THE NATURAL HISTORY OF
INBREEDING AND OUTBREEDING

THE
NATURAL HISTORY
OF
INBREEDING
AND
OUTBREEDING

Theoretical and Empirical Perspectives

EDITED BY
NANCY WILMSEN THORNHILL

The University of Chicago Press • Chicago and London

Nancy Wilmsen Thornhill is research assistant professor in the Institute of Social Research, assistant professor in the Department of Communications, and an associate in The Evolution and Human Behavior Program at the University of Michigan.

The University of Chicago Press, Chicago 60637
The University of Chicago Press, Ltd., London
© 1993 by The University of Chicago
All rights reserved. Published 1993
Printed in the United States of America
02 01 00 99 98 97 96 95 94 93   5 4 3 2 1

ISBN (cloth): 0-226-79854-2
ISBN (paper): 0-226-79855-0

Library of Congress Cataloging-in-Publication Data

The Natural history of inbreeding and outbreeding : theoretical and
    empirical perspectives / edited by Nancy Wilmsen Thornhill.
        p.   cm.
    Includes index.
    ISBN 0-226-79854-2 (cloth).—ISBN 0-226-79855-0
    (pbk.)
    1. Sex (Biology)   2. Inbreeding.   3. Evolution (Biology)
I. Thornhill, Nancy Wilmsen.
QH481.N38   1993
575.1'33—dc20   92-38497
                                                          CIP

CONTENTS

v

Acknowledgments

Charles Darwin suggested in 1868 that "nature abhors perpetual self-fertilization." Since then the occurrence and fitness consequences of inbreeding, and the nature of breeding patterns more generally, have been foci of interest in disciplines as diverse as anthropology, behavioral ecology, and population genetics. Apparently taking Darwin's dictum to heart, workers in these disciplines generally have assumed that inbreeding is rare or pathological. As empirical evidence has accumulated, however, it has become clear that the true picture is much richer and more varied.

The literature on breeding patterns is widely scattered in different, often taxonomically oriented journals that appeal primarily to specialists, and thus it is difficult for a more general audience to appreciate. One intention of this book is to bring together a broad and deep review of the literature for a diversity of taxa. In so doing, the book also brings together workers whose perspectives are primarily theoretical or primarily empirical, and provides a comparison of these perspectives. The book addresses several central questions: (1) Do organisms in many taxa facultatively inbreed, and if so, at what level and with what frequency? (2) What are the fitness consequences of inbreeding and outbreeding? (3) What are the ecological and genetic costs and benefits associated with different levels of inbreeding and outbreeding? (4) How would observed breeding patterns affect the evolution of social systems and the process of speciation? (5) What are the most fruitful avenues for future research on the causes and effects of inbreeding and outbreeding?

This book has been four years in the making. It was conceived, along with an associated symposium, in collaboration with William Shields. Shields also co-authored a National Science Foundation proposal to support the symposium and originally served as co-editor of the book. Other commitments necessitated his withdrawl from the latter responsibility, but his input into the content and general plan of the volume has been major and very critical.

Preparation of the book has benefited from the attention and assistance of many people, principally the authors, who contributed their knowledge, skill, and time to writing and revising. Each author also served as a reviewer for at least one other chapter, an effort, in conjunction with reviews of the entire manuscript by the University of Chicago Press, that has promoted integration among chapters.

The National Science Foundation and the Animal Behavior Society provided financial support for the symposium associated with the book, held at the 1989 ABS meetings at Northern Kentucky University. I am grateful to the NSF, the ABS, and to Tom Rambo, the local host, for the opportunity to bring all the authors together at an early stage of the book's gestation.

The Department of Biology at the University of New Mexico provide me with space and an atmosphere conducive to scholarly activities. Professor Peter Sauer and the Fakultat für Biologie, Universitat Bielefeld, also provided assistance when I was in Germany in 1990. Professor Peter Weingart and the Centrum für Interdisciplinäre Forschung, Universitat Bielefeld, provided similar support in 1991 and 1992.

I am most grateful for the superior technical assistance of L. Jegerlehner. Lynn Fullerton's eleventh-hour help in compiling the bibliography is sincerely appreciated. Anne Rice's expert word processing made my editorial burden far lighter.

Randy Thornhill has provided much support throughout a decade and a half of collaboration. Finally, my children belong prominently on my list of debts. Patrick, Aubri, and Margo Thornhill each have precious input into every final product. I thank them for sharing their lives with me.

1

Sex, Mating Systems, Inbreeding, and Outbreeding

Nickolas M. Waser

"Historically [the effects of inbreeding and outbreeding] are . . . practical problems of considerable significance bound up with man's gravest affairs, his marriage customs and his means of subsistence" (East and Jones 1919; 13).

Sex in higher organisms involves syngamy, the fusion of gametes. Thus sex immediately implies the mating of individuals. Beyond our lay fascination with sex—we belong to a sexual species—lies a scientific fascination with the diversity of sexual expression within and across taxa, including the diversity of mating systems.

Matings are likely to deviate from a random expectation in several ways. First, individuals sharing a phenotypic trait may mate with each other more or less often than expected at random; these deviations are referred to respectively as positive and negative assortative mating (Roughgarden 1979). Second (and related to assortative mating insofar as phenotypic similarity reflects common ancestry), individuals may mate based on degree of genetic similarity derived from kinship and shared versus separate evolutionary history. That is, they may *inbreed* or *outbreed*, terms I strive below to define more exactly. This volume addresses the theoretical treatment and natural history of these latter deviations from random mating.

DEFINITION OF TERMS

Mating System

Mating is defined biologically not by copulation, spawning, pollination, or other events that place male and female gametes in proximity, but by the production of a zygote, i.e., by syngamy (Bateson 1983; Lyons et al. 1989). The mating system describes patterns of syngamy, or equivalently, who shares parentage with whom in a population. Measuring it ideally takes the form of a complete enumeration of matings.

Inbreeding

From what might be called an "absolute" perspective, inbreeding is the mating of relatives (kin)—individuals (or one individual in the case of self-fertilization) that carry alleles identical by descent (*ibd*) from a common ancestor or ancestors. A "relative" perspective is that inbreeding is the mating of kin more often than expected at random, given the population size in question (Roughgarden 1979; see also Shields [chapter 8]; Knowlton and Jackson [chapter 10]). The coefficient of kinship ("coancestry") of two individuals is the probability that allele copies drawn at random from each one are *ibd*, or alternatively, the genetic correlation between the two allele copies (Michod [chapter 5]). This is equivalent to the expected value of Wright's (1922) inbreeding coefficient, f, of an offspring produced by mating of these individuals, and to half the genetic correlation, r, of the individuals unless they themselves are inbred (Crow and Kimura 1970; Michod and Hamilton 1980). It also is useful to consider the average inbreeding coefficient, F. For the "relative" definition given above, F is defined as a deviation from the random mating expectation of some reference population. In fact, one can define a hierarchical series of F values for a series of population units each nested within the next larger unit (Wright 1943). These are called "fixation indices" to acknowledge the influence of evolutionary factors other than the mating system. The most commonly estimated parameters are F_{IS}, the average fixation of individuals relative to subpopulations, and F_{ST}, the average of subpopulations within a total population (equivalently, the standardized allelic variance among subpopulations; Crow and Kimura 1970; Hartl and Clark 1989; Michod [chapter 5]). When natural selection is negligible, F_{IS} and F_{ST} respectively estimate contributions to inbreeding from nonrandom mating within subpopulations and from division of the population into subpopulations each with a finite pool of potential mates (Waser [chapter 9]; Knowlton and Jackson [chapter 10]).

Connotations of "inbreeding" actually vary considerably among researchers. One perspective downplays the population genetics definition of inbreeding and the contributions of population structure to mating opportunities, focusing instead on individual behavior and fitness. Anthropologists and zoologists with this focus sometimes adopt a threshold, at about the level of first-cousin mating, between "close inbreeding" (or "incest") and trivial inbreeding (Shields [chapter 8];

Rowley, Russell, and Brooker [chapter 13]; Smith [chapter 14]; Moore [chapter 17]), and furthermore distinguish inbreeding from outbreeding at some second threshold kinship level (which may be as high as first cousins; Smith [chapter 14]). Technically speaking, however, inbreeding forms a continuum (and is continuous with outbreeding; see below and Shields [chapter 8]), rather than falling into discrete classes separated by thresholds. This continuum is recognized in a second, population genetics perspective, which would consider Smith's "outbreeding" to be substantial inbreeding! Replacing the continuum with discrete classes does have some practical advantages. For example, fitness consequences of inbreeding may be detectable only within matings above some threshold f. Grouping matings into classes may be necessary to achieve statistical power when the sample of matings is limited (the usual case!). Furthermore, the value of f in a mating may be impossible to measure precisely, so that grouping is necessary (e.g., Packer and Pusey [chapter 16]). Finally, grouping may logically reflect distinctions that the organisms themselves appear to make, for example, when they behave differently toward close kin than toward all other individuals (Waldman and McKinnon [chapter 11]; Smith [chapter 14]). However, the use of threshold definitions does contribute to miscommunication with those interested in population consequences who adopt definitions precisely grounded in population genetics. To reduce this problem, researchers should keep in mind that discrete classes are a simplification, should clearly define those they use, and should articulate their reasons for adopting them.

In botany there is a different "threshold" focus on self-fertilization as the only form of inbreeding. An example is the recent definition of *inbreeding* depression as fitness decline under *selfing* (e.g., Lande and Schemske 1985; Waller [chapter 6]), which contrasts with a more traditional and general definition (Falconer 1989, 248ff.). Such usage may have several historical roots. First, individual flowering plants usually express both sexes (i.e., are "cosexual"). Selfing is easily recognized and distinguished from "crossing" or "outcrossing" with other plants (terms sometimes taken as synonyms for outbreeding). Furthermore, self-pollination is straightforward to achieve experimentally, whereas other forms of inbreeding require estimates of parental kinship. The focus on selfing is reinforced by dramatic "self-incompatibility" reactions in many angiosperms (Fryxell 1957). Second, the simplest mating system that yields a stable intermediate F value is a mixture of selfing and outbreeding (Haldane 1924). A "mixed mating" model forms the basis for almost all discussions of plant mating systems (Fyfe and Bailey 1951; Brown 1990). Whereas it may be approximately correct for some crop plants, the model ignores the kinship structures that often develop in natural populations due to restricted seed and pollen dispersal (Waser [chapter 9]). This limitation is now being addressed with more elaborate models (e.g., Ritland 1985).

Outbreeding

Outbreeding refers to the opposite of inbreeding, i.e., to mating of nonkin (in an "absolute" perspective), or to kin mating less often than expected at random given the population size (in a "relative" perspective). What must be kept in mind is that kinship always is defined relative to some ancestral generation. With N ancestors ($2N$ ancestral alleles), the smallest possible average inbreeding (F) of descendants converges on $1/2N$. Furthermore, this assumes no sampling error in transmission of alleles across generations, i.e., no genetic drift. With a pure drift process (or directional selection), the lineages of all allele copies in a population derive from (or *coalesce* in) a single ancestral allele (Hartl and Clark 1989). Under these conditions matings that are outbred from either an "absolute" or a "relative" perspective (i.e., $f = 0$ or $f_{IS} < 0$, respectively), within one frame of reference, will not be if the frame is moved far enough back in time. One reaches the somewhat perplexing conclusion that all allele copies in a species ultimately share descent, implying no scope for outbreeding (East and Jones 1919: 80–81; Shields 1982, and chapter 8, this volume). If shared descent is not traceable to the time of species formation, it should be in principle to more distant ancestors. As Darwin recognized, this also means that an ultimate form of *ibd* occurs across species, so that, for example, it is reasonable to construct a phylogeny based on an ancient enzyme such as cytochrome c (Fitch and Margoliash 1970).

These thought experiments (or mental machinations?) may suggest that the continuum of genetic similarity embodied in an inbreeding coefficient ranging from one (*ibd*) to zero (outbreeding) is illusory. Consider an isolated population (as a conservation geneticist may). Rare matings that involve individuals from outside the population might at first be considered outbred (or at least less inbred than those within the population), whereas from a larger perspective all may appear inbred. What is missing here is the recognition that alleles themselves undergo "descent with modification." The structural similarity of alleles depends not only on shared ancestry, but also on evolutionary change in separate lineages with different histories of mutation, selection, gene flow, and drift. When one expands the scope of sufficiently large geographic, historical, and taxonomic scales, considerations of shared ancestry in a strict sense (i.e., all alleles traceable to a single ancestral copy) are overshadowed by evolutionary divergence among lineages. Thus there is indeed a continuum of possible genetic similarity. Although a judgement of *ibd* in a strict sense becomes more all-inclusive as a more distant ancestral generation is chosen, individuals mating in nature or in captivity will differ genetically to an extent not adequately described by this choice. The "hybridity" measures of Templeton and Read (1984) and Lynch (1991) attempt to capture some of this increased range of possible genetic dissimilarity with respect to coadapted gene complexes and other differences between parental genomes.

REASONS FOR INTEREST IN MATING SYSTEMS

Why should we be interested in mating systems, and in the degree of inbreeding and outbreeding they encompass? A biologist specializing in one taxon is likely to discover a variety of mating systems. A generalist is certain to discover a variety so rich that it demands questions.

The biologist first notices the near ubiquity of sexual reproduction and recognizes the challenge of explaining the maintenance of sex via short-term individual selective advantage (Maynard Smith 1971, 1978; Williams 1975; but see Nunney 1989). The wide taxonomic distribution of sex indicates that it arose early in the history of life (G. Bell 1982). Its characteristic features of recombination, segregation, and syngamy suggest that sex is on the one hand an adaptation in part to a constantly changing biotic environment and on the other a cause of the evolution of organic diversity (Ghiselin 1974; Williams 1975; Maynard Smith 1978; G. Bell 1982; Law and Lewis 1983; Hamilton [chapter 18]).

Moving beyond the basic features of sex, the biologist next is struck by the diversity of individual sex expression (G. Bell 1982). Why should hermaphroditism be virtually absent in vertebrates, rare in arthropods, more common in echinoderms, and the norm in urochordates (Ghiselin 1974)? Why should approximately 70% of higher plants be cosexual (Yampolsky and Yampolsky 1922)? Sex expression (the "packaging" of sexual function into individuals) obviously constrains mating patterns, with attendant fitness consequences insofar as fitness is related to the degree of inbreeding and outbreeding. The evolution of sex expression is thought to be driven in part by such fitness consequences (Charnov 1982).

Variation in mating systems, in the form of monogamy, polygyny, promiscuity, and the like, also occurs within taxa that exhibit uniform sex expression (e.g., birds; Emlen and Oring 1977; Rowley, Russell, and Brooker [chapter 13]; and mammals; Smith [chapter 14]). In seeking to understand variation at this level, biologists once again look to the genetic and fitness consequences of different degrees of inbreeding and outbreeding, and to other ecological features of populations, as discussed below.

In parallel with pure scientific curiosity about sex and mating systems are several more practical concerns. We can immediately list plant and animal husbandry. The recognition that fitness and variation in trait expression are affected by inbreeding, outbreeding, and ultimately hybridization of subspecies and related species is of long standing, derived in large part from selective animal and plant breeding (e.g., Kölreuter 1761; Darwin 1868, 1876; East and Jones 1919). Sewall Wright's (1922) development of a tractable theory of inbreeding, for which he invented the method of path coefficients, was motivated by questions from animal breeding. Those managing captive populations in zoos and botanical gardens, and more recently, conservation biologists managing dwindling natural populations, share concerns about inbreeding and outbreeding effects on fitness (e.g., Soulé 1986; Rowley, Russell, and Brooker [chapter 13]; Lacy, Petric, and Warneke [chapter 15]).

The empirical study of sex, mating systems, inbreeding, and outbreeding has not proceeded in isolation, but has coevolved with increasingly sophisticated theoretical efforts. The consensus that Mendelian genetics explains both discrete and quantitative phenotypic traits, reached early in this century after much dispute, followed from the development of mathematical genetics. In turn these advances fostered the "New Evolutionary Synthesis" of the 1930s and 1940s (Mayr and Provine 1980). The stage was then set for explicit theoretical consideration of the evolution and maintenance of sex, sex expression, and mating systems, beginning with Fisher (1930).

MAJOR QUESTIONS, AND THE SCOPE OF THIS VOLUME

Given these reasons for interest in mating systems, what are the major theoretical and empirical issues facing us? I offer the following taxonomy of questions and their coverage in this volume.

What Are the Patterns of Inbreeding and Outbreeding?
One major task researchers face is simply to characterize mating systems, including patterns of inbreeding and outbreeding. The word "simply" does not imply that this is easy. At least four approaches (and their combinations) are found in this volume. Observing copulations is the most direct, but is a daunting assignment even for organisms with relatively visible behavior, such as diurnal mammals and birds (Row-

ley, Russell, and Brooker [chapter 13]; Packer and Pusey [chapter 16]), and next to impossible for organisms that are small or secretive (Reichert and Roeloffs [chapter 12]; Smith [chapter 14]). Direct observation is also intractable with organisms that shed gametes into a medium where they are quickly diluted and lost from sight (Knowlton and Jackson [chapter 10]; Waldman and McKinnon [chapter 11]). Even if one can enumerate copulations, the production of offspring is not guaranteed; thus a second approach is to assign parentage based on offspring genotypes (Rowley, Russell, and Brooker [chapter 13]). Here the kinship of mates also must be assessed with pedigree information (Rowley, Russell, and Brooker [chapter 13]; Smith [chapter 14]; Lacy, Petric, and Warneke [chapter 15]) or some measure of genetic similarity (Waldman and McKinnon [chapter 11]; Packer and Pusey [chapter 16]). A third, less precise approach is to use information on population structure and dispersal patterns to infer mating opportunities (Waser [chapter 9]; Knowlton and Jackson [chapter 10]; Waldman and McKinnon [chapter 11]; Reichert and Roeloffs [chapter 12]; Rowley, Russell, and Brooker [chapter 13]; Smith [chapter 14]; Packer and Pusey [chapter 16]; Moore [chapter 17]). A final approach is to assess offspring genotypes (or fixation indices), with the assumption that these primarily reflect matings and that indistinguishable alleles indicate *ibd* (Michod [chapter 5]; Waser [chapter 9]; Knowlton and Jackson [chapter 10]; Reichert and Roeloffs [chapter 12]; Rowley, Russell, and Brooker [chapter 13]). The empirical chapters that follow use these approaches to document varying degrees of inbreeding across a range of taxa. One obvious pattern that emerges is the importance of population structure. Sessile organisms generally experience substantially more opportunity for inbreeding than vagile ones.

What Are the Consequences for Individual Fitness?

A second task is to ask what consequences inbreeding and outbreeding carry for survival and reproduction, i.e., individual fitness. This has long-term evolutionary implications, and also short-term ones insofar as inbreeding or outbreeding depression increase the chance of population extinction due to "demographic stochasticity" (Goodman 1987).

Inbreeding depression, a decline under inbreeding in the mean value of traits, including fitness components, was first explored systematically in the late nineteenth century (East and Jones 1919; Wright 1977). Contributors to this volume present evidence for inbreeding depression in plants (Waller [chapter 6]; Waser [chapter 9]), fishes, amphibians, and reptiles (Waldman and McKinnon [chapter 11]), zoo animals (Lacy, Petric, and Warneke [chapter 15]; but see Shields [chapter

8]), and primates (Moore [chapter 17]; but see Shields [chapter 8]); but not in birds (Rowley, Russell, and Brooker [chapter 13]) and some small mammals (Smith [chapter 14]). Again, these patterns may in large part reflect constraints on mating opportunities imposed by restricted effective population size. Although such constraints play a role even for organisms with natal or adult dispersal (Riechert and Roeloffs [chapter 12]; Rowley, Russell, and Brooker [chapter 13]), they surely are more severe for sessile organisms.

As Shields ([chapter 8]) and Waser ([chapter 9]) stress, what ultimately is needed for rigorous evaluation of inbreeding depression is information on lifetime fitness (not just fitness components), assessed under natural conditions. The reason is that expression of genetic load may be concentrated late in the life cycle, and fitness expression in artificial environments may not reflect that in nature. Waller ([chapter 6]) also points out the theoretical importance of documenting fitness effects across successive generations of inbreeding in order to predict mating system evolution.

What Causes Inbreeding and Outbreeding Depression?

With the rediscovery of particulate inheritance, it was recognized that inbreeding depression has a likely explanation in the nonadditive expression of alleles at a locus, involving either dominance (East and Jones 1919) or overdominance (Crow 1948). Most discussions in this volume conclude that the evidence favors a dominance model (Werren [chapter 3]; Waller [chapter 6]; Hamilton [chapter 18]), but Mitton (chapter 2) invokes overdominance on empirical grounds and from biochemical considerations, and Uyenoyama (chapter 4) assumes overdominance in models of the evolution of incompatibility systems.

If one accepts that inbreeding and outbreeding are simplified terms describing a continuum of genetic similarity, and that outbreeding logically extends to the level of differentiation represented by subspecies and closely related species, it follows that postzygotic reproductive barriers in crosses at these taxonomic levels represent "outbreeding depression" in fitness. In this sense outbreeding depression is part and parcel of a biological species concept. Controversy over outbreeding depression would wisely be directed not to questions of its existence, but instead to questions of the taxonomic and spatial scale on which it is found, its magnitude, and its mechanism (see Lynch 1991).

Chapters in this volume summarize evidence for outbreeding depression in some plants (Waser [chapter 9]), marine invertebrates (Knowlton and Jackson [chapter 10]), and zoo populations (Lacy, Petric, and Warneke [chapter 15]), and Shields (chapter 8) surveys cases more broadly. Although disruption of coadapted gene complexes is emphasized in most chapters, another cause of outbreeding depression, the disruption of adaptation to local environments, may be more likely in sessile organisms (Waser [chapter 9]).

What Are The Consequences for Mating System Evolution?

Fitness consequences of inbreeding and outbreeding presumably exert selection on a number of traits that in turn affect mating patterns. One such trait is the dispersal behavior of vagile animals, and the dispersal abilities of larvae, seeds, gametes, or gametophytes of organisms with sessile adults. Dispersal, along with selection and genetic drift, affects the kin structure of a population, and thus the kinship of potential mates (Waser [chapter 9]; Knowlton and Jackson [chapter 10]; Waldman and McKinnon [chapter 11]). To the extent that there is heritable variation in dispersal ability, we expect the coevolution of dispersal and the mating system (see also Campbell and Waser 1987). This reasoning is implicit in the argument that dispersal has evolved as an inbreeding avoidance mechanism in birds and mammals (Rowley, Russell, and Brooker [chapter 13]; Smith [chapter 14]; Packer and Pusey [chapter 16]; but see Moore [chapter 17]). One must consider also whether the genetic value of matings available after dispersal is not counterbalanced by what zoologists call the "somatic" fitness costs to dispersers, i.e., risk of starvation and predation and loss of mating opportunities (Shields [chapter 8]; Smith [chapter 14]; Moore [chapter 17]).

Final mating patterns also will be affected by behavioral or physiological discrimination among potential mates, so that only some are "chosen." In animals this can involve recognition of traits that reflect overall heterozygosity (Mitton [chapter 3]), and recognition of kin, which is reported both for invertebrates (Knowlton and Jackson [chapter 10]) and vertebrates (Waldman and McKinnon [chapter 11]; Smith [chapter 14]; Moore [chapter 17]). Some, but not all, kin recognition reduces to recognition of natal associates (e.g., Rowley, Russell, and Brooker [chapter 13]). In plants, discrimination may involve prezygotic inhibition of certain male microgametophytes and a possible (still controversial!) contribution from postzygotic abortion of some embryos (Waser [chapter 9]). Discrimination in plants must occur after dispersal of pollen; but in animals discrimination may either drive dispersal (i.e., animals deprived of mating opportunities disperse from the natal site) or follow it. Discrimination is not universal or perfect, and this matches theoretical expectations. Uyenoyama (chapter 4) explores necessary conditions for the evolution of one form of discrimination, self-incompatibility, with multilocus models.

Beyond the coevolution of dispersal and mate discrimination with the mating system lies the possibility for evolution at loci determining the fitness effects of inbreeding or outbreeding themselves. The sessile adult habit of plants and numerous aquatic invertebrates, with its potential for high inbreeding, sets the stage for substantial selection on genetic systems that give rise to inbreeding depression. Waller (chapter 6) contrasts simple "static" models of plant mating system evolution with more complex "dynamic" models in which inbreeding depression coevolves with population structure. Dynamic models predict a much richer range of final mating systems, a pleasing result given the evolutionary diversification of mating systems and sex expression in plants. Werren (chapter 4) constructs genetic models of the consequences of diploid versus haplodiploid sex determination for genetic load under inbreeding, and argues that haplodiploidy (and the correlated ability of females to skew offspring sex ratio) facilitates the evolution of extreme inbreeding, a commonly derived mating system in hymenopterans.

Are There Other Evolutionary Consequences?

As dispersal behavior, mate discrimination, and fitness effects of matings coevolve with the mating system, so must the genetic makeup of a population. The frequencies and spatial distribution of alleles and genotypes within a population (i.e., its genetic structure) are constrained largely by the dispersion of individuals in space (i.e., the ecological structure), with important influences from the mating system and its fitness consequences. Michod (chapter 5; see also Moore [chapter 17]) constructs models of how inbreeding, through its effect on kin structure in populations, influences the evolution of various forms of altruism (i.e., social evolution). Genetic structure of populations also has macroevolutionary implications. Howard (chapter 7) sketches an engaging conceptual picture of the importance of small population size, inbreeding, and genetic drift to a class of speciation models dating back half a century, and summarizes the (mixed) support for each model.

THE FUTURE

Since the advent of mathematical genetics we have made impressive progress in understanding patterns of inbreeding and outbreeding and their causes and consequences. This is evident in comparing the present volume to one with a similar title published 70 years ago (East and Jones 1919). However, what we know remains small compared to what we wish we knew. What follow are my musings on some profitable trends for future research. Other pleas for theoretical and empirical effort are found throughout this volume.

The first issue is empirical characterization of mating systems. Regardless of whether one eventually classifies levels of inbreeding and outbreeding into discrete categories for analysis, it is desirable to improve precision in estimating who produces zygotes with whom. Molecular techniques hold great promise here. If sufficient genotypic information is obtainable from allozymes or newer techniques (e.g., Burke 1989; Williams et al. 1990), one can assign paternity and maternity of offspring relatively unambiguously; otherwise one can apply maximum likelihood methods that have become rapidly more powerful (e.g., Devlin, Roeder, and Ellstrand 1988). One also can use the improved genetic information to estimate more accurately kinship of parents in the absence of pedigree information (e.g., Queller and Goodnight 1989; Weir 1990). These approaches open up new possibilities for studying temporal and spatial variation in mating systems, fitness consequences, and selection on related traits. The difficulty is that they are tedious, especially when populations are large.

Despite more than half a century of study, we still are uncertain of the relative contributions of various genetic mechanisms to inbreeding depression, especially in natural populations. Causes of outbreeding depression are even less well understood (Lynch 1991). Multigeneration experimental crosses (Bulmer 1985), using species with short generation times that can be bred in captivity, should help us separate the roles of overdominance versus dominance in inbreeding depression and detect pseudo-overdominance due to linkage disequilibrium. Studies of fitness over a range of inbreeding levels will be required to look for nonlinearities that may indicate epistasis and that play important roles in models of mating system evolution (Waller [chapter 6]). The issue here involves gene expression, and we should be aware of growing evidence for nontraditional expression, for example, differential parental "imprinting" of alleles (i.e., Reik et al. 1987). What role might such an effect play in inbreeding and outbreeding depression? Might disruption of parental coordination of imprinting (e.g., Surani 1987) be a source of outbreeding depression, expressed during F_1 offspring development, in addition to (or instead of) disruption of multilocus allelic associations (usually considered to be expressed primarily in the F_2 generation), and disruption of local adaptation? Finally, we need more careful assessment of possible ecological contributions to inbreeding depression. Such mechanisms include frequency-dependent selection through resource competition or pathogen attack (Price and Waser 1982; Charnov 1987), inferior competitive ability of inbred individuals (Schmitt and Ehrhardt 1990), and disproportionate loss of adaptation to local conditions by such individuals (Waser and Price 1989; Schmitt and Gamble 1990).

The fact that mating patterns affect fitness implies that they often evolve in an adaptive manner. This does not mean that mating systems are necessarily "optimal" in more than a restricted sense. Some restrictions on achieving more adaptive mating patterns are relatively absolute, for example, restrictions on the dispersal biology of a species, its kin recognition abilities, or on the evolution of multilocus genetic systems that determine the fitness consequences of matings. Other restrictions involve fitness trade-offs with the quality of differently inbred or outbred offspring, such as somatic costs of dispersal and searching for mates, and loss of mating opportunities from overly strict mate discrimination (e.g., Bateman 1952; Pomiankowski 1987; Motro 1991). These restrictions apply to plants as well as animals. For example, the behavior of animal pollinators and seed dispersers may place severe limits on plant gene flow and thus population structure. As dispersal distance increases, pollen "mortality" may increase, because pollen is more likely to fall off the pollinator, be deposited on an inappropriate flower, or lose viability. Progress in assessing the adaptive nature of mating systems requires that workers lay out such costs and benefits in a clearheaded way, and estimate them all (and the sensitivity of any conclusion to variance in the estimates), rather than focusing on one or two. In addition, we need to estimate the heritability of traits that influence mating patterns, and genetic and phenotypic correlations among traits. There are no shortcuts to gaining a more rigorous adaptive understanding of mating systems.

Finally, we can anticipate continued theoretical progress, yielding tools for understanding mating system evolution that are more powerful than those available previously. As already noted, some recent models allow simultaneous coevolution at gene loci affecting fitness and loci involved in kinship discrimination and/or mating behavior. Other models include biparental inbreeding in addition to selfing and outbreeding (e.g., Uyenoyama 1986; M. Uyenoyama, K. Holsinger, and D. Waller, unpublished data). This realism yields exciting new predictions. We should be able to progress even further by including more of the ecology of pollen, seed, larval, and/or adult dispersal. A start in this direction has been made with Holsinger's (1991) models that include pollen delivery. Additional efforts are needed that incorporate somatic costs of different mating options in models tailored for both plant and animal systems.

ACKNOWLEDGMENTS

Thanks for the thoughtful comments of Norm Ellstrand, Rick Grosberg, Arlee Montalvo, Mary Price, Ruth Shaw, Peter Waser, Rick Williams, and one anonymous reader, and for the support of NSF Grant BSR 8905808.

PART ONE

Theoretical Perspectives

2

Theory and Data Pertinent to the Relationship between Heterozygosity and Fitness

Jeffry B. Mitton

The primary purposes of this chapter are to review the empirical literature that relates to the correlations between components of fitness and individual heterozygosity and to relate this recent literature to the vast literature on inbreeding depression and heterosis. In general, components of fitness such as growth rate, physiological efficiency, and fecundity increase with individual heterozygosity. Data sets from ryegrass, killifish, and cod are examined with the adaptive distance model of Smouse (1986) to provide further insight into the correlations. These analyses suggest that selection discriminates among outcrossed genotypes, providing an incentive that may influence the evolution of mating systems. Finally, a model of the evolution of female choice is described. In this model, females bearing a dominant allele influencing mating behavior seek a highly heterozygous male to enhance the fitness of their offspring. Empirical data consistent with this model are presented.

INBREEDING DEPRESSION AND HETEROSIS

Inbreeding depression, heterosis, and the performance of randomly mating populations can be presented as fitnesses on a continuum of individual heterozygosity (figure 2.1). Here fitness is defined as relative reproductive success, but in most empirical studies fitness is estimated with a component of fitness, such as viability or fecundity (Hartl and Clark 1989). Individual heterozygosity is the number of loci heterozygous in an individual (Mitton and Pierce 1980), and it can be esti-

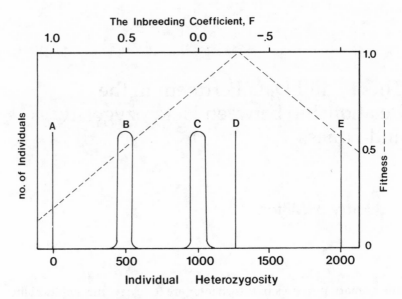

FIGURE 2.1   The distribution of individual heterozygosity, the inbreeding coefficient, F, and postulated fitness as a function of the mating system. The genome consists of 2,000 polymorphic loci, each with two alleles with frquencies of 0.5. A is one of many possible inbred strains, retaining no genetic variability. C is an outcrossing population in Hardy-Weinberg equilibrium. B is a sample of individuals produced by one generation of selfing in population C. D and E are crosses between inbred strains; D exhibits heterosis, while E exhibits outbreeding depression. The variance in individual heterozygosity in populations B and C is $S^2_H = [(L - H) * H]/L$, where L is the number of polymorphic loci and H is the number of heterozygous loci. $S^2_H = 0.0$ in A, D, and E.

mated with protein polymorphisms. Inbreeding decreases heterozygosity, and relative to the benchmark of randomly mating populations, the fitness of inbred individuals is typically depressed. Inbreeding depression is manifest as a depression of growth rate, viability, developmental stability, fecundity, and fertility (Lerner 1954; Wright 1977; D. Charlesworth and B. Charlesworth 1987). Inbreeding depression can be severe in species that are typically outcrossing (Crumpacker 1967; Simmons and Crow 1977; Franklin 1972; Sorensen and Miles 1982; Crow and Simmons 1983), and it may be less intense (Lande and Schemske 1985) but not necessarily absent (D. Charlesworth and B. Charlesworth 1987) in selfing species. Crosses between populations or between differentiated strains produce offspring with levels of heterozygosity higher than those observed in typical outcrossing populations. Heterosis (Shull 1948; Lerner 1954; Frankel 1983) is the enhanced fitness produced by crosses between differentiated populations and strains. Fitnesses do not increase indefinitely with heterozygosity; several observations indicate that when the populations pass some critical degree of differentiation, the fitnesses of offspring decline. This decline in fitness, called outbreeding depression, has been attributed to the breakup of coadapted gene complexes (Moll et al. 1965; Price and Waser 1979; Shields 1982; Waser and Price 1983, 1985; Campbell and Waser 1987; Frei, Stuber, and Goodman 1986).

The axis of individual heterozygosity can be used to illustrate the distribution of heterozygosity within populations and the mean fitness of a population as a function of the mating system (figure 2.1). From the outcrossed population (C), fitness and heterozygosity decline with increasing levels of inbreeding. But it does not necessarily follow that heterozygosity per se is the most important variable influencing fitness. Inbreeding depression and heterosis have been under investigation for decades (Wright 1977; Frankel 1983), but there is still no consensus concerning the genetic mechanisms underlying these phenomena. Two contending hypotheses, the dominance hypothesis and the overdominance hypothesis, are compatible with many of the empirical results.

The Dominance Hypothesis

The dominance hypothesis focuses upon the phenotypic expression of recessive alleles that are either lethal or detrimental. Some groups of species, such as conifers and *Drosophila*, have relatively high frequencies of deleterious alleles, and the rate of mutation to deleterious alleles per gamete is surprisingly high (Lynch 1988c). Inbreeding causes rare recessive alleles to occur more frequently in the homozygous condition, increasing the frequency of aberrant phenotypes. From the perspective of the dominance hypothesis, inbreeding depression is the expression of deleterious alleles that had been masked in the heterozygous condition. Because completely inbred strains tend to be fixed for deleterious alleles at different loci, crosses between strains produce genotypes bearing deleterious alleles in the heterozygous condition, masking the great majority of deleterious alleles and producing heterosis.

The Overdominance Hypothesis

The overdominance hypothesis supposes the performance of heterozygotes to be superior to that of homozygotes. From this perspective, inbreeding depression results when inbreeding decreases the frequency of heterozygous genotypes and increases the frequency of homozygous genotypes. Heterosis is the phenotypic manifestation of enhanced heterozygosity resulting from crosses between differentiated populations.

From a single-locus perspective, it is difficult to imagine that modest changes in genotypic frequency could produce the dramatic consequences that we recognize as inbreeding depression. For example, imagine a population of ponderosa pine segregating the alleles A and a, each with a frequency of 0.5. The frequencies of the genotypes AA, Aa, and aa in a typical outcrossing population would be 0.25, 0.50, and 0.25. Now consider the consequences of selfing, which decreases the level of heterozygosity by 50% each generation. The frequencies of the genotypes in selfed seeds collected from the same population would be 0.375, 0.25, and 0.375. Selfing produces only a moderate change in genotypic frequencies, but inbreeding depression, expressed as decreased germinability, growth rate, and viability, is dramatic. There is a serious problem with this single-locus perspective.

Consider the same problem, but with a pair of unlinked polymorphic loci, segregating alleles A and a at the first locus, and B and b at the second locus (figure 2.2). Once again, all allelic frequencies are 0.5. The frequency of double heterozygotes is expected to be 0.25 under random mating, but this frequency would fall to 0.0625 with just one generation of selfing. Similarly, the frequency of double homozygotes would be 0.25 under random mating, but one generation of selfing would increase it to 0.56. Although the proportion of heterozygotes drops by 50%, the proportion of double heterozygotes drops by 75%.

The disparity between the heterozygosities of outcrossed and selfed genotypes increases with the number of loci under consideration. For example, in figure 2.1, the axis of individual heterozygosity is based upon 2,000 independently assorting polymorphic loci, each segregating two alleles with frequencies of 0.5. Average heterozygosity in the outcrossed group is 1,000, but one generation of selfing drops the mean heterozygosity to 500, far below the range expected in the outcrossed group. Thus, a single generation of selfing produces genotypes never seen in the outcrossing population (Mitton and Jeffers 1989). This disparity between selfed and outcrossed genotypes is simply not apparent with a single-locus perspective.

Concerns about the level of segregational load have led some biologists to reject the overdominance hypothesis. Briefly, the reasoning goes as follows: Imagine an overdominant locus with the AA, Aa, and aa genotypes conferring fitnesses of 0.9, 1.0, and 0.9. These fitnesses incur a segregational load of 0.05 at equilibrium (when $p = q = 0.5$), reducing the average fitness to 0.95. How many loci may remain polymorphic under this form and intensity of selection? One answer is that

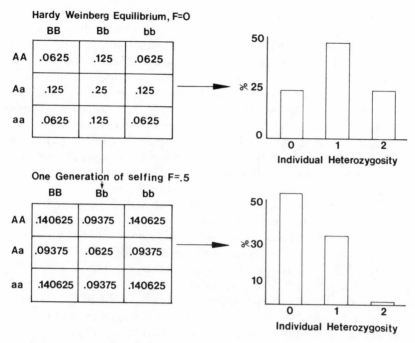

FIGURE 2.2   The influence of one generation of selfing upon the distribution of individual heterozygosity for two polymorphic loci. All allele frequencies are equal to 0.5.

the segregational load will be multiplicative across loci (Lewontin 1974), so that the average fitness will be equal to

$$W = (0.95)^L$$

where L is the number of polymorphic loci experiencing this intensity of selection. The average fitness of a population with 100 overdominant loci would be 0.0059. This extremely low estimate of fitness led many biologists to conclude that very few overdominant loci could sustain this intensity of selection, and that most of the genetic variation detected in electrophoretic surveys of proteins was selectively neutral (Kimura and Ohta 1971; Nei 1975; Kimura 1983). However, the magnitude of genetic load is highly dependent upon the model of fitness determination; truncation selection (Milkman 1978, 1982; Wills 1978, 1981) is a form of natural selection that essentially eliminates the perceived problem with genetic load (Crow and Kimura 1979; Kimura and Crow 1978; Wills 1978, 1981). This model of natural selection first ranks individuals by fitness potential, which may be constructed with additive, multiplicative, or epistatic interactions between loci. Then the environment and/or carrying capacity of the environment imposes a threshold below which fitness is constrained to zero, and above which fitness is determined by an interaction of environmental and genetic factors. The biological reality of this model is in the placement of the threshold upon the rank-ordered fitnesses. A low threshold imposes little selection, as might be the case in a salubrious environment, while a high threshold imposes the selection intensity that may occur in extreme environments. Because this is a multilocus model of selection of individuals, selection against an individual imposes selection upon many loci simultaneously, and therefore the genetic load drops dramatically. Models of truncation selection reveal that thousands of loci can be balanced by multilocus overdominance (Wills 1978, 1981).

A general theoretical result is consistent with the overdominance hypothesis. Whether loci contributing to fitness interact additively, multiplicatively, or epistatically, average fitness will generally increase with the number of heterozygous loci (Ginzburg 1979, 1983; Turelli and Ginzburg 1983). Therefore, if we could genotype individuals for 100 polymorphic loci contributing to fitness, we would see that individuals heterozygous for 80 loci had higher fitness than individuals heterozygous for 60 loci, and the fitnesses of both of these groups would exceed that of individuals heterozygous for 20 loci. This result does not rely solely upon strict overdominance, the unconditional superiority of heterozygous genotypes, but it would also apply to marginal overdominance. Imagine selection that is variable through the life cycle, favoring first one homozygote, then another, with the heterozygote always intermediate. The product of fitnesses over time or through various life stages (Mitton and Grant 1984) assigns the highest fitness to heterozygotes. This type of selection, as well as other forms of selection in heterogeneous environments (Hedrick, Ginevan, and Ewing 1976; Hedrick 1986), may be common and important for the maintenance of genetic variation (Gillespie and Turelli, 1989).

The Controversy Gone Stale

The mechanisms underlying inbreeding depression and heterosis will probably never be attributed exclusively to either the dominance hypothesis or the overdominance hypothesis. First of all, this simple dichotomy does not encompass all forms of fitness variation and selection (D. Charlesworth and B. Charlesworth 1987). Other forms of selection, such as frequency-dependent selection and selection in environments heterogeneous in space and time (Gillespie 1978; Hedrick 1986) may be important as well. Although the bulk of the evidence appears now to be most compatible with the dominance hypothesis

(Sprague 1983; Jinks 1983; D. Charlesworth and B. Charlesworth 1987), it is most likely that heterosis and inbreeding depression are caused by a combination of mechanisms. There is no doubt that mutation regularly produces deleterious alleles (Lynch 1988c), but it is also likely that overdominance or marginal overdominance contributes to fitness differentials (Turelli and Ginzburg 1983). It is possible that all of these mechanisms may contribute to inbreeding depression and heterosis, as well as to the typical fitness differentials seen in natural populations.

Although the lingering debate is unlikely to be resolved according to the simplistic dichotomy in which it is presented here, questions about the basic nature of inbreeding depression, heterosis, and fitness differentials are certainly worth pursuing. One profitable way to proceed is to examine whether fitnesses are associated with morphological characters or with single loci (Endler 1986). This treatment will focus upon allozyme loci, and more specifically, variation in levels of heterozygosity among individuals within populations.

Difficulties of Measuring and Interpreting Individual Heterozygosity

Several characteristics of the distribution of individual heterozygosity have led field biologists to be pessimistic about its application to studies of natural populations.

1. The range of individual heterozygosity is narrow in natural populations. This point can be seen most clearly in population C (figure 2.1), which has 2,000 independently assorting polymorphic loci, all segregating two alleles with frequencies of 0.5. This population is randomly mating, and is in Hardy-Weinberg equilibrium. Mean heterozygosity in this population is 1,000, and 95% of the population falls between 955 and 1,045 heterozygous loci.

2. Alterations in the mating system can profoundly affect the level of heterozygosity, compared with the limited level of variation found within an outcrossing population. Just a single generation of selfing decreases heterozygosity by half (population B, figure 2.1), dropping the selfed population far below the limits of the outcrossing population. Continued inbreeding ultimately depletes the population of all heterozygosity, producing a totally homozygous, inbred population (population A, figure 2.1). Thus, variation in individual heterozygosity among individuals might be due to the segregation of polymorphic loci in an outcrossing population, variation in the degree of inbreeding among individuals within a population, or some mixture of those two factors.

3. Within outcrossing populations, individual heterozygosity of the

entire genome is difficult or effectively impossible to measure. The genome of *Drosophila* contains approximately 5,000 genes, while the human genome contains approximately 50,000 genes. Electrophoretic surveys may examine 30–50 genes, but the most comprehensive studies can collect data on only 12–20 polymorphic loci. We can rank individuals by their heterozygosity at a dozen or so polymorphic loci, but this ranking cannot reasonably estimate the ranking of individual heterozygosity across the entire genome (Mitton and Pierce 1980; Chakraborty 1981).

These problems have led population biologists to consider that the vast literature on inbreeding depression and heterosis is perhaps irrelevant and probably inapplicable to the study of genetic variation and fitness differentials within natural populations. Although this conclusion seems to follow from the points above, it is incorrect. Just as fitness appears to parallel individual heterozygosity among the populations in figure 2.1, fitness also follows individual heterozygosity within outcrossing populations. Application of appropriate sampling schemes and statistical analyses can reveal important insights into fitness differentials in natural populations.

ASSOCIATIONS BETWEEN COMPONENTS OF FITNESS AND HETEROZYGOSITY

Viability

Many studies have reported associations between individual heterozygosity for protein polymorphisms and viability in animals. Viability differentials favoring heterozygotes have been documented for the leucine aminopeptidase locus in the blue mussel (*Mytilus edulis*; Koehn, Milkman, and Mitton 1976) and for the tetrazolium oxidase locus in the ribbed mussel (*Modiolus demissus*; Koehn, Turano, and Mitton 1973). Viability selection favored heterozygous genotypes in the American oyster (*Crassostrea virginica*; Singh 1982; Zouros et. al. 1983), the Pacific oyster (*Crassostrea gigas*; Fujio 1982) and *Macoma balthica* (Green et al. 1983). Overwintering mortality increased heterozygosity at twelve polymorphic loci in the common killifish (*Fundulus heteroclitus*; Mitton and Koehn 1975). Survival under laboratory conditions favored heterozygosity in the guppy (*Poecilia reticulata*; Beardmore and Shami 1979), and differential survival of captive cod (*Gadus morhua*) favored heterozygotes at the lactate dehydrogenase and phosphoglucose isomerase loci (Mork and Sundnes 1985; see below). Mortality throughout the life cycle enhanced heterozygosity at the phosphoglucose isomerase locus in *Colias* butterflies (Watt 1979).

Reports of differential survival favoring heterozygous genotypes are also found in plants. For example, viability selection favoring heterozygotes was reported in the slender wild oat (*Avena barbata;* Clegg and Allard 1973). Similarly, selection favoring heterozygous genotypes was seen in annual ryegrass (*Lolium multiflorum;* Mitton 1989). Viability selection also favored heterozygous genotypes in *Liatris cylindracea* (Schaal and Levin 1976), corn (Kahler, Gardner, and Allard 1984), ponderosa pine (*Pinus ponderosa;* Farris and Mitton 1984), and yellow poplar (*Liriodendron tulipifera;* Brotschol, Roberds, and Namkoong 1986).

When viability selection favoring heterozygous genotypes is observed in species with mixed mating systems and documented inbreeding depression, such as ponderosa pine and corn, it is most parsimonious to consider the selection to be against inbred individuals. Because selfed individuals are more homozygous than outcrossed individuals, and because marked inbreeding depression is commonly seen in selfed progeny, it is most parsimonious to assume that the homozygous individuals that died were produced by selfing. But in some studies it is clear that selection against inbred individuals is not sufficient to explain the changes observed within a life cycle—one must postulate selection favoring heterozygous genotypes within the outcrossed portion of the population. Such observations are most abundant in forest trees (summarized in Mitton and Jeffers 1989), and examples can be taken from several studies of ponderosa pine. When seeds or seedlings from natural populations of ponderosa pine are sampled, values of F vary from 0.17 in a low-density stand (Farris and Mitton 1984) to 0.02 in typical populations (Linhart et. al. 1981). Seeds collected from the low-density stand were grown in a greenhouse, where mortality removed predominantly homozygous individuals, altering F from 0.17 to -0.03 (Farris and Mitton 1984). This shift in the value of the inbreeding coefficient is most simply explained as selection against inbred individuals. However, the mean value of F is -0.11 in mature individuals in a typical stand, indicating a significant excess of heterozygotes (Linhart et al. 1981; Mitton and Jeffers 1989). Similarly, studies of forest trees that focus upon old or exceptionally large trees reveal excesses of heterozygotes in black spruce (*Picea mariana*), Monterey pine (*Pinus radiata*), Douglas fir (*Pseudotsuga menziesii*), balsam fir (*Abies balsamea*), and jack pine (*Pinus banksiana*) (Mitton and Jeffers 1989). Selection against inbred individuals can decrease values of F to zero, but selection against selfed genotypes cannot produce negative values of F. Negative values of F indicate excesses of heterozygotes, and can only be produced by selection favoring heterozygous genotypes in the outcrossed portion of the population.

Even when selection within the life cycle decreases values of F toward zero, selection against selfed and inbred individuals may not suffice to explain the change in genotypic frequencies. This point can be illustrated with a study of the dynamics of genetic variation in annual ryegrass (*Lolium multiflorum*; Mitton 1989). Seeds were sampled from two natural populations at Lake Berryessa, in Napa County, California, and genotypes from progeny arrays were used to infer maternal genotypes and to estimate components of the mating system (table 2.1). Annual ryegrass employs a mixed mating system, with approximately 20% of its seed produced by selfing. Consequently, values of the inbreeding coefficient (F) in seedlings vary from 0.025 to 0.190, with a mean of 0.103. Reproductively mature plants, however, have values of F that vary from 0.066 to -0.067, with a mean of 0.005. Selection within the life cycle removes predominantly homozygous individuals, reducing the apparent level of inbreeding to essentially zero. However, examination of the selection coefficients associated with the genotypes reveals that the pattern of selection is not explained simply by selection against homozygotes. The value of χ^2 measures the significance of the difference between the genotypic distributions in seedlings and in adults. In each case where the difference is highly significant, the same pattern in relative viabilities appears—the heterozygous genotype exhibits the highest viability, followed by the most common homozygote, with the less common homozygote exhibiting the lowest viability. This pattern of fitnesses is not expected by simple selection against inbred genotypes (the dominance hypothesis), but it is predicted by the adaptive distance model of Smouse (1986), which models the overdominance hypothesis (see below).

Growth Rate

The first studies linking protein heterozygosity to rates of growth were conducted with the American oyster (*Crassostrea virginica*; Singh and Zouros 1978; Zouros, Singh, and Miles 1980). Oyster spat were placed in a cage anchored in a bay for almost a year. The oysters were then harvested, and size (growth rate) was plotted as a function of enzyme heterozygosity at five or seven loci. The rate of growth increased, and the variance of growth decreased, with heterozygosity. In the second experiment, the mean weights of the most homozygous and the most heterozygous oysters were 4.28 grams and 7.33 grams, respectively. A similar study was conducted with the blue mussel, and once again growth rate increased with heterozygosity (Koehn and Gaffney 1984). Growth rate has also been reported to increase with heterozygosity in the coot clam (*Mulinia lateralis*; Koehn, Diehl, and Scott 1988), the tiger

TABLE 2.1 Frequencies, F Values, and Selection Coefficients in Annual Ryegrass

| | | | Frequencies | | | | | Relative Fitness | | |
| | | | Genotype | | | | | Genotype | | |
	Enzyme[a]	Sample	11	12	22	F^b	χ^{2c}	11	12	22
Site A	PER	Seedlings	257	1,355	2,954	0.088	***	0.59	1	0.87
		Adults	19	171	323	−0.027				
	AP	Seedlings	4,247	411	46	0.137	***	0.64	1	0
		Adults	455	68	0	−0.069				
	PGM	Seedlings	4,425	251	35	0.190	***	0.66	1	0.19
		Adults	481	41	1	0.006				
	PGI	Seedlings	583	1,997	1,958	0.031	*	0.94	0.98	1
		Adults	62	222	221	0.024				
	GOT	Seedlings	47	737	3,872	0.026	NS	0.78	0.86	1
		Adults	4	72	441	0.024				
Site B	PER	Seedlings	278	626	758	0.178	***	0.57	1	0.72
		Adults	22	87	76	−0.028				
	AP	Seedlings	1,529	123	14	0.147	***	0.85	1	0.54
		Adults	168	16	1	0.066				
	PGI	Seedlings	132	644	890	0.025	NS	0.83	0.86	1
		Adults	13	66	106	0.045				

Source: From Mitton 1989.
[a]Insufficient genetic variation at GOT and PGM at Site B to infer maternal genotypes.
[b]Inbreeding coefficient.
[c]χ^2 tests homogeneity of genotypic frequencies in seedlings and adults with rows by column test of independence. * = $P < .05$; *** = $P < .001$; NS = nonsignificant.

salamander (*Ambystoma tigrinum;* Pierce and Mitton 1982), the white-tailed deer (*Odocoileus virginianus;* Cothran et al. 1983), sheep (Baker and Manwell 1977), pigs (Makaveev, Venev, and Baulov 1978), and humans (Bottini et al. 1979). However, there have been other studies that detected no relationship between heterozygosity and growth, with perhaps the most exhaustive study reporting no relationship in humans (Ward et al. 1985).

Physiological Efficiency

It is now apparent that hemoglobin and enzyme polymorphisms can have a major effect on physiological variation (reviewed in Koehn, Zera, and Hall 1983). Furthermore, a variety of empirical studies suggests a general relationship: physiological efficiency increases with protein heterozygosity. These studies employ enzyme polymorphisms to estimate individual heterozygosity and use oxygen consumption at rest to estimate basal metabolic cost. Basal metabolic cost is the energetic cost of keeping the metabolic machinery running. Individuals with the lowest basal metabolic costs can apportion more energy to growth and reproduction. The first data to suggest this relationship came from the American oyster (*Crassostrea virginica*; Koehn and Shumway 1982). Weight-corrected oxygen consumption rates decreased with heterozygosity at five enzyme polymorphisms. This result was consistent in both optimal and stressful environments, but the variation in oxygen consumption among genotypes was much greater in the stressful environment, where the oxygen consumption of the most homozygous genotype was 2.5 times greater than that of the most heterozygous genotype. Basal oxygen consumption also decreased with heterozygosity in the gastropod *Thais hemostomata* (Garton 1984), the coot clam (*Mulinia lateralis*; Garton et al. 1984), the blue mussel (Diehl et al., 1985, Diehl, Gaffney, and Koehn 1986), and the rainbow trout (Danzmann, Ferguson, and Allendorf 1987, 1988). Weight loss in starving clams was least severe in highly heterozygous individuals (Rodhouse and Gaffney 1984).

Enzyme heterozygosity at eight loci was related to both resting and active oxygen consumption in the tiger salamander (*Ambystoma tigrinum*; Mitton, Carey, and Kocher 1986). In resting tiger salamanders, oxygen consumption decreased with heterozygosity, just as it did in marine mollusks and trout (above). This observation is consistent with the superior growth rates observed in heterozygous individuals (Pierce and Mitton 1980). However, the relationship between oxygen consumption and heterozygosity was reversed when the salamanders were forced to exercise (Mitton, Carey, and Kocher 1986). Consequently, the scope for activity, defined as the difference between resting and active oxygen consumption, increased with heterozygosity. Thus, highly heterozygous individuals appear to enjoy two advantages: their lower basal metabolic rates provide greater growth rates, and their greater scopes for activity may provide either greater endurance or higher burst speed during interactions with predators, conspecifics, and prey.

Although the mechanism providing greater physiological efficiency in heterozygotes has not yet been identified, greater insight into the problem has been revealed by studies of protein turnover rate (Hawkins, Bayne, and Day 1986; Hawkins et al. 1989). The half-life of soluble enzymes is on the order of 16 hours, so the majority of the thousands of enzymes needed to control metabolism must be replaced each day. Consequently, a substantial portion of basal metabolic cost, perhaps 20% to 40%, is spent in breaking down damaged proteins and synthesizing new proteins. The protein turnover rate in the blue mussel decreases with enzyme heterozygosity, with the turnover rate of the most homozygous class exceeding the rate of the heterozygous class by 25%. The energy saved in heterozygous individuals can then be invested in growth or reproduction.

Fecundity

If heterozygous individuals are, indeed, more physiologically efficient than homozygous individuals, then heterozygous individuals may have more energy to put into reproduction, and may be more fecund than homozygotes. Data consistent with this hypothesis come from a study of patterns of mating and fecundity in *Drosophila melanogaster* (Serradilla and Ayala 1983) (table 2.2). Reproductively mature females were exposed to males and allowed to lay eggs for several days. Each female was then removed, and the number of eggs in the medium was counted. Whether the data are analyzed by the genotype of the α-glycerophosphate dehydrogenase locus, the alcohol dehydrogenase locus, or the acid phosphatase locus, heterozygotes deposited 40% to 50% more eggs than homozygotes. Data consistent with these were obtained in a study of components of fitness associated with the alcohol dehydrogenase polymorphism in *D. melanogaster* (Bijlsma-Meeles and Bijlsma, 1988). Relative fecundities estimated for slow homozygotes (*SS*), heterozygotes (*SF*), and fast homozygotes (*FF*) were 0.88, 1.00, and 0.61, respectively.

Fecundity also increases with heterozygosity in both the blue mussel (Rodhouse et. al. 1986) and the guppy (Beardmore and Shami 1979). Fecundity increases regularly with individual heterozygosity at six enzyme loci in the brine shrimp (*Artemia franciscana;* Gajardo and Beardmore 1989). Reproductive performance of individuals was examined by counting the lifetime production of nauplii and cysts. The number of nauplii was not significantly related to individual heterozygosity, but the average number of cysts increased from 61 in the most homozygous females to 527 in the most heterozygous females. The total number of offspring (nauplii + cysts) produced was 580, 608, 703, and 900 in females with zero, one, two, and three or more heterozygous loci, respectively.

Associations between enzyme heterozygosity and fecundity are not common, but this does not necessarily indicate that fecundity and heterozygosity are unrelated. Because variation among genotypes has little impact upon the equilibrium distribution of genotypes, associations between heterozygosity and fecundity may not be apparent in anything but direct counts of offspring. Another reason for the paucity of associations is that the apportionment of energy between growth and reproduction may vary among individuals and among genotypes.

TABLE 2.2 Fecundity of *Drosophila melanogaster* as a Function of
Enzyme Genotype

Enzyme	Genotype	No. Eggs ± SE
α-Glycerophosphate	F/F	101.6 ± 3.0
dehydrogenase	F/S	138.9 ± 3.0
	S/S	101.8 ± 3.0
Alcohol	F/F	100.0 ± 3.4
dehydrogenase	F/S	149.3 ± 2.5
	S/S	102.5 ± 3.0
Acid	F/F	102.4 ± 3.3
phosphatase	F/S	160.7 ± 3.8
	S/S	118.0 ± 3.8

Source: Serradilla and Ayala 1983.
Note: For each locus, fecundities are heterogeneous among female genotypes ($P < .001$).

For example, highly heterozygous ponderosa pine tend to grow more
regularly, but to produce cones more irregularly, than more homozygous genotypes (Linhart and Mitton 1985). Similarly, there is a complex relationship between fecundity and heterozygosity in the blue
mussel. The relationship is not seen throughout the life cycle, for
young mussels put more energy into somatic tissue, while older and
larger mussels put most of their energy into gamete production. But if
a sample is restricted to those mussels in which gamete production exceeds somatic tissue production, fecundity increases with heterozygosity (Rodhouse et al. 1986).

INTERPRETING ASSOCIATIONS BETWEEN HETEROZYGOSITY AND COMPONENTS OF FITNESS

While the studies above are a necessary first step in describing the patterns of variation in fitness in populations, there is more than one interpretation of the results. One interpretation is that fitness increases
with heterozygosity (Ginzburg 1979, 1983; Turelli and Ginzburg 1983).
Another interpretation relies upon variation in the degree of inbreeding. Because individuals tend to mate with individuals in close proximity (Shields 1982), natural populations are composed of individuals
with a variety of degrees of inbreeding. Inbred individuals will tend to
be more homozygous at all of their polymorphic loci (figure 2.1). Are
the accounts of selection listed above simply selection against inbred
individuals? Or is selection actually favoring heterozygotes at the protein polymorphisms (or at loci in linkage disequilibrium with the protein polymorphisms)? Decisive evidence for the selective maintenance

of a few protein polymorphisms comes from kinetic, physiological, and demographic studies of polymorphic enzyme loci (Koehn, Zera, and Hall 1983; Koehn 1987), but these programs are labor intensive, and for that reason such studies will never be common. We need a simple test to indicate whether selection is acting upon a polymorphism.

The Adaptive Distance Model

A recent theoretical development (Smouse 1986) provides a test to make inferences concerning the presence or absence of selection at polymorphic loci. This test is particularly useful when correlations between heterozygosity and components of fitness cause an investigator to wonder whether selection is acting directly at the locus in question. The test determines whether the pattern of fitnesses estimated for the genotypes is consistent with the allele frequencies.

Three assumptions lead to a clear relationship between allele frequencies and relative fitnesses:

1. Selection is acting directly upon the polymorphism.
2. There is overdominance or heterozygote superiority at the locus.
3. Allele frequencies are at equilibrium.

If natural selection is acting upon genotypes to produce overdominance, then the fitnesses of those genotypes are predicted by the allele frequencies. Consider a locus with two alleles, A and a, whose frequencies are p and q.

	AA	Aa	aa
Frequency	p^2	$2pq$	q^2
Fitness	$1 - s$	1	$1 - t$
Adaptive distance (X)	$1/p$	0	$1/q$

If we consider the classic case of overdominance, in which the fitness of the AA genotype is depressed by s and the fitness of the aa genotype is depressed by t relative to the heterozygote, then the equilibrium allele frequencies are $p = t/(s + t)$ and $q = s/(s + t)$, and the segregational load is $L = st/(s + t)$. Clearly, the fitness of the more common homozygote exceeds the fitness of the rarer homozygote. Smouse (1986) defined the adaptive distance X of a homozygous genotype as the inverse of the frequency of its allele, and showed that fitnesses were related to segregational load and adaptive distances in the following way:

ln (fitness AA) $= -LX = -[st/(s + t)][1/p]$
ln (fitness Aa) $= -LX = -[st/(s + t)][0]$
ln (fitness aa) $= -LX = -[st/(s + t)][1/q]$

Furthermore, because fitnesses are assumed to be multiplicative across loci, the fitnesses of multilocus genotypes for loci 1 and 2 can be represented as:

$$ln \text{ fitness} = -L_1X_1 - L_2X_2$$

The adaptive distance model generally predicts that fitness will increase with the number of heterozygous loci, but the predictions are much more precise than a ranking of individuals by individual heterozygosity. In addition to the highest fitness falling to the multilocus heterozygote, the common multilocus homozygote should have relatively high fitness, and the rarest multilocus homozygote should have the lowest fitness.

These relations permit us to test the fit of estimated fitnesses to fitnesses predicted from allelic frequencies. If the locus is not influenced by natural selection, there should be no significant differences among the observed fitnesses, and there should be no relationship between observed fitnesses and fitnesses predicted from allele frequencies. Although there will always be the possibility that the marker locus is neutral but in linkage disequilibrium with a locus directly influenced by selection, linkage disequilibrium appears to be a rare phenomenon in outbreeding species (Mukai, Mettler, and Chigusa 1971; D. Charlesworth and B. Charlesworth 1976; Hedrick, Jain, and Holden 1978; Clegg, Kidwell, and Horch 1980), and there should be no predictable relationship between fitnesses at a selected locus and allelic frequencies at a linked marker locus. Therefore, this test may be used to make a weak inference concerning the action of selection upon the locus.

The adaptive distance model has been applied to empirical data on pitch pine to study protein polymorphisms (Bush, Smouse, and Ledig 1987). The radial growth rate of pitch pine increased with heterozygosity of proteins, although this relationship was only observed in mature stands. The initial report (Ledig, Guries, and Bonefield 1983) presented growth rate as a function of heterozygosity class with regressions run on class means rather than on individuals. When the regression was run with individuals, the positive relationship between growth and heterozygosity was no longer statistically significant (Bush, Smouse, and Ledig 1987). The adaptive distance model was then applied to these data to determine whether a higher proportion of the variation in growth rate might be explained. Using multiple regression, the researchers tested the relationship of age-standardized growth rate to adaptive distance at eight polymorphic loci. This analysis explained higher proportions of the variances in growth rates than did linear regressions on the number of heterozygous loci. For example, the proportion of variance of growth explained in two of the populations was $R^2 = 0.49$ and 0.28. The authors concluded that the data were not consistent with selection solely against selfed genotypes, but that specific genotypes played a role in the growth of pitch pine.

Analyses of Genetic Variation in Killifish

The common killifish (*Fundulus heteroclitus*) is abundant in the marshes and bays of the North Atlantic from the Matanzas River in Florida to Newfoundland. It has abundant genetic variation (Mitton and Koehn 1975), and at least some of this genetic variation helps adapt populations to variations in temperature (Mitton and Koehn 1975; Powers 1987). Two polymorphisms, serum esterase (EST) and phosphoglucomutase-1 (PGM) were chosen for this analysis for their high levels of genetic variation. Dilocus genotypes were available from 574 fish collected in 1971 and 1972 from Flax Pond, Long Island, New York. The sample was broken into 107 young, reproductively immature individuals and 467 reproductively mature individuals on the basis of size (Mitton and Koehn 1975).

With the assumption that the sampled population was at equilibrium, the transition from young to mature individuals yielded both single-locus and dilocus fitnesses (table 2.3). Fitness differentials at the PGM locus were moderate, with the heterozygote having the highest viability. The heterozygote at the EST locus also exhibited the highest viability, and the common homozygote had a relative viability of 0.94. The proportion of the less common homozygote (*BB*) dropped substantially in the transition between young and mature, yielding a relative viability of 0.46. These single-locus fitnesses and the allele frequencies in the mature group were used to predict (Smouse 1986) dilocus fitnesses (table 2.3b). The observed and predicted frequencies were in good agreement ($r = 0.91$, $P < .001$). The double heterozygote had the highest relative fitness, as predicted. The most common (*AABB*) and least common (*BBCC*) double homozygotes were predicted to have fitnesses of 0.90 and 0.76, and their fitnesses were estimated to be 0.84 and 0.56.

Analyses of Genetic Variation in Cod

A study of viability of Atlantic cod (*Gadus morhua*) was conducted with a sample of young cod collected with a seine at Trondheimsfjorden, Norway (Mork and Sundnes 1985). The fish refused to eat in captivity. The experiment was terminated when approximately one-third of the animals remained alive; the genotypes of all the animals were obtained for lactate dehydrogenase (LDH) and phoshopglucose isomerase (PGI)

TABLE 2.3 Genotypic Frequencies and Selection Coefficients for PGM and Serum Esterase in Killifish

A. SINGLE LOCUS

	PGM Genotypes		
	BB	BC	CC
Population samples			
Young	36	48	23
Mature	153	214	100
Fitnesses	0.95	1.00	0.97

	EST Genotypes		
	AA	AB	BB
Population samples			
Young	62	35	10
Mature	278	167	22
Fitnesses	0.94	1.00	0.46

B. DILOCUS FITNESSES, observed and (expected)

	PGM Genotypes		
EST Genotypes	BB	BC	CC
AA	0.84	0.89	0.84
	(0.90)	(0.93)	(0.90)
AB	0.85	1.00	0.86
	(0.97)	(1.00)	(0.96)
BB	0.51	0.34	0.56
	(0.77)	(0.79)	(0.76)

Source: Data from Mitton and Koehn 1975.
Note: r, observed and expected fitnesses = 0.91; $P < .001$.

(table 2.4). The observed fitnesses were estimated from the genotypic distributions of surviving and deceased cod.

Selection was intense in the starving cod. The LDH heterozygote had the highest relative fitness, but the fitnesses of the homozygotes were both 0.60. The PGI heterozygote also had the highest relative fitness, and the fitnesses of the common and rare homozygotes were 0.95 and 0.69, respectively.

The correlation between observed dilocus fitnesses and fitnesses predicted by the adaptive distance model was $r = 0.69$ ($p < .05$). The double heterozygote was predicted to have the highest fitness, but it was the second highest with a fitness of 0.96. The fitness of the rare

TABLE 2.4 Genotypic Frequencies and Selection Coefficients for LDH and PGI in Cod

A. SINGLE LOCUS

	LDH Genotypes		
	100/100	100/70	70/70
Population samples			
Dead	72	68	27
Alive	24	48	9
Fitnesses	0.60	1.00	0.60

	PGI Genotypes		
	100/100	100/135	135/135
Population samples			
Dead	77	74	16
Alive	37	39	5
Fitnesses	0.95	1.00	0.69

B. DILOCUS FITNESSES, observed and (expected)

LDH Genotypes	PGI Genotypes		
	100/100	100/135	135/135
100/100	0.43	0.70	0.37
	(0.68)	(0.73)	(0.63)
100/70	0.90	0.96	1.00
	(0.94)	(1.00)	(0.86)
70/70	0.95	0.50	0.00
	(0.56)	(0.59)	(0.51)

Source: Data provided by J. Mork.
Note: r, observed and expected fitnesses $= 0.69$, $P < .05$.

double homozygote (70/70 135/135) was predicted to be 0.51, but not a single individual of this genotype survived, so the fitness was estimated to be 0.00.

Implications for the Overdominance and Dominance Hypotheses

Because these data are consistent with the adaptive distance model, selection appears not only to favor the most heterozygous genotype, but also to discriminate between homozygous genotypes, favoring common homozygotes over rare homozygotes. Thus, these analyses suggest that selection does not simply eliminate inbred individuals, but discriminates among genotypes in the outcrossed portion of the population. Furthermore, these data reveal intense selection in natural populations of ryegrass (table 2.1) and killifish (table 2.3) and in a laboratory experiment utilizing cod (table 2.4). Observed dilocus fitnesses vary from 1.00 to 0.34 in killifish, and from 1.00 to 0.00 in cod. The average observed fitnesses of doubly heterozygous, singly heterozygous, and doubly homozygous genotypes in cod were 0.96, 0.78, and 0.44, respectively. If these fitness differentials could be extended to more loci, the major impact of the mating system upon multilocus genotypes (figure 2.1) could easily produce the severe inbreeding depression characteristic of some species. To the extent that these analyses are relevant to hypotheses concerning the mechanisms underlying inbreeding depression and heterosis, the data are more consistent with the overdominance hypothesis. These data certainly cannot deny the role of deleterious alleles in the continuum between inbreeding depression and heterosis, but they do suggest a role for some form of overdominance.

This tentative conclusion suggests a question: are the differences among genotypes sufficiently reliable and detectable to influence the behavior of individuals? More specifically, could the choice of mates with high fitness potential drive the evolution of female choice?

THE EVOLUTION OF FEMALE CHOICE
Models

The evolution of female choice is a contentious issue among evolutionary biologists. Although the incentive of "good genes" and the choice of highly heterozygous individuals has always had intuitive appeal (Borgia 1979), many analyses of the evolution of female choice have failed to find the conditions that would allow female choice of adaptive phenotypes to evolve (Arnold 1987; Kirkpatrick 1987). Two factors work against the evolution of female choice by selection of mates for the fitness potential of their genotypes. First, when the most fit genotypes are homozygotes, selection will drive the population toward monomorphism, removing the initial incentive for female choice. If the most fit genotypes are heterozygotes, selection drives the system toward the equilibrium allele frequencies. As the system approaches genetic equilibrium, the heritability of fitness tends toward zero (Partridge 1983), once again removing the incentive for female choice. The empirical data as well as the theoretical studies summarized here point to a general advantage of heterozygous genotypes, but choice of heterozygous genotypes cannot sustain the evolution of female choice unless something prevents the system from coming to equilibrium. Oscillations through time driven by frequency-dependent selection imposed by a parasite upon a host might provide the necessary conditions (Hamilton and Zuk 1982). Although this model may not be able to initiate the evolution of epigamic sexual selection (Kirkpatrick 1986; Pomiankowski 1988), once female choice arises, it is possible that natural selection could favor the choice of the currently fittest male genotype (Eshel and Hamilton 1984; Charlesworth 1988). Environmental heterogeneity could also produce the perturbations that prevent a system from coming to equilibrium.

A "good genes" or adaptive choice model for the evolution of female choice (Charlesworth 1988) relies upon a diallelic locus affecting fitness in a population inhabiting a fluctuating environment. Environmental fluctuations are simulated by reversing the fitnesses of the homozygotes. A second locus, segregating independently from the first locus, has an initially rare, dominant allele that causes females to choose heterozygous males. When selection coefficients are in the range of 0.01 to 0.10, female choice evolves. Charlesworth's model takes more than a thousand generations for the choice allele to reach fixation, but a multilocus version of this model (Mitton and Thornhill, in preparation) may drive the choice allele to fixation in 50 to 200 generations.

Another model of female choice relies upon perturbations inherent in the genetic system to prevent the genetic system from coming to equilibrium. This model employs 5 to 100 independently segregating, overdominant loci affecting viability. An additional independent locus segregates a rare, dominant allele that drives females to mate with the most heterozygous male available. The perturbations in the system arise not from environmental heterogeneity but from the fact that perfect males are simply not available (a biological reality!). This point is most easily made with an example. Imagine that 10 overdominant loci affect viability, and that at each of these loci, the fitness potential of homozygotes is 0.90, and the fitness potential of the heterozygote is 1.0. At equilibrium, allele frequencies would equal 0.50, and in each generation, before selection, heterozygosity at each locus would equal 0.50. The probability that an individual would be heterozygous at all 10 loci would be 0.5^{10}, or .00098. The average male would be heterozygous for 5 loci, and individuals heterozygous for less than 2 loci or more than 8 loci would be extremely rare in finite populations. Three examples of the most heterozygous males are presented in figure 2.3. Imagine that male 1 appears in generation 1, male 2 appears in generation 2, and so on. If there is a substantial proportion of choosy females in the population (>0.30), these males will have very high fitness. But they will perturb the population from equilibrium at the loci for which

| | Optimal | | Best Available | |
| | | 1 | 2 | 3 |

Optimal	1	2	3
Aa	AA	Aa	aa
Bb	Bb	bb	Bb
Cc	cc	Cc	Cc
Dd	Dd	DD	Dd
Ee	Ee	Ee	Ee
Ff	Ff	Ff	Ff
Gg	GG	Gg	Gg
Hh	Hh	hh	Hh
Ii	Ii	Ii	II
Jj	Jj	Jj	jj

$$(.5)^7 = .0078$$
$$(.5)^{10} = .00098$$

FIGURE 2.3   The optimal genotype and the best available genotypes in a finite population. Fitness potentials are highest for genotypes heterozygous at genes A through J. When allele frequencies are equal to 0.50, and heterozygosity at each locus is equal to 0.50, the probability of finding an individual heterozygous for all ten loci is .00098. In populations of a few hundred individuals, we do not expect to find individuals heterozygous for more than seven of the ten selected loci.

they are homozygous. Each locus will experience this recurrent perturbation, simply because the optimal genotypes are not available in finite populations. If the genetic system is not at equilibrium, the offspring of the highly heterozygous males will be more heterozygous than the offspring of a random sample of males, and therefore the offspring of highly heterozygous males will have a higher probability of survival. Consequently, choosy females will have higher fitness, and the allele causing choice will increase in frequency.

Empirical Evidence of Superior Mating Success of Heterozygotes

Models can help us determine what is possible, or perhaps even what is likely. But it is empirical data that allow us to decide which models are most germane to natural populations. The models discussed here describe the evolution of female choice via selection of heterozygous males (Charlesworth 1988; Mitton and Thornhill, in preparation; Mitton and Boyce, in preparation), but there are many other models of the evolution of female choice (Arnold 1987; Kirkpatrick 1987; Pomiankowski 1988). Are there data that indicate that heterozygous males have superior mating success?

The data most directly pertinent to the mating success of male genotypes comes from work on the butterflies *Colias eurytheme* and *C. philodice eriphyle*. The data are most extensive for the phosphoglucose isomerase (PGI) locus, for which there are enzyme kinetic data, demographic data, behavioral data, and data on mating success (Watt 1979, 1983; Watt et al. 1983; Watt, Carter, and Blower 1985; Watt, Carter, and Donohue 1986). There are biochemical differences among the enzyme genotypes of the PGI locus, with some (but not all) heterozygous genotypes being the most kinetically efficient. In general, these favored genotypes are kinetically efficient over a broader range of temperatures than the homozygous genotypes are. The heterozygotes have greater viability, and they fly over a greater range of temperatures than the homozygotes do. But the most remarkable aspect of this story is the mating success of the heterozygotes. Because females use only the sperm from a single male, an analysis of the genotype of the female and the genotypic array of her offspring reveals the genotype of a successfully reproducing male. For comparison, a sample of genotypes flying in the field reveals the male genotypes available for mating. The proportion of heterozygotes in males flying in the field varies from 40% to 56%, but 67% to 85% of the females mate with heterozygous males (table 2.5). This is a marked mating advantage for heterozygotes at this locus. Furthermore, heterozygotes enjoy similar mating advantages at the phosphoglucomutase and glucose-6-phosphate dehydrogenase loci. The pattern of heterozygous advantage for male mating success is relatively consistent across loci, across species, and over years. When the data are combined across years, loci, and species, 46% of the available males were heterozygous, but 69% of the mating success was captured by heterozygotes. If the mating success of heterozygotes is standardized to 1.00, the mating success of homozygotes is only 0.38.

CONCLUSIONS AND RECOMMENDATIONS FOR FURTHER RESEARCH

Selection differentials in natural populations are surprisingly large. This statement is most generally supported by studies of morphological variation (Endler 1986), but it may apply to other classes of genetic variation as well. Examples are presented here to indicate that selection, as measured with protein polymorphisms, can be intense (see tables 2.1, 2.3, 2.4). Further estimates of fitnesses are needed for biochemical and molecular genetic markers, particularly in natural populations.

Several theoretical analyses of fitness determination point to the general result that fitness will generally increase with heterozygosity at

TABLE 2.5 Mating Success in *Colias* Butterflies

		% Heterozygotes		
		Flying[a]	Mating[b]	Probability[c]
Colias eurytheme				
Locus	Date			
PGI	1984	40	77	* * *
	1985	52	67	*
PGM	1984	56	74	+
	1985	46	55	NS
G6PD	1984	46	85	* * *
	1985	47	72	* *
Colias philodice eriphyle				
Locus	Date			
PGI	1984	52	74	* * *
	1985	56	85	* * *
PGM	1984	44	63	* *
	1985	47	63	+
G6PD	1984	33	61	* * *
	1985	38	52	+
Average		46	69	

Source: Data from Carter and Watt, 1988.
[a]Percentage of males in the field heterozygous for the enzyme
[b]Percentage of males successfully siring broods. These genotypes were inferred from the mother's genotype and the distribution of genotypes in her brood.
[c]Probability that the percentages are the same: $+ = P < .10$, $* = P < .05$, $** = P < .01$, $*** = P < .001$.

selected loci. Analyses of multilocus allozyme genotypes have produced data consistent with this expectation; physiological efficiency (measured with resting oxygen consumption), growth rate, and viability often increase with the heterozygosity of allozyme loci. These results are consistent with both the overdominance hypothesis and the dominance hypothesis. The adaptive distance model of Smouse (1986), however, makes specific predictions from the overdominance hypothesis, providing a test that utilizes both allele frequencies and estimates of fitness to determine whether selection is acting upon the locus. Data presented here (see tables 2.1, 2.3, 2.4) reveal patterns of fitness differentials consistent with the overdominance hypothesis. Selection favors heterozygous genotypes, and is most severe in rare homozygous genotypes. Further studies are needed to determine how frequently fitness differentials conform to the predictions of the adaptive distance model. We need to know more about fitness differentials among individuals, the extent to which fitness differentials are genetic, and the models of fitness determination that are most useful. While there are advantages to conducting these studies in the laboratory, intensities of selection and fitness differentials in natural populations are of greater interest.

If heterozygotes are reliably more fit, and if selection differentials in natural populations are sufficiently large to be perceived, then common genetic variation may have an effect on behavior. One possibility is that reliable fitness differentials could drive the evolution of female choice. This possibility is consistent with several models, and is supported by empirical data from *Colias* butterflies. Genetic analyses of male mating success in *Colias* reveal intense sexual selection (see table 2.5). While heterozygotes constitute only 46% of the males available for mating, they achieve 69% of the matings. These data certainly justify further genetic studies of sexual selection.

Acknowledgments

Pat Carter and Anna Goebel contributed comments on earlier drafts of the manuscript. Jarle Mork provided the data needed to conduct the analyses of cod. This effort was substantially helped by a fellowship from the Committee on Research and Creative Work at the University of Colorado. I also wish to acknowledge the hospitality and creative atmosphere of Hopkins Marine Station.

3

The Evolution of Inbreeding in Haplodiploid Organisms

John H. Werren

Chronic inbreeding occurs in many haplodiploid species (Hamilton 1967; Adamson 1989; Kirkendall 1983). For example, Hamilton (1967a) lists species typifying extreme levels of inbreeding, and most of those species are male haplodiploid hymenopterans and mites. Hamilton described a set of characteristics found in many inbreeding species, which includes (1) a subdivided mating structure composed of the progeny of only one or a few parents; (2) small, short-lived and/or poorly dispersing males who are most likely to mate only within their local mating population; and (3) female-biased primary sex ratios, which are selected for under these population structures. Chronically inbreeding species can also be identified by the absence of inbreeding depression (Hoy 1977).

Many parasitoid hymenopterans exhibit these characteristics. They range from extreme inbreeders such as *Mellitobia acasta* (Schmeider and Whiting 1947) to "facultative inbreeders" such as *Nasonia vitripennis* (Werren 1980, 1983). Inbreeding is also likely to be common in haplodiploid oxyuridans (pinworms), xyleborids (ambrosia beetles), and mites (Kirkendall 1983; Adamson 1989; Oliver 1962; Hoy 1977).

Although it has been argued that chronic inbreeding will facilitate the evolution of haplodiploidy (Borgia 1980), for many haplodiploid species inbreeding is likely to represent a derived condition. In the Hymenoptera, many parasitoid chalcidoids have life histories that indicate inbreeding. However, many (if not most) hymenopterans, including the primitive sawflies, are likely to be outbreeders based upon their sex-determining mechanism (Whiting 1943; Crozier 1971; Smith and Wallace 1971). Similarly, in the haplodiploid nematode order Oxyurida (pinworms), life histories and strongly female-biased sex ratios typical of inbreeding are often found; however, many other species have out-breeding population structures (Adamson 1989).

The purpose of this paper is to investigate factors likely to promote the evolution of inbreeding in haplodiploids relative to diploids. Two factors, the role of mutational load and sex ratio control, are investigated. The mutational loads of haplodiploids and diploids are compared under conditions of chronic outbreeding, chronic inbreeding, and during the transition from outbreeding to inbreeding. It is concluded that under a wide range of biologically reasonable values, haplodiploid species will suffer less inbreeding depression than diploid species, and therefore will make the transition from outbreeding to inbreeding more readily. In addition, it is concluded that the ability of haplodiploid species to control sex ratio provides them with a competitive advantage relative to diploid species when inbreeding is common.

GENETIC LOAD IN HAPLODIPLOIDS

Because recessive lethal and deleterious genes are exposed to selection in haploid males, it has been widely recognized that the equilibrium frequencies of such alleles will be lower in haplodiploid species than in diploid species (Crozier 1985). Derivations for haplodiploid and X-linked genes are virtually identical, and the equilibria under mutation-selection balance have been derived by a number of researchers (Haldane 1926; Crozier 1976, 1985; Avery 1984). For example, Crozier (1976, 1985) has contrasted the equilibrium frequencies in haplodiploid versus diploid systems for a fully recessive deleterious allele, where μ = mutation rate per haploid gene and $1 - s$ = fitness of aa recessives (fitnesses of AA and Aa are 1); and for partial recessives, where $1 - s$ = fitness of the aa genotype and $1 - sh$ = fitness of the Aa heterozygotes. Equilibrium frequencies of the deleterious alleles in zygotes (q_z) are shown below.

	Haplodiploids	Diploids
Recessive	$q_z = 3(\mu/s)$	$q_z = (\mu/s)^{1/2}$
Partial recessive	$q_z = 3\mu/s(1 + h)$	$q_z = \mu/sh$
Female-limited recessive	$q_z = (3\mu/2s)^{1/2}$	$q_z = (2\mu/s)^{1/2}$

Based upon the derivation for recessive alleles, it has been argued that the genetic load, i.e., the reduction in fitness due to deleterious alleles maintained in mutational/selection equilibrium (Crow 1970), is lower in haplodiploid species. However, Crozier (1985) has recently ar-

gued that inbreeding depression resulting from genetic load may not
be very different in haplodiploids, particularly if codominant and sex-
limited loci make a major contribution to this load. Similarly, Avery
(1984) has suggested that haplodiploids may suffer greater genetic
load.

The derivations above assume either (implicitly or explicitly) equal
frequencies in males and females for the allele in question. This assumption is not necessarily valid for male haploids, and therefore the
formulations can only be applied to cases of weak selection (small s)
where the deviations in male and female frequencies will be slight.

The equilibrium frequencies and genetic loads are derived below
without the assumption of weak selection. Results show that frequencies for many deleterious genes can be lower in male haploids than
previously expected. The following additional definitions are used:

q_{mz} = frequency of the allele in male zygotes
q_{fz} = frequency of the allele in female zygotes
q_{ma} = frequency of the allele in male adults
q_{fa} = frequency of the allele in female adults
l_m = per locus genetic load in males, or loss of males due to selec-
      tion at the locus in question
l_f = per locus genetic load in females, or loss of females due to
      selection at the locus in question

Assume that selection occurs between the zygote and adult stages
and that mating occurs randomly among surviving adults in a large
outbred population. For a recessive deleterious allele in a haplodiploid,
loss of the allele by selection will occur almost exclusively in males,
since the recessive is expressed in the haploid state at approximately q
versus in the homozygous state at approximately q^2. Because q is itself
assumed to be very small, the effects of selection in the female upon
the equilibrium frequency can be neglected. At equilibrium, the loss of
a recessive deleterious allele will be balanced by the production of new
copies of the allele by mutation, and therefore, the following condition
is approximately satisfied:

$$(q_{fa} + \mu)s = 3\mu$$

Since q_{mz} closely approximates $q_{fa} + \mu$, and q_{fz} closely approximates
$(q_{ma} + \mu + q_{fa} + \mu)/2$, the frequencies for a recessive deleterious allele
are

$$q_{fz} = \mu(3 - s)/s \qquad\qquad q_{mz} = 3\mu/s$$
$$q_{fa} = \mu(3 - s)/s \qquad\qquad q_{ma} = 3\mu(1 - s)/s$$

The per locus genetic load for X-linked genes and haplodiploids, here defined as loss of fitness due to selection against deleterious alleles (Crow 1970), is distributed differently in males and in females. In males, $l_m = q_{mz}s$ and in females $l_f = (q_{ma} + \mu)(q_{fa} + \mu)s$; therefore,

$$l_f = 3\mu^2(3 - 2s)/s \qquad l_m = 3\mu$$

Several interesting conclusions result from these derivations. First, previous derivations for X-linked genes and haplodiploids correctly indicate the frequency of recessive deleterious alleles in male zygotes, but do not accurately reflect frequencies in females. The differences can be as large as 1/3 (for $s = 1$) from the derivations presented earlier. Second, the frequencies of recessive deleterious alleles will differ between males and females, even in the zygotic stage prior to the action of selection. Third, the genetic load in males is independent of s, as is also the case for autosomal genes in diploid species (Haldane-Muller principle [Hartl and Clark 1989]). However, genetic load in females is not independent of s. Fourth, the genetic load is dramatically lower in females than in males, by approximately a factor μ. In a sense this last point is obvious. However, the significance of the sexual asymmetry in load in haplodiploids has not been widely considered, and will be discussed in detail below.

Using a similar approach, the equilibrium frequencies for a partial recessive allele can be derived. Assuming the following fitnesses, $AA = 1, Aa = 1 - sh, aa = 1 - s, A = 1, a = s$, then an equilibrium will be closely approximated when the following is satisfied:

$$(q_{fa} + \mu)s + (q_{fa} + q_{ma} + 2\mu)hs = 3\mu$$

The first term represents loss of the allele through males and the second term represents loss of the allele through females. The formula can be solved by recalling that $q_{ma} = (q_{fa} + \mu)(1 - s)$ and $q_{mz} = q_{fa} + \mu$. Similarly, the formulae for female-limited recessives and female-limited partial recessives are, respectively,

$$s(q_{fa} + \mu)(q_{ma} + \mu)2 = 3\mu$$

$$sh(q_{fa} + q_{ma} + 2\mu) = 3\mu$$

Table 3.1 contrasts the equilibrium frequencies and per locus genetic loads in haplodiploid and diploid species. Based upon the formulae, it can be concluded that equilibrium frequencies for deleterious alleles are lower in haplodiploids than in diploids for recessives, partial recessives, and female-limited genes over a wide range of biologically reasonable values. For partial recessive alleles, the per locus mutational

TABLE 3.1 Equilibrium Frequencies and Genetic Loads for Deleterious Genes in Diploid and Haplodiploid Species

	q_z	Load
Recessive		
Diploid	$(\mu/s)^{1/2}$	2μ
Haplodiploid		
Female	$(3 - s)\mu/s$	$3\mu^2(3 - 2s)/S$
Male	$3\mu/s$	3μ
Variable Dominance ($h > 0$)		
Diploid	μ/hs	2μ
Haplodiploid		
Female	$(3 - s)\mu/s\,[1 + h(2 - s)]$	$2h\mu(3 - s)/[1 + h(2 - s)]$
Male	$(3 - sh)\mu/s[1 + h(2 - s)]$	$(3 - sh)\mu/[1 + h(2 - s)]$
Female-Limited Recessive		
Diploid	$(2\mu/s)^{1/2}$	2μ
Haplodiploid		
Female	$(3\mu/2s)^{1/2}$	$3\mu/2$
Male	$(3\mu/2s)^{1/2}$	$\varnothing$
Female-Limited Variable Dominance ($h > 0$)		
Diploid	$2\mu/sh$	4μ
Haplodiploid		
Female	$3\mu/2sh$	3μ
Male	$3\mu/2sh$	$\varnothing$

Note: Frequencies (q_z) are those in zygotes prior to the action of selection. Diploid frequencies are the same in males and females. Haplodiploid frequencies for males and females are shown separately.

load is higher in males of haplodiploid species, but significantly lower in females of haplodiploid species, than it is in diploids.

Studies in *Drosophila* indicate that lethal alleles detected in natural populations typically are only partially recessive, with estimates of h being approximately 0.02 (Simmons and Crow 1977; D. Charlesworth and B. Charlesworth 1987). Figure 3.1a shows equilibrium frequencies and per locus mutational loads for $h = 0.02$ as a function of selection (s). Results are normalized with respect to the mutation rate (μ). The mutation load in diploids is 2μ and independent of s, as shown by Haldane (1926). However, mutational load in male haplodiploids is considerably higher than this, and increases slightly with increasing s ($l = 3\mu$ when $s = 1$). In female haplodiploids, the opposite pattern results: mutational load is considerably less than in diploids and decreases with increasing s ($l_f = .078\mu$ when $s = 1$). Thus, the pattern for primarily recessive alleles indicates a male load approximately 1.5 times

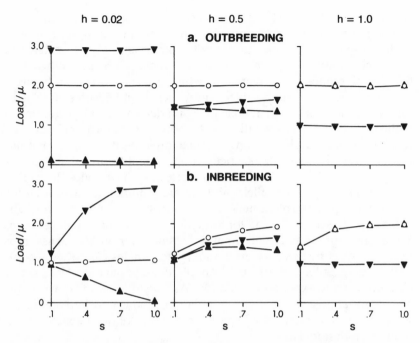

FIGURE 3.1 Per locus genetic loads for diploids ○ and haplodiploids (male ▼, female ▲) are shown as a function of s for different values of dominance (h). Results are present for (a) outbreeding versus (b) chronic inbreeding.

that of diploids, but a female load ranging from 1/10 to 1/25 that found in diploids.

Alleles with additive fitness effects at a locus ($h = 0.5$) continue to show differences between male and female loads, although the differences are less dramatic. Diploid mutational load is independent of s and h over a wide range of these values (Haldane 1926), and remains 2μ. However, both male and female haplodiploid load is lower than 2μ over the range of s. Male load ranges from 75% ($s = 0.1$) to 83% ($s = 1.0$) of diploid load, whereas female load ranges from 74% ($s = 0.1$) to 67% ($s = 1.0$) of diploid load. For dominant alleles ($h = 1.0$) the situation is totally reversed. Both female haplodiploids and diploids have a mutational load of 2μ, whereas haploid males have μ mutational load.

The rather curious changes in mutational load in haplodiploids with changing h and s makes some intuitive sense (retrospectively) if one considers to what extent selective elimination of alleles occurs in males and females depending upon these parameters. In cases of nearly re-

cessive alleles ($h = 0.02$) virtually all selection against a deleterious allele occurs in males, and hence they suffer an increased, and females
a decreased, mutational load. With increasing s, the extent to
which elimination of the allele occurs in males increases. When $h =$
0.5, the difference in the degree to which selection occurs in hemizy-
gous males and heterozygous females is less, but still present. Therefore female load is less than male load and decreases with increasing s,
while male load increases with s. Note, however, that load in both male
and female haplodiploids is lower than in diploids at $h = 0.5$. For dominant alleles, the loss rate is strictly proportional to probability of carrying the allele. Since females are diploid, they have twice the probability of carrying a (rare) allele, and therefore mutational load is 2μ for
females and μ for haploid males. It is interesting to note that for haplodiploids, $l_f + l_m = 3\mu$ and for diploids $l_f + l_m = 4\mu$, for all values of
h and s (except for extremely small values of h, for which the assumption of neglecting selection in homozygote females cannot be made).
This finding reflects the fact that haplodiploids have one less set of
chromosomes to experience mutation. As a result, the "effective" number of mutations entering the population is lower in haplodiploids,
which helps to explain why haplodiploids (and X-linked genes) have a
reduced mutational load.

Calculating Total Mutational Load

The derivations above reflect the *per locus* equilibrium frequencies and
genetic loads for different types of alleles. To estimate the *total* load of
an organism requires several assumptions concerning the number of
genes involved, interactions among alleles, and interactions among
loci. Empirical studies of genetic load in *Drosophila* and other organisms have been reviewed by Simmons and Crow (1977) and D. Charlesworth and B. Charlesworth (1987). From these studies it can be concluded that total genetic load can be partitioned into two major
components, lethal load (L) and detrimental load (D). Estimates indicate that for lethals, $h = 0.02$ and $\mu = 2.5 \times 10^{-6}$. The average fitness
decrement for detrimentals in natural populations is $s = 0.03$, degree
of dominance is $h = 0.2$, and $\mu = 6 \times 10^{-5}$. Finally, a reasonable estimate for the number of loci that can mutate to lethal or deleterious
genes (N) in *Drosophila melanogaster* is 5,000 (Wills 1981).

Most estimates of total genetic load (L) assume multiplicativity, i.e.,
the fitness decrement of a deleterious allele is independent at each locus. Based upon this assumption, total lethal load (L), total detrimental
load (D) and total load (T) are approximately

$$L = 1 - (1 - l)^N$$
$$D = 1 - (1 - d)^N$$
$$T = 1 - [(1 - l)(1 - d)]^N$$

where l and d are the per locus lethal and detrimental loads. Using this approach, the total recessive lethal load in haplodiploids and diploids can be contrasted. Figure 3.2 shows L and D in diploids (male or female), male haplodiploids, and female haplodiploids as a function of mutation rate. As can be seen, total lethal and total detrimental load is strongly influenced by mutation rate. However, over a wide range of mutation rates, lethal loads in male haplodiploids and (male and female) diploids are comparable, whereas lethal loads in haplodiploid females are significantly lower. Total female lethal loads in haplodiploids range from 1/20 to 1/10 of the diploid load. For detrimentals ($s = 0.03$, $h = 0.2$, $\mu = 6 \times 10^{-5}$), female load is 40–60% of the diploid load.

Both empirical and theoretical studies suggest that detrimentals rather than lethals cause most of the total genetic load in a population (D. Charlesworth and B. Charlesworth 1987). Indeed, this is expected because deleterious alleles appear to have higher mutation rates than lethals. Table 3.2 presents the total load in outbreeding diploids and haplodiploids based upon the assumptions of mutation rate, dominance, selection, and number of loci described above. As can be seen, over 90% of the total genetic load is caused by detrimentals in both diploids and haplodiploids. However, the total load (T) for female haplodiploids is approximately 50% that observed in diploids, whereas male haplodiploid total load is only slightly higher than in diploids. It should be noted that these total loads are in reference to an "ideal" fitness in the absence of any deleterious genes. Therefore, a load of 0.461 does not mean that 46.1% of individuals die in the population, but that average fitness is 46.1% of its theoretical maximum in the absence of load. Clearly, it can be concluded that the genetic load in haplodiploids (especially females) will be lower than that in diploids for reasonable biological values. As shown in figure 3.2, it can be concluded that the result is likely to be robust for a wide range of mutation rates for lethals and detrimentals, so long as those mutation rates are comparable in haplodiploids and diploids.

Significance of Sexual Asymmetry in Load

In species with little or no male parental care (e.g., most insects) and under conditions of exponential growth, the growth potential of a pop-

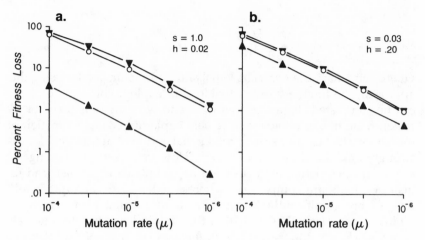

FIGURE 3.2   Total lethal load (a) and total detrimental load (b) are shown as a function of mutation rate for diploids and haplodiploids. Symbols are the same as in Figure 3.1. Load is represented as percent fitness lost. Results assume that $s = 1$, $h = 0.02$, $\mu = 2.5 \times 10^{-6}$ for lethals and that $s = 0.03$, $h = 0.2$, $\mu = 6.0 \times 10^{-6}$ for detrimentals. Number of loci is 5×10^{3}.

ulation is likely to be more strongly linked to the number of females in the population than to the number of males. This generalization will be true so long as female reproduction is not limited by access to sperm. All other things being equal, the intrinsic rate of increase of a haplodiploid species would be 1.43 times greater than that of a diploid species due to its reduced genetic load, based upon the estimates above. Even if only the total lethal load is considered, haplodiploid population growth rate would be 1.02 times greater. Naturally, many assumptions go into these calculations. However, the population growth advantages are significant, and the analysis therefore suggests that haplodiploid species may have certain competitive advantages relative to diploids due to a reduced female genetic load. This interpretation is directly opposite that of Avery (1984), who concluded that haplodiploid systems should be at a competitive disadvantage due to increased expression of deleterious alleles in males.

Mutational Load Under Chronic Inbreeding

It is not immediately obvious whether chronic inbreeding would result in a lower genetic load than outbreeding. Inbreeding results in lower equilibrium frequencies of deleterious alleles due to their near immediate expression in the homozygous state; however, this very fact also could result in increased genetic load. A common approach for genetic

TABLE 3.2 Genetic Loads in Diploid and Haplodiploid Species

	L	D	T
Outbred			
Diploid (M or F)	0.025	0.451	0.461
Haplodiploid (F)	0.001	0.226	0.227
Haplodiploid (M)	0.034	0.475	0.493
Inbred			
Diploid (M or F)	0.014	0.267	0.277
Haplodiploid (F)	0.001	0.260	0.261
Haplodiploid (M)	0.035	0.284	0.310

Note: Total lethal load (L), total detrimental load (D), and total load (T) are shown for diploid and haplodiploid species. Results assume that for lethals $s = 1$, $h = 0.02$, $\mu = 2.5 \times 10^{-6}$; for detrimentals $s = 0.03$, $h = 0.2$, $\mu = 6.0 \times 10^{-5}$. It is further assumed that both species types have the same functional genome size (5×10^3 loci) and that the diploid species does not have sex chromosomes.

analysis of inbreeding is to use Wright's coefficient of inbreeding (F); however, calculations of F typically assume neutral alleles and therefore are not applicable to an analysis of deleterious genes.

One complication of analyzing genetic load under inbreeding is that selection could be quite different in males and females due to the mating structure. Under inbreeding, male fitness is influenced by competition with other males who are relatives. The degree of competition among male relatives who share a deleterious allele will depend in part on the family size. For example, in a sibling mating system, if only one male is produced per family, then a male carrying a newly arisen deleterious mutation would not compete with nonmutant relatives, and the fitness detriment (s) of the allele would not occur in males (assuming of course that $s < 1$, and the male is healthy enough to adequately inseminate his sisters). On the other hand, if family size is large, then mutant males will compete for mates with nonmutant siblings, and fitness effects will be relevant among males.

A second major consideration is the nature of the fitness effects. For example, if $s = 1/2$ means that mutant males have half the competitive ability of nonmutant males for procuring mates, then the consequences in an inbred system will be quite different from $s = 1/2$, meaning 50% mortality. This is again especially true if family size is small because of the influences upon mate competition between mutant and nonmutant males and the probability that females remain uninseminated. As a result, in order to calculate genetic load under inbreeding, certain explicit assumptions must be made concerning the nature of mate competition and fitness effects.

Having recognized these complications, I will now ignore them by assuming a very large family size in a species with complete sib mating. For haplodiploid species under this regime, six different mating types are possible ($A \times AA$, $A \times Aa$, $A \times aa$, $a \times AA$, $a \times Aa$, $a \times aa$) with frequencies ($G_1 \ldots G_6$) and fecundities ($W_1 \ldots W_6$). The transmission matrix for a deleterious gene is approximately

$$
\begin{bmatrix}
1-s & 0 & \dfrac{(1-s)^2}{2(2-s)} & 0 & 0 & 0 \\[2ex]
0 & 0 & \dfrac{(1-s)}{2(2-s)} & 0 & 0 & 0 \\[2ex]
0 & 1-sh & \dfrac{(1-sh)(1-s)}{2(2-s)} & \dfrac{(1-sh)(1-s)}{2(2-s)} & 0 & 0 \\[2ex]
0 & 0 & \dfrac{1-sh}{2(2-s)} & \dfrac{1-sh}{2(2-s)} & 1-sh & \dfrac{2\mu(1-sh)}{1-\mu s} \\[2ex]
0 & 0 & 0 & \dfrac{1-s}{2(2-s)} & 0 & \dfrac{\mu(1-s)}{1-\mu s} \\[2ex]
0 & 0 & 0 & \dfrac{1}{2(2-s)} & 0 & \dfrac{(1-2\mu)(1-\mu)}{1-\mu s}
\end{bmatrix}
\dfrac{1}{\Sigma G_i W_i}
$$

where $W_1 = 1 - s$, $W_2 = 1 - sh$, $W_3 = (2 - s - sh)/2$, $W_4 = (2 - sh)/2$, $W_5 = 1 - sh$, and $W_6 = 1 - 2\mu sh$.

The transmission matrix for a deleterious gene in a diploid species under sib mating is

$$
\begin{bmatrix}
1-s & \dfrac{(1-s)^2}{2(2-s-sh)} & 0 & \dfrac{(1-s)^2}{4(4-s-2sh)} & 0 & 0 \\[2ex]
0 & \dfrac{(1-s)^2}{2-s-sh} & 0 & \dfrac{(1-s)(1-sh)}{4-s-2sh} & 0 & 0 \\[2ex]
0 & 0 & 0 & \dfrac{(1-s)}{2(4-s-2sh)} & 0 & 0 \\[2ex]
0 & \dfrac{(1-sh)^2}{2(2-s-sh)} & 1-sh & \dfrac{(1-sh)^2}{4-s-2sh} & \dfrac{(1-sh)^2}{2(2-sh)} & 0 \\[2ex]
0 & 0 & 0 & \dfrac{1-sh}{4-s-2sh} & \dfrac{1-sh}{2-sh} & \dfrac{4\mu(1-sh)}{1-2\mu sh} \\[2ex]
0 & 0 & 0 & \dfrac{1}{4(4-s-2sh)} & \dfrac{1}{2(2-sh)} & \dfrac{(1-2\mu)^2}{1-2\mu sh}
\end{bmatrix}
\dfrac{1}{\Sigma G_i W_i}
$$

where $W_1 = 1 - s$, $W_2 = (2 - s - sh)/2$, $W_3 = 1 - sh$), $W_4 = (4 - s - 2sh)/4$, $W_5 = (2 - sh)/2$, and $W_6 = 1 - 2\mu sh$.

This approach is similar to that of Ohta and Cockerham (1974), who derived mutational loads under partial selfing in plants. However, rather than attempting to derive the analytical solutions for these matrices, I will determine equilibrium frequencies by computer calculation. Solutions were determined by iterating for five hundred generations (or until equilibrium frequencies were achieved) for different values of h and s. Once the equilibrium was determined, it was confirmed by reinitializing the starting frequencies first above and then below the equilibrium and determining whether values converged on the equilibrium as expected. Genetic loads were simultaneously determined.

Per locus mutational loads under chronic inbreeding are shown in figure 3.1. The contrast between diploids and haplodiploids is interesting. For nearly recessive genes ($h = 0.02$), mutational load in diploids remains nearly constant at μ. In male haplodiploids, load increases rapidly with increasing s, from μ for small s to 3μ for large s. Female haplodiploid load, on the other hand, declines with increases in s, from μ to 0. This makes intuitive sense, since the stronger selection is, the more likely a nearly recessive allele will be eliminated in males prior to being expressed in females. Thus, males "absorb" the mutational load. The pattern is quite different for codominant alleles ($h = 0.5$). Here, load increases with s for both diploids and haplodiploids. However, load is greater in diploids than in male haplodiploids, and greater in male haplodiploids than in female haplodiploids. For dominant alleles, the situation is reversed; male haplodiploids have the lowest load (μ), whereas load in diploids and female haplodiploids increases with s from μ to 2μ.

Determining the total mutational load under partial inbreeding is complicated, since heterozygosity between loci can be correlated (Bennet and Binet 1956; Weir and Cockerham 1973). However, under the assumption of complete sibling mating made here, the "identity disequilibrium" (Weir and Cockerham 1973) is likely to be small. Table 3.2 shows the calculated values of total genetic load under inbreeding, assuming independence between loci. As can be seen, total genetic loads are dramatically lower (40%) in inbred than in outbred diploids. In contrast, the genetic load in haplodiploid females has not changed appreciably. Male load has declined 40% from the outbred condition. Thus, diploid species appear to "benefit" more than haplodiploids from *chronic* inbreeding, in terms of a reduction in genetic load.

Transition from Outbreeding to Inbreeding

Since outbred haplodiploid species have dramatically lower frequencies of deleterious alleles than do diploids, it seems reasonable that inbreeding depression will be lower in such species (but see Crozier 1985). Therefore, we expect that haplodiploids will more easily complete the transition from outbreeding population structures to inbreeding population structures.

To investigate this effect, let us assume an extreme case. Suppose an outbreeding species immediately switched to obligatory sibling mating. What would be the level of inbreeding depression? To investigate this on a per locus basis, the transmission matrices described above are initialized with the equilibrium frequencies for an outbred population and then iterated over successive generations. Using the standard assumption of multiplicativity, the *total* load can then be determined for each successive generation following the onset of inbreeding. The relative population mean fitness (W_g) at generation g during the transition from outbreeding to inbreeding is then defined as

$$W_g = (1 - T_g)/(1 - T_o)$$

where T_o and Tg are the total genetic loads in the outbred population and at generation g after onset of inbreeding. Inbreeding depression is $1 - W_g$. Most treatments of inbreeding depression compare the fitness of inbreeders to that of outbreeders in an otherwise outbreeding population. Here, inbreeding is population-wide, and therefore the average fitness of individuals in an inbreeding population is compared to average fitness in an outbred population.

Figure 3.3 shows the fitness changes for the first fifty generations following a change from outbreeding to sibling mating in diploid and haplodiploid species. As before, the standard values for mutation rate, selection, dominance, and number of loci are applied to computations of total load. Several conclusions can be drawn. First, the per generation reduction in fitness was significantly lower in haplodiploids than in diploids. Male haplodiploids suffered no inbreeding depression, as is to be expected. Indeed, by the fiftieth generation, male haplodiploids had greater fitness than under outbreeding due to reduced mutational load under inbreeding. Female haplodiploids had a significantly lower fitness loss than their female diploid counterparts. At its lowest point (generation 10) female haplodiploid fitness was near 75%, whereas diploid fitness was 55%. The fitnesses of both types begin to rebound after the tenth generation, due to the elimination of existing deleterious genes and the reduced mutational load under chronic inbreeding. By the fiftieth generation, relative fitness of female diploids was higher than that of female haplodiploids. It should be kept in mind, however, that this is fitness relative to outbreeding. In absolute terms, female haplodiploid fitness is still higher than that of diploids (due to lower genetic load). These results clearly suggest that haplodiploid species can make the transition from outbreeding to inbreeding more easily than can diploid species.

Crozier (1985) has argued that inbreeding depression may not be lower in haplodiploids than in diploids, if a major component of the depression is due to the expression of sex-limited lethals and deleteri-

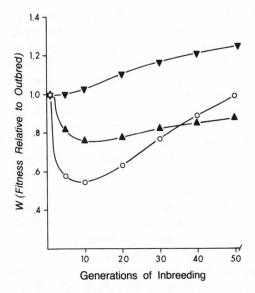

FIGURE 3.3   Change in fitness in diploids and haplodiploids during the transition from outbreeding to inbreeding. Assumptions about lethal and detrimental load previously described apply. Symbols are the same as in Figure 3.1.

ous genes. however, the derivations for sex-limited genes shown in table 3.1 indicate that per locus sex-limited mutational load in haplodiploids is three-fourths the comparable load in diploids. Therefore, although the computations have not been performed, it is clear that the total load from sex-limited genes will be lower in haplodiploid species, assuming equal numbers of loci and equal mutation rates with diploid species.

Sex Determination Load

Current evidence suggests that a single sex-determining locus (Whiting 1943) is the ancestral mode for the hymenopterans. This very tentative conclusion is based upon the observation that the single sex locus mechanism has been identified for diverse hymenopteran taxa, including the honeybees, a braconid (Whiting 1943), and a sawfly (Smith and Wallace 1971), and is also indicated for other species (Crozier 1977). The single sex locus system is not compatible with inbreeding, since inbreeding results in diploid (homozygous) males and triploid females that are typically sterile or have reduced fitness. In contrast, mechanisms compatible with inbreeding have been documented within the Hymenoptera in some parasitoid wasp taxa (e.g.,

Eulophidae and Pteromalidae). More species should be examined before any firm conclusions are drawn about the distribution of sex determination modes within the Hymenoptera. It would be very useful to know whether the single-locus system has evolved many times into mechanisms more compatible with inbreeding, or only a few times within a limited number of taxa (Crozier 1977).

All the foregoing analyses strongly support the view that haplodiploid species will suffer less mutational load and will experience less inbreeding depression due to deleterious genes during a transition from outbreeding to inbreeding than will diploid species. However, arrhentokous species with Whiting's single-locus sex determination experience a genetic load specific to this mode of sex determination (Yokoyama and Nei 1979). Normal males result from hemizygosity at the sex locus and females from heterozygosity. However, diploid males result from homozygosity. These diploid males (or their progeny) have severely reduced fitness. As a result, large breeding populations will suffer a sex determination load under equilibrium in which $1/z$ proportion of males produced will be diploid, given z different sex alleles. Since allele frequencies in a finite population can deviate significantly from equilibrium (also $1/z$), then sex determination load can often be larger (Yokoyama and Nei 1979). Inbreeding with this form of sex determination leads to the generation of large numbers of (homozygous) diploid males, thus resulting in a severe inbreeding depression. Therefore, contrary to the conclusions above, those haplodiploid species with this mode of sex determination will have great difficulty occupying ecological niches requiring inbreeding. Since many haplodiploid species do inbreed, either the single-locus sex determination mechanism is evolutionarily labile, or it is restricted only to certain taxa in the Hymenoptera.

Sex Ratio Control

The apparent ability of haplodiploid species to evolve female-biased sex ratios could provide a major advantage for haplodiploids relative to inbreeding diploid species. Once a species makes the transition to inbreeding, the resultant local mate competition selects for female-biased sex ratios (Hamilton 1967a; Werren 1987). This gives haplodiploid species a greater intrinsic rate of increase relative to diploid species, all other things being equal. For example, if a diploid producing a 50% female sex ratio had an intrinsic rate of increase of 1.2, the haplodiploid equivalent producing 80% females would have a rate of 1.92. Furthermore, haplodiploids with facultative sex ratio adjustment increase their proportion of males under outbreeding conditions (Werren 1983).

If outbreeding is associated with greater population densities, then this response could potentially reduce the probability of resource overexploitation and unstable population cycles (Hassell, Waage, and May 1983). These are, of course, species-level selection arguments; sex ratio control is selected for because of an individual fitness advantage within a species, but species in which individuals exercise such control have a competitive advantage relative to those that do not under inbreeding circumstances.

Another obvious advantage of haplodiploid species under inbreeding is that uninseminated females can produce sons with which to mate (Borgia 1980). Routine mother-son mating in haplodiploids has been documented in mites (Oliver 1962), hymenopterans (Schmeider and Whiting 1947), xyleborids (Kirkendall 1983), and oxyuridan nematodes (Adamson 1989).

DISCUSSION AND CONCLUSIONS

Based on this analysis, it can be concluded that (1) in outbred haplodiploid species, female genetic load is significantly lower than, and male load is similar to, that of diploid species; and (2) haplodiploid species suffer less inbreeding depression resulting from mutational load than do diploids and therefore will more easily make the transition from outbreeding to inbreeding. These conclusions appear to be robust for a range of biologically reasonable values of mutation rate, selection coefficients, and dominance. The reduced genetic load in haplodiploids is due primarily to two factors. First, the "effective" mutation rate is lower in haplodiploid populations (3μ) than in diploids (4μ). Second, increased elimination of partial recessives in haploid males results in lower equilibrium values of deleterious alleles. This second factor also results in the sexual asymmetry in genetic loads found in haplodiploids.

Several complications have not been addressed in this study. The genetic load effects of overdominant alleles have not been investigated here. Empirical studies in *Drosophila* suggest that overdominant mutation load is probably not important (D. Charlesworth and B. Charlesworth 1987). The complications of truncation selection (Kondrashov and Crow 1988) and other epistatic interactions have been ignored. It is unclear how these factors might influence genetic loads in haplodiploids relative to diploids.

Results suggest that chronic inbreeding actually increases mean fitness by reducing genetic load; nevertheless, for diploid species it may be difficult to make the transition from outbreeding to inbreeding, due to the transient inbreeding depression. For haplodiploids the situation is different. Genetic load for females actually increases slightly under chronic inbreeding, although it decreases for males. This is particularly true for mildly deleterious alleles (see figure 3.1), which account for most of the load. Nevertheless, the transition from outbreeding to extreme inbreeding is easier for haplodiploid populations, due to a considerably smaller transient inbreeding depression.

Several authors have investigated the population genetics of modifiers for inbreeding in the presence of mutational load–induced inbreeding depression (Campbell 1986; Lande and Schemske 1985; Charlesworth, Morgan, and Charlesworth 1990). As expected from "cost of sex" arguments (Williams 1975), modifiers for promoting inbreeding are selectively favored under a wide range of circumstances. Because many species outbreed rather than evolving to obligatory inbreeding, the suggestion is that the short-term costs of inbreeding must be very large.

An alternative approach to studying the dynamics of modifiers for inbreeding level is to consider what happens when inbreeding is "imposed" upon a population by ecological circumstances. For example, this could occur in plants at very low density or when pollinators are scarce. Under those circumstances, modifiers allowing inbreeding (e.g., breakdown of self-incompatibility), would clearly be selected for, but the population might not survive the consequences of inbreeding depression. Similarly, in animals, low population density, or few opportunities for mating outside of small natal populations, are circumstances in which inbreeding is imposed upon the population rather than actively selected for. The argument presented in this paper is that haplodiploid populations are more likely to survive this transition than are diploid populations, and therefore may occupy niches requiring inbreeding more frequently than diploids.

Although in the medium term (e.g., after fifty generations) chronic inbreeding may be advantageous relative to outbreeding in diploids because of reduced genetic load, it is likely to become detrimental quickly for a number of reasons. Classic arguments for sex concern the advantages of novel beneficial gene combinations (Stebbins 1957; Maynard Smith 1977) and the ability to adapt to changing environments. However, chronic inbreeders also face the problem of accumulating genes of mild deleterious effects within lineages. Every lineage would accumulate such alleles at different loci via drift in a process analogous to Muller's ratchet (Muller 1964). It would be interesting to determine what levels of outbreeding are necessary to prevent the accumulation of deleterious genes in inbreeders. Haplodiploids are likely to be less prone to an accumulation process because of greater elimination of deleterious genes prior to fixation in a lineage, and because of lower "effective" mutation rates.

Finally, it is worth reemphasizing that the derivations for haplodiploids also apply to sex-linked genes in diploids. The total mutational loads indicated in this paper are for a diploid species with only autosomal genes. Total load for a species with sex chromosomes would simply be $XT_{HD} + (1 - X)T_D$, where X = proportion of loci on the X chromosomes. This suggests the prediction that the probability of making a transition from outbreeding to inbreeding in diploid species will increase as the proportion of (coding) genes on the X increases. Similarly, taxa in which the female is the heterogametic sex (e.g., snakes, birds, and butterflies) will have increasing outbred genetic loads in females as the proportion of genes on the sex chromosome increases, but will have reduced inbreeding depression in females. Thus, we would expect that female heterogametic species in which the sex chromosome represents a large component of the (coding) genome will readily make the transition to chronic inbreeding.

ACKNOWLEDGMENTS

The following individuals are thanked for helpful comments and discussion: L. Beukeboom, H. Breeuwer, D. Charlesworth, R. Crozier, N. Johnson, U. Nur, M. Palopoli and R. Stouthammer. The research here was supported in part by a grant from the NSF.

4

Genetic Incompatibility as a Eugenic Mechanism

Marcy K. Uyenoyama

INTRODUCTION

Reproductive success varies within broods: certain offspring produce many offspring of their own, while others fail even to survive to reproductive age. If it were possible for parents to identify with complete accuracy those offspring destined to be most successful, natural selection would favor parents that discriminate among their own offspring. Selective advantages would accrue to parental discrimination even if the variation in offspring quality had no genetic basis: the critical issue concerns the success of the offspring themselves as carriers of parental genes, with heritability of quality of secondary importance (see also Lloyd 1980; Seavey and Bawa 1986; Stearns 1987).

The problem immediately arises that key components of offspring fitness, particularly viability or fertility, are generally not expressed before conception or sufficiently early in development to permit parents to invest preferentially in certain offspring. A strategy that has met with success in artificial selection involves the improvement of characters expressed late in life by selection on correlated characters that are expressed at an earlier stage (see Falconer 1981: chapter 19). In an analogous fashion, parents may preferentially direct resources toward offspring destined to achieve high reproductive success by discriminating on the basis of a distinct set of signals that can be monitored sufficiently early in development.

In this chapter, I explore the hypothesis that the common evolutionary function of diverse systems of incompatibility in mammals, including self-incompatibility in fungi and flowering plants and maternal-fetal incompatibility in mammals, is to provide a mechanism for the assessment and differential support of offspring. I recognize as a com-

ponent of incompatibility any process that permits preferential investment by parents in offspring of high quality (Uyenoyama 1988a). Such
processes include active termination of certain offspring, encouragement of competition among prospective mates, gametes, or offspring,
and amplification of already expressed variation in offspring viability.
Various kinds of criteria, including characters expressed in prospective
mates as well as in the offspring themselves, may serve as predictors of
offspring quality (see Uyenoyama 1988c). This discussion is restricted
to mechanisms of incompatibility that evolve in response to associative
overdominance at specific antigen loci. Under this hypothesis, heterozygosity at antigen loci serves as a predictor of offspring quality because it provides an indication of general genomic heterozygosity,
which in turn is associated with higher viability or fertility.

Whether or not selection on correlated characters represents a general eugenic mechanism that operates under natural selection depends
upon three issues. Given the existence of an association between offspring reproductive success and characters with early expression, will
parental discrimination based on the early characters evolve? How
does parental discrimination based on early characters affect the association between early and late characters? What conditions generate
associations between early and late characters? I review results derived
from quantitative models (Uyenoyama 1988d; 1989a, 1989b, 1989c) that
were designed to provide preliminary answers to these questions. The
last section appeals for further experimental and theoretical work crucial for the development of an understanding of the structure and evolutionary function of genetic incompatibility systems.

Mechanisms of Genetic Incompatibility

The primary examples of incompatibility systems that influence mate
formation or the fertility of mating combinations occur in fungi, flowering plants, and mammals. Self-incompatibility loci in fungi and angiosperms define mating types, with mating within types completely
restricted. In mammals, the sharing of histocompatibility antigens can
retard mate formation (reviewed by Boyse, Beauchamp, and Yamazaki
1987) or fail to stimulate maternal recognition of the fetus (reviewed by
Rodger and Drake 1987).

Mating Types in Fungi. The genetic mechanisms controlling the determination of mating type are best known in baker's yeast (*Saccharomyces
cerevisiae;* reviewed by Herskowitz 1988, 1989). Resolution of the molecular basis of mating incompatibility is also proceeding rapidly in *Schi-*

zosaccharomyces (Kelly et al. 1988), *Neurospora* (Glass et al. 1988), and *Ustilago* (Schultz et al. 1990). *S. cerevisiae* cells express one of three morphologically and physiologically distinct forms: the haploid types **a** and α, which undergo mating, and the diploid type **a**/α, which undergoes meiosis and sporulation. Mating types express distinct physiological functions, in particular, the production of a mating pheromone (**a**-factor or α-factor). Mating requires expression of one factor and the receptor for the opposite pheromone type. A single locus, *MAT*, determines mating type. Allele *MAT*α encodes both α1, an activator of α-specific genes, and α2, a repressor of **a**-specific genes. Products of the mating type locus interact with a protein encoded by a distinct gene (*MCM1*) that activates transcription of **a**-specific and α-specific genes (see figure 2 in Herskowitz 1989). Allele *MAT***a** encodes **a**1, which, in association with α2 in **a**/α cells, regulates expression in the diploid cell; in particular, **a**1 and α2 together permit meiosis and sporulation by inhibiting a negative regulator of meiosis. Mating requires expression of a battery of genes distinct from the mating type locus itself. Herskowitz (1988) lists five that are α-specific (essential for mating by α-cells but not **a**-cells), six that are **a**-specific, and six that are required by both cell types.

The extensive body of work on the determination of mating type in yeast provides a wealth of information on the genetic structure of mating systems. The *S*-locus has traditionally been portrayed as a supergene that encodes at least three physiological functions associated with self-incompatibility: antigen expression, antigen recognition, and inhibition of fertilization (see de Nettancourt 1977). This view is consistent with early classical genetic studies on induced and spontaneous mutations (Lewis 1949; Lewis and Crowe 1954; Pandey 1956). In contrast, the model that emerges from the yeast system indicates that although mating type on a gross phenotypic level segregates as a single locus (*MAT*), the locus is not a supergene in the sense of serving as the structural locus for several physiological functions, including pheromone production and reception. Rather, the mating type locus regulates transcription at distinct genes, which in turn control the production of precursors and modifiers that ultimately generate the physiological effects.

Maternal-fetal Incompatibility in Mammals. The major histocompatibility complex (MHC) may interact with reproduction in mammalian systems in a fashion that is functionally analogous to self-incompatibility systems in fungi and higher plants. Rather than partitioning the population into compatible and incompatible mating types, mammalian systems generate partial reductions in the rate of mate formation or fertility between individuals that share MHC or related antigens. Histocompatibility between mother and fetus in the mouse generates reproductive incompatibility: contrary to the reproductive incompatibility expressed in response to Rh factors, the rate of successful gestation appears to *decline* as antigenic similarity between mother and fetus increases.

In humans, low fertility appears to be associated with sharing by mates of HLA or related (TLX) antigens (see review by McIntyre and Faulk 1983). Østergård et al. (1989) reviewed the literature on commercial livestock, which demonstrates that fertility declines as antigenic similarity between mates increases. In pigs, backcrosses involving mates that shared one but not both SLA (swine lymphocyte antigen) haplotypes revealed a strong segregation distortion favoring the unshared SLA haplotype if it carried a particular antigen type (Philipsen and Kristensen 1985; Fredholm and Kristensen 1987). Østergård et al. (1989) noted that antigen sharing between mates was confounded with offspring homozygosity in the studies on swine because transmission of the shared haplotype produced homozygosity in the studies on swine because transmission of the shared haplotype produced offspring that were inbred at the SLA and associated regions. This caveat applies in fact to all conclusions based solely on correlations between fertility and antigen similarity between mates.

To address the problem of distinguishing between inbreeding depression in the offspring and maternal-fetal interactions, direct experiments involving stimulation of the maternal immune system by transfusing the mother with foreign antigens have been conducted in mice and humans. The first experiments of this kind (James 1965; 1967) could not be confirmed by later studies, even though the same laboratory lines were used (Clarke 1971; McLaren 1975). Recent studies, employing more refined techniques, indicated that immunization of pregnant mice reduced the rate of abortion, and that this effect could be transferred to nonimmunized mice through serum alone (Chaouat et al. 1985). These observations have again become controversial due to difficulties in interpreting and repeating these experiments (see review and discussion by Chaouat et al. 1987; Bobé et al. 1986; Tartakovsky 1987). Taylor and Faulk (1981) reported that transfusions of foreign leukocytes during pregnancy resulted in live births in three of four women with histories of spontaneous abortions. Beer et al. (1987) described a larger study, involving the immunization of seventy-seven women, that showed an increase in births in immunized women over a nonimmunized group (which cannot be regarded as a proper control). To con-

trol for any placebo effects of the immunization treatment itself, Mowbray et al. (1985) conducted an experiment involving ninety-seven
couples, in which the women were transfused either with their own
cells or with their partner's cells. This study indicated that transfusion
of paternal cells significantly increased the rate of live births (Mowbray
et al. 1985; Mowbray 1987).

While still controversial, these studies suggest the existence of a system involving interactions between mother and fetus that promotes
the gestation of antigenically dissimilar offspring. Unlike the self-
incompatibility systems described in flowering plants and fungi, mammalian incompatibility systems inhibit but do not preclude mating between individuals that share antigens.

MODELS OF THE ORIGIN OF INCOMPATIBILITY SYSTEMS

At minimum, genetic incompatibility systems of the kind hypothesized
comprise three sets of loci, which encode variation in offspring quality
(genetic load), antigen expression, and maternal response to recognized antigens. The three central issues raised in the introductory section address evolutionary interactions among loci encoding these functions. To identify the key evolutionary phenomena involved and to
develop preliminary answers to the three questions, I constructed and
analyzed a series of simple quantitative models of the origin of integrated incompatibility systems.

Effect of Existing Associations between Genetic Load and Antigen Expression

The MHC/t-complex system. The region on chromosome 17 in the
house mouse that encompasses the MHC and the *t*-complex appears to
possess all of the fundamental features of incompatibility systems of
the kind under discussion: genetic load at the *t*-complex maintained by
segregation distortion, associations between the *t*-haplotype and antigens encoded by the MHC loci, and implication of antigen sharing in
interactions that affect conception or gestation (see Silver [1985] for a
review of the genetic structure; and Uyenoyama [1989a] for an interpretation within the present conceptual framework). The findings that
recombination between + - and *t*-haplotypes occurs at a very low rate
(see Silver 1985) and that such recombinants are likely to be nonfunc-
tional (Sarvetnick et al. 1986) permit restriction of the quantitative
model to the case involving recombination suppression in + /*t* individuals; this restriction formally transforms the three-locus analysis into a

two-locus analysis by fusing the antigen and load loci. My results (Uyenoyama 1989b) indicate that enhancement of the expression of incompatibility in response to antigen sharing evolves if the number of $+$-specific antigens (a) exceeds the number of t-specific antigens (b) by twofold ($a/b > 2$). This result, together with the observation that several $+$-specific antigens and a single t-specific antigen generally occur in natural populations (Nadeau et al. 1981), suggests that modifier loci situated virtually anywhere in the genome would be expected to increase the expression of incompatibility.

Incompatibility serves a eugenic function in $+/t$ mothers by reducing the formation of t/t offspring. Each $+/t$ female recognizes one $+$-specific and one t-specific antigen; the expression of incompatibility inhibits transmission through sperm of $1/a$ $+$-haplotypes and $1/b$ t-haplotypes. Exclusion of t-bearing sperm becomes more effective as the ratio of $+$-specific to t-specific antigens (a/b) increases. This eugenic advantage favors the enhancement of incompatibility, even though the ultimate consequence of sheltering the expression of t-haplotypes is to increase genetic load and depress mean fitness (Uyenoyama 1989a).

Incomplete Associations between Load and Antigen Sharing. To explore further the issue of the evolution of the expression of incompatibility in response to antigen sharing, I studied changes at a modifier locus controlling the response to antigen sharing that is distinct from the locus encoding antigen expression itself (Uyenoyama 1988d). In this model, the third locus, encoding variation in offspring quality, is represented only implicitly through its association with the antigen locus. I assumed that each antigen allele is associated with a particular recessive lethal allele at the load locus; different defective alleles are noncomplementing in their effects on viability. The magnitude of association between the load and antigen loci determines the expected viability of offspring that are heterozygous or homozygous at the antigen locus, given whether they were derived by selfing or by random mating.

I obtained an expression for the minimal association between the antigen and load loci that would permit the enhancement of incompatibility (see equation 17 in Uyenoyama 1988d). For any level of association, there exists some selfing rate that is sufficiently low to ensure the adaptiveness of incompatibility. This result reflects that the expression of incompatibility improves the viability of offspring derived by random outcrossing, but not by selfing, under which gametes bearing either maternal antigen are equally inhibited. Modifiers that improve the viability of their carriers by enhancing heterozygosity at both the antigen locus and the load locus increase in frequency in the population.

Effect of Incompatibility on Associations between Antigen and Load Loci

Selection imposed on a secondary character generates a response in the primary character under improvement, the magnitude of change in which depends on the heritabilities of the two characters and their additive genetic correlation (Falconer 1981: chapter 19). Falconer argued that selective pressures imposed on both characters simultaneously would rapidly exhaust genetic variation that contributes to both characters in the favored directions, resulting ultimately in a negative genetic correlation generated by negative pleiotropy. To explore the effect of maternal preference for antigenically dissimilar offspring on the association between the MHC and the t-complex in the house mouse, I examined the erosion of the state of strong disequilibrium observed in natural populations (Uyenoyama 1989a).

The restriction of t-haplotypes to certain MHC antigens suggested their common origin (Hammerberg and Klein 1975; Levinson and McDevitt 1976), an inference that was confirmed by direct sequence comparisons of the t-complex region itself (Silver et al. 1987). By considering the fate of rare recombinants bearing formerly t-specific antigens on the $+$-haplotype or formerly $+$-specific antigens on the t-haplotype, I examined two questions: whether recombination suppression alone maintains this restriction, and, in particular, whether the expression of mating incompatibility tends to reduce it. My analysis indicates that the expression of incompatibility actively maintains a state of association between the MHC and the t-complex, with the strength of the association expected to decline in its absence. High numbers of t-specific antigens are not maintained within single populations, presumably because formation of the t/t genotype eliminates t-specific antigens. A single t-specific antigen can be maintained, provided that the number of $+$-specific antigens falls below a threshold that depends on the intensity of segregation distortion. Similarly, $+$-specific antigens remain restricted to the $+$-haplotype for sufficiently low ratios of the numbers of $+$-specific to t-specific antigens. For values of the rate of segregation distortion falling in the lower end of the range observed in natural populations ($0.90 < k < 0.95$; see Dunn 1957; Gummere, McCormick, and Bennett 1986), the results of my analysis suggest that one t-specific antigen and up to five $+$-specific antigens can be maintained in single populations. Estimates of the numbers of t-specific and common $+$-specific antigens obtained from surveys of natural populations agree roughly with these figures (Nadeau et al. 1981; Klein and Figueroa 1981); however, the presence of several rare +-specific antigens represents a general feature, the significance of which is not explained by the model.

Origin of Associations between Antigen Expression and Offspring Quality

This summary of analytical models (Uyenoyama 1988d; 1989b) indicates that gametic level associations between particular antigens and particular alleles influencing offspring quality encourage the evolution of prezygotic and postzygotic incompatibility. In the case of the house mouse, the high specificity of the association between the MHC and the *t*-complex is probably historical, with its persistence reflecting strong recombination suppression in the region and the derivation of new *t*-haplotypes from existing *t*-haplotypes (Shin et al. 1982; Silver 1982; Figueroa et al. 1985; Silver et al. 1987). The nature of associations between components of genetic load and antigens that can serve as the basis for the origin of incompatibility, and whether such associations can arise by means other than historical accident, need exploration.

Incompatibility systems may evolve in response to associative overdominance conferred on antigen loci by loci affecting viability or fertility. This view suggests that in addition to disequilibria on the gametic level, incompatibility may arise as a response to higher-order associations, including identity disequilibrium (Cockerham and Weir 1968), an association in heterozygosity among loci. Ohta and Cockerham (1974) showed that purifying selection confers associative overdominance on a neutral locus through identity disequilibrium generated by inbreeding (partial selfing).

To address whether incompatibility systems can evolve in response to associative overdominance generated by identity disequilibrium, even in the absence of special assumptions that ensure preexisting gametic disequilibria, I analyzed a model of the origin of sporophytic incompatibility in a partially selfing population subject to overdominant viability selection at a locus distinct from the incipient S-locus (Uyenoyama 1989c). Individuals that are heterozygous at the viability locus, at which an arbitrary number of alleles segregate, survive to reproductive age at a higher rate than do homozygotes. Initially, all alleles segregating at the incipient S-locus are entirely inactive, neither expressing nor eliciting an incompatibility reaction on any stigma. An incomplete form of sporophytic incompatibility arises upon the introduction of an active S-allele, which causes stigmas carrying the active allele to inhibit weakly fertilization by pollen produced by individuals that share the active allele.

Gametic phase disequilibrium, or linkage disequilibrium, describes associations between alleles (A or a at the first locus, and B or b at the second) held on the same chromosome:

$$D = f(AB)f(ab) - f(Ab)f(aB),$$

in which $f(\cdot)$ denotes the frequency of the corresponding chromosome. Genotypic phase disequilibrium, or within-individual disequilibrium, describes the difference in the frequencies of the two kinds of double heterozygotes:

$$G = f\left(\frac{AB}{ab}\right) - f\left(\frac{Ab}{aB}\right)$$

(see Cockerham and Weir 1977a). In the absence of selection, both D and G converge to zero at a rate determined by the recombination fraction and the level of selfing (Karlin 1968). Selfing reduces the effective rate of recombination, but does not itself generate associations on these levels. However, partial selfing does in fact generate a third kind of association, called identity disequilibrium, which reflects associations between loci in the level of heterozygosity (Weir and Cockerham 1973; Cockerham and Weir 1977a). At equilibrium, the genotypic frequencies for two neutral loci converge to the values given in figure 4.1, in which the departure from random association between loci is proportional to identity disequilibrium (η):

$$\frac{e}{4p_A q_A p_B q_B} = \eta = \frac{4s(1 - s)[1 + \lambda^2(1 - s)]}{(2 - s)^2[4 - s(1 + \lambda^2)]},$$

in which p_A and q_A denote the frequencies of A and a, p_B and q_B the corresponding frequencies at the B locus, s the rate of selfing, and λ is equal to $(1 - 2r)$ for r the rate of recombination (Weir and Cockerham 1973). Positive identity disequilibrium (positive e in figure 4.1) implies an excess of double heterozygotes and double homozygotes, and a deficiency of single heterozygotes, relative to expectation based on the genotypic frequencies at each locus separately. What this means for the question of the evolution of incompatibility is that the incompatibility locus can serve as a predictor of heterozygosity at other loci under partial selfing, even in the absence of gametic level associations.

Identity disequilibrium increases the probability that offspring heterozygous at the S-locus are heterozygous throughout the genome as well. This association, even without prior gametic level disequilibria, promotes the origin of incompatibility (Uyenoyama 1989c). I found that

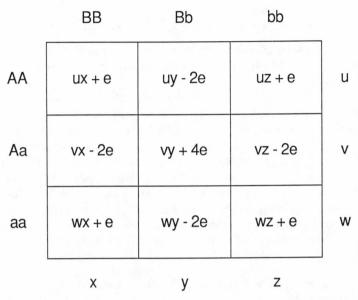

FIGURE 4.1 Genotypic frequencies at equilibrium for two biallelic loci in the absence of gametic and genotypic phase disequilibria. The genotypic frequencies within locus A are denoted u, v, and w; with x, y, and z the corresponding frequencies within locus B.

the twofold cost of outcrossing (see Williams and Mitton 1973; Maynard Smith 1974) is accommodated in part by the reduced average viability among offspring produced by selfing relative to offspring produced by outcrossing; associations evolving between the S-locus and the viability locus, particularly identity disequilibrium, can compensate for the residual cost. High numbers of alleles at the viability locus encourage the evolution of incompatibility by increasing the relative average viability of offspring produced by random outcrossing. Tight linkage (low rates of recombination) between the S-locus and the viability locus encourages the origin of incompatibility by increasing identity disequilibrium; however, incompatibility can arise even in the absence of linkage under sufficiently large levels of genetic load at the viability locus.

This analysis suggests that inbreeding need not be regarded merely as a burden that the population must devise various means to reduce. Rather, inbreeding represents a preadaptation that promotes associations among components of multilocus incompatibility systems.

A Call for Experimental and Theoretical Work

Classical analyses of self-incompatibility in flowering plants (see de Nettancourt 1977) have provided information about the several physiological processes involved, but very little about the genetic mechanisms of its origin. Similarly, although theoretical studies derive from a distinguished tradition (see Wright 1969: chapter 14; Fisher 1958: chapter 4; Ewens 1969: chapter 6), they have for the most part addressed the consequences of the expression of well-developed incompatibility, as opposed to the origin of primitive incompatibility. This appeal for experimental and theoretical work is directed toward improving our understanding of the genetic basis of mating incompatibility, developing a means of distinguishing between inbreeding depression and partial incompatibility, and better characterizing the forms of associative overdominance that can promote the evolution of incompatibility systems.

Gene Regulation in Self-Incompatibility Systems in Flowering Plants

Modern genetic analyses of mating type determination in yeast have revealed a complex and tightly coordinated multilocus system underlying what appears on a gross morphological level to be a monogenic sex determination system. Although theoretical arguments can be contrived to support the classical view that all physiological effects of self-incompatibility in flowering plants derive from a single polypeptide chain encoded by the S-locus (Charlesworth 1982), the genetic basis of self-incompatibility is an empirical question whose resolution appears to be forthcoming.

Nucleotide and amino acid sequences associated with S-locus-specific glycoproteins have been analyzed in both sporophytic (Nasrallah et al. 1985; Takayama et al. 1987; Nasrallah et al. 1987) and gametophytic (M. A. Anderson et al. 1986, 1989) self-incompatibility systems. In *Brassica*, the appearance of the glycoproteins in the seed parent coincides with the onset of the expression of self-incompatibility (see review by Nasrallah and Nasrallah 1986). The glycoproteins appear to function as receptors or sites of recognition of S-alleles in germinating pollen. At present, the question of whether the same locus that encodes glycoproteins in the stigma or style also controls specificities expressed by pollen has not been resolved.

Knowledge of the genetic structure of S-locus systems is crucial for the resolution of questions surrounding the origin of self-incompatibility. An alternative to assuming that the S-locus arose as a

supergene entails the evolution of the coordinated expression of multiple loci that encode the several physiological functions associated
with well-developed self-incompatibility (Bateman 1952). Under such a
scenario, the problem of understanding how a supergene can arise by
mutation is replaced by the problem of understanding the generation
of associations among interacting loci and the evolutionary responses
to those associations.

Distinguishing between Inbreeding Depression and Partial Incompatibility

From a functional as well as an evolutionary point of view, one may
find it useful to distinguish between prezygotic and postzygotic reproductive barriers (cf. N. O. Anderson et al. 1989). A simple assay used
in detecting self-incompatibility in hermaphroditic plants involves
comparing seed set in selfed and outcrossed flowers. While a reduction in viability in inbred offspring would not easily be confused with
the complete inhibition of fertilization in well-developed self-
incompatibility systems, the distinction between the expression of inbreeding depression in offspring and the expression of partial or postzygotic incompatibility in parents is less clear. Moreover, because the
central hypothesis in the present discussion portrays parental discrimination as a response to variation in offspring quality, the two phenomena are expected to exhibit a close association (Seavey and Bawa 1986).

A major strength of the immunization experiments in humans and
mice described in the introductory section is the ability to manipulate
the intensity of the interactions between mother and fetus independently of the level of inbreeding in the offspring. This aspect avoids the
confounding of inbreeding depression and incompatibility that
plagues the interpretation of statistical associations between reduced
fertility and antigen sharing between mates. Similar experiments that
involve the modification or disabling of partial or complete incompatibility in plants appear feasible, particularly as our understanding of the
genetic basis of incompatibility improves.

Theoretical Studies of the Evolution of Incompatibility in Response to Associative Overdominance

That the S-locus arose as a supergene is implausible on theoretical as
well as functional grounds (Bateman 1952). Analysis of theoretical
models that incorporate this assumption indicates that the introduction of functional S-alleles into self-compatible populations requires extreme levels of inbreeding depression (Charlesworth and Charlesworth 1979; Uyenoyama 1988b, 1988c). One-locus models serve an

essential function by establishing a framework within which the complex evolutionary phenomena observed in multilocus models can be interpreted. Models of systems of multiple interacting loci must now be developed, not because such models are likely to provide better descriptions of genetic incompatibility systems (although this is probably true), but because the study of phenomena, such as epistasis and genetic associations, that are crucial to the evolutionary hypothesis under investigation requires a multilocus approach.

The theoretical work summarized in this discussion addressed the coevolution of three components of incompatibility systems: variation in offspring quality, antigen expression, and modification of the expression of incompatibility in parents. The evolutionary response to two forms of genetic association between a locus controlling offspring quality and a modifier of incompatibility was investigated. Gametic level disequilibrium reflects nonrandom associations within haplotypes of particular alleles at loci controlling antigen expression and offspring viability or fertility, and identity disequilibrium associations among the levels of homozygosity at those loci. To address whether incompatibility represents an evolutionary phenomenon of general significance, theoretical investigations are needed to determine the minimal restrictions on the nature and magnitude of associations between offspring quality and the criteria available for its assessment that promote the evolution of incompatibility. Are low levels of genetic load at each of many loci throughout the genome equivalent to high levels of genetic load at a few loci with respect to providing a basis for the origin of incompatibility? Do incompatibility systems evolve to respond to particular antigens because they are encoded by loci that exhibit close associations with offspring quality? Do components of incompatibility systems, including loci that influence offspring quality, antigen expression, and response to antigen sharing, tend to evolve closer linkage and higher levels of genetic association?

Inbreeding generates associations that serve as preadaptations for the evolution of incompatibility. Incompatibility responding to particular antigen loci is favored because the antigen loci provide an indication of general genomic heterozygosity. Self-incompatibility in flowering plants may be regarded as more than a mechanism for the avoidance of selfing; in particular, it may represent only an extreme manifestation of a general evolutionary process that improves offspring quality by exploiting associative overdominance.

ACKNOWLEDGMENTS

I thank Hanne Østergård and Ira Herskowitz for discussion and for providing an introduction to their work. This study was supported by U.S. Public Health Service Grant GM-37841.

5

Inbreeding and the Evolution of Social Behavior

Richard E. Michod

In *Sociobiology,* Wilson (1975) referred to the effects of inbreeding on social behavior as one of the "central problems" of sociobiology. In his classic paper on kinship theory, Hamilton (1964a) noted that "it does seem necessary to invoke at least mild inbreeding if we are to explain some of the phenomena of the social insects—and indeed of animal sociability in general—by means of this theory." During the past twenty or so years, there have been a number of analyses of inbreeding and social behavior from a variety of different perspectives. However, the field is still a long way from a complete understanding of this important but complicated topic.

There are two perspectives concerning the relationship of inbreeding to social behavior. The first is the view that social structure creates conditions conducive to inbreeding. The emphasis is then on the concomitant costs of inbreeding depression (for review see D. Charlesworth and B. Charlesworth 1987) and the evolution of mechanisms, such as juvenile dispersal and kin recognition, by which these costs can be avoided. The second perspective concerns the implications of the population and relatedness structures created by inbreeding for the evolution of social behavior. Although both perspectives are related and necessary for a complete understanding of inbreeding and social evolution, this chapter will primarily focus on the second perspective.

Social evolution is interpreted here in a narrow but important sense as the evolution of behaviors that benefit the group but are costly to the individuals that perform them. The term altruism has been applied to such behaviors. However, it should be realized that most of the theory to be discussed applies to the evolution of intraspecific interactions generally. In other words, the effects of the behavior under study, on the fitnesses of both the donor and the recipient, can take any sign.

It is possible to distinguish three main mechanisms for social evolution: kin selection, group selection, and reciprocation. Parental manipulation has been offered as a separate hypothesis for social evolution (see, for example, Alexander 1974). Parental manipulation and kin selection differ in the life stage at which the allele causing sibling altruism is assumed to be expressed and in the distribution of the costs of altruism among the siblings. This gives rise to more relaxed conditions for the spread of sibling altruism under parental manipulation than under offspring control. However, both processes share a number of fundamental correlates, especially from the population genetics point of view, and so parental manipulation will not be considered separately here (see Michod 1982 for further discussion).

The three mechanisms given above are often not mutually exclusive, and conditions promoting the operation of one process often promote the operation of another. The similarities between kin selection and group selection have been extensively studied. The population structures that promote continued contact among the same individuals, as is necessary for the operation of reciprocation, may also promote genetic relatedness. The list of overlap and similarities can go on. Nevertheless, it seems reasonable to begin exploring the effects of inbreeding on social evolution by studying the effects of inbreeding on the above three processes.

INBREEDING

General Comments

Before beginning, it may be worth reviewing some obvious and, perhaps, not so obvious, points about inbreeding. If an individual is inbred, its parents must have been related. Inbreeding reflects the genetic relatedness of mating pairs. As discussed in more detail below, inbreeding also increases genetic variance at the population level. Consequently, it is not obvious that interacting individuals are more related than they would be in a population without inbreeding, although the parents must have been related. Thus, there are two separate issues, relatedness of parents and relatedness of interactants, and these do not necessarily go hand in hand. Indeed, we will see that in some situations, interactants can actually be less related under inbreeding than they are in a randomly mating population.

Alleles may be identical in a variety of different senses. They may be identical in nucleotide sequence or they may have identical effects on the phenotype. The central concept in the study of inbreeding is "identity by descent" (*ibd*), which refers to the property that two alleles are copies of the same DNA sequence in a common ancestor.

The different inbreeding coefficients measure the probability of *ibd* between alleles that are picked in different ways. However, many inbreeding coefficients can be interpreted in another sense, as the correlation in genetic value between two gametes. In this chapter, I will mainly use the probability of *ibd* interpretation, since it is more common and is easily interpretable. However, the correlation interpretation has advantages in some situations, since it can be negative or positive.

Inbreeding Coefficients

At an arbitrary gene locus in diploid organisms there are nine possible states of identity by descent between the two alleles in one organism and the two alleles in another organism, if no distinction is made concerning the parent from which each allele came. Corresponding to each of these nine states of identity by descent are nine "condensed identity coefficients," each giving the probability of one of the nine states. These condensed identity coefficients can be calculated from pedigree information by a variety of procedures (Cockerham 1956; Denniston 1974; Harris 1964). The two allele coefficients discussed below can be calculated in terms of these nine condensed identity coefficients.

The coefficient of inbreeding, f, is the probability that, at an arbitrary gene locus, alleles in two uniting gametes are *ibd*. Consequently, f is also the probability that the alleles in the resulting zygote are *ibd*. As already noted, f may also be interpreted as the genetic correlation between uniting gametes.

The coefficient of consanguinity, f_{AR}, is defined as the probability that a randomly chosen gamete from A is *ibd* to a randomly chosen gamete from R. This coefficient has other names in the literature, including "coefficient of kinship" (Jacquard 1974) and "coefficient of coancestry" (Malecot 1969).

F-Statistics and Group Structure

Populations, especially of social species, are not homogeneous but are usually highly structured. Inbreeding coefficients are useful tools in the study of population structure. One of Sewall Wright's great accomplishments was the development of a series of inbreeding statistics, termed *f*-statistics, for the description of population structure (Wright 1943, 1951, 1965). As we shall see later, this description is especially useful in understanding the effects of inbreeding on social evolution.

From the point of view of *ibd* between alleles, there are three natural hierarchical levels of focus: individuals, subpopulations, and the total population or species. A single inbreeding coefficient cannot begin to completely describe this situation, and so Wright proposed a hierarchical series of three basic coefficients, f_{IS}, f_{IT}, and f_{ST}. At an arbitrary gene locus, alleles can be "picked" from individuals and their *ibd* can either be compared to alleles picked at random from the subpopulation, f_{IS}, or from the total population, f_{IT}. Alleles picked at random from subpopulations can be compared to randomly picked alleles from the total population, f_{ST}.

In subpopulations S, let Q_S be the frequency of an allele and H_S the frequency of heterozygotes. Let Q_T and H_T be the same respective variables in the total population. f_{IS} is usually calculated from the deviation of the actual heterozygosity in subpopulations from that expected if there was random mating, or,

(1) $$f_{IS} = 1 - H_S/2Q_S(1 - Q_S).$$

F_{IT} can be calculated from the deviation of the actual heterozygosity in the total population from that expected if alleles combined randomly, or,

(2) $$f_{IT} = 1 - H_T/2Q_T(1 - Q_T).$$

F_{ST} can be shown to be equal to the ratio of the actual variance in frequency between groups divided by the maximal possible variance if all local subpopulations were fixed for one type or the other, or

(3) $$f_{ST} = \text{Var}[Q_S]/[Q_T(1 - Q_T)]$$

The three f-statistics can be related by the formula

(4) $$f_{IT} = f_{ST} + f_{IS}(1 - f_{ST}).$$

Using equation 4, if any two of the f-statistics are known, the other can be calculated.

F-statistics can be calculated for a variety of situations. They have been used to study kin selection in family-structured populations (Michod 1980) and will be used below to study the effect of inbreeding on group selection.

Group structure is reflected in correlations between alleles picked at different levels and in different ways. As discussed in the last section, inbreeding is a natural tool for analyzing group structure.

Wright (1951) studied how different hierarchical population structures and situations of inbreeding are reflected in the f-statistics given above. First, consider the case in which each subpopulation mates randomly ($f_{IS} = 0$). With slight differentiation among subpopulations, f_{ST} and f_{IT} will both be positive but small. With extreme differentiation among subpopulations, f_{ST} and f_{IT} will both be close to 1. Second, consider the case in which each subpopulation is highly inbred ($f_{IS} = 1$). With slight differentiation among subpopulations, f_{ST} will be positive but small, whereas $f_{IT} = 1$. With extreme differentiation among subpopulations, f_{ST} will be close to 1 and $f_{IT} = 1$. The effect of population subdivision into family groups on f-statistics has been discussed by Wright (1965) and Michod (1980).

The effect of inbreeding on the statistical properties of populations has been explored extensively by Sewall Wright (see, for example, Wright 1951). Let f be the correlation between uniting gametes. The exact effect of f on the statistical properties of populations depends upon dominance, but here we consider only the special case of semidominance. All comparisons made are to an identical population under random mating ($f = 0$). The population mean is not affected by f in this case, although if there is dominance, the mean is decreased by f. The total genetic variance increases by a factor $1 + f$ and the frequency of heterozygotes decreases by a factor of $1 - f$.

INBREEDING AND KIN SELECTION

We discuss the effects of inbreeding on kin selection in two ways. First, we consider the effects of inbreeding on Hamilton's cost/benefit rule, and second, we consider the effects of inbreeding on the between- and within-family components of selection in family-structured populations. As discussed later, Hamilton's cost/benefit rule can be related to the within-family and between-family components of selection in family-structured populations. Consequently, these two approaches are not independent by any means. Cost/benefit rules have been derived using inclusive fitness models as well as family-structured models. The relation between inclusive fitness models and family-structured models has been studied for outcrossing (see Michod 1982 for review), but not for systems of inbreeding. Before considering these issues we will discuss the major underlying factor in both cases, the genotypic distribution of interactions.

Genotypic Distribution of Interactions

To evolve by kin selection, a genetic trait expressed by one individual, the actor, must affect the genotypic fitness of one or more other individuals who are genetically related to the actor in a nonrandom way at the loci determining the trait (Michod 1982). Consequently, one way to analyze the effects of inbreeding on kin selection is to study the effects of inbreeding on genotypic fitness in kin selection models. Recall that although inbreeding reflects the increased relatedness among parents, inbreeding does not necessarily imply that interactants are more related. By considering genotypic fitness, we will see that the main effect of inbreeding is on the genotypic distribution of interactions (Michod 1979, 1980; Wade and Breden 1981; Breden and Wade 1981; Uyenoyama 1984). We will make this explicit in the case of the interaction of altruism, although the points made hold for other interactions.

The fitness of a focal genotype depends upon the costs and benefits of an altruistic act as well as upon the number of altruistic acts the focal genotype donates and receives. The number of acts donated to others depends in most kin selection models only on the genotype of the donor. Thus this component of a genotype's fitness is unaffected by inbreeding in most models. However, the number of altruistic acts received depends upon the genotypes of the other individuals with whom the focal individual interacts and whether those individuals perform altruistic acts. With most kinds of population structures, the genotype distribution of interactants is not independent of the genotype of the focal individual. Furthermore, this genotypic distribution of interactions is profoundly affected by inbreeding (Michod and Anderson 1979; Michod 1979; Elston and Lange 1976; Jacquard 1974). Under inbreeding, pedigree information can be used to obtain the genotypic distribution of interactions as detailed in the references just given.

Hamilton's Cost/Benefit Rule

Hamilton's cost/benefit rule,

$$(5) \qquad\qquad c/b < r,$$

provides the conditions for the increase of an altruistic trait that decreases the fitness of individuals performing the altruistic act by c and increases the fitness of individuals receiving the act by b. The donor and recipient of the altruistic act are assumed to have a coefficient of genetic relatedness, r. What r is and how it should be calculated is a topic in which there has been considerable interest, as discussed in more detail below.

The effect of inbreeding on Hamilton's rule (equation 5) has been studied in two ways. First, workers have assumed the form of the rule derived for outbred populations and then asked how r is affected by inbreeding (see, for example, Hamilton 1972; Michod and Anderson 1979; Bartz 1980; Tyson 1984). Hamilton's rule was first derived for outbred populations, and it was not clear that the form of the rule would apply to inbred populations. Consequently, a second approach was taken to derive a cost/benefit rule for inbreeding populations (Michod 1979; Michod 1980; Uyenoyama 1984; Pamilo 1984). The results of this latter approach will be discussed shortly, after we consider the various coefficients of relatedness and their implications for inbreeding.

Relatedness Coefficients. A variety of coefficients have been proposed for use in the right-hand-side of equation 5, some of which are obvious and others not so obvious functions of inbreeding. This topic has been given considerable attention in the literature and so will not be discussed in detail here except to define the various coefficients for use later in the chapter (see, for overviews, Michod and Hamilton 1980; Michod 1982; Uyenoyama 1984).

The most general coefficient for use in Hamilton's rule is the regression coefficient of relatedness,

$$(6) \qquad\qquad b_{AR} = \text{Cov}(A_P R_G)/\text{Var}(A_G).$$

In this coefficient A and R denote altruist and recipient respectively, but more generally refer to any two individuals; and P and G denote phenotype and genotype, respectively. In Michod and Hamilton (1980), R_G is the frequency of one of the alleles in individuals and takes values of 0.0, 0.5, or 1.0 in diploids. In Uyenoyama (1984), R_G is the additive genotypic or breeding value (Falconer 1981), calculated from the average effects of gene substitution, which minimize the dominance variance associated with the altruism phenotype. An advantage of Uyenoyama's approach is that it handles multiple alleles easily.

In earlier work, Hamilton suggested several other coefficients for use in the study of inclusive fitness and altruism (Hamilton 1971, 1972), which can be related to b_{AR} under simplifying assumptions (Michod and Hamilton 1980). These coefficients are especially relevant here, since they explicitly include inbreeding, and the effects of inbreeding have often been explored using them (see, for example Hamilton [1972], discussed below). These coefficients are,

$$(7) \qquad\qquad h_{AR} = 2f_{AR}/(1 + f_A),$$

where f_{AR} and f_A are the coefficients of consanguinity and inbreeding, respectively, and,

$$(8) \qquad\qquad f_S = 2f_{ST}/(1 + f_{IT}),$$

for use in structured populations characterized by Wright's f-statistics discussed above.

Finally Wright defined the coefficient of relationship,

$$(9) \qquad\qquad r_{AR} = 2f_{AR}/[(1 + f_A)(1 + f_R)]^{1/2}.$$

Although cautioning against the use of such coefficients in populations undergoing selection and inbreeding, Hamilton (1972) used modifications of equation 7 to study the effect of inbreeding on the evolution of eusocial behavior in the social insects. A variety of workers have followed suit. Unfortunately, the overall effect of inbreeding on the coefficients proposed in equations 7 and 8 is ambiguous. These coefficients are inversely related to the inbreeding coefficient of the actor. The rationale for this is that the more inbred an individual is, the more likely that individual will pass on a copy of a particular allele of interest (say the altruistic allele). However, the numerator of both coefficients may be expected to increase with inbreeding if the recipients of the behavior are drawn from the same population as the actor. Consequently, the overall effect of inbreeding is not clear. We will return to this ambiguity later on.

Models that Derive Hamilton's Rule with Inbreeding. Since Hamilton's original rule (equation 5) was derived for outbreeding populations, a more satisfying and explicit approach to the study of inbreeding and kin selection than those given above is to rederive Hamilton's rule for inbred populations.

Michod (1979) derived a cost/benefit rule for the increase of altruism in a population genetic model in which interactants were assumed to be related by the nine condensed identity coefficients discussed above. The nine condensed identity coefficients were used to generate the genotypic distribution of interactions, which is a major component of genotypic fitness in any population genetic model of kin selection, as discussed above. The population processes giving rise to the relatedness situation described by the condensed identity coefficients were not addressed by the model. This approach has the disadvantage of implicitly assuming weak selection so that the neutral pedigree values can be used. However, an advantage of this approach is that arbitrary genetic relationships can be studied.

From the basic gene frequency equation, a cost/benefit rule was derived. The right-hand side (RHS) of the rule was later shown (Michod and Hamilton 1980) to equal b_{AR} (equation 6). The RHS of this inbred rule differed from the RHS of the corresponding outbred rule for weak selection in several important respects. The inbred rule depended on gene frequency and dominance in addition to the identity coefficients. This made stable polymorphisms possible so that altruistic alleles that increased initially would not necessarily increase to fixation. It was also shown that the RHS of the cost/benefit rule was not some simple function of the inbreeding coefficient, f, since the condensed identity coefficients that appeared there could not be expressed in terms of f. In the case in which the interactants were equally inbred, the RHS equaled h_{AR} (equation 7), as originally suggested by Hamilton (1972). However, if the individuals were inbred to different extents, as is common in termite pedigrees, then h_{AR} was not a sufficient predictor of selection. It was also pointed out that inbreeding increases homozygosity, which can dilute out asymmetries in genetic relatedness necessary for kin selection.

Michod (1980) studied the effects of inbreeding in the context of family-structured models for explicitly specified mating systems such as mixed-sib and mixed-selfing. The intensity of inbreeding was altered by changing the relative frequencies of, for example, sib mating and random mating. It was found that in the case of an additive model of altruism, in which costs and benefits combine in an additive fashion, the cost/benefit ratio necessary for the increase of a rare altruistic allele itself increased with the level of sib mating or selfing. Therefore, inbreeding facilitated the increase of altruism in this case. When the costs and benefits combined in a multiplicative fashion, inbreeding still facilitated the spread of altruism for low costs. However, for high costs, such as in the case of total altruism in which the altruist has no individual fitness, the cost/benefit threshold decreased with the level of inbreeding. We shall consider this case next, since it shows how inbreeding affects the genotypic distribution of interactions through altering the family frequencies.

Let a denote the altruistic allele. In the case of total altruism there are only three families possible: $Aa \times Aa$, $Aa \times AA$, $AA \times AA$. Inbreeding increases the frequency of homogenotypic matings at the expense of heterogenotypic matings, which, of course, changes the genotypic distribution of interactions in the total population. The effect of this on the evolution of altruism depends upon how the altruistic a allele does in the families affected, through a relationship that can be explicitly specified (see, for example, Michod 1980, equation 17). From this relationship it can be shown that inbreeding usually decreases the prospects of a completely altruistic gene. We find again, however, that the important effect of inbreeding is on the genotypic distribution of interactions.

Uyenoyama (1984) studied the evolution of sibling altruism in inbreeding diploid and haplodiploid populations, which were specified by explicit mating systems such as mixed parthenogenesis, selfing, and sib mating. She varied the intensity of inbreeding by changing the proportions of random mating and, for example, sib mating. She assumed that the heterozygote was intermediate between the two homozygotes in its propensity to perform altruism (no over- or underdominance). Interactions between siblings were assumed to either take place in "combined broods," in which inbred and outbred offspring interacted freely, or in "separate broods," in which inbred and outbred offspring were kept segregated from each other and interacted only with like types.

Uyenoyama calculated b_{AR} (equation 6), using additive genotypic values for G, by assuming no selection and the asymptotic mating type distribution for the mating system under study. She showed that b_{AR} calculated in this way could be used in Hamilton's rule to determine the initial increase conditions for altruistic alleles under the different mating systems studied.

The covariance between fitness and additive genotypic value was partitioned into between-family and within-family components. Fitness was defined as the ratio of the number of individuals of a certain genotype at the adult stage to the corresponding number at birth. b_{AR} was found to be proportional to the between-family component, with $1 - b_{AR}$ proportional to the within-family component (Uyenoyama 1984, equations 28–33).

If the cost/benefit ratio necessary for altruism assuming no inbreeding held for the genetic system under study (say, diploidy), then inbreeding was said to facilitate altruism if the same ratio was sufficient for the increase of altruism under all levels of inbreeding (fraction of population which, say, sib-mates). In the cases of partial selfing or partial parthenogenesis, it was found that increased inbreeding facilitates altruism in the case of separate broods. However, in the case of combined broods, conditions were discovered in which inbreeding retarded the spread of altruism. Here, high levels of inbreeding facilitated the spread of altruism; however, for low levels of inbreeding, increased inbreeding led to lower levels of altruism than with no inbreeding. In these cases, relatedness values of b_{AR} were less than one-half, the usual value for diploid genetic systems. In the case of partial sib-mating it was generally found that inbreeding facilitated the spread of altruism for haplodiploids and diploids.

Uyenoyama (1984) also found that inbreeding made unequal cost/benefit thresholds more likely, so that the threshold for the increase of an allele that increased the intensity of altruism was different from that for an allele that decreased the intensity of altruism. She emphasized that with inbreeding, fitness becomes inherently multiplicative in situations in which males and females are distinguished. In a family-structured situation with separate sexes, the productivity of families depends upon the female genotype. Inbreeding males mate with a restricted and nonrandom portion of the female population. This is not the case with randomly mating males. Therefore inbred males have different expected fertilities than do outbred males. As a consequence, male fitness in the inbred portion of the population becomes a complicated product of male and female fitness even in the case of "additive" models of altruism. This does not happen in randomly mating populations, since the family fitnesses exactly cancel. This effect of inbreeding on male fitness is further evidence of the profound effect inbreeding can have on the genotypic distribution of interactions.

Inbreeding and Between- and Within-Family Selection

Kin selection can be viewed, and explicitly modeled, as a selection process involving within-family and between-family selection. In such models, individuals are assumed to interact within sibships. It can be shown for random mating and an additive model of weak selection that the total change in gene frequency can be partitioned into two components (Wade 1980), a between-family component, or group selection, and a within-family component, or individual selection. This partitioning has been extended to more complex situations by Uyenoyama (1984). The between-family component of selection is proportional to relatedness (Michod 1980; Uyenoyama 1984) and the within-family components is proportional to one minus relatedness (Uyenoyama 1984).

By definition, altruism is selected against by within-family selection (individual selection) and favored by between-family selection, since the average fitness of a family is directly proportional to the frequency of altruists within the family. Indeed, it can be shown that Hamilton's rule translates into the condition in which between-family selection overrides within-family selection (Wade 1980). Consequently, one may understand the effects of inbreeding on kin selection by studying its effects on the within- and between-family components of selection in family-structured populations.

Family Frequencies. Wade and Breden (1981) and Breden and Wade (1981) used family-structured models to study inbreeding by varying the proportion of within-sibship mating and random mating. They assumed that the fitness interactions take place within families with additive fitness effects and weak selection. The primary effect of inbreeding was on the frequencies of different family types, and, consequently, on the genotypic distribution of interactions. This effect was studied for altruistic alleles that begin in appreciable frequency and for intermediate fitness schemes of altruism.

Breden and Wade pointed out that inbreeding eventually fixes the population for two family types resulting from matings between like homozygotes (*AA* × *AA*, *aa* × *aa*). Matings between like homozygotes result in families that have no genetic variation in them. Consequently, they concluded that inbreeding generally decreases the average within-family variance. The two types of families produced by matings between like homozygotes are also maximally different from each other in frequency of the altruistic allele and in average fitness. Consequently, inbreeding increases the between-family variance. Both of these factors, that is, the decrease of within-family selection and the increase of between-family selection, have the effect of making the evolution of altruism easier. This effect was also emphasized by Boorman and Levitt (1980: 350). They concluded that ". . . a high degree of inbreeding may be expected to favor sib altruism over the random mating case."

During this process of fixation of the two family types, *AA* × *AA* and *aa* × *aa*, the frequencies of other family types are also altered. Breden and Wade found that in the interim, inbreeding increased the frequency of families in which one of the parents is homozygous for the altruistic allele. This facilitates the increase of altruism, since in these families altruism is not selected against *as much* as in the other families, and these families also have the highest family fitnesses. However, for extreme forms of altruism (e.g., the altruist has zero individual fitness), parental homozygosity for the altruistic allele does not exist, removing this effect of inbreeding (Michod 1980; Breden and Wade 1981).

Initial Increase Conditions for Rare Alleles. There are exceptions to the general rule that inbreeding facilitates evolution of altruism by kin selection. These exceptions have been discovered through the study of the initial increase conditions for rare altruistic alleles, as discussed above (Michod 1980; Uyenoyama 1984). Inbreeding produces rare homozygote classes which can, under some circumstances, increase the within-family variance and make altruism less likely than with outbreeding (Uyenoyama 1984). This effect will reappear below in our study of group selection. As already mentioned, extreme forms of altruism, in which the fitness of altruists is zero, can lead to inbreeding retarding the increase of altruism. The main beneficial effect of inbreeding for the evolution of altruism in the studies of Breden and Wade (1981) stemmed from families in which at least one of the parents is homozygous for the altruistic allele, and these families do not exist if altruism is lethal to the altruists.

Conclusions. In conclusion, the evolution of altruism is favored by between-family selection and disfavored by within-family selection. Inbreeding can make the evolution of altruism easier by increasing the between-family variance and decreasing the within-family variance. However, exceptions to this rule exist concerning the increase of rare altruistic alleles or extreme forms of altruism.

INBREEDING AND GROUP SELECTION

Overview

Group selection can be defined as a process of genetic change that is caused by the differential extinction or proliferation of groups of organisms (Wade 1978a). Three distinct phases of group dynamics can be distinguished: extinction, dispersion, and colonization. Differences in extinction or dispersion based on genetic differences between groups must exist for group selection to occur.

Wade (1978a) has distinguished two classes of group selection models, the "traditional" and the "intrademic." The "traditional group" selection models are approximations of Wright's hierarchical model, which can be described by f-statistics as discussed above. In Wright's hierarchical model, a species consists of many local subpopulations within which nonrandom mating (assortative or disassortative mating) may occur and between which limited migration occurs. However, most traditional group selection models do not allow the full richness of factors that can be described by f-statistics. In particular, and of special relevance here, these models focus primarily on migration and do not allow for nonrandom mating within subpopulations. In these models, group selection occurs by the differential extinction and recolonization of the partially isolated local subpopulations. The second class of model, termed the "intrademic group" model, assumes a single panmictic population whose members are distributed into isolated neighborhoods or "trait groups" (Wilson 1980). In this model group selection occurs by the differential dispersion of the trait groups back to the mating pool.

There are a variety of assumptions common to both models (Wade 1978a), and several of these are relevant to the topic of inbreeding. Most important for the study of inbreeding is the assumption of a migrant pool, which is often made in both classes of mathematical models of group selection. The migrant pool assumption requires that individuals leave the local subpopulations to mate in a large panmictic pool, termed the migrant pool, which is formed by random mixing of the local subpopulations. Genetic variation between groups is severely limited in such models, since it can only be created by random sampling from the migrant pool. The migrant pool assumption prohibits local mating, either by assortative or disassortative mating, within subpopulations. As a consequence of this assumption, most group selection models as presently formulated are inappropriate for the study of inbreeding.

The approach taken here is to study the effects of local mating through the use of f-statistics, discussed above, and Price's covariance approach to group selection, discussed next. In this discussion, it is important to distinguish among the various biological factors that cause inbreeding and affect the f-statistics. The primary biological factors involved are migration between subpopulations and assortative mating within subpopulations. Lower migration rates mean greater differentiation of gene frequency between subpopulations, as subpopulations diverge from one another due to genetic drift. At equilibrium between genetic drift and migration,

(10) $$f_{ST} = 1/4(N_e m + 1),$$

where m is the migration rate or probability that a gamete comes from outside the subpopulation and N_e is the effective population size.

Assortative mating refers to a phenotypic or genotypic correlation between mates, which may be positive or negative. A variety of factors can give rise to assortative mating. Specific systems of mating can give rise to assortative mating, including brother-sister mating, selfing, or mixtures of these inbred systems with outcrossing, as in the mixed mating models discussed in the section on kin selection. Mate choice based on genotype, such as the rare male mating advantage in *Drosophila* populations, also produces assortative mating. Because of these multiple factors, the effects of assortative mating are more difficult to represent in general equations without specification of the breeding system.

Covariance Approach

The effect of different hierarchical structures can be explored with the covariance approach to natural selection (Price 1972, 1970; Hamilton 1975; Wade 1985). These references should be consulted, as only a brief development of the main equations will be given here. Let E[x], Cov[x,y], and Var[x] denote, respectively, the expected value, covariance, and variance of variables x and y.

Again let Q_T, Q_S, and Q_I and W_T, W_S, and W_I be, respectively, the frequency of an allele of interest (Q), and fitness (W), in the total population T, subpopulation S, and individual I. Define $b_{W_S Q_S}$ as the regression coefficient of the average fitness of subpopulation S on the frequency of an allele in subpopulation S (this regression will be positive for altruism),

(11) $$b_{W_S Q_S} = \text{Cov}[W_S, Q_S]/\text{Var}[Q_S].$$

The equation for total gene frequency change can be shown to be

(12) $$W_T \Delta Q_T = b_{W_S Q_S} \text{Var}[Q_S] + E[W_S \Delta Q_S],$$

in which ΔQ_T is the change in gene frequency in the total population. The first term on the RHS represents the effect of the between-group variance on selection, while the second term on the RHS is the average of the within-group change in gene frequency due to individual selection.

Equation 12 can be applied at any hierarchical level. Let us focus on the level of individuals I within a particular subpopulation S and the next lower level, which is that of alleles within these individuals. Define $b_{W_I Q_I}$ as the regression of individual fitness on the frequency of a particular allele within individuals (this regression will be negative for altruism),

(13) $$b_{W_I Q_I} = \text{Cov}[W_I, Q_I]/\text{Var}[Q_I].$$

Applying equation 12, we have

(14) $$W_S \Delta Q_S = b_{W_I Q_I} \text{Var}[Q_I] + E[W_I \Delta Q_I].$$

Ignoring mutation, transposition, and the like, the second term of equation 14 will be zero. Note that equation 14 is to be substituted in the last term of the RHS of equation 12.

Effect of Inbreeding on Group Selection

The effect of inbreeding on group selection may now be investigated through the use of Wright's f-statistics and Price's equations 12 and 14. The opportunity for individual selection is proportional to the genetic variance between individuals within groups (first term on RHS of equation 14 and the opportunity for group selection is proportional to the genetic variance between groups (first term on RHS of equation 12). The effect of inbreeding on the within-group and between-group variance may then be studied by expressing these variances in terms of Wright's f-statistics and then using what we know of the effect of inbreeding on f-statistics (see above). This approach is not completely rigorous, since the effects of inbreeding on f-statistics discussed above did not include selection. Clearly, a more rigorous approach is needed using explicit systems of mating. Nevertheless, as a first approach, we hope that the insights gained are accurate, at least for weak selection.

Consider, first, the between-subpopulation variance. From equation (3), we have

$$(15) \qquad \text{Var}[Q_S] = f_{ST}Q_T(1 - Q_T),$$

which can be substituted into equation (12). This shows that the total change in gene frequency is directly proportional to f_{ST}, through its positive effects on between-group selection. What are the effects of inbreeding on f_{ST}? With low migration and the effects of finite effective population size (equation 10), f_{ST} will increase. However, f_{ST} is relatively unaffected by the inbreeding effects of consanguineous matings within subpopulations. Consanguineous matings within subpopulations do not change gene frequency within subpopulations, and so cannot affect f_{ST}.

Now, consider the between-individual variance in gene frequency within a subpopulation in which there are consanguineous matings, $\text{Var}[Q_I]$. As discussed above, inbreeding will tend to increase the within-subpopulation variance in equation (14) to $\text{Var}[Q_I](1 + f_{IS})$, where f_{IS} is the correlation of uniting gametes within subpopulations. This increase in variance is a consequence of the fact that inbreeding produces homozygous individuals that have extreme gene frequencies, that is, more individuals either with gene frequency 0 (say, in the case of aa homozygotes) or 1 (in the case of AA homozygotes).

Putting these two factors together, we see that inbreeding will increase the opportunity for selection, either by increasing the between-individual/within-subpopulation variance in gene frequency, or the between-subpopulation variance in gene frequency, or both. However, it is extremely important to keep in mind the different biological causes of inbreeding. The correlation between gametes picked at random within a subpopulation (relative to the total population), f_{ST}, is a result of low migration and the random differentiation of subpopulations due to genetic drift. The correlation between gametes within individuals relative to randomly picked gametes in a subpopulation, f_{IS}, is a result of assortative mating within subpopulations producing consanguineous matings. The total correlation between uniting gametes, f_{IT}, is a result of both of these factors through equation 4. It is possible for either of these two biological causes of inbreeding to exist without the other. For example, in extremely large subpopulations, the effects of genetic drift will be small, and f_{ST} will likewise be small. Still, consanguineous matings within subpopulations produce $f_{IS}(= f_{IT})$. Likewise, even with random mating within subpopulations ($f_{IS} = 0$), small effective population size produces large $f_{ST}(= f_{IT})$.

The overall effect of these two factors, genetic drift and assortative mating, on the total change in gene frequency depends on the regression coefficients of fitness upon frequency at the two hierarchical levels at which these factors are manifest, the subpopulation level, $b_{W_S Q_S}$, and the individual level, $b_{W_I Q_I}$ (equations 11 and 13). As discussed in more detail in the next section, in the case of altruism, these coefficients have different sign, so there appears to be no clear effect of "inbreeding" in general on the spread of altruism by selection in hierarchically structured populations. To understand the effect of inbreeding on the evolution of altruism in structured populations we must specify which of its biological causes are in effect.

Group Selection of Altruism

The gene frequency change caused by group selection may be in the same or in a different direction than that caused by individual selection. Altruism, by definition, is favored by selection at the group level but disfavored by selection at the individual level. As discussed above, inbreeding can increase both the between-group and the between-individual/within-group variances and hence facilitate the operation of selection at both the group and individual levels. However, since the selection of altruism is favored at the group level but disfavored at the individual level, the effect of inbreeding on the evolution of altruism by group selection is ambiguous unless we specify the biological causes of inbreeding.

Thus it appears that inbreeding could have different effects on the evolution of altruism by kin selection (see above) than by group selection in hierarchically structured populations. Breden and Wade (1981) argued that in family-structured populations, inbreeding tends to increase the frequency of homogenotypic matings and that this tends to decrease the within-family variance. In the family-structured models studied, families function as groups. The decrease in within-group (family) variance is favorable to the evolution of altruism, since this weakens the selection against altruism within families. Although counterexamples can be found (Michod 1980; Uyenoyama 1984), Breden and Wade's point seems to be generally correct, at least in the limit as the inbreeding process proceeds to completion. How can this be reconciled with the apparently different conclusion we have reached for selection in hierarchically structured populations, in which inbreeding, by increasing the between-individual/within-group variance, retards the evolution of altruism?

In family-structured populations, there is a direct relationship between the inbreeding process and the foundation of the groups within which social interactions take place (families). Family groups are a direct product of a single mating. To the extent that matings are between relatives, the resulting offspring arrays can be expected to be less variable. However, this direct relationship between inbreeding and the foundation of groups in family-structured populations is absent in group selection in hierarchically structured populations. In hierarchically structured populations, interactions take place within groups (subpopulations) that include many families. The subpopulation in a hierarchically structured species is more analogous to the total population of a family-structured population, in which the total variance does increase with inbreeding. By studying inbreeding models of multifamily groups (Wade and Breden 1987), the apparent difference between kin selection and group selection concerning the prospects of altruism within groups should be resolved. The beneficial effect of inbreeding on the prospects of altruism due to individual selection should decrease as more and more families contribute to a group, so long as the offspring of families within the same group interact at random.

Hamilton (1971) argued that for the purposes of calculating inclusive fitness in a hierarchically structured population, relatedness could be defined by f_S (equation 8). f_S is directly proportional to f_{ST} and inversely proportional to f_{IT}; recall that f_{IT} measures the inbreeding coefficient of an individual relative to the total population. Since equation 8 is inversely related to f_{IT}, using f_S as relatedness implies that the more inbred individuals are, the less likely altruism is to evolve. On the other hand, the greater the between-subpopulation variance, the more similar are individuals within subpopulations relative to the total population. As discussed above, reduced migration results in inbreeding with an increase in f_{ST}. Likewise, consanguineous matings within subpopulations will increase f_{IT}. Again we see that inbreeding can have conflicting effects on the prospects of altruism. These conflicting effects are reflected in the gene frequency equations 12 and 14 above.

Hamilton's coefficient for use in studying the evolution of altruism in structured populations (equation 8) implies that relatedness should be directly proportional to f_{ST}. This can be confirmed by a dynamic model based on Wilson's (1980) structured deme model of group selection in which mating is random ($f_{IS} = 0, f_{IT} = f_{ST}$). In such "interdemic group" models, individuals interact in local subpopulations, termed "trait groups." It can be shown, using Wilson's own conditions for selection of altruism, that altruism will be selected for if $c/b < f_{ST}$ (Michod and Sanderson 1985), where f_{ST} is defined as in equation 3. The effects of inbreeding within groups are not addressed by this model.

These themes concerning inbreeding are also played out in Sewall right's shifting balance model of evolution (Wright 1932). In this model the total population is composed of partially isolated inbreeding groups. Wright distinguished three phases in the shifting balance process: random drift, mass selection within subpopulations, and interdemic selection. The phase of random drift generates between-subpopulation variation, which allows the total population, or species, to explore the adaptive topography. This process is enhanced by the inbreeding resulting from reduced migration between subpopulations. Likewise, the phase of mass selection should be enhanced by the increased genetic variance within subpopulations resulting from consanguineous matings. Concerning the evolution of altruism, interdemic selection favors altruists, since the average subpopulation fitness is an increasing function of the number of altruists in the subpopulation. However, the phase of mass selection within subpopulations selects against altruists, and consanguineous matings within subpopulations will tend to enhance selection against altruism, as already discussed.

INBREEDING AND RECIPROCATION

Reciprocation represents a conditional strategy by which individuals modify their behavior on information from past interactions. A simple but powerful such strategy is tit-for-tat (TFT), which has figured prominently in theory in this area (Trivers 1971; Axelrod and Hamilton 1981; Brown, Sanderson, and Michod 1982; Michod and Sanderson 1985). The TFT strategy assumes the following: if an individual recognizes its present partner from a previous interaction, then the individual behaves now as its partner did during that previous encounter. On the other hand, if an individual does not recognize its partner, then the individual behaves altruistically. The TFT strategy is often compared within the all D strategy, in which individuals never behave altruistically.

Although the TFT model has been extensively studied, there are no studies of the role of inbreeding in the evolution of social behavior by reciprocation. One approach is to ask what the key variables in TFT models are and how might they be affected by inbreeding.

There are two basic variables in TFT models of reciprocation, which are either implicitly or explicitly defined: α, the total number of interactions per generation, and β, the number of those interactions that are with strangers (Brown, Sanderson, and Michod 1982). Consequently, $\alpha - \beta$ is the number of interactions with recognized individuals. Brown et al. showed that so long as the ratio α/β can be made large, reciprocation may increase when it is arbitrarily rare. This conclusion holds for various multipartner models, in addition to the single-partner model considered by Axelrod and Hamilton (1981).

Other formulations of reciprocation in the literature can be related to the variables α and β. For example, Axelrod and Hamilton's (1981) single-partner model is based on w, defined as the probability an encounter with a partner continues during the next interval of time. Thus, w is the probability of yet another interaction. Since in this model an individual has only one partner per generation and consequently must always "recognize" this partner on all interactions but the very first, $\beta = 1$. Since the interactions with a partner take place in some finite period of time, usually a single generation, there must be some upper limit on the number of interactions possible. Let this limit be u. It is then a simple matter to show that

$$(16) \qquad \alpha = (1 - w^u)/(1 - w).$$

What factors will make α/β large, and how might inbreeding relate to those factors? There is no obvious relation between inbreeding and α, the total number of interactions per generation. What about the number of strangers, β? To the extent that inbreeding and the capacity of recognizing other individuals are based on a common cause, such as population viscosity, the two might be correlated. Yet then the population setting would seem to involve elements of kin selection. It is difficult to see any direct casual role for inbreeding in the evolution of social behavior by reciprocation, which is distinct from its role in kin selection and group selection.

Conclusions

It is important to stress the limitations of this chapter. As discussed in the Introduction, inbreeding is both a cause and an effect of social evolution. On the one hand, social structures are conducive to inbreeding, and an obvious biological effect of inbreeding is inbreeding depression. This may in turn create selection for mechanisms for avoiding inbreeding. On the other hand, inbreeding may change the population and relatedness structure and, as a result, affect the evolution of social behaviors. This chapter reflects the latter perspective.

Social behavior has been interpreted here in the narrow but important sense of altruism, that is, behaviors that benefit other individuals yet are costly to the individuals that perform them.

My focus has been almost entirely on the theory of inbreeding and social evolution with almost no consideration of data. In part, my emphasis on theory reflects the paucity of data that bear directly on the issues raised.

Genotypic Distribution of Interactions

Social behavior requires interactions between individuals, and selection of social behavior requires differences in the distribution of interactions among genotypes. Thus it is not surprising that inbreeding can have profound effects on social evolution, since it directly changes the genotypic distribution of interactions. This is the case in both kin selection and group selection models. In family-structured kin selection models, inbreeding alters the family frequencies, which directly alters the genotypic distribution of interactions. In kin selection models based on identity coefficients, the identity coefficients are used to restructure the genotypic distribution of interactions, which is integral to the calculation of individual fitness.

In hierarchically structured groups, inbreeding increases both the within- and between-subpopulation variance in gene frequency. As a result, inbreeding increases the opportunity for selection at both levels. Since interactions are assumed to take place within subpopulations, and since the altered genetic variance under inbreeding ultimately reflects new genotype frequencies, inbreeding alters the genotypic distribution of interactions.

Effect of Inbreeding on Altruism

Although, by definition, inbreeding reflects genetic relatedness among parents, inbreeding does not necessarily produce greater genetic relatedness among interacting individuals. Inbreeding can both increase and decrease genetic variance, depending upon the level at which we are looking and what kind of population structure is envisioned. For example, the genetic variance of a population is increased by inbreeding, while the offspring arrays of inbred matings are less genetically variable, that is, less heterozygous.

In family-structured kin selection models, inbreeding ultimately produces families that result from matings between like homozygotes. This effect is attained asymptotically. These homogenotypic families have no genetic variation within them, so there is no opportunity for individual selection against the altruistic genotype. These families also differ maximally in both the frequency of the altruistic allele and in average family fitness. As a consequence of both the decreased within-family selection and the increased between-family selection under inbreeding, inbreeding enhances the opportunities for the evolution of altruism in many cases.

However, this effect is most clearly understood for altruistic alleles in appreciable frequency. There are explicit counterexamples for rare altruistic alleles. These counterexamples either stem from extreme forms of selection, in which the altruistic genotype has no individual fitness, or from the generation by inbreeding of normally rare homogenotypic classes in appreciable frequencies. In addition, there may be other, yet undiscovered, counterexamples.

In hierarchically structured populations, the overall effect of inbreeding on the evolution of altruism is unclear. Inbreeding increases both the opportunity for individual selection within groups and group selection by increasing the variance at both levels. However, since the prospects for altruism are negatively related to selection at the individual level but positively related to selection at the group level, the effects of inbreeding on the evolution of altruism are unclear.

Continued inbreeding increases the general level of relatedness within groups, and this should lower the selective scope for selfish behavior within the group. However, at the same time, inbreeding will dilute out the asymmetries in relatedness within the group, thereby reducing the selective scope for differences in altruistic behavior within groups.

Origin versus Maintenance of Social Behavior

There were several instances in the theory reviewed here in which inbreeding had different effects depending upon whether the altruistic allele was rare or common. For example, Wade and Breden (1981) and Breden and Wade (1981) argued that as inbreeding proceeded, homogenotypic matings should become more common, and this should facilitate altruism by decreasing the within-family variance but increasing the between-family variance. The counterexamples to the expectation that inbreeding facilitates social evolution (Michod 1980; Uyenoyama 1984) were usually confined to rare altruistic genes. In this situation inbreeding can generate the usually rare homozygous classes in appreciable frequency and thereby increase the genetic variance within families. These examples are suggestive and raise the general issue of whether inbreeding can have different roles to play in the initial stages of social evolution than it does in the maintenance of established social structures. However, a complete understanding of this issue awaits further analysis.

Directions for Future Work

There is much that we do not know about inbreeding and social behavior from the theoretical point of view. Understanding the effect of inbreeding on group selection has been impeded by the almost universal use of the migrant pool assumption in group selection models, which assumes a random mating pool of migrants each generation. The use of f-statistics in the covariance approach to selection indicates that inbreeding increases the opportunity for selection at both the individual and the group level. However, the net effect of this on the evolution of altruism is unclear. Explicit models of the effects of inbreeding on group selection in hierarchically structured populations are needed. By studying inbreeding models of multifamily groups (Wade and Breden 1987), the apparent difference between kin selection and group selection concerning the prospects of altruism within groups should be able to be resolved. The beneficial effect of inbreeding on the prospects of altruism due to individual selection found in the models of Wade and Breden (1981) and Breden and Wade (1981) should decrease as more and more families contribute to a group, so long as the offspring of families within the same group interact at random.

We need to know more about the joint effects of inbreeding and social behavior. A start in this direction has been made by Wade and Breden (1987) and Breden and Wade (1991), who studied a two-locus model with one locus controlling the amount of sib mating and the second the propensity to be altruistic. In part, the structure of this chapter reflects this lack of integration at the level of theory. Inbreeding depression, the effect of inbreeding on genetic structure, and the effect of social behavior on the evolution of inbreeding all need to be investigated together. Whether inbreeding has different roles to play in the origin of social behavior than in its maintenance needs further study.

6

The Statics and Dynamics of Mating System Evolution

Donald M. Waller

Plants usually pay third parties for the privilege of engaging in the chancy process of pollen transfer. Not surprisingly, a great deal of variation often exists in the quality of these matings. Perhaps in response to this variation, plant species have evolved a remarkable diversity of breeding systems (Darwin 1877; Bristow 1978; Richards 1986). This diversity makes plants excellent subjects for studies of mating system evolution.

The most useful set of species in which to study mating system evolution are those that simultaneously mix two genetically distinct reproductive modes. The mixed mating systems found in many plant species (Allard 1975; Jain 1976) stand as both an enigma and an opportunity in evolutionary biology. The enigma arises because we expect natural selection to favor whichever mode has even a slight selective advantage. The opportunity arises because it may be feasible to directly examine the costs and benefits associated with each reproductive mode. Williams (1975) stressed this advantage by noting that their coexistence implies a balance between the short-term evolutionary forces maintaining each mode of reproduction.

The most common mixed mating system involves outcrossing with variable levels of self-fertilization (Fryxell 1957; Stebbins 1970; Jain 1976). Selfing may occur by any of a number of mechanisms, ranging from occasional selfing between different flowers on the same plant (geitonogamy), through "fail-safe" contrivances to ensure selfing if no cross pollen is received, to consistent selfing effected via reduced flowers. Plant populations also undergo additional inbreeding as the result of assortative mating or of crosses with near neighbors, which tend to be related. Such inbreeding resembles the situation in many animal species, particularly sessile marine invertebrates (Knowlton and Jackson, this volume), lending the review developed here some generality.

Despite more than a century of interest in the evolutionary dynamics of plant breeding systems, no consensus has yet emerged on the evolutionary mechanism(s) responsible for maintaining their diversity. Indeed, many authors emphasize the enigma posed by outcrossed sexuality in general (Williams 1975; Maynard Smith 1978) and the potential instability of mating systems that combine inbreeding and outbreeding in particular (Williams 1975; Lloyd 1979; Lande and Schemske 1985; Charlesworth, Morgan, and Charlesworth 1990).

Here, I first present a general phenotypic model of mating system evolution for species that produce both inbred and outcrossed progeny. This model portrays evolutionary "statics" by representing the relative costs of progeny, the faithfulness with which they transmit maternal genes, and differences in fitness between them as constants. After reviewing the classic mechanisms of inbreeding depression, I then discuss evolutionary "dynamics": how key genetic parameters may covary with differences in population structure. This covariation will determine whether mixed reproductive systems can persist.

COMPONENTS OF PLANT MATING SYSTEM EVOLUTION

The genetic structure of a population, together with patterns of pollen and seed movement, establishes the context in which selection acts on the mating system. Conversely, the mating system reciprocally affects the genetic structure of a population by influencing patterns of mating and gene flow (figure 6.1). Such interactions complicate studies of the evolution of mating systems (Ennos and Clegg 1982; D. A. Levin 1984; Ritland and Ganders 1985). They also raise questions regarding how genetic variation is maintained and how such variation, in turn, affects selection on the mating system (Lewontin 1974; Maynard Smith 1978; Bell 1982; Mitton and Grant 1984; Lewontin 1985; and papers in Michod and Levin 1988). Much of this work has been theoretical, resulting in a multitude of models and continuing debate. Broad surveys of patterns of genetic variation among species (Hamrick, Linhart, and Mitton 1979; Gottlieb 1981; Hamrick 1983; Loveless and Hamrick 1984) have revealed certain patterns, but have not resolved these issues.

Models of mating system evolution can be either simple or complex, depending on their assumptions. Even complex models, however, often contain simplifying assumptions that restrict their generality. A complete model of mating system evolution in a particular species would require assumptions or estimates of a wide variety of parame-

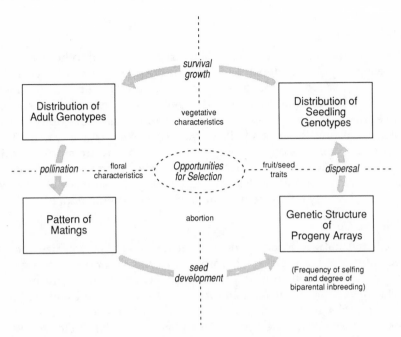

FIGURE 6.1   The reciprocal interactions between population genetic structure and patterns of genetic variability in the progeny. The pattern of mating in a population results from the distribution of adult genotypes in combination with patterns of pollination. The distribution of seedling genotypes, in turn, derives from patterns of seed development and dispersal. Those seedlings that survive to reproduce then constitute the next generation's distribution of genotypes. Selection acts indirectly on the mating system by modifying the characteristics listed along the dotted lines.

ters. Lloyd (1979, 1988) provides epic catalogs of relevant variables under a wide range of botanical scenarios. Self-fertilization offers the immediate rewards of high genetic representation in the next generation (Fisher 1941) and assured reproduction (Baker 1955). In addition, many plant species realize further advantages by reallocating floral resources away from male function in favor of female function. This reallocation is particularly conspicuous in the tiny, closed, and therefore obligately self-fertilized "cleistogamous" flowers that Darwin (1877) recognized as "wonderfully efficient."

Here, I present a simple model of mating system evolution based on three factors of general significance in the evolution of a mixed mating system. The particular form of this model was developed to describe the cleistogamous plant *Impatiens capensis*, and therefore describes a mixture of self- and cross-fertilization. It is general enough, however, to apply to any species producing more and less inbred progeny.

General Model

The invasion and persistence of self-fertilization in an otherwise out-crossing population depends on four factors: (1) the relative costs of producing selfed and outcrossed seeds; (2) the "fidelity" with which each type of seed passes on maternal genetic information, (3) their relative fitnesses, and (4) the genetic associations that develop between alleles affecting fitness and alleles modifying the mating system. The first factor depends on whether outcrossed progeny are more "expensive" to produce, while the second and third depend on genetic differences between more and less inbred progeny. These three factors are described briefly before being combined into a general model.

Relative Costs

There is usually a close relationship between an organism's size and its fecundity, implying that total reproductive output for plants of a given size is limited. Such limits imply that progeny within an organism compete for resources, making the relative costs of inbred and outbred progeny relevant to mating system evolution. These costs (C_S and C_O, for selfed and outcrossed progeny) will generally reflect biological details such as the efficiency with which outcrossed pollen is transferred. In animals with separate sexes, they might reflect the costs of dispersal necessary for outbreeding (Bengtsson 1978). In hermaphroditic plants, they depend on the extent of male costs, pollinator effectiveness, lost opportunities for donating pollen, and whether specializations for self-fertilization exist (Charnov 1982, 1987; B. Charlesworth and D. Charlesworth 1987).

What are these costs and how might they be measured? Clearly, plants require energy, material, and time to produce flowers and ripen fruit. Ecologists have spent some time trying to define the "correct" currency with which to measure these costs (for plants, see, e.g., Bazzaz and Reekie 1985; Bazzaz et al. 1987). Perhaps the best measure from an evolutionist's perspective is one that represents the trade-off between competing components of fitness. Ideally, the (presumably negative) genetic correlation could be measured between competing modes of reproduction (Antonovics 1980; Bell and Koufopanou 1986; Uyenoyama 1988a). To avoid the controlled crosses and progeny testing this approach would require, it is often possible to approximate this correlation using a phenotypic partial correlation, controlling for the size of individuals. In practice, the various methods may not diverge significantly in their results. In two species of *Impatiens*, each outcrossed seed costs its mother 1.5–3 times as much material, energy, and time as a selfed seed derived from her cleistogamous flowers (Schemske 1978; Waller 1979). These relative costs play an important role in mating system evolution, but have often been ignored in theoretical models. They reflect both external ecological circumstances (e.g., pollinator effectiveness) and internal constraints (e.g., the degree to which resources spent on selfing decrease or "discount" the amount of resources that can be allocated to outcrossing; Holsinger, Feldman, and Christiansen 1984).

The Genetic "Fidelity" of Progeny

The rate at which a seed transmits maternal genes may be termed its "fidelity." That is measured by the "relatedness" between parents and progeny, or, more specifically, the regression coefficient of the additive genotypic value of offspring on the additive genotypic value of the parent (b) (Uyenoyama and Bengtsson 1982). It reflects the "cost of meiosis" (Williams 1975; Charlesworth 1980a) that accompanies the decreased relatedness of outcrossed progeny to their maternal parent. For uniparental progeny, $b_S = 1$, since each allele is, on average, represented once in a selfed offspring. For nonselfed progeny, b_O is usually assumed to be 1/2. This, however, will only be true in a randomly mating population. More generally, when mating can occur between relatives, the probability that a particular maternal allele will be passed on is the sum of the probability that the individual will itself pass it through meiosis (1/2) and the probability that an allele identical by descent is present in the individual's mate (m) times the probability that the mate passes it on (1/2):

$$(1) \qquad\qquad b_O = (1 + m)/2$$

Here, m is the genotypic correlation between the additive genetic value of mates, formally equivalent to Wright's "coefficient of relationship" (Wright 1921; Wright 1978: 367; Uyenoyama 1986). Fisher (1941) first noted that the high genetic fidelity of selfed progeny would produce "automatic" selection for self-fertilization. The size of this advantage clearly depends on m.

Relative Fitness

Inbreeding depression constitutes a third determinant of mating system evolution and the likely source of any advantage to outcrossing. It may be defined as the relative decrease in fitness of selfed relative to nonselfed progeny [$(W_O - W_S)/W_O$]. Its existence has been appreciated for more than a century (Darwin 1876), but its exact genetic mechanism is still debated. It has classically been attributed to the increased expression of deleterious and recessive alleles and/or the loss of overdominance that accompanies the higher frequency of homozygosity among inbred progeny (see below). Clearly, the extent of the fitness difference depends on the relative levels of inbreeding in the two groups of progeny. Such effects make estimates of the levels of inbreeding in each progeny group essential in studies of inbreeding depression (D. Charlesworth and B. Charlesworth 1987).

These three factors may be combined into a static model to examine how selection might act on the mating system. We expect selection to favor that type of progeny in which the fitness "return" (the product of fitness and fidelity) per unit cost is greater than the return per unit cost for an alternative type of progeny. To favor outcrossed over selfed progeny, then,

(2) $$W_O b_O / C_O > W_S / C_S$$

Although this static inequality cannot represent the evolutionary changes that can occur in all of its parameters, it represents a useful benchmark for evaluating further mating system dynamics (see below).

This model is based on what maximizes overall individual fitness, making it a "phenotypic" model. More complex models that contain the genetic details of how alternative alleles that influence the mating system change in frequency are "genotypic" models. The predictions made by the two classes of models are quite often similar (Nei 1967; Feldman 1972; Maynard Smith 1982). Only genotypic models can incorporate the associations among loci that have important effects on the evolution of selfing (Holsinger 1988a; Uyenoyama and Waller 1991a, 1991b, 1991c). I use a phenotypic model here to explore how the first three elements of mating system evolution interact and respond to evolving differences in population structure.

The Theory of Inbreeding Depression

The superiority of outcrossed over inbred progeny has classically been attributed to either dominance (the masking of deleterious recessive alleles) or overdominance (the superior fitness of heterozygous individuals) expressed at many loci (Crow and Kimura 1970; Wright 1977; Falconer 1981: 227). Both types of models predict that inbred progeny suffer reduced fitness due to their greater expression of the hidden genetic load. In either type of model (assuming independence and multiplicative fitness interactions among the loci), the logarithm of fitness should decline linearly with the degree of inbreeding in the progeny (as estimated by their fixation index, f; Wright 1922; Morton, Crow, and

FIGURE 6.2   The serial inbreeding effect. In an original outcrossing population, progeny are only slightly inbred at a level attributable to biparental inbreeding. Subsequent generations of selfing progressively increase individual fixation (f_{S1}, f_{S2}, etc.), causing corresponding decreases in fitness (W) and increases in observed inbreeding depression (brackets). Inbreeding load is B where $-B$ is the slope of the line.

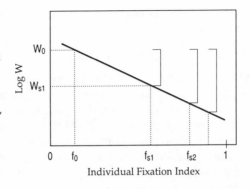

Muller 1956; Latter and Robertson 1962). Deviations from linearity occur if there are consistent epistatic interactions among the loci affecting fitness (Crow and Kimura 1970, 1979).

In the formulation of Morton, Crow, and Muller (1956), the load resulting from mutation-selection balance can be represented as a line with slope $-B$ when log fitness is plotted against f (figure 6.2). The expected total decline in fitness with complete inbreeding ($-B$) is termed the inbreeding load. It can be estimated from the amount of inbreeding depression and progeny f values:

(3) $\log (W_O/W_S)/(f_S - f_O).$

If the load is due to recurrent deleterious mutation, its magnitude will depend on the per genome mutation rate, the fitnesses of the inferior homozygotes, and the extent of dominance at these loci. The greater the mutation rate and the smaller the dominance, the greater the decline in fitness with inbreeding will be. Load maintained by heterozygote advantage has $A = B = st/(s + t)$, where s and t are the decreases in fitness of the two homozygotes. D. Charlesworth and B. Charlesworth (1987) recently reviewed the overdominance and partial dominance models of inbreeding depression and associated empirical data.

The classic mechanisms of dominance and overdominance appear sufficient to account for observed inbreeding depression in plants. Other more exotic mechanisms, such as frequency-dependent selection and/or more intense sibling competition (reviewed by Bell 1988), have yet to be demonstrated in self-fertilizing species (Willson et al. 1987; Schmitt and Ehrhardt 1987; McCall, Mitchell-Olds, and Waller 1989; Karron and Marshall, in preparation) although they could occur in other situations (Antonovics and Ellstrand 1984; Schmitt and Antonovics 1986).

Effects of Biparental Inbreeding

When two mates are more genetically similar than two individuals taken at random from a reference population, a correlation exists between their additive genotypic values, and "biparental inbreeding" is said to occur (Uyenoyama 1985). Such biparental inbreeding is often caused by (and can contribute to) population substructuring. Restricted dispersal leading to such structure is likely to be present in many animal populations (e.g., sessile marine invertebrates) and pervasive in plant populations (Ehrlich and Raven 1969; Levin and Kerster 1974; Levin 1981; Hamrick 1982). Biparental inbreeding mimics selfing in producing homozygous progeny and can therefore seriously bias estimates of the outcrossing rate (Schoen and Clegg 1984). This sort of inbreeding has been detected even in normally outcrossing species like maize (Bijlsma, Allard, and Kahler 1986) and has been also labeled "apparent selfing" (Ritland 1984). Such effects may explain why outcrossed taxa generally have less heterozygosity than expected (Brown's [1979] "heterozygosity paradox") (Hedrick and Cockerham 1986).

Biparental inbreeding has two important contrasting effects on mating system evolution: it increases the genetic fidelity of nonselfed progeny while simultaneously reducing genetic and fitness differences between selfed and nonselfed progeny (Uyenoyama 1986). Such opposing effects make it difficult to predict exactly how mating system evolution will be affected. Uyenoyama (1988a) suggested that biparental inbreeding could favor an optimal intermediate genetic relationship between outcrossed mates (cf. Bengtsson 1978; Shields 1982; Waser and Price 1983). Like inbreeding due to other sources, biparental inbreeding reduces local genetic variation while enhancing among-population variation (Wright 1978). Hedrick (1985) noted with regard to equilibrium heterozygosity and the potential for selection that "If there is more than one type of inbreeding, the cumulative effect is greater than the sum of the individual effects." Clearly, biparental inbreeding needs to be incorporated into realistic models of mating system evolution.

Given the frequency of biparental inbreeding among both plants and animals and its importance in theoretical models of mating system evolution, it would be useful to be able to estimate it in natural populations. Hedrick (1985) stated only a few years ago that "estimates of inbreeding from between-relative matings are generally unknown." Recently, however, the situation has improved. Because biparental inbreeding biases single-locus estimates of the outcrossing rate (Ritland

and Jain 1981), differences between single-locus and less biased multi-locus estimates of the outcrossing rate can be interpreted as evidence for mating with relatives (Ellstrand, Torres, and Levin 1978; Kesseli and Jain 1985).

Quantitative estimates of biparental inbreeding are also possible using routine electrophoretic data. Ritland and Ganders (1985) interpreted the regression of pollen allele frequencies on ovule genotype as an estimate of the genotypic correlation between mates, m, in *Bidens menziesii*. They also used the correlation between observed effective selfing rates and population and individual fixation indexes to estimate levels of biparental inbreeding in *Mimulus guttatus* equivalent to 11% selfing (Ritland and Ganders 1987a).

Waller and Knight (1989) used the ratio of fixation indices in selfed and "outcrossed" progeny in *Impatiens capensis* to estimate m. Theory predicts an increase in the inbreeding coefficient following a generation of selfing according to the formula

$$(4a) \qquad f_S = (1 + f_P)/2,$$

where f_P is the fixation in the parental generation. The inbreeding coefficient following outcrossing should be

$$(4b) \qquad f_O = (1 + f_P)m/2$$

(Wright 1921), suggesting that the ratio of these expectations provides an estimate of m. In populations of *Impatiens capensis*, these estimates varied widely over populations, but were significantly greater than zero in more than half of the populations (table 6.1).

These techniques are more general than might at first be expected. They provide straightforward estimates of m without the lengthy and detailed analyses of ancestry often applied in studies of animal populations. All that is required is a number of moderately variable loci to genotype reasonably sized arrays of naturally crossed progeny. Selfed progeny make it easier to infer the maternal genotypes, but this can be done without selfed progeny with sufficient numbers of outcrossed progeny (Ritland 1983; Schoen and Clegg 1984). Thus, data from a single generation of outcrossed progeny can be used to estimate the outcrossing rate, the level of inbreeding in the outcrossed group, and the expected f for selfed progeny, allowing direct estimation of the important evolutionary parameter, m.

Evolutionary Dynamics

If the terms in the general model of mating system evolution presented above (equation 2) are replaced with the theoretical expectations pre-

TABLE 6.1 The Extent of Biparental Inbreeding in Six Populations of *Impatiens capensis*

Population	m_t mean	m_t S.E.	m_b
HF	0.58	0.16	0.10
GI	0.48	0.17	−0.19
AT	0.43	0.18	−0.92
UWM	0.60	0.25	0.41
MZ	0.61	0.13	0.37
SF	0.72	0.22	0.42
Mean:	0.57		0.03

Source: Waller and Knight 1986, table 3.
Notes: Total genotypic correlations (m_t) and their standard errors were estimated via a bootstrapping procedure and include a component due to geitonogamous selfing. The residual genotypic correlation between truly outcrossed mates (m_b) removes the correlation due to selfing, as estimated via the multilocus procedure of Ritland and Jain (1981). Populations HF, GI, and AT were sampled at a finer spatial scale (2 × 5 m) than the other populations.

sented in equations 1, 3, and 4, the expression favoring the evolution of outcrossing over self-fertilization becomes:

(5) $$(B/2) (1 + f_P) (1 - m) > \ln [2 \, C_O/C_S (1 + m)]$$

This formulation makes clear how increased levels of inbreeding (f_P) tend to favor outcrossing. High inbreeding loads (B) and low relative costs of producing outcrossed seeds (C_O/C_S) also favor outcrossing, while biparental inbreeding (m), as noted before, has mixed effects.

EVOLUTIONARY STATICS AND DYNAMICS

Simple models like this only reflect evolutionary statics, i.e., the immediate trade-offs involved when genetic fidelities and inbreeding depression remain constant. Such models (e.g., Bengtsson 1978; Wells 1979; Feldman and Christiansen 1984; Holsinger, Feldman, and Christiansen 1984) generally generate "neutrally stable" predictions in which only a razor's edge balance of returns per unit of investment permit the (precarious) maintenance of simultaneous selfing and outcrossing. Yet all three quantities vary demonstrably among natural populations (and perhaps even among individuals). Campbell (1986) notes that "a significant oversimplification in these models is the assumption that inbreeding depression is a constant parameter which characterizes populations." Clearly, evolutionary dynamics depend on how inbreeding depression (B) and genetic fidelities (dependent on m) respond to variation in population inbreeding history and structure.

The existence of species with stable mixed mating systems suggests at least the occasional presence of dynamically stable equilibria (Waller 1986). The existence of a preexisting genetic load determines the immediate, short-term fitness effects of inbreeding. Over several generations, however, mutation and selection modify the extent of the load and consequent inbreeding depression. Thus, inbreeding depression is expected to coevolve with population structure. The exact effect of a history of inbreeding on subsequent mating system evolution depends on both the mechanism of inbreeding depression and the population structure. This section reviews the role the mechanism plays and examines competing hypotheses regarding how inbreeding depression might coevolve with population structure.

Dominance versus Overdominance

Both dominance and overdominance mechanisms predict that fitness should decline with inbreeding and homozygosity, but they have contrasting implications for how inbreeding depression should coevolve with population structure. If heterosis via overdominance is important, inbreeding depression increases with selfing when homozygote fitnesses are symmetrical, and will also increase initially when fitnesses are asymmetrical (D. Charlesworth and B. Charlesworth 1987; Campbell 1986). Further increases in selfing with asymmetrical homozygote fitnesses lead to losses of the polymorphism(s) and consequent eventual erosion of inbreeding depression. If recurrent deleterious recessive mutations cause inbreeding depression, the outcome is less clear (see below).

Although empirical data exist to support both mechanisms, most reviews have concluded that partial dominance plays a greater role in generating inbreeding depression than overdominance does (Jinks 1983; Sprague 1983; D. Charlesworth and B. Charlesworth 1987b; but see Mitton, chapter 2, this volume). For example, estimates in *Fagopyrum esculentum* suggest that many loci of small deleterious effect contribute to load (Ohnishi 1982). In reviewing proximity-dependent genetic loads in *Picea* and *Phlox*, Klekowski (1988) concludes that these are consistent with the mutational load hypothesis, and suggests that recurrent somatic mutations would continuously replenish the load in long-lived plant species.

Methods potentially capable of resolving the mechanism of heterosis present difficulties (Lewontin 1974). Crosses between inbred lines can be used to infer levels of dominance, but linkage causes "associative overdominance," which inflates estimates of dominance (Robinson and Comstock 1955; Bulmer 1985: 70). Smouse (1986) proposed using the relationship of (log) fitness to "adaptive distance" (0 for heterozygotes and $1/p_i$ for homozygotes, estimated for each locus) to resolve the mechanism (Mitton, this volume). Application of this method is constrained, however, by its assumption that mating is random.

The Serial Inbreeding Effect

The effects of selfing and outcrossing are asymmetrical in a way that can strongly affect selection on the mating system. One generation of random outcrossing suffices to eliminate inbreeding ($f = 0$) under any circumstances. Less complete outcrossing will still rapidly restore progeny to an equilibrium level of inbreeding dependent on m:

$$f_O = m/(2 - m)$$

In contrast, the genetic consequences of selfing depend critically on the preexisting level of parental inbreeding (cf. Campbell 1986). As noted above (equation 4a), selfed progeny have an expected fixation index of $(1 + f_P)/2$ (Wright 1921). Thus, the level of inbreeding in a selfing lineage is cumulative, increasing with the number of generations of selfing. As fitnesses decline with f, the relative advantage of outcrossing will increase with progressive inbreeding (see figure 6.2).

The fitness difference between selfed and outcrossed progeny (inbreeding depression) will depend on three factors: (1) the steepness of the line (B), (2) the number of generations of selfing, and (3) the relative level of inbreeding in the "outcrossed" group. If a fixed cost differential exists between selfed and outcrossed seeds, and m is relatively constant, further increases in selfing could cease to be favored after one or some other discrete number of generations of selfing, regardless of which mechanism of inbreeding depression (dominance or overdominance) applies (cf. Maynard Smith 1978: 127).

This serial inbreeding effect could provide stabilizing selection on the mating system if realized inbreeding depression increases with the level of inbreeding in the parents. Strong selection against homozygotes would retard the increase of f with selfing (or bow the line slightly), but would not change the qualitative result or the prediction that inbreeding depression should increase with f_P. An evolutionary stable alternation of selfed and outcrossed generations could also be favored, as seemingly occurs in *Amphicarpaea bracteata* (Leguminosae), where only subterranean selfed cleistogamous seeds produce plants large enough to outcross (Schnee and Waller 1986).

Purging the Genetic Load through Inbreeding

Assuming that the mutational load and not overdominance causes inbreeding depression, selection against deleterious alleles is expected to purge inbred populations of some fraction of their load (Stebbins 1950; Bengtsson 1978; Shields 1982; Lande and Schemske 1985). This should result in a population whose homozygotes would not suffer as much inbreeding depression as the ancestral, outcrossed population.

Lande and Schemske (1985) modeled this purging process based on the mutation/selection balance at many loci contributing to quantitative variation in fitness. This model was the first to make inbreeding depression a dynamic variable rather than a static parameter. At equilibrium, inbreeding depression due to recessive lethals and sublethals was substantially reduced by even a small rate of selfing. Once a population was purged of enough of its load, selection in favor of increasing self-fertilization could then occur. These results led Lande and Schemske to predict that populations with intermediate rates of selfing should undergo disruptive selection for either more or less selfing, and that mostly or exclusively selfing populations could be at evolutionary equilibria with reduced inbreeding depression (Schemske and Lande 1985).

Although this model represents a significant improvement over earlier models, its conclusions rest on several assumptions that appear questionable for many plant populations. First, it assumes that no biparental inbreeding occurs. A model incorporating such matings (Uyenoyama 1986) allows a stable mixed mating system to evolve. In addition, the purging of deleterious recessive alleles necessarily involves a "cost of selection" (Haldane 1957; Maynard Smith 1968) which limits the rate at which deleterious alleles can be eliminated in small populations. Lande and Schemske assumed that selection could operate independently and simultaneously across all loci, yet this seems unlikely given the forms of genetic association expected to arise in inbred populations (see below). Finally, the degree of purging has been shown to depend critically on both the degree of dominance of the deleterious alleles and the strength of selection operating against them. If the deleterious alleles have even the slight penetrance found in experiments with *Drosophila* (4%–5%), most of the selection against them will occur in heterozygotes rather than in homozygotes (Morton, Crow, and Muller 1956). This makes purging in inbred lines much less effective (D. Charlesworth and B. Charlesworth 1987). Similarly, small selection coefficients greatly reduce the speed with which purging occurs. D. Charlesworth and B. Charlesworth (1987) conclude that "even populations with vary high levels of selfing retain a substantial fraction of the inbreeding depression found with random mating, as far as mutations of small effect are concerned."

Possible Accumulations of Load with Inbreeding

While purging of some magnitude seems likely to occur in populations subject to continued inbreeding, it is also conceivable that inbred populations might accumulate genetic load under certain circumstances. Factors causing inbreeding (small effective population size, selfing, and limited gene flow) create conditions that could retard, or even reverse, the purging of deleterious alleles. For example, small populations tend to accumulate mutational load due to genetic drift and/or correlated responses to selection at other loci. Completely selfing populations, like asexual populations, can accumulate genetic load irreversibly via "Muller's ratchet" (Heller and Maynard Smith 1978). The efficacy of purging relative to fixation via random processes hinges on the intensity of selection. For strongly deleterious or lethal alleles, selection alone is probably enough to counter their accumulation, even under close inbreeding. For that substantial part of the load due to genes with individually small effect, however, selection coefficients less than $1/N_e$ cannot be expected to be effective. Thus, even with selection, small populations are expected to accumulate and eventually fix many mildly deleterious alleles in each subpopulation (Feldman, Christiansen, and Brooks 1980; Kondrashov 1982, 1985; Hedrick 1985).

We might also expect inbreeding and restricted effective population sizes to hamper the ability of selection to eliminate deleterious alleles at different loci. Both inbreeding and genetic drift increase levels of randomly generated linkage disequilibrium and so reduce the ability of selection to operate independently at different loci (Hill and Robertson 1966; Hedrick, Jain, and Holden 1978; Birky and Walsh 1988). Such associations violate Lande and Schemske's (1985) assumption that genotype frequencies are independent at different loci. These effects could also cause moderately deleterious alleles to "hitchhike" to high frequency as selection against strongly deleterious alleles occurs (Maynard Smith 1978, 1988; Crow 1988; Felsenstein and Yokayama 1976; Felsenstein 1988).

With such complex processes, it is difficult to determine exactly how selection will operate. If linked sets of deleterious alleles are eliminated together, small inbred populations might actually experience accelerated purging. This appears plausible, since inbreeding does tend to increase associations among homozygosities at different loci ("identity disequilibrium"; Weir and Cockerham 1973; Cockerham and Weir

1977a). New deleterious mutations, however, must arise as often in repulsion as in coupling with existing deleterious alleles, restricting such economies of selection for the newer part of the load. In addition, if the load is due largely to many alleles of individually small effect or to polymorphisms maintained by frequency dependence or overdominance, it seems unreasonable to assume that selection can operate simultaneously and efficiently at many loci. We should expect random fixation of a substantial fraction of these alleles in small inbred populations.

Such processes raise the possibility that group selection might be occurring among inbred and substructured populations and that it could be purging the general population of some fraction of its genetic load (D. Schemske, personal communication). The effectiveness of such a process clearly rests on the variance in genetic load among subpopulations. While the possibility of such processes deserves investigation, it is unclear how effective such selection could be if each population becomes fixed for some similarly sized, but different, subset of moderately deleterious alleles. It is also unclear whether selection on this scale could operate quickly enough to counter within-population processes.

In summary, whether genetic load dissipates or accumulates with a history of inbreeding has obvious and important implications for subsequent selection on the mating system. Although we can expect a history of selfing to purge a population of its load due to lethal and semilethal recessive alleles, such selection will be less effective against that substantial fraction of the load due to less deleterious and/or more penetrant alleles. Furthermore, associations among loci and the ratchet may slow or reverse this purging.

Effects of Population Structure on Inbreeding Depression

As mentioned above, we expect most plant populations to be substructured due to restrictions on pollen and seed dispersal. The theory of population structuring is relatively well developed, originally by S. Wright (1965, 1978), whose hierarchical f-statistics and ideas of isolation by distance and genetic neighborhoods provide key tools. These methods have now been applied enough to show that population substructuring depends on many biological characteristics and that it varies greatly among taxa (Hamrick, Linhart, and Mitton 1979; Loveless and Hamrick 1984). It also varies somewhat among related species (Levin 1978) and even within species (Linhart et al. 1981; Schoen 1982). Estimates of neighborhood sizes in herbaceous species typically range from one to three hundred individuals, with areas of 1–50 m². Genetic differentiation may occur quite locally over only a few meters (see, e.g.,

Brown and Clegg 1984; Ellstrand and Marshall 1985; Calahan and Gliddon 1985; Bos, Harmens, and Vrieling 1986; Smyth and Hamrick 1987; Fenster 1991).

A population's substructure strongly affects both patterns of mating and the measurement of f. Estimates of the apparent inbreeding load require comparing the fitness of progeny at various levels of inbreeding (see figure 6.2). Fixation (f), however, is always relative, making inbreeding depression "a dynamic concept for a population" (Campbell 1986). This has several interesting implications for tests of how inbreeding depression should vary with a history of inbreeding.

Let us consider what will happen to an ancestral, outcrossing population that starts with a large neighborhood area but then begins to experience localized mating and dispersal. As demes contract, biparental inbreeding should begin to occur (figure 6.3), as has been observed among clumped plants in certain populations (Ennos and Clegg 1982; Ritland and Ganders 1985). Progeny resulting from crosses occurring within these neighborhoods will therefore be partially inbred (i.e., f_O will increase) and suffer some depression in fitness. This will reduce the fitness difference between selfed and outcrossed progeny, causing a decrease in observed inbreeding depression (figure 6.3b). Such an effect will occur even without any change in the inbreeding load.

Localized inbreeding and drift due to reduced effective population size will also tend to deplete local genetic variation while increasing among-neighborhood variation. If f is measured relative to a set area, such samples will contain more neighborhoods in the more subdivided population (figure 6.4a). Thus, plants derived from crossing within these neighborhoods will display a higher apparent f than a similar plant from an unstructured population, since each deme will likely be fixed for different alleles (the Wahlund effect). If, instead, f is always measured relative to a single neighborhood area, no Wahlund effect will occur, and the results of random mating within the neighborhood will not appear inbred. The progeny of such crosses might, however, be expected to show reduced fitness relative to the result of outcrosses within a less structured population (figure 6.4b). For example, this might occur if the substructured population had lost genetic variation due to drift and selection. Such an effect could cause an apparent decrease in both inbreeding depression and inbreeding load.

Population inbreeding and local substructuring might also be expected to increase the probability that some outcrossing will occur among distinct genetic neighborhoods (figure 6.5). If neighborhoods are small, due to a high rate of selfing, for example, the few outcrosses that do occur might well involve other demes. These will likely be fixed

A) Large Neighborhood Size With Effective Outcrossing

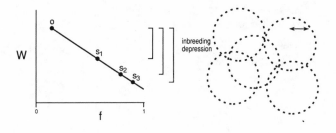

B) Local Mating With Biparental Inbreeding

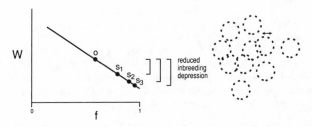

FIGURE 6.3   Effect of reduced neighborhood size on observed inbreeding depression. An ancestral outcrossing population (A) is assumed to have low f, large genetic neighborhoods, and little population substructure. Inbreeding depression is large. As neighborhoods contract due to restricted pollen and seed movement, biparental inbreeding will begin to occur (B). This will reduce f_o and therefore reduce observed inbreeding depression.

for deleterious alleles at different loci, causing crosses between them to produce progeny with enhanced fitness. This could cause an apparent increase in inbreeding load in more inbred, structured populations (figure 6.5). Such a process mirrors the practice common in plant and animal breeding of crossing inbred lines to obtain hybrid progeny of high vigor. Such an effect, moreover, should only occur in some natural outcrosses, causing increases in the variance of genotypic correlations between mates and consequent outcrossed progeny quality in more inbred populations. Thus the inbreeding depression and inbreeding load we observe in a population will depend critically on the existence

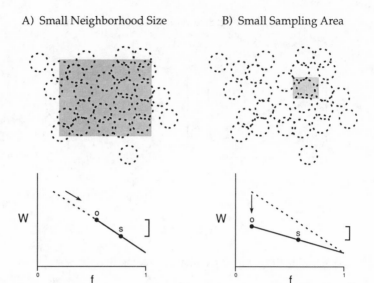

FIGURE 6.4 Effect of population subdivision and scale of sampling. If the sampled population extends across several genetic neighborhoods (A), a plant derived from out-crossing within a neighborhood from a subdivided population will exhibit an elevated apparent f due to the Wahlund effect relative to an unstructured population (dotted line). Alternatively, if f is measured relative to the neighborhood (B), drift and selection might eliminate genetic variation within neighborhoods. This could reduce the effect of f on fitness relative to the unstructured population.

Among-Neighborhood Crosses Can Occur

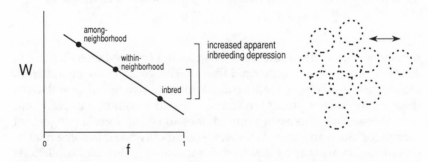

FIGURE 6.5 Effect of among-neighborhood crossing. Because different neighborhoods tend to (randomly) fix different deleterious alleles, crosses between them should result in progeny with enhanced fitness. If only a few outcrosses occur and these occur be-tween neighborhoods, this effect could increase the observed inbreeding depression.

of subpopulation structure, how we measure it, and the patterns of mating within and among neighborhoods. This makes investigations of population substructure and mating patterns central for testing predictions regarding mating system evolution.

An Appeal for Empirical Data

Lacunae still exist in the theory of inbreeding depression. Maynard Smith (1988), for example, provides a list of important outstanding questions needing theoretical work, which includes: "How much outcrossing is needed to arrest the ratchet in selfing species?" Nevertheless, the greater current need is for more empirical information to distinguish among the competing hypotheses reviewed above. Most existing studies have not paid attention to all the potentially relevant factors, making it impossible to unambiguously test the competing ideas presented here.

A central question here is: Does inbreeding depression increase or decrease with a history of population inbreeding? If load accumulates in more inbred populations, there is potential for stable mixed mating systems to evolve. The purging hypothesis instead suggests a decrease in load and disruptive selection for full outcrossing or complete selfing. The existing data are quite sparse, but three studies deserve mention. In a study of lima beans, Harding, Allard, and Smeltzer (1966) found the fitness of heterozygotes relative to homozygotes at a marker locus to decline as they increased in frequency. While not a test of the purging or load accumulation hypotheses, this frequency dependence could provide a mechanism that would favor outcrossing in more inbred populations. In *Impatiens capensis*, populations that are more inbred appear to suffer greater inbreeding depression (figure 6.6a). These data are preliminary, flawed by the small number of populations and the fact that estimates of fixation and of inbreeding depression came from different years. Finally, Ritland (1990) developed a technique to infer relative selfed and outcrossed fitnesses from levels of heterozygosity in natural populations. Although resulting estimates of relative selfed fitness in *Mimulus guttatus* were uncorrelated with selfing rates, they are positively correlated with population f (figure 6.6b), supporting the purging hypothesis.

Further studies are clearly needed to examine how levels of inbreeding depression covary with population inbreeding and population structure across many plant and animal species. Although the effects of inbreeding depression have been the subject of intense study in applied plant and animal breeding, surprisingly little is known about their magnitude and variability in natural populations. Furthermore, studies of inbreeding depression have traditionally been divorced from

A

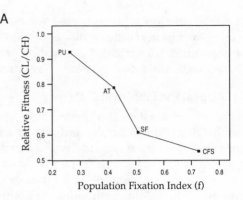

FIGURE 6.6 Relation between inbreeding depression and population fixation. The fitness of selfed progeny relative to naturally outcrossed progeny plotted as a function of f. (Higher f values are assumed here to reflect a history of inbreeding.)

(A) The fitness of progeny derived from selfed (CL) flowers relative to outcrossed (CH) progeny fitness decreased with increased f across four populations of *Impatiens capensis*. Data on relative fitnesses from Waller 1984; f values from Knight and Waller 1987.

(B) The inferred relative fitness of selfed progeny increased with population fixation across thirteen populations of *Mimulus guttatus*. Data from Ritland (1990).

B

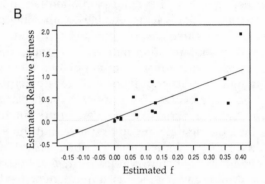

studies of population structure. This specialization is readily apparent in the literature: a search of the extensive "Agricola" data base from 1979 to 1989 revealed 121 references concerning "inbreeding depression," but only one of these (Sorensen and White 1988) was cross-listed under "population structure."

CONCLUSIONS

Many plant species have a mating system that consists of a mixture of self- and cross-pollination, presenting a model system for understanding the evolutionary dynamics of inbreeding. Whether selfed progeny will be favored over more outbred progeny depends on four factors: (1) the relative costs of progeny derived from selfing and outcrossing; (2) the fidelity with which selfed and outcrossed progeny transmit maternal genes; (3) differences in fitness between selfed and outcrossed progeny; and (4) genetic associations that arise between alleles influencing fitness and alleles modifying the mating system. Ignoring any of these factors may lead us to misinterpret evolutionary dynamics. Furthermore, we are still some way from understanding how these factors covary and coevolve. The degree to which outbred progeny are related to their parents hinges on the relatedness between "outcrossed" mates. Similarly, observed inbreeding depression depends both on how inbred the various progeny are and on how quickly fitness declines with inbreeding. All these parameters are labile, and reflect differences in pollination ecology, inbreeding history, and population structure.

Whether mixed reproductive systems are evolutionarily stable depends on how these genetic parameters coevolve with the changes in population structure that accompany inbreeding. Alternative theoretical models make contrasting predictions, necessitating further empirical studies to resolve how the relatedness between mates and inbreeding depression covary with changes in population structure. Such studies will reveal whether genetic loads tend to accumulate in more inbred populations or are purged via selection. This understanding should illuminate the genetic hazards posed by genetic isolation and inbreeding in rare and threatened species.

ACKNOWLEDGMENTS

The ideas presented here were originally worked out while I was a visitor to the Institute for Ecology of the Technical University of Berlin. I am grateful to the Alexander von Humboldt Foundation for providing fellowship support. M. Uyenoyama provided encouragement and many critical discussions. I also wish to thank D. Charlesworth, J. Crow, N. Knowlton, M. Kuchenreuther, K. Ritland, D. Schemske, and C. R. Williams for their thoughtful comments on a version of the manuscript.

7

Small Populations, Inbreeding, and Speciation

Daniel J. Howard

The study of inbreeding and outbreeding in natural populations has become one of the growth industries of the environmental sciences as ecologists and evolutionary biologists become increasingly aware that an understanding of population structure and mating systems is critical for reaching an understanding of the evolution of sex, the maintenance of viable populations, and the evolutionary dynamics of natural populations. Perhaps less well appreciated by many biologists is the important role that population structure and mating structure, especially small population size and inbreeding, play in models of speciation. In this chapter I will review models of speciation that emphasize the consequences of small populations and inbreeding for genetic divergence and the acquisition of reproductive isolation. I will then examine the effects attributed to small population size and inbreeding, and I will compare the population structures of two groups (lower vertebrates and mammals) that appear to differ in rates of speciation and chromosomal evolution. One of my conclusions will echo that of other authors in this volume; there is much work, especially fieldwork, yet to be done.

HISTORICAL OVERVIEW AND MODELS

It can be argued that the 1940s was the last period in which there was general agreement on ideas about animal speciation. The best statement of those ideas can be found in Ernst Mayr's 1942 book, *Systematics and the Origin of Species*. In some ways, the book was one long argument in favor of what Mayr (1942: 155) called "orthodox ideas on species formation." The basic orthodox idea was that geographic speciation was the predominant, if not the only, mode of speciation among animals.

The course of geographic speciation was as follows: "A new species develops only if a population which has become geographically isolated from its parental species acquires during this period of isolation characters which promote or guarantee reproductive isolation when the external barriers break down" (Mayr 1942: 155). According to Mayr, geographic speciation was thinkable only if subspecies are incipient species.

One may wonder why Mayr felt compelled to write a book defending orthodox ideas. What was the motivating force if most biologists were in general agreement about the ideas presented? The force was Richard Goldschmidt, who had published *The Material Basis of Evolution* in 1940. In this book, Goldschmidt argued that subspecies of polytypic species do not differ enough to represent incipient species. He contended that there are bridgeless gaps between true species, and that these gaps can only be crossed by a fundamental reordering of chromosomal pattern or by mutations that produce marked phenotypic effects by acting upon developmental rates in early ontogeny. The result of such changes are not subspecies, but hopeful monsters: monsters that will start a new evolutionary line if they can find an appropriate and empty environmental niche.

The orthodox view of species formation as described by Mayr in 1942 did not include population size or mating system as a factor. The important factors were geographic variation and the isolating action of geographic barriers. This decoupling of population structure and speciation was common at the time. Even Wright and Dobzhansky, who both emphasized the importance of limited population size and partial isolation for the dynamics of the evolutionary process, considered these as factors in the evolution of a species as a whole. The splitting of a species demanded more complete isolation (Dobzhansky 1937, 1941; Wright 1940, 1949).

The clearest voice invoking population structure as a component of geographic speciation during this time was that of Julian Huxley. Clearly influenced by Wright, Huxley (1942) pointed out that nonadaptive differentiation of small populations owing to "drift" explained many observations that puzzled earlier evolutionists, such as the greater phenotypic divergence of island versus mainland forms. Spatial isolation remained the fundamental initiating factor of speciation for Huxley (1942), but he envisioned that isolation might be linked in small populations with drift.

Consideration of the conspicuous differences of many peripherally isolated populations eventually had a major impact on Mayr's thinking as well, and led to a dramatic change in his ideas about the nature and evolutionary potential of geographic variation. In his 1954 paper, "Change of environment and speciation," Mayr broke geographical variation down into two types, ecotypic and typostrophic. Ecotypic variation was the variation typical of populations that are members of a continuous series of populations. Because of the homogenizing effects of gene flow, these populations are merely variations on a single theme, although they are sufficiently distinct to be recognized as subspecies. Ecotypic subspecies as incipient species was the notion that Goldschmidt (1940) had attacked and that Mayr had defended in his 1942 book. In one of the most remarkable conversions in modern evolutionary thought, Mayr (1954) now conceded that Goldschmidt was correct. Subspecies that are subdivisions of a widespread array of continuous populations are not incipient species, unless they become isolated subsequent to differentiation. Even then they are not the stuff of evolutionary novelty.

Typostrophic variation was the sort of variation found in peripherally isolated populations. This variation broke from the "type" of the parental species and sometimes represented something entirely new. Mayr (1954) argued that peripherally isolated populations are not only incipient species, but they are the site of origin of evolutionary novelties.

Whereas Huxley (1942) attributed the dramatic differences in peripheral isolates to random genetic drift in small populations, Mayr (1954) attributed the dramatic differences to isolation and selection. Invoking the importance of coadapted gene complexes and the continuous stream of gene flow washing through contiguous populations, Mayr (1954) contended that in such populations selection will favor alleles that produce heterozygotes of high fitness and alleles that have high selective values on a large number of different genetic backgrounds (good mixers). When a few individuals are isolated from such a population, Mayr (1954) maintained that there is a dramatic change in the genetic environment. Genetic variability decreases, and this change in genetic background may alter the relative selective values of different alleles so that the "soloist" allele is favored rather than the "good mixer." Thus, the "mere change of the genetic environment may change the selective value of a gene very considerably" (Mayr 1954: 169). The result may be a genetic revolution in which changes at one locus affect the selective values of many other loci, until finally a new equilibrium is reached.

In an important early variant of Mayr's model, Carson (1959) advocated the importance of ecologically marginal, peripheral populations in the formation of species. According to Carson, ecologically marginal, peripheral populations are characterized by isolation, small size, a high coefficient of inbreeding due to small size, and homozygosity. Homozygosity evolves because a small population cannot afford to support a lot of heterozygotes segregating less fit homozygotes and because the relatively few ecological niches available in marginal areas favor well-adapted homozygotes. Inbreeding and random genetic drift bolster levels of homozygosity. Carson (1959) argued that the combination of inbreeding and homoselection in small populations can produce extreme phenotypes and a high level of genetically encoded, specific adaptations. If given the opportunity to expand, such a population could form the basis of a new species.

In 1957, Brown offered a third model of allopatric speciation that contrasted sharply with those of Mayr (1954) and Carson (1959). Brown (1957) contended that continental species are subject to contractions and expansions of species range. The more important genetic changes tend to be incorporated in the central populations during the contraction phase, when these populations are isolated from those on the periphery. If the genetic changes are so great that they result in reproductive isolation, it is likely that the "new central" species will overwhelm the "old peripheral" species during the next expansion phase. Coexistence is possibly only if the two species specialize and exploit separate ecological niches. Brown (1957) invoked some of Wright's (1956) ideas about interdemic selection to support his contention that central populations are the principal source of "potent" new species and evolutionary novelty. He assumed that species, even within their central range, are organized into small, partially isolated demes, especially during the contraction phase. Thus, he argued, mutations are more likely to arise among the central demes (because there are more of them), and central demes will harbor the largest number of genetic variants (because they exchange genes with a greater number of demes).

For reasons of space and competence, I am confining my attention in this paper to the models and literature of animal speciation. However, I will present one model of speciation largely oriented toward plants, H. Lewis's model of saltational speciation, because one of the major pieces of evidence used to support the model was drawn from a study of animals. Lewis (1962, 1966) was impressed by the finding that adjacent plant populations that are very similar in morphology and ecology sometimes differ greatly in chromosome arrangement and basic chromosome number. Given that the frequency of a single chromosome rearrangement replacing the original arrangement throughout a population is low, Lewis (1966) considered it extremely unlikely that multiple chromosome rearrangements occurred and independently

marched to fixation in a population. Instead, he postulated that saltational reorganization of the chromosomes sometimes occurs following
a unique event that isolates one or a few individuals in an open habitat
free from conspecific competition. The chromosomal reorganization is
triggered by the intensive inbreeding that accompanies the isolation of
extremely small populations. As evidence for the relationship between
a sudden shift to inbreeding and chromosomal change, H. Lewis (1966)
cited a study by K. R. Lewis and John (1959). These investigators reported a high number of chromosomal abnormalities in the germ line
within individuals of an inbred culture of the locust *Pyrgomorpha
kraussi.*

White (1968) pointed out that even the most closely related species
of animals usually differ in karyotype and that closely related taxa often
occupy contiguous areas, with an extremely narrow zone of overlap.
Based on these observations and on detailed work on morabine grass-
hoppers, White (1968) proposed a "stasipatric" model of speciation.
According to this model some types of chromosomal rearrangements
function as strong primary genetic isolating mechanisms. Because the
heterozygotes are less fit, these rearrangements are selected against
when they first arise in a population. White (1968, 1978) suggested four
factors that alone or in combination would enhance the probability of
fixation of a chromosomal variant: random genetic drift in small populations, inbreeding, meiotic drive, and selective advantage of the new
arrangement as a homozygote. White considered it likely that the rearrangements initially arise and go to fixation in a small, isolated population in the interior of the species range. Subsequent spread of the
nascent species occurs at the expense of the parental species, and the
border between the two taxa is characterized by a narrow hybrid zone.

Motivated by some of the same observations as Brown (1957), Carson (1968) developed a model of allopatric speciation based on the genetic consequences of a population flush for organisms that have a
high recombination index and carry extensive genetic variability. He
postulated that during a population flush, natural selection for survival
is less than normally encountered, and genetic combinations that
would not ordinarily survive and reproduce are successful. Widespread dispersal ensues, carrying with it the possibility of colonization
of new areas by one or a few individuals. Following the dispersal
phase, the population may crash, isolating many demes and reinstitut-
ing strong selection. At this point, many genotypes will be eradicated,
as will demes with unfit gene pools. Those demes surviving may have
novel gene pools as a result of the many gene combinations present in
the flush phase of the population cycle, coupled with random genetic

drift and inbreeding. Some of the genetic changes are likely to modify reproductive behavior in some way. The result of this modification could be reproductive isolation and the formation of a new species.

In 1971, clearly influenced by new results from his work on Hawaiian *Drosophila*, Carson modified his 1968 model to emphasize the importance of a single founder individual in the colonization of new areas. Carson further modified his 1968 model in 1975 by breaking the genome down into two parts: an open part that is relatively free of epistatic interactions and able to vary without having a major effect on the phenotype, and a closed part, probably regulatory in nature, that is highly coadapted and balanced. Carson (1975) postulated that the most important aspect of founder-induced speciation is the disruption of the closed part of the gene pool and its subsequent reorganization. The end result is a new, highly coadapted gene pool reproductively isolated from the ancestral gene pool.

Until recently, geography and the physical and biotic environment were regarded as the crucial factors controlling population size and levels of gene flow among populations. Bush et al. (1977) emphasized a new element, the importance of intrinsic factors, especially social structuring, in reducing effective population size (N_e), increasing inbreeding, and therefore increasing rates of chromosomal evolution and speciation. Bush et al. (1977) were led to this hypothesis by the finding that some large, highly mobile mammals display rapid rates of karyotypic evolution and speciation. These patterns appear inconsistent with the mobility of the organisms, but are perhaps understandable if the small social units typical of many mammals limit gene flow and N_e.

In 1980, Templeton introduced a third model of founder-induced speciation. Like Mayr's (1954) model, this model emphasized the change in genetic environment caused by the act of isolation. Unlike Mayr, Templeton (1980) did not regard the loss of genetic variation as critical; rather, he stressed the stochastic effects of the founder event on the frequencies of a few major genes. A drastic change in the pattern of variation at one of these major genes was envisioned as having cascading fitness effects on modifier loci in a strongly epistatic genetic system. The resulting trait differences between the founder population and the parental population could be great enough to cause reproductive isolation. Empirical studies on parthenogenesis in *Drosophila mercatorum* (Templeton 1979) inspired the development of this model.

This brief survey of models of speciation suggests that isolation, small population size, and inbreeding were incorporated into models of speciation as a response to the discovery of dramatic phenotypic divergence in peripherally isolated populations and to the realization that the only way to arrive at some of the differences between closely related species is for populations to go through a maladaptive phase. In most of the models, inbreeding results from small population size rather than from nonrandom mating within demes. The only model that incorporates mating structure is that of Bush et al. (1977), but even this model emphasizes deme size rather than incestuous matings within demes. This survey also reveals that despite the similarities between the models (e.g., most involve allopatry), there are important differences between them, especially in the effects ascribed to small population size and inbreeding. Because of these differences it is not reasonable to lump some of the models together, as has commonly been done with the three models of founder-induced speciation. Moreover, any assessment of the significance of the various models must include an evaluation of the evidence for and against the effects each model attributes to inbreeding and small population size. Such an evaluation will serve as the major focus of the remainder of this chapter.

NON-ADAPTIVE DIFFERENTIATION DUE TO RANDOM GENETIC DRIFT

Random genetic drift is an important component of several of the small-population models of speciation, but the only model to rely exclusively on nonadaptive differentiation in small isolated populations to explain genetic divergence and the onset of reproductive isolation was that of Huxley (1942). Theoretically, there is no doubt that random genetic drift is more important in small populations than in large populations, all other factors being equal (Wright 1931, 1932, 1969). These theoretical expectations have been confirmed by experimental work with *Tribolium* populations (Rich, Bell, and Wilson 1979; Wool 1987) and by computer simulations (Chesser and Baker 1986). Small, isolated natural populations often display much greater morphological and genetic differentiation than populations of greater size or populations exposed to greater levels of gene flow. For example, in a widely cited study, Kramer and Mertens (1938) reported that Adriatic lizards (*Lacerta sicula*) from small islands showed greater divergence from the mainland phenotype than Adriatic lizards from large islands. More recently, Berry, Jakson, and Peters (1978) have demonstrated rapid allozymic and phenotypic differentiation in house mice of the Faroe Islands compared with that of mainland populations. They attributed the differentiation to the stochastic nature of founder events rather than to adaptation because newer populations on the islands were as distinct as older populations. Populations of house finches on the California Islands also displayed greater phenotypic divergence than mainland populations, and some of the most highly differentiated populations occurred on smaller islands (Power 1979).

One of the difficulties with the study of divergence on islands is determining its cause. Random genetic drift due to small population size is a possibility, but so too is natural selection as a result of differences in the physical, biotic, and genetic environment. Distinguishing among the possible explanations for divergence requires detailed knowledge of the trait and its variation and genetic control, as well as a thorough, in-depth understanding of the group under consideration. Population size, behavior, and ecological factors should be studied and taken into consideration. Only one long-term study that I am aware of has come close to meeting these objectives: the investigation of Darwin's finches by Grant and his colleagues (summarized in Grant 1986). In this case, selection appears to be of overriding importance in driving phenotypic and genetic divergence.

LOSS OF GENETIC VARIATION

A decrease in genetic variability associated with a founder event or with small population size plays a critical role in two models of speciation, that of Mayr (1954) and that of Carson (1959). In addition, the centrifugal speciation model of Brown (1957) predicts that peripherally isolated populations will harbor less genetic variation than central populations. Mayr (1954) attributed the loss of genetic variation in founder populations to sampling effects, to the elimination of deleterious recessive alleles exposed to natural selection by inbreeding, and to a change in the selective environment. Carson (1959) attributed the loss of genetic variation in peripheral populations to the effects of selection for homozygosity coupled with the effects of inbreeding and random genetic drift. Brown (1957) believed that mutations were more likely to arise and spread in central populations than in peripheral populations, leading to a central-peripheral decline in genetic variation.

It has long been clear on theoretical grounds that small populations should harbor less genetic variation than large populations (Wright 1931). What was less clear for many years was how much genetic variation would be lost if a population went through a severe but temporary bottleneck in size. Mayr (1954) assumed that the loss of variation due to sampling effects would be substantial if a population was founded by one or a few individuals. Nei, Maruyama, and Chakraborty (1975) demonstrated through the use of neutral models that a population founded by ten or more individuals loses very little variation, and that a population founded by two individuals suffers a great loss of heterozygosity only if the population grows slowly. If the population grows rapidly, the reduction in heterozygosity may be small—35% or less of the original level. On the other hand, a population founded by two individuals will lose many alleles, especially rare alleles, regardless of how fast the population grows subsequently (Nei, Maruyama, and Chakraborty 1975; Maruyama and Fuerst 1985). Thus, the consequences of a population bottleneck for the level of genetic variation depends not only on the size of the bottleneck and how rapidly the population grows following the bottleneck, but also on the measure of genetic variation considered to be relevant.

Several investigators have used enzyme electrophoresis to assess changes in variability associated with a known bottleneck. In all cases, the population(s) involved in the bottleneck displayed less variability than source populations or conspecific populations presumed not to have undergone a recent reduction in size (table 7.1). The loss of genetic variation is sometimes substantial and sometimes small. North American gypsy moth populations, which grew out of an accidental introduction from France in 1869, demonstrated the most dramatic reduction of genetic variation associated with a founding event (Harrison, Wintermeyer, and Odell 1983). In France, more than half of the twenty loci that were examined exhibited at least some variability (average observed heterozygosity = 0.06). In the United States, seven of fourteen populations displayed no variation, and the other seven populations only varied at a single locus. The average observed heterozygosity was less than 0.01. In contrast, founder populations of *Anolis* lizards (Taylor and Gorman 1975), face flies (Bryant 1981), and Eurasian tree sparrows (St. Louis and Barlow 1988) did not differ significantly from source populations in measures of genetic variation. The reasons for the different outcomes of different founder events are hard to determine. The size of the initial founder population was known in only three of the seven studies cited in table 7.1, and no investigation attempted to assess differences in the environments of the parental and daughter populations.

No group of organisms is understood better than *Drosophila* with regard to the genetic characteristics of central and peripheral populations. Geographic patterns of genetic variation in this group differ depending on the type of genetic variation under consideration. Most species of *Drosophila* exhibit a reduction in inversion heterozygosity in peripheral populations compared with that of central populations (da Cunha, Burla, and Dobzhansky 1950; da Cunha and Dobzhansky 1954; Dobzhansky 1957; Krimbas and Loukas 1980). On the other hand,

TABLE 7.1   Genetic Variability Measures for Founder Populations and Source Populations

Species	Mean Sample Size/ Locus	Av. Obs. Het.	Mean No. Alleles/ Locus	Proportion of Loci Polymorphic
Anolis grahami[a]				
Source (Jamaica)	38.0	0.078	1.75	0.50
Founder (Bermuda)	43.0	0.064	1.50	0.29
Musca autumnalis[b]				
Source (Europe)	134.6	0.053	1.55	0.36
Founder (North America)	110.3	0.038	1.46	0.29
Lymantria dispar[c]				
Source (France)	216.0	0.053	2.10	0.50
Founder (United States)	354.4	0.002	1.05	0.05
Acridotheres tristis[d]				
Source (India)	28.4	0.05	1.43	0.31
Founder (South Africa)	38.3	0.03	1.15	0.13
Rhagoletis pomonella[e]				
Source (Illinois)	57.1	0.189	2.8	0.59
Founder (Utah)	56.4	0.095	1.5	0.24
Passer montanus[f]				
Source (W. Germany)	30	0.093	1.5	0.36
Founder (Illinois)	93	0.079	1.3	0.28
Theba pisana[g]				
Source (W. Australia)	20–33	0.083	1.24	0.22
Founder (E. Rottnest Isl.)	20–33	0.056	1.16	0.16

[a]24 loci; Taylor and Gorman 1975
[b]14 loci; Bryant, van Dijk, and van Delden 1981
[c]20 loci; Harrison, Wintermeyer, and Odell 1983
[d]39 loci; Baker and Moeed 1987
[e]17 loci; McPheron, Jorgenson, and Berlocher 1988
[f]39 loci; St. Louis and Barlow 1988
[g]25 loci; Johnson 1988

there is no indication of a loss of allozyme heterozygosity in peripheral populations of *Drosophila* (Soulé 1973; Brussard 1984). In a recent review article, Brussard (1984) attributed the lack of central-peripheral decline in allozyme heterozygosity to the selective neutrality of allozyme variants combined with the effect of high dispersal rates in marginal environments. A high dispersal rate will increase the overall effective population size of a peripheral population and the level of gene flow into a peripheral population from more central areas. Like Carson, Brussard (1984) invoked selection favoring homozygosity, albeit via a somewhat different mechanism, to explain the commonly observed central-peripheral reduction in inversion heterozygosity. However, considering the current lack of understanding of the ecology, structure, and dynamics of *Drosophila* populations in nature, explanations of the allozyme data and the chromosomal inversion data must be regarded with some caution.

Given that some founder populations experience a dramatic loss of genetic variation and that peripheral populations of *Drosophila* usually exhibit a lessening of inversion heterozygosity, the question of evolutionary consequences becomes important. Brown (1957) took a decidedly dim view of the future of genetically depauperate peripheral isolates, contending that they represented evolutionary dead ends. Mayr (1954) and Carson (1959) argued that new ecological adaptations would sometimes accompany the loss of genetic variation associated with peripheral populations and founder events. These ecological differences would hasten the onset of reproductive isolation. Carson (1959) also postulated that a reduction of inversion heterozygosity would free whatever variation is present in a peripheral population to recombine, thus permitting a more rapid response to selection and aiding in the synthesis of evolutionary novelty. Carson tested his ideas about the consequences of greater recombination in peripheral populations by exposing *D. robusta* from a central and a peripheral population to selection for mobility toward light. Eight of ten lines from the peripheral population responded positively, whereas only eight out of fifteen lines from the central population demonstrated a positive response (Carson 1958). The response was also more exaggerated in the peripheral population lines.

Besides Carson's (1958) study, there are very few data from natural populations that are pertinent to the question of the evolutionary consequences of a loss of genetic variation in founder and peripheral populations. It has been argued from population genetic considerations that a rapid response to selection is unlikely in a population with a low level of variation (Carson and Templeton 1984). The relevance of this criticism to founder-induced speciation is unclear because Mayr envisioned the population "making the ecological shift during the 'genetic revolution' and during the period of relaxed selection accompanying the phase of rapid expansion" (Mayr, 1954: 173). Moreover, while it is clear that bottlenecks cause a loss of variation at single loci with additive genetic effects, recent work on the housefly demonstrates that additive genetic variances of morphological traits can actually increase as the result of a bottleneck (Bryant, McCommas, and Combs 1986). Nonadditive effects within and among loci (i.e., dominance, overdominance, and epistasis) seem to be responsible for the increases. Bottlenecks can also disrupt the additive genetic interrelationships among morphometric traits in the housefly, potentially allowing a founder population to move in an evolutionary direction unavailable to the parental population (Bryant and Meffert 1988, 1990). It is not clear whether the results from the housefly can be extended to other organisms, but if so, the rate of evolution in a newly founded population may be enhanced if it goes through a bottleneck.

GENETIC REVOLUTION

Mayr (1954, 1963) postulated that some populations would experience a genetic revolution upon going through a founder event. At the center of this hypothesis was Mayr's conviction that the gene pool of a species was highly coadapted and cohesive. Mayr thought that the loss of variability associated with a founder event would change the genetic environment so dramatically that the selective value of alleles at many loci would be affected. Carson's (1968, 1971, 1975) model of founder-induced speciation was similar in emphasizing the highly coadapted and stabilized closed part of the gene pool. Templeton's (1980) model of founder event speciation also focused on coadapted gene complexes, but in a much more limited way. Whereas Carson (1968, 1971, 1975) and Mayr (1954) stressed the coadaptation of the gene pool, or at least a significant portion of it, Templeton (1980) emphasized the importance of epistatic polygenic systems with a few major genes.

Using Wallace's definition of coadaptation: "Genes are said to be coadapted if high fitness depends upon specific interactions between them" (Wallace 1968: 305), there is no doubt that coadapted gene complexes involving a limited number of genes exist in populations (for comprehensive reviews of the literature on coadaptation see Hedrick et al. [1978] and Shields [1982]). Among the more compelling examples are the segregation distorter system in *Drosophila melanogaster* (Hartl 1977) and the alpha-glycerol phosphate dehydrogenase–alcohol dehydrogenase system, also in *D. melanogaster* (Cavener and Clegg 1981). On the other hand, there is considerably less empirical evidence to support the concept of comprehensive gene pool coadaptation (Hedrick et al. 1979; Nei 1980; Bush 1982). The most widely cited evidence comes from investigations of the relative fitness of offspring from crosses between geographically separated populations. In many cases, the F_1 hybrids demonstrate an increase in various components of fitness compared with the offspring of intrapopulation crosses, but this is followed by a significant decline in the fitness of F_2, F_3, and backcross individuals (Brncic 1954, 1961; Vetukhiv 1954, 1956, 1957; Wallace and Vetukhiv 1955; Anderson 1968; Ohta 1980). Although these results point to the presence of coadapted gene complexes within a population, they give little indication of the number of genes involved. Thus, they do not provide evidence for the comprehensive gene pool integration envisioned by Mayr (1954, 1963) and to a more limited extent by Carson (1975, 1982). (Other discussions of outbreeding depression can be found in Knowlton and Jackson, chapter 10; Shields, chapter 8; Smith, chapter 14; Waldman and McKinnon, chapter 11; and Waser, chapter 9; all in this volume.)

As pointed out by Bush and Howard (1986), the concept of a highly coadapted and cohesive gene pool hails from a time when evolutionary biologists perceived organismal development as an intricate interplay among all the genes in a genome, with each gene regarded as a member of a successful team (Mayr 1963). At least in part, this point of view grew out of a rejection of Goldschmidt's (1940) ideas about the importance of "systemic mutations" in speciation and macroevolution. Mayr (1954) argued that the well-balanced gene pool of a species would resist dramatic alterations produced by mutations affecting the timing and pattern of developmental events. This left Mayr with the problem of explaining how an incipient species broke out of the straitjacket of a highly integrated gene pool. His solution was the idea of genetic revolutions in peripherally isolated populations. He believed this solution to be more consistent with the evidence from genetics and natural populations than the idea of "systemic mutations." Ironically, an ever-increasing amount of evidence from the field of developmental biology indicates that changes in key developmental regulatory genes can have a major effect on the phenotype, and that such changes can be viable and are potentially capable of explaining the origin of evolutionary novelties (Tompkins 1978; Hunkapiller et al. 1982; Raff and Kaufman 1983; Shaffer 1984a, 1984b; Raff et al. 1987).

For many years after Mayr's (1954) original enunciation of the concept of a genetic revolution in founder populations, there were few attempts to test the hypothesis. Part of the problem was the lack of attention the paper generated initially (Mayr 1976), and part of the problem was the lack of an adequate means of sampling the gene pool. The latter constraint was seemingly overcome in the middle 1960s with the introduction of electrophoretic techniques to the study of natural populations by Lewontin, Hubby, and Harris (Harris 1966; Hubby and Lewontin 1966; Lewontin and Hubby 1966).

The 1970s witnessed a wealth of investigations devoted to measuring allozymic differences between populations at various stages of evolutionary divergence. The purpose of many of these studies was the elucidation of the "genetics of speciation"; that is, the determination of the amount of genetic differentiation that accompanies the acquisition of reproductive isolation (Johnson and Selander 1971; Ayala et al. 1974; Carson et al. 1975; Avise and Smith 1977; Johnson, Clarke, and Murray 1977; Avise 1978; Johnson 1978; Zimmerman, Kilpatrick, and Hart 1978; Benado et al. 1979; Craddock and Johnson 1979; Ryman, Allendorf, and Stahl 1979; Turner, Johnson, and Eanes 1979). No single answer emerged from these investigations: varying amounts of change at genes coding for soluble proteins were associated with speciation events (table 7.2). But the impression left on evolutionary biologists was that reproductive isolation could evolve with relatively little allozymic divergence, even when the reproductive isolation was accompanied by dramatic morphological alterations such as those exhibited by *D. heteroneura* and *D. silvestris* (Sene and Carson 1977), two sister species of Hawaiian *Drosophila* presumed to have speciated via a founder event (Carson 1986).

The electrophoretic approach to the genetics of speciation reflected a belief common in the 1960s and 1970s that speciation was a by-product of genetic differences that accumulated between two populations when they were geographically separated for some period of time. The by-product aspect of the acquisition of reproductive isolation was especially emphasized by Mayr (1963), and formed a cornerstone of his allopatric models of speciation. As it became evident that reproductive isolation and morphological evolution were not closely coupled to the amounts and patterns of protein divergence, evolutionary biologists became disillusioned with the electrophoretic approach to the genetics of speciation. A number of evolutionary biologists, particularly Allan Wilson and his colleagues (Wilson, Maxson, and Sarich 1974; King and Wilson 1975; Wilson 1976), argued that changes in soluble proteins did not form the basis of anatomical evolution. These workers postulated that a variable process, alterations in gene regulatory systems, accounted for behavioral, ecological, and morphological evolution. Whether or not Wilson and his colleagues were correct (and we still do not know the answer to that question), the electrophoretic results implied that speciation is not merely the by-product of overall genetic change, nor is the genetics of speciation simply some measure of genetic differentiation between closely related species. Rather, the genetics of speciation is the genetics of the traits responsible for reproductive isolation between closely related species.

The clarity with which this message has reached evolutionary biol-

TABLE 7.2 Genetic Distances between Closely Related Species

Taxa	Number of loci	Nei's D	Reference
Drosophila setosimentum/			
ochrobasis	14	0.108	Carson et al. 1975
Drosophila athabasca	17	0.105	Johnson 1978
Drosophila heteroneura/			
silvestris	25	0.063	Sene and Carson 1977
			Ryman, Allendorf, and
Salmo trutta (brown trout)	54	0.025	Stahl 1979
Partula taeniata/mirabilis			Johnson, Clarke, and
(land snails)	20	0.061	Murray 1977
Geomys bursarius/			
tropicalis (pocket			Penney and Zimmerman
gophers)	22	0.308	1976
Albula vulpes (bonefish)	40	1.3	Shaklee and Tamaru 1977

ogists is apparent in the recent flush of animal hybridization studies devoted to understanding the genetic basis of reproductive isolation between closely related species. The resolving power of these studies is low and the results are variable, but a number of investigators have reported that changes in traits affecting reproductive isolation are controlled by a small number of genes (Huettel and Bush 1972; Oliver 1979; Grula and Taylor 1980). Especially dramatic have been the reports that a mutation of a single gene can rescue an otherwise inviable interspecific hybrid (Watanabe 1979; Hutter and Ashburner 1987). In a series of studies, Coyne (1983, 1984, 1985) specifically addressed the question of the genetic differences between two species that may have split via a founder event (*Drosophila mauritiana/D. simulans*) compared with the genetic differences between two more distantly related species (*D. simulans/D. melanogaster*) presumed not to have undergone founder-induced speciation. He demonstrated that differences between *D. mauritiana* and *D. simulans* in four traits related to reproductive isolation were, in each case, under the control of the largest number of genes that could be detected by his method of analysis. He also reported that all measures of developmental anomaly were more advanced in hybrids between the more distantly related species pair than in hybrids between the more closely related species pair. Coyne (1985) concluded that single genes of large effect were not important in the speciation event separating *D. mauritiana* and *D. simulans,* and that the acquisition of reproductive isolation is merely one step in a continuous process of genetic differentiation between isolated populations.

Coyne's investigations have been excellent, but his conclusions can be disputed. A number of authors, even those arguing that speciation can occur through changes at one or a few loci of large effect (i.e., Bush and Howard 1986), have agreed that species will continue to accumulate genetic differences after a speciation event and that these differences will influence traits responsible for reproductive isolation. Thus, Coyne's (1985) demonstration that isolation increases with time does not invalidate the idea that speciation can be initiated by changes at a few critical genes. Moreover, hybridization studies of closely related species, unless divergence was very recent, will tend to overestimate the number of genetic changes needed for the evolution of reproductive isolation. This has been a consistent problem in animal hybridization investigations, including those of Coyne. Even his closely related species pair, *D. mauritiana* and *D. simulans*, are estimated to have split 2.7 to 2.9 million years ago (Bodmer and Ashburner 1984; Cohn, Thompson, and Moore 1984). Obtaining a true picture of the genetics of speciation demands that investigators focus on taxa that have very recently acquired reproductive isolation, so that the accumulation of genetic differences after the speciation event is less of a confounding factor.

The importance of studying taxa in an early stage of speciation is underscored by the recent work of Orr (1989), who reported that the large effect of the X chromosome on hybrid male fertility in crosses between the subspecies *Drosophila pseudoobscura pseudoobscura* and *D. pseudoobscura bogatana* is due to a relatively small section of the X chromosome, "suggesting the involvement of one or a few loci" (Orr 1989: 187). This finding contrasts sharply with hybridization studies of the older species pair, *D. pseudoobscura-D. persimilis.* In that case, sterility genes are spread on both arms of the X chromosome, indicating the presence of more genes (Orr, 1987).

The genetics of speciation is still in a very rudimentary stage of development, but at present no evidence exists that the acquisition of reproductive isolation is accompanied by a genetic revolution, that is, a massive reorganization of the genome. There is a growing amount of evidence that differences in traits responsible for reproductive isolation can be controlled by a few loci with major effect, and even some evidence of epistatic interactions among genes involved with mate-recognition systems (Ahearn 1989), providing support for Templeton's (1980) model of founder event speciation. However, the majority of genetic investigations of speciation have found that species differences are under polygenic control (Dobzhansky 1936, 1974; Spencer 1944; Patterson and Dobzhansky 1945; Spieth 1949; Weisbrot 1963; Coyne 1983, 1984, 1985; Orr 1987; Coyne and Charlesworth, 1989). Determin-

ing whether this reflects biological reality or a tendency to study ancient species pairs awaits further studies of species in early stages of
divergence.

INBREEDING AND INCREASED MUTATION

H. Lewis's (1966) model of catastrophic speciation was based on the
hypothesis that saltational reorganization of the chromosomes is
sparked by the intensive inbreeding that accompanies the isolation of
extremely small marginal populations. As evidence that a shift from
outbreeding to inbreeding might cause a high rate of chromosomal mutations, he cited a study by K. R. Lewis and John (1959). These investigators sampled twenty-seven males and fifteen females from an inbred
culture of the locust *Pyrgomorpha kraussi*, and reported that the germ
line of twelve males exhibited abnormal chromosomal behavior indicative of genotypic unbalance. K. R. Lewis and John (1959) interpreted
this result to mean that the change in the breeding system had upset
the balance of the genome with a consequent increase in the rate of
spontaneous mutation. Unfortunately, K. R. Lewis and John (1959) did
not study the mating system of this species in nature, nor did they
examine the germ line of individuals from natural populations. Thus,
there is no way of knowing whether inbreeding in the laboratory represented a dramatic change in the breeding system, or whether the
chromosomal mutation rate in the laboratory was higher than in nature.

    H. Lewis's ideas about the relationship between a shift to inbreeding
and a higher mutation rate attracted relatively little attention from animal evolutionists; nevertheless, a study relevant to the issue was that
of Fitch and Atchley (1985), who reported that inbred strains of mice
appear to evolve rapidly and suggested that one potential explanation
for this finding was an elevated mutation rate. Fitch and Atchley (1985)
did not endorse this explanation and cited substantial evidence unfavorable to it. One of the pieces of evidence was a detailed investigation of mutagenesis in mice by Johnson et al. (1981), which found that
inbred strains of mice used as controls exhibited a mutation rate too
low to explain an excess of genetic variation.

    Very recently, a renewed interest in mutation rates of founder pop-
ulations has been generated by the study of transposable elements. In
particular, McDonald (1989) has speculated that because retroviruslike
transposable elements (RLEs) can be regulated by host-encoded suppressor and enhancer genes, a founder event could lead to a population with substantially different suppressor allele frequencies and

therefore an increased rate of transposition. Because the insertion and excision of RLEs from genes and their flanking sequences is an important source of mutation in *Drosophila melanogaster* (Green 1988; Sankaranarayanan 1988), an increased rate of transposition could have a dramatic effect on a population's mutation rate and levels of phenotypic variability. McDonald's speculations are provocative, but, as he acknowledges, there is a dearth of direct data bearing on the issue.

Perhaps the most intriguing data are those of Biemont, Aouar, and Arnault (1987), who studied the chromosomal distribution of *mdg-1* and *copia* RLEs in 17 highly inbred lines of *D. melanogaster* over 69 generations of sib mating. There were no changes in the distribution of the elements in most of the lines, but one line exhibited a dramatic reshuffling of the *copia* element between generations 52 and 69. Biemont, Aouar, and Arnault (1987) point out that such reshuffling events provide a mechanism by which inbred strains can rapidly create new genetic variability and remain responsive to the demands of the environment. The question that still needs to be addressed is whether these reshuffling events are more common in inbred than in outbred populations.

Chromosomal Change and Speciation

The centerpiece of White's (1968, 1978) model of stasipatric speciation is the contention that chromosomal rearrangements can serve as strong primary genetic isolating mechanisms because of the diminished fecundity of heterozygotes. Faced with the difficulty of explaining how a rearrangement that decreased the fitness of heterozygotes could go from low to high frequency in a population, White (1978) suggested four factors that alone or in combination could overcome negative heterotic selection: random genetic drift in small populations, inbreeding, meiotic drive, and selective advantage of the new arrangement as a homozygote. Bush et al. (1977) emphasized that social structuring could also accelerate chromosomal evolution and speciation by reducing effective population sizes, restricting gene flow, and increasing levels of inbreeding.

The importance of small effective population sizes for the spread of negatively heterotic chromosomal variants has received abundant support from mathematical investigations. Wright (1941) calculated that the chance of fixation of a reciprocal translocation causing a 50% loss of relative fertility in heterozygotes is on the order of 1 in 1,000 if N_e is 10, on the order of 2×10^{-6} if N_e is 20, and on the order of 3×10^{-14} if N_e is 50. Emphasizing the interdependence of population size and the level of selection against heterozygotes, Bengtsson and Bodmer (1976) pointed out that a chromosomal mutation causing a relative fertility decrease in heterozygotes of 0.001 has almost no chance of making it to fixation in a population of ten thousand animals, whereas a mutation reducing the fertility of heterozygotes by 25% would have a reasonable chance of reaching fixation in a population of ten animals. Chesser and Baker (1986) reached similar conclusions from a computer simulation study.

Turning to gene flow, Lande (1979) calculated that the effective migration of only a few individuals per generation ($N_e m > 1$) greatly reduces the probability of fixation of a new chromosomal arrangement in a population. This finding led Lande to emphasize the importance of geographic isolation for the establishment of chromosomal variants.

The most extensive mathematical investigation of White's ideas was carried out by Hedrick (1981). His most dramatic finding was that even a small amount of meiotic drive can substantially increase the probability of fixation of a chromosomal rearrangement that suffers a heterozygote disadvantage. Other factors that had a large effect on the probability of fixation were meiotic drive in combination with genetic drift, inbreeding in combination with genetic drift, and inbreeding in combination with a selective advantage for the new variant in the homozygous state. In contrast, the probability of fixation of the new variant was relatively low when the only force operating was a selective advantage for the new homozygote.

Given the importance White (1968, 1978) attributed to the meiotic difficulties caused by chromosomal rearrangements, one is struck by the small number of examples of chromosomal aberrations known to influence reproductive success presented in the 1973 edition of his classic book, *Animal Cytology and Evolution*. Although such examples certainly exist, for instance, in mice (Tettenborn and Gropp 1970; Capanna et al. 1976), in cattle (Gustavsson 1969), and in Australian stick insects (White 1978), it is becoming abundantly clear that meiotic difficulties do not automatically accompany chromosomal rearrangements. For example, the Przewalski horse and the domestic horse are separated by a single Robertsonian translocation; however, hybrids are fertile and there is normal meiosis and gametogenesis in both sexes (Short et al. 1974). Porter and Sites (1985) reported that when males of the iguanid lizard *Sceloporus grammicus* were heterozygous for one to three centric fissions, they exhibited normal disjunction and produced chromosomally balanced gametes. Apparently, the genetic background of a species has a large influence on the meiotic effects of chromosomal rearrangements (Patton and Sherwood 1983; Baker, Qumsiyeh, and Hood 1987).

There are data from sheep suggesting that even when chromosomal rearrangements lead to the production of aneuploid gametes, these gametes degenerate or are outcompeted for fertilization, allowing the heterozygous carriers to maintain normal fertility (Bruere 1974; Bruere, Scott, and Henderson 1981). Aspects of reproductive biology such as intrauterine competition between zygotes may also mitigate the effects of the production of aneuploid gametes because the death of zygotes is a normal affair, and lethality caused by chromosomal imbalance will not severely reduce the number of progeny at birth (Turner 1983). Thus, evolutionary biologists cannot assume that a chromosomal rearrangement will have a negatively heterotic effect. The actual effect of a chromosomal variant on reproductive success will have to be evaluated for each group of animals that comes under scrutiny.

Although meiotic drive is capable of a strong influence on chromosomal change (Hedrick 1981), and known examples of meiotic drive exist in natural populations (such as the *t* alleles in *Mus:* Lewontin and Dunn 1960; Young 1967; Silver 1985), the only known case of an association between meiotic drive and chromosomal change did not explain the direction of the chromosomal change (White 1978). The lack of examples may be explained by the rapidity with which a driven chromosome will reach fixation in a population if unopposed by selection (Hedrick 1981; Hartl and Clark 1989). Alternatively, meiotic drive may be a truly rare phenomenon in the context of chromosomal evolution.

Bush et al. (1977) wanted to directly examine whether small population size promoted rapid speciation and chromosomal change due to inbreeding and drift, but were thwarted by the dearth of N_e estimates for natural populations. Instead, they examined the relationship between rate of speciation and rate of chromosomal evolution in vertebrates, and pointed out that if both processes depend on the existence of small demes, the two rates should be correlated. Indeed, the predicted correlation was found, and Bush et al. urged that subdivision of species into small demes be considered as a possible explanation. An independent study by Bengtsson (1980) confirmed the results of Bush et al. (1977); however, contrary evidence also exists. Tegelstrom, Ebenhard, and Ryttman (1983) found no correlation between the rate of karyotype evolution and rate of speciation in birds, and Baker and Bickham (1980) reported a similar result in bats.

A specific prediction made by Bush et al. (1977) was that N_e values of lower vertebrates, such as salamanders and frogs, that exhibit low rates of karyotypic evolution and speciation should be larger than the N_e values of most mammals. Twelve years later, N_e estimates for verte-

TABLE 7.3 Estimates of Effective Population Size for Vertebrates

Taxon	N_e	Reference
Odocoileus virginianus (white-tailed deer)	45–80	Chepko-Sade et al. 1987
Equus cabullus (horse)	50	Chepko-Sade et al. 1987
Ursus americanus (black bear)	550–850	Chepko-Sade et al. 1987
Helogale parvaula (dwarf mongoose)	1–12	Chepko-Sade et al. 1987
Canis lupus (wolf)	804–1, 662	Chepko-Sade et al. 1987
Cynomys ludovicianus (black-tailed prairie dog)	23–31	Chepko-Sade et al. 1987
Dipodomys spectabilis (kangaroo rat)	7–16	Chepko-Sade et al. 1987
Ochotona princeps (pika)	4–59	Chepko-Sade et al. 1987
Acris crepitans (cricket frog)	4–111	Gray 1984
Notophthalmus viridescens (red-spotted newt)	25	Gill 1978
Uta stansburiana (iguanid lizard)	17	Tinkle 1965
Sceloporus olivaceus (rusty lizard)	225–270	Kerster 1964
Rana pipiens (leopard frog)	46–112	Merrell 1968
Bufo woodhousei fowleri (Fowler's toad)	38–152	Breden 1987

Note: The majority of the estimates are for the inbreeding effective population size

brates are still in short supply, and the reliability of the estimates is doubtful. However, a preliminary assessment of the prediction is possible (table 7.3). Contrary to the prediction, the N_e estimates for lower vertebrates do not differ significantly from the N_e estimates for mammals (Kruskal-Wallis test using midpoint values, $P > .5$). Thus, the available data on effective population sizes do not support the contention that the rapid chromosomal evolution of mammals is the result of subdivision into small demes.

The N_e estimates given in table 7.3 were obtained through direct field studies of each species. Estimates of N_e have also been obtained for vertebrates, especially lower vertebrates (i.e., Taylor and Gorman 1975; Larson, Wake, and Yanev 1984; Easteal 1985), by the application of indirect methods to allozyme data. In general, the N_e values for lower vertebrates estimated by these indirect methods are much higher than the values obtained through direct field studies. Indirect estimates of N_e were not included in table 7.3 because comparisons with direct estimates of N_e obtained from the same group of organisms are lacking, so we do not know whether the two estimation procedures should be expected to give different answers. However, indirect estimates of gene flow ($N_e m$) are consistently higher than direct estimates of gene flow for the same species (Slatkin 1987).

Sweeping comparative studies have only limited usefulness in providing insight into the process of chromosomal change and speciation. Such studies have low resolution and cannot bring to light the unique aspects of an individual speciation event—unique aspects that may hold the key to understanding the role of population structure in the event. Unfortunately, rigorous studies of population structure are rarely incorporated into individual investigations of chromosomal evolution and speciation. In one of the few examples, Thompson and Sites (1986) compared the population structure of two closely related iguanid lizards: *Sceloporus grammicus*, a chromosomally polytypic species, and *S. graciosus*, a chromosomally monotypic species. The results indicated that the monotypic species possessed a more subdivided population structure than the species characterized by chromosomal races. On the surface, this result appears very damaging to the position that population subdivision promotes the fixation of underdominant chromosomal variants. Scratching below the surface a bit, it is not known whether the fission rearrangements that distinguish the chromosome races of *S. grammicus* have a negative effect on the reproductive success of heterozygotes. Work performed in Sites's laboratory indicates that these fission rearrangements either have no effect (Porter and Sites 1985) or a very small effect (Porter and Sites 1987) on segregation patterns in meiosis. A rigorous evaluation of the role of population structure in the spread of underdominant chromosomal mutations can only come from the study of chromosomal mutations clearly shown to be negatively heterotic.

Analyses of population structure are critical for the evaluation of the ideas of White (1968, 1978) and Bush et al. (1977), but the uncertainties associated with such studies should be kept in mind. Many parameters influence N_e and gene flow. These parameters, such as long-range migration rates, can be extraordinarily difficult to measure (Breden 1987; Chepko-Sade et al. 1987). An additional weakness of most studies of population structure is that they concentrate on a single population over a short period of time, but population dynamics are known to vary over space and time (Begon 1977; Weatherhead and Boak 1986; King 1987; A. T. Smith 1987; Rowley, Russell, and Brooker, chapter 13;

Smith, chapter 14; Knowlton and Jackson, chapter 10; Waldman and McKinnon, chapter 11; all in this volume). Finally, even when population structure comes under careful scrutiny, it is difficult to eliminate the possibility that a rare event, such as a bottleneck in size, in an isolated part of the range of a species has promoted chromosomal change.

Undoubtedly, the most original idea proposed by Bush et al. (1977) was that intrinsic factors, especially social structuring, could reduce N_e, increase inbreeding, and therefore increase rates of chromosomal evolution and speciation. Surprisingly, this hypothesis has received little direct evaluation. However, in a synthesis of the available data on rodent social systems, Lidicker and Patton (1987) tentatively concluded that social level and genetic structuring are not closely correlated and that the amount of gene flow among demes cannot be predicted from the level of social complexity. Lidicker and Patton (1987) regarded their conclusions as tentative because of the small number of species examined and the inadequate quality of the relevant data.

Thus, twenty-two years after White (1968) began promulgating a stasipatric model of speciation, a lack of critical in-depth studies prevents judgement on the merits of the model and the modifications of it that have followed (e.g., Bush et al., 1977). Still at issue is the importance of chromosomal rearrangements as strong primary genetic isolating mechanisms, as well as the frequency with which certain aspects of the structure of natural populations have promoted the fixation of negatively heterotic chromosomal variants.

CONCLUSIONS

For many years evolutionary biologists interested in speciation bemoaned the extended time scale required for species formation, a time scale that precluded direct human observation and measurement. As attention has shifted in recent years to small-population models of speciation, the time scale problem has seemingly diminished because of the rapidity with which genetic divergence and the onset of reproductive isolation is believed to occur. But this sets up a new series of questions: in particular, how does one study a process that occurs rapidly in small, isolated populations?

One approach is to study the process in the laboratory. The feasibility of a laboratory approach for small, rapidly reproducing animals has been demonstrated by Powell (1978), Templeton (1979), and Galiana, Ayala, and Moya (1989) in studies of founder effect speciation. A second approach is to study the effects of a known founder event or a known population bottleneck on the evolution of reproductive isolation. As discussed earlier, populations resulting from known founder events have been used by a number of investigators to evaluate the impact of a founder event on levels of genetic variability. But until now, there has been no effort to ascertain whether a known founder event or population bottleneck has led to the onset of reproductive isolation between parental and daughter populations. The great advantage of the two approaches outlined above is that the scourge of speciation studies—history—is understood.

A third approach to the investigation of small-population models of speciation is to find a case of incipient speciation and to analyze it in great detail, combining genetic, ecological, and behavioral studies. The goals of such a study should be to identify the specific trait differences responsible for reproductive isolation, the genetic bases of the trait differences, and the ecological, geographic, and selective conditions that promoted the evolution of the trait differences. The multidisciplinary approach advocated here is suitable, and indeed, necessary, for the study of any speciation event. By advocating the relatively indiscriminate study of apparent cases of incipient speciation, I am taking the stand that the single greatest challenge evolutionary biologists face in reaching an understanding of the process of speciation (gradual or rapid) is the full documentation of some cases of speciation. If enough well-documented cases are accumulated, the importance of population structure, the isolation of peripheral populations, and founder events should be evident.

The importance of a multidisciplinary approach to the study of speciation cannot be overemphasized. As is apparent from the body of this chapter, few conclusions about the influence of small populations and inbreeding on speciation are yet possible. To a large degree, this is because studies of genetic divergence, chromosomal change, etc. are rarely coupled with studies of behavior and population structure. The outstanding exception to this state of affairs is the work of Patton and his students on pocket gophers. In-depth studies of population structure, ecology, and genetic differentiation (e.g., Patton and Yang 1977; Patton and Feder 1981; Patton and Smith 1981; Patton and Sherwood 1982; Sherwood and Patton 1982; Hafner et al. 1983; Smith et al. 1983; Daly and Patton 1986) have allowed Patton (1985) to argue persuasively that chromosomal speciation in *Thomomys* does not occur within the stasipatric framework of White (1968, 1978), but as a result of the fixation of rearrangements in geographic isolation due to founder events. We need more work like Patton's, especially more investigations of speciation that emphasize studies of population structure. This will entail forging stronger links between field biologists (who are gathering the population structure data, but largely for other purposes) and evolutionary biologists. These links have eroded during the past decade as a growing number of evolutionary biologists have turned their attention to molecular biology and to the laboratory bench rather than to the study of natural populations.

ACKNOWLEDGMENTS

The writing of this chapter was partially supported by NSF Grant BSR 8600429. I thank William Shields, Nancy Thornhill, two anonymous reviewers, and especially Kerry Shaw for their advice and criticisms.

8

The Natural and Unnatural History of Inbreeding and Outbreeding

William M. Shields

Just as the fitness consequences of inbreeding and outbreeding are conditioned by the breeding history of any species, so too do concepts about what inbreeding and outbreeding actually are, and whether they are good or bad, depend on one's conceptual history. Many biologists, especially vertebrate zoologists, assert that inbreeding is a rare and aberrant event that should be, and usually is, avoided by organisms in nature. As evidence for the harmful nature of inbreeding they cite numerous studies and reviews of inbreeding depression in laboratory organisms (e.g., Lerner 1954), domestic plants, birds, and mammals (e.g., Wright 1977; Falconer 1981), and in zoos (e.g., Ralls, Brugger, and Ballou 1979; Ralls and Ballou 1983; Lacy, Petric and Warneke, chapter 15, this volume). In the founding volume of conservation biology, Soulé (1980: 158) suggested that "it is obvious that conservationists ought to view inbreeding as anathema." Following his lead, but on the basis of their more extensive review of studies of numerous bird and mammal populations, Ralls, Harvey, and Lyles (1986: 55) concluded that "matings between close relatives are uncommon in most natural populations of birds and mammals . . . and that individuals often avoid mating with their closest relatives." This line of argument, which I have elsewhere (Shields 1982, 1987) called the detrimental inbreeding or inbreeding avoidance (IA) hypothesis, currently permeates vertebrate population biology.

One result of acceptance of a view that inbreeding is unconditionally detrimental has been a cottage industry of studies attempting to demonstrate that organisms have a variety of adaptations, including self-sterility systems, sex-biased dispersal, and kin recognition, that evolved explicitly to allow avoidance of mating with kin (for reviews see Greenwood 1980; Ralls, Harvey, and Lyles 1986; Pusey 1988). Jour-

nals like *Zoo Biology* and *Conservation Biology* contain numerous articles
that trumpet the importance of genetic diversity and what is presented
as the concomitant dangers of small populations and inbreeding
depression in planning for the conservation or reintroduction of rare
or endangered species. A plethora of management plans for zoos (e.g.,
American Association of Zoological Parks and Aquaria, Species Survival Plans) and reserves calling for maximum avoidance of inbreeding,
equalization of family sizes, and even the transfer of individuals between zoos on different continents or between geographically distant
populations in nature (e.g., transferring cheetahs from South Africa to
Kenya; P. P. G. Bateson, personal communication) either have been
proposed or implemented. All of this effort and expense stems from
the good intentions of many who wish to combat the dangers, if not
the downright evils, of inbreeding.

In contrast to many vertebrate biologists, many entomologists, and
especially botanists, view intense inbreeding as a fact to be explained
rather than abjured or avoided (e.g., Hamilton 1967a; Grant 1975; Jain
1976). In addition, a new generation of botanists (e.g., Price and Waser
1979; Campbell and Waser 1987; Waser, chapter 9, this volume), some
zoologists (e.g., Bateson 1978; 1983) and some evolutionary geneticists
(e.g., Shields 1979, 1982; Lande and Schemske 1985) have resurrected
or elaborated earlier suggestions that *some* level of inbreeding might be
beneficial in some circumstances. In particular, Price and Waser, Bateson, and Shields, independently, and on the basis of different kinds of
reasoning and evidence, proposed hypotheses that breeding with
distantly related conspecifics (e.g., members of different local populations) might be as bad or even worse than close inbreeding. Like
interspecific hybridization, such conspecific outbreeding, if with sufficiently distant kin, was thought to be capable of generating a significant depression in fitness, that is, an outbreeding depression. Many
who entertained the resulting "optimal inbreeding" or "optimal outbreeding" hypotheses cited the same studies of mating systems and
population structure as did proponents of the detrimental inbreeding
hypothesis but reached an opposite conclusion. Many suggested that
the data were consistent with a relatively high incidence of inbreeding
and failed to provide clear evidence of inbreeding depression in natural
populations (e.g., Shields 1982; Moore and Ali 1984).

As Lacy, Petric, and Warneke (chapter 15, this volume) note, "the
generally (almost universally) deleterious effects of inbreeding have become widely recognized, and almost dogma," resulting in the conventional wisdom of the deleterious inbreeding school. Other lines of reasoning suggesting that inbreeding at some levels may not be all that

bad, and indeed at times might actually be beneficial, are then seen as "heretical." Such heresy, of course, has almost as long and distinguished an intellectual pedigree as the detrimental school (reviewed in Wright 1978; Shields 1982). In one telling contrast, Soulé himself (in Frankel and Soulé 1981: 153–56) warned managers about the dangers of outbreeding depression. Since he suggested that inbreeding ought to be anathema, the manager is left in a quandary.

It is a conundrum that the same underlying theory and data have led to such radically different conclusions about what is going on in nature. To me it appears that two likely candidates that permit and might even promote the disagreement are: (1) semantic and conceptual differences between the two schools of thought, with the result that they may not always be discussing the same phenomenon, and more importantly, (2) a paucity of critical data on the actual levels of inbreeding and outbreeding characterizing natural populations and their actual effects on fitness in plants and animals.

If Kuhn's (1970) description of revolutionary science is reasonable, the opposing camps (if that is what they are) could result from the two schools being immersed in wholly different paradigms. Members of different paradigms are defined as possessing different research strategies and even incommensurate definitions of the same words (wholly different theoretical structures). The semantic soup that results can do nothing but generate misunderstanding. By reviewing the conceptual history surrounding these issues and providing common definitions for inbreeding and outbreeding, I hope that all of the theoreticians and empiricists contributing to this book and the interested reader, whether ornithologist, anthropologist, botanist, or invertebrate zoologist, will find a common ground on which to explore the occurrence, causes, and consequences of the mating systems actually observed in nature.

While the critical importance of mating systems and population structure to understanding many aspects of the evolutionary process has been discussed extensively since Wright's (1922, 1931, 1946) pioneering work, very little is known about these topics in nature (for reviews, see Wright 1978; Shields 1982; Chepko-Sade and Halpin 1987, and any chapter in this volume). The paucity of data stems from the difficulty and expense of doing the kind of detailed and long-term studies needed to determine how large and how isolated the random mating groups characterizing a species might be (Chepko-Sade and Halpin 1987; chapters 9–18, this volume). Despite the difficulties, if population biology is to provide a sufficient description of the demographic structure of a species to illuminate the genetic structure observed in that species, then such studies will be necessary. Only then will they serve as a basis for understanding the evolutionary process or assist in the wise management of our natural resources as one foundation of conservation biology. To achieve this end, we must stop interpolating from theory and artificial systems (domestic stock, laboratory populations, or zoos) and gather the critical data in nature. Only then can we be reasonably certain whether inbreeding or outbreeding are good, bad, or indifferent in particular circumstances.

SEMANTIC CONFUSIONS THAT REFUSE TO DIE

One of the knottiest problems facing someone interested in population structure is the use of different and often implicit definitions of inbreeding and outbreeding by various parties in the debate (for reviews, see Shields 1982, 1987; Ralls, Harvey, and Lyles 1986). This brief discussion is intended to illustrate how controversy about the evolutionary role of inbreeding and outbreeding may be rooted in the semantic morass caused by their multiple and often changeable meanings. As Jacquard (1975) noted, inbreeding is "one word, several meanings." Most people think of inbreeding simply as the mating of relatives, with the result that when siblings or cousins mate, they talk about inbreeding. Few people would intuitively conclude that if siblings mated, they might be outbreeding. On the same level, if mates do not share a common great-grandparent, most would conclude they were not inbreeding. Such thinking would be based on an implicit definition of inbreeding based on an *absolute* concept of relatedness.

In contrast, when discussing whether a population or species is regularly inbreeding or outbreeding, more technical definitions make reference to *nonrandom* mating based on kinship; that is, inbreeding is the mating of relatives more frequently than expected by chance, and outbreeding is the opposite (e.g., Crow and Kimura 1970). If siblings or third cousins mate more frequently than would be expected, based on their frequencies in the pool of all potential mates, then a population or species would be considered inbred. If mating is random with respect to kinship, the population is considered outbreeding regardless of the observed frequency of close-kin mating. This population perspective on inbreeding relies on a *relative* measure of relatedness. Here the intuitive possibilities include the notion that some "inbred" matings are likely to happen in any otherwise "outbred" population purely by chance.

There are two problems with fully grasping and using such intuitive and at times conflicting notions of inbreeding. The first stems from the fact that relatedness, and hence inbreeding and outbreeding, is a continuous property. As a continuous variable relatedness results in an inbreeding/outbreeding continuum. Only a more or less arbitrary decision can divide this continuum into inbreeding and outbreeding zones. As I put it, (Shields 1982: 33), "Inbreeding intensity can range from a maximum with continuous self-fertilization, to a theoretical minimum with perfect outbreeding (Table 3). Although the latter can only be approximated for short periods of time, the remainder of the continuum is often arbitrarily (and at times implicitly) divided into 'inbreeding' and 'outbreeding' levels. These cryptic divisions are then used in subsequent discussion." In a remarkably similar statement, Ralls, Harvey, and Lyles (1986: 51) stated, "Inbreeding is clearly a continuous variable and it is not possible to adequately describe the continuum using the dichotomy 'inbred' versus 'noninbred.' Since all members of a species have common ancestors and are therefore related, the question becomes, how close should the relatedness between mates be for their offspring to qualify as inbred? Different authors dichotomize the continuum at different (and often unspecified) points, which results in considerable confusion in the literature."

A second source of confusion stems from the relativity of relatedness, which results in relative measures of inbreeding and outbreeding. Here, the algorithms estimate inbreeding by the average relatedness of actual mates relative to the average relatedness of all theoretically possible mates in some larger reference group. Ambiguity and controversy emerge from a tendency to choose whatever reference group one wishes and from failures to make the choice explicit when discussing whether a group is inbreeding or outbreeding. At times, some forget that an explicit reference group is necessary to discuss the problem coherently, and that even then it can remain a problem.

For example, consider a breeding population (< 50) of mice on an island. Since the population is small, and all mates share numerous common ancestors in every preceding generation, someone might assert that *this* deme is inbred. Looking at the same group, however, someone else might argue, no, it is outbred, because in each generation individuals actively avoid mating with their closest kin; that is, *close* kin mate *less* frequently than expected by chance, given their frequency in the population. Of course, in the context of their own definitions, both are correct. They only conflict if one forgets that they are using two related, but different, definitions of inbreeding.

Ralls, Harvey, and Lyles (1986: 52) noted, "If we assume that all individuals of a particular species with an effectively infinite population size are descended from a single mated pair, and that pair is used as our reference population, then the species has an inbreeding coefficient of one, and is therefore totally inbred, irrespective of its mating system. However, if the reference generation for the same species is taken as one generation back from the present and mating is at random, the inbreeding coefficient is zero, and the species is therefore totally outbred. Even at any single point in time we can measure inbreeding with reference to a variety of base populations. For example, consider a species that consists of only two small populations on two different isolated islands between which migration is almost impossible. Our measure of the degree of inbreeding will be much smaller if we use one of the populations instead of both as the base population."

Using similar assumptions, I (Shields 1982: 34) noted, "If we define inbreeding as nonrandom mating, and then use the deme as reference group, we define away the very possibility of inbreeding regardless of observed patterns of coancestry." This can generate the paradoxical conclusion that exclusive full sib-mating, if it occurs via random mating in large nuclear families (as actually might occur within the bodies of some female mites whose progeny mate before their birth; Elbadry and Tawfik 1966; Werren, chapter 3, this volume), is, by dint of the random mating, actually outbreeding. Although it would be logically correct to reach this conclusion, it could be misleading if one were interested in the potential consequences of mating with kin versus nonkin. Since inbreeding is relative, a reference group is necessary before meaningful estimates of the rates of occurrence or comparisons of inbreeding intensity can be made (for review, see Wright 1965; Jacquard 1975; Shields 1982).

If the recent controversy surrounding questions about mating and kinship has done nothing else, it has served to force everyone to sharpen their definitions and assumptions and make them more explicit. For example, Soulé's (1980) initial assertion that "it is obvious that conservationists ought to view inbreeding as anathema," is more precisely focused in his more recent reflections, "If 'Nature knows best,' and Nature abhors *close* inbreeding (chapter 3), then managers should too." (Soulé 1986, emphasis added). The addition of "close" represents broader recognition that the relatedness of mates reflects a continuum and so requires some qualification if we are to accurately describe the patterns observed in nature.

Unfortunately, it is not enough to recognize that semantic and conceptual problems exist so that everyone at least qualifies what they mean by inbreeding and outbreeding. It is also necessary that parties to the debate make sure that they are actually criticizing each other's positions and not misunderstanding or misrepresenting them. For example, Ralls, Harvey, and Lyles (1986: 36) compared inbreeding definitions like the one in "Webster's *New Collegiate Dictionary* of 'mating between close relatives' to mating between relatives 'that share a greater common ancestry than if they had been drawn at random from an entire species' (Shields, in preparation)." They also stated that my definition (Shields 1987), "seems of little value when discussing inbreeding and outbreeding in birds and mammals, because almost all species of birds and mammals would then be described as inbreeding."

The value of any definition of inbreeding does not depend only on what Ralls and her colleagues wish to discuss or conclude, but rather on whether birds and mammals, or any other organisms, possess sufficiently structured populations, and hence are inbreeding intensely enough, to influence their evolution. In essence, the usefulness of a definition depends on the goals of the user. Standard population genetic treatments repeatedly note that one can define the level of inbreeding as inversely proportional to the size of the deme. As Falconer (1981: 57) put it, "pairs mating at random are more closely related to each other in a small population than in a large one." The resulting continuum can be divided, more or less arbitrarily, into inbreeding and outbreeding levels depending on the goals of the investigator.

The extreme level of relatedness (incest) defined as inbreeding by Ralls, Harvey, and Lyles (1986) is more than would be necessary to affect the genetic structure (e.g., Waller, chapter 6; Werren, chapter 3; or Mitton, chapter 2; all in this volume), the tempo and mode of evolution (e.g., Wright 1978), the likely modes and rates of speciation (White 1978; Howard, chapter 7, this volume), or the kinds of possible social evolution (e.g., Wade and Breden 1987; Werren, chapter 3, Michod, chapter 5, Riechert and Roeloffs, chapter 12, all in this volume) characterizing a particular taxon, and perhaps even the costs and benefits of sexual reproduction (e.g., Shields 1982, 1988; Uyenoyama, chapter 4, this volume; Hamilton, chapter 18, this volume). Thus, their definition might not be the best in discussions of evolutionary phenomena that could be driven by much lower intensities of inbreeding. At the same time, however, their division of the spectrum into extreme inbreeding and everything else just might be the most appropriate dichotomy for exploring the question of whether organisms actively avoid mating with their *closest* kin. It would not, however, permit discussion of whether inbreeding, as an unmodified term, were frequent or rare and good or bad.

For a variety of reasons, discussed at length in genetics and population biology texts (e.g., Wright 1978; Shields 1982; Chepko-Sade and

Halpin 1987; Waser, chapter 9, this volume), so long as a species breeds in structured demes, the degree of inbreeding entailed can be sufficient to have significant evolutionary effects. As Falconer (1981: 73) noted about drift, "The conclusion to which the neighborhood model leads is that a large amount of local differentiation will take place if the effective number in the neighborhoods is of the order of 20, a moderate amount if it is of the order of 200, but a negligible amount if it is larger than about 1,000" (see also Wright 1978).

On the basis of this discussion, and accepting criticism of my earlier definitions as being too broad for some purposes, by using standard expectations of the consequences of different population sizes as a guideline, I hope the following definitions would allow for anyone's needs.

Deme. The random mating population. It is the breeding group from within which adult parents are drawn at random, that is, in which all breeding individuals have equal opportunities of mating with any other individual of the appropriate gender. A deme has an areal extent, a neighborhood area, and a population or neighborhood size, determined by the number of individuals living in that area (Mayr 1963; Dobzhansky 1970).

Census population size. The number of individuals capable of breeding censused in the delimited area or social group thought to coincide with a deme (synonym: N; Crow and Kimura 1970).

Effective population size. The size of an ideal randomly mating population (deme) that would produce the same level of inbreeding and opportunity for drift as any real population with a specific census size, areal extent, and characteristics known to affect kinship within the group (synonym N_e; Crow and Kimura 1970). Many factors, including dispersal distances, breeder sex ratios, variance in reproductive success, and variance in census size over time, would affect N_e (reviewed in Wright 1978; Shields 1982; Chepko-Sade et al. 1987). For example, even if there were 100 males and 100 females in a deme, if only 10 males bred, that is, the species was polygynous, and if only 2 of the females produced all the progeny, that is, the variance in reproductive success were large, then the relatedness of mates, and independently, the opportunity for drift, would be much higher. This would result in an effective population size considerably smaller than the census size of 200. There are two estimates of N_e: an inbreeding and a variance size. The former relates to the number and relatedness of the parents in any generation, while the latter depends on the number and relatedness of the progeny they produce in that generation. They would differ substantially only in increasing or decreasing populations (Crow and Kimura 1970; Kimura and Ohta 1971).

TABLE 8.1 The Inbreeding/Outbreeding
Continuum

Extreme inbreeding	$N_e \leq 4$
Moderate inbreeding	$5 \leq N_e \leq 100$
Mild inbreeding	$101 \leq N_e \leq 1,000$
Mild outbreeding	$1,001 \leq N_e \leq 10,000$
Moderate outbreeding	$10,001 \leq N_e$
Extreme outbreeding	Interpopulation or interspecies

Inbreeding. The mating of relatives that share greater common ancestry than if they had been drawn at random from a *large* deme. Here,
large is considered to number greater than 1,000. Below this population size, the opportunity for drift is real, the correlation between uniting gametes will be high, the expected rate of loss of heterozygosity is
moderate to large ($1/2N_e$ per generation), and mutation cannot replenish variability lost to drift, as it is expected to in populations larger than
10,000. The latter expectation stems from 10,000 being the inverse of
conservative estimates of the per-locus per-generation mutation rate
for large eukaryotes (10^{-4}). Inbreeding intensity would increase as the
size of the breeding population decreases (table 8.1). As population
size declines, then, the population might be more likely to evolve in
the manner suggested by Wright's shifting balance (Wright 1978), until
it becomes so small (< 10) that drift could overpower selection, whatever selection's form or intensity, and nonadaptive differentiation en-
sued.

Outbreeding. As the mating system is a continuum, outbreeding is
defined as random mating in demes large enough to avoid the levels of
relatedness that would result in what is defined as inbreeding ($N_e >$
1,000, table 8.1). Careful distinctions must be made here, since avoidance of extreme inbreeding (e.g., parent-offspring or full-sibling incest)
does *not* necessarily reduce the absolute level of inbreeding very much,
except in the smallest populations (about $N_e < 10$). As Falconer noted
(1981: 64), "The exclusion of closely related matings, however, does not
make a great deal of difference to the rate of inbreeding, for the following reason. The progeny of a closely related mating have a higher coefficient of inbreeding than those of less closely related matings. Their
presence therefore raises the average coefficient of inbreeding of the
population at any time. But their higher inbreeding is not permanent:
mating at random, they themselves are likely to mate with less closely
related individuals, and so their higher-than-average inbreeding is not
passed on to their progeny. Thus, the exclusion of closely related matings reduces the average coefficient of inbreeding throughout, but it
does not much affect the rate at which the inbreeding accumulates."

Contrary to Ralls, Harvey, and Lyles (1986) and others, then, it does not follow that if incest is avoided or rare that a population is outbreeding. It will be less inbred than if it mated at random, but if it remains small, it remains inbred, at least with respect to the expected consequences of the mating system for other evolutionary processes and patterns (see also Moore and Ali 1984).

Tentatively accepting these definitions still leaves us with critical questions about the occurrence, causes, and consequences of inbreeding and outbreeding in nature. Even if our primary interest were in understanding how mating systems influence evolution, we would still have to agree on a conceptual framework within which we could assess the frequency of mating among individuals of various degrees of kinship. Only when we agree that we know how much inbreeding and outbreeding is occurring in a particular species or population can we move on to discussions of (1) what proximate and ultimate factors might influence the observed patterns of occurrence (causes), and (2) what consequences the observed pattern of kinship might have for the inbred or outbred individuals or populations.

THE FREQUENCY OF INBREEDING AND OUTBREEDING

In order to fruitfully discuss the roles of inbreeding and outbreeding in evolution, we must have an objective and explicit method for calculating their absolute and relative frequencies. Current discussions often examine the same data and reach opposite conclusions. For example, the conclusion reached by Ralls, Harvey, and Lyles (1986) that close inbreeding is avoided and as a result is uncommon in most birds and mammals was based on the same data that led me to the (on the surface) contradictory conclusion that inbreeding is more common than many might expect (Shields 1982).

Ralls, Harvey, and Lyles (1986: 55) emphasized this conflict in their claim that, "In sharp contrast to assertions made from weak evidence, a number of studies are beginning to provide data from which clear conclusions can be drawn." This implies that my conclusions were, at best, in error. In my opinion, our differences might as easily be attributed to the different definitions of inbreeding on which they were based (i.e., their recent restriction of inbreeding to synonymy with close or extreme inbreeding versus my original broader definition), different methods of determining the frequency of inbreeding and outbreeding, and insufficient criteria to objectively determine whether something is uncommon or common.

In order to explore any of these questions we must first have (1) a set of criteria with which to classify particular matings as inbred or outbred and (2) an estimate of how frequent such matings would be based purely on chance encounters. There are two standard methods for determining the degree of inbreeding characterizing a population. First, if an investigator has information on effective population size, this can be translated directly into an estimate of degree of inbreeding, with F, the inbreeding coefficient for the generation in question, being approximated by

(1) $1/2N_e$.

In this scheme, perpetual self-fertilization is represented as random mating in a deme with $N_e = 1$, continuous full-sibling incest as random mating in a deme with $N_e = 2$, and so on (Allen 1965; Wright 1978; Falconer 1981). If the data on dispersal, mating systems, and variance in reproductive success needed to estimate N_e (Chepko-Sade et al. 1987) is difficult or impossible to obtain, pedigree data would still permit estimation of degree of inbreeding or outbreeding (Crow and Kimura 1970). Here observational or genetic data, and preferably both, on paternity and maternity within a population are used to generate known pedigrees. The historical data from such pedigrees are then used to estimate the relatedness of mates (e.g., Waldman and McKinnon, chapter 11, this volume; Packer and Pusey, chapter 16, this volume).

The ratio of the observed frequency of mating between one class of kin and that of mating between individuals that are known not to be that class of kin are the data that Ralls, Harvey, and Lyles (1986) assert permit their clear conclusion that close inbreeding is rare in the populations they examined. They (1986: 38) noted that, "A clear pattern seems to emerge. With two exceptions, parent-offspring and full sib matings are between 0 and 6 percent of those observed. Over half of the studies put that figure at below 2 percent." Even if their assertions were true, and the data as presented unequivocal, it would not imply that the species in question were not inbred. Even if close-pedigree kin do mate infrequently, if the entire deme is small and relatively isolated, then all members will share numerous common ancestors further back in the pedigree, where it is more difficult for the investigator to gather the needed information. Thus, immediate close-pedigree kin may mate infrequently, while the frequency of mating with relatively close kin remains high (e.g., see the discussion of Hoogland's data on black-tailed prairie dogs in Smith, chapter 14, this volume).

Pedigree analyses require objective and consistently applied methods for determining the relative frequency of different classes of matings. Intuitively, and essentially, the question being addressed is how often individuals of different degrees of relatedness mate with one another. First, we must assume that genetic and not social mates can be identified reasonably well, and that paternity and maternity can be attributed with little error (but see Rowley, Russell, and Brooker, chapter 13, this volume). This would allow for the building of pedigrees and estimation of the relative frequency of matings between particular classes of kin. Alternatively, with molecular techniques with the appropriate resolution (e.g., Waldman and McKinnon, chapter 11, this volume; Packer and Pusey, chapter 16, this volume), molecular estimates of kinship would permit similar classification of individuals into kinship groups as checks on the pedigree.

Ideally, then, to estimate the frequency of a particular class of mating, say extreme inbreeding (parent-offspring or full-sib incest), one would count the number of mated pairs *known* to be that related and the number of pairs *known* not to be that related (i.e., only pairs with individuals with complete enough pedigrees to insure that they are not mated with parents, offspring, or full sibs would be included in the latter sample). Pairs with members having too little pedigree information to unequivocally assign kinship should be removed from the sample. The relative frequency of extreme inbreeding, then, would be the number of incestuous pairs divided by the total number of pairs of *known* pedigree relatedness.

In a table reporting their version of such frequencies, Ralls, Harvey, and Lyles (1986: 39, reproduced in part in table 8.2 here), illustrate some of the dangers associated with this approach. First, for various species they present frequencies based on (1) the number of copulations, (2) the number of young born, or (3) the number of pairs multiplied by the number of breeding seasons the pairs were together. These measures are not strictly comparable, could bias the analysis, and are offered with no rationale for the choices given.

One example of the dangers of mixing these methods is illustrated by their report that, based on a personal communication from W. D. Koenig, and using the number of pairs multiplied by the number of breeding seasons, they estimated that 3.2% of the matings observed in acorn woodpeckers were close inbreeding. In contrast, Koenig, Mumme, and Pitelka (1984: 269) reported that of about 49 different pairs observed where relatedness could be assigned, "10 cases of known or probable inbreeding have been recorded" (parent-offspring or full-sib), implying that the amount of close inbreeding was about 20%. Which is the better estimate is not clear from the data.

Pedigree data become more complete over time. Early in any study

TABLE 8.2 Observed Frequency of Close Inbreeding and Estimated Effective Population Sizes in Selected Natural Populations of Birds and Mammals

Species	Frequency of Close Inbreeding	N_e
White-fronted bee eater	0.000	>1,000
Florida scrub Jay	0.004	750
Arabian babbler	0.006	500
Yellow-eyed penguin	0.006	500
Purple martin	0.007	429
Great tit	0.015	200
Cliff swallow	0.017	176
Great tit	0.030	100
Acorn woodpecker	0.032	94
Mute swan	0.098	31
Splendid wren	0.194	15
Belding's ground squirrel	0.000	>1,000
African wild dog	0.000	>1,000
Red deer	0.000	>1,000
Yellow-bellied marmot	0.012	250
Beaver	0.013	231
Chimpanzee	0.017	176
Black-tailed prairie dog	0.025	120
Mountain gorilla	0.036	83
Horse	0.039	77
Dwarf mongoose	0.058	52

Note: Estimates of frequency of inbreeding taken from Ralls, Harvey, and Lyles (1986: table 2). N_e is estimated using equation 2 ($P(I) = 3/N_e$).

many pedigrees will be incomplete, so that the relatedness of many pairs will be unknown. Only after many generations of study will the pedigrees become complete enough to estimate the relatedness of most of the breeding population. Thus, only later in a study will many pairs that are closely related be identified as such. Under such conditions, inbred pairs will become more frequent later in a study, and so have been known as breeders for fewer seasons than other pairs classified as not closely related. Under such conditions, "noninbred" pairs will have been observed to be together for more breeding seasons than "inbred" pairs, even if both types of pairs actually stay together for the same number of seasons on average.

Potential biases in such analyses are further exacerbated if investigators misclassify pairs of unknown relatedness (e.g., all pairs in the first generation of a study) as noninbred. Ralls, Harvey, and Lyles (1986) provide an example of this error with their report that 9.8% of the matings in a population of mute swans were inbreeding closely.

They (Ralls, Harvey, and Lyles 1986: 41) cited Reese (1980), who reported that, "of 32 banded pairs recorded, two were siblings from the same brood, three were siblings from different broods, and one was a male paired with his daughter following the loss of his mate."[1] Arithmetically, 6 of 32 does not compute to 9.8%, but rather to about 19%. Reading their footnote sheds light on this inconsistency as they note that, "We are grateful to Dr. Reese for providing us with his full data set, including information on the pairs in which only one member was banded." Obviously, they included pairs with an unmarked bird in their noninbred sample. Since no one knows how related members of such pairs actually were, it is inappropriate to classify them as either noninbred or inbred. They should have been removed from the sample.

Finally, care must be taken to insure that the data reported reflect the actual mating patterns observed in nature. Both Ralls, Harvey, and Lyles (1986) and I (Shields 1987) agreed that perhaps the best evidence for avoidance of inbreeding had been reported in studies of two marsupial species in the genus *Antechinus* for which no inbreeding was reported (Cockburn, Scott, and Scotts 1985). The crucial evidence we found so convincing was that juvenile females remained on their natal ranges, while *all* juvenile males dispersed. Since all adult males die after breeding, juvenile male dispersal could not be the result of competition for breeding opportunities with older dominant males. We reached an identical conclusion that "the only benefits to the mothers appear to be the exchange of sons to mate with their daughters" (Ralls, Harvey, and Lyles 1986: 46).

Our hidden, and as it turns out, incorrect, assumption was that mating took place on the females' ranges and thus with the strange males observed on their ranges after dispersal of related natal males. More recent work has demonstrated that *Antechinus stuartii* actually practices a form of lek promiscuity in which *all* males disperse from their natal ranges to a few of the area's nest trees. Females then leave their ranges briefly and visit trees with male aggregations in order to mate (Lazenby-Cohen and Cockburn 1988). Since all males must join, and move among, these mating aggregations, and all females visit them to mate, additional evidence will be required to determine whether they are inbreeding closely or not, though there is opportunity for both inbreeding and outbreeding. Additional study is necessary before this group can be counted as a good example of anything (see Smith, chapter 14, this volume, for a discussion of new data on the genus).

A NULL MODEL FOR INBREEDING AND OUTBREEDING

Given that we do a reasonable job of assigning matings to different classes of inbreeding or outbreeding, we are left with the question of what our results tell us about their relative frequencies. It is not enough to know that the number of closely inbred matings is 0%, 2%, or 19% of the total. In order to decide whether close, or any other degree of, inbreeding or outbreeding is rare or common, and especially whether certain kin are avoided or preferred as mates, we must estimate the frequency of inbred and outbred pairs we would expect to occur by chance. While I agree with Ralls, Harvey, and Lyles (1986) that generating appropriate and accurate null models is likely to be difficult, owing to the various assumptions that can be changed during the process, I believe that *some* null model is absolutely necessary to provide objectivity to the discussion.

Any model of random mating requires an estimate of the total number of individuals of appropriate age and gender in a breeding population, together with their relatedness to one another. In essence, and ideally, if we have an accurate measure of N, the size of the potential randomly mating group, together with appropriate pedigree data, we can estimate directly the expected frequencies of different kinds of pairings under random mating. To do this formally and rigorously is a problem in combinatorial probabilities. The difficulty here is that we are seeking the expected number of matches based on kinship, and kinship is a relational rather than an absolute and discrete property of individuals, and finally that mating often occurs without replacement.

It is still possible to generate an approximation that should be useful for many purposes. For example, we can assume that: (1) breeding populations have equal sex ratios, (2) everyone breeds at most twice and then dies, (3) the population size is at a steady state, and (4) all breeders produce two surviving progeny, one of each sex (i.e., there is no variance in breeding success). All of these assumptions are conservative with respect to our estimates of inbreeding frequency because violation of *any* of them would *increase* the null frequency of inbreeding expected under random mating. Similarly, increasing longevity would increase the overlap in generations and hence the expectation of inbreeding (Charlesworth 1980b). Under these simple conditions, then, an average adult breeder in a population would have one parent, one progeny, and one full sibling of the opposite sex, alive and available as potential mates. Since some of these individuals might be mated already (e.g., both parents might still be alive), the actual expected frequency of available first-degree (i.e., nuclear family) kin as mates would decline.

Ideally, we would like sufficiently detailed life tables and pedigrees to determine K, defined as the exact number of kin of specified relatedness available as potential mates in a population. In stable populations, the average pair will have no more than two descendants. For the sake of argument, then, we can assume that for any individual, the expected (mean) K for first-degree kin will range between two and four in populations of *any* size. If all matings occur independently, and the probability of a single individual closely inbreeding by chance was $P(I)$, then

(2) $$P(I) = K/N_e,$$

and so would decline exponentially with increasing population size (figure 8.1). While the assumption that matings are independent will rarely if ever be true (e.g., it would require mating with replacement), its violation is not likely to change the expected trends in most real populations.

In an ideal population of 200 monogamous individuals, 100 females would be available as potential mates for each male, but only 3 would be first-degree kin (his mother, daughter, and sister). Since we know there are 100 females in this population, $P(I)$ equals 0.03 for each male. Given a metapopulation, subgrouped into demes with $N_e = 200$, we would expect that 3% of the sampled pairs would be inbreeding closely, regardless of sample size. If the same metapopulation were breeding in larger demes, then we would expect fewer of our sampled pairs to be inbreeding closely (e.g., if $N_e = 1,000$, then $P(I) = 0.003$, and the expected number of closely inbred pairs in samples of most sizes would be zero; figure 8.1). If deme size was much smaller (e.g., $N_e = 50$), then $P(I) = 0.12$, and we would expect 6 closely inbred pairs, if our sample consisted of say 50 pairs (figure 8.1).

Given that we have a good estimate of deme size, a null model would permit us to determine whether there were fewer or more inbred matings than expected by chance, and hence whether inbreeding or outbreeding were being avoided, preferred, or were irrelevant in a study population. This would be especially true in regions of parameter space where the model is robust to changes in its assumptions. In our case, and under *any* reasonable assumptions, the models converge on an implication that in *any* deme large enough to be classified unequivocally as outbreeding (i.e., $N_e > 1,000$), we would expect to see no inbred matings in any but the largest of samples (figure 8.1). Conversely, observing significant proportions of pairs of first-degree kin

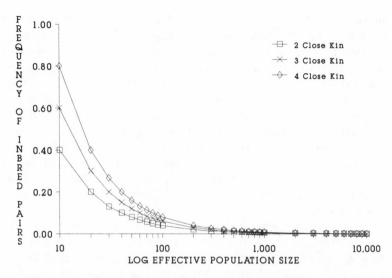

FIGURE 8.1 The relationship between expected proportion of inbred pairs in a sample and the log of effective population size using different assumptions about how many close kin are alive and available as potential mates in a randomly mating population.

(> 0.5%) would be consistent with an assumption that the organism is breeding in demes small enough (N_e < 500) to insure significant levels of background inbreeding or that mating with close kin is preferred (figure 8.1). In this region, however, changes in assumptions would change the probability of incest to a sufficient degree to make tests of inbreeding avoidance or preference more difficult (figure 8.1).

A second property of this null model is that it can work both ways. Gathering all of the empirical data needed to determine effective population size accurately and then predicting the frequency of inbreeding can be exceedingly difficult in the field (for reviews see Wright 1978 and Chepko-Sade et al. 1987). On those occasions when field biologists can develop reasonable pedigrees and have behavioral or genetic data on mating patterns, but do not have data on dispersal or variation in reproductive success, for example, they can use observed frequencies of close inbreeding to generate a reasonable estimate of effective population size.

This will sound circular to some readers and therefore generate skepticism. In this case the skepticism is unwarranted, as the ability to do this is directly deducible from the definition of effective population size. Remember that N_e is *defined* as the size of the ideal randomly mating population that produces a given amount of inbreeding and opportunity for drift. If we document that 20% of the matings between individuals with known pedigrees occur between close kin and that the remaining 80% occur between individuals that are known not to be close kin, then we can conclude that if they *had been* mating at random, it would have to have been in a small deme ($20 > N_e < 40$, depending on demography, figure 8.1). If the local breeding population were larger, we would expect fewer matings between close kin. While the convergence of all of the curves to an asymptote near zero at about N_e = 1,000 (figure 8.1), means that we cannot generate quantitative estimates of N_e in an absence of inbred pairs in a sample, it would be consistent with a qualitative conclusion that the population is randomly mating in demes large enough to be unequivocally classed as outbreeding (see table 8.1).

This null model permits a reassessment of Ralls, Harvey, and Lyles's (1986) conclusions about the rarity of inbreeding in populations of birds and mammals. The observed frequencies, which by their estimates (leaving aside disagreements about the accuracy of those estimates) ranged from 0%–19% (table 8.2), can be translated directly into estimates of effective population size, using the assumption that each individual had three first-degree kin available as potential mates in these populations (table 8.2). In these circumstances, most of the species in question appear to be randomly mating in sufficiently small demes ($N_e \leq 500$) that for most purposes they could be classified as at least mildly inbred. It is certainly open to question whether the amount of close inbreeding observed in these studies should be classified as rare. This is especially true if Ralls, Harvey, and Lyles's (1986) assumption that inbreeding was being avoided were true, since that would decrease the expected incest relative to the random null model. With inbreeding avoidance, then, estimates of N_e would be even smaller than those reported here (table 8.2). A safe conclusion would be that most of the species in question bred in small enough demes ($N_e < 1,000$) for inbreeding to have had some influence on their population structure and evolutionary potential.

Finally, we can test the suitability of this probabilistic model by comparing the estimates of N_e it generates (table 8.2), with those calculated for the same populations in standard fashion using data on dispersal, population density, sex ratios, and variance in reproductive success. For two of the species in Ralls, Harvey, and Lyles (1986), wild horses and dwarf mongoose, sufficient data were available to estimate the frequency of incest using pedigrees and to generate an independent estimate of effective population size (Chepko-Sade et al. 1987). For horses, Berger (1987) reported that 3.9% of the observed matings were incestuous, yielding an estimated $N_e = 77$, while the standard population techniques yielded an estimate of $N_e = 50.2$ (Chepko-Sade et al. 1987: table 19.2). For the dwarf mongoose, Rood (1987) reported that 4 of 69 pairs were close kin (5.8% and not 5.5% as reported in Ralls, Harvey, and Lyles, 1986), yielding an $N_e = 51.8$, while the standard method yielded an N_e between 11 and 87 depending on estimates of local population density (Chepko-Sade et al. 1987). These estimates are similar enough to warrant continued development of this method.

THE CAUSES OF INBREEDING AND OUTBREEDING

Given reasonably objective and consistent methods for defining mating systems and for determining their patterns and frequencies of occurrence, we are then left with how best to develop and test hypotheses about their causes and their consequences. Standard teleonomic methodology (Williams 1966) implies that there are always at least two major classes of potential explanations, a nonadaptive and adaptive case, for any phenotype. A nonadaptive explanation for an observed mating system might assume that some factor(s) forced an organism to inbreed or outbreed as an unselected side effect of another adaptive behavior, or that the mating system stems from historical constraints. For example, it could assume that some genetic, morphological, physiological, behavioral, or ecological factors constrain the organism's opportunities to freely choose mates. With such a model there need be no obvious connection between the causes and consequences of inbreeding or outbreeding. An adaptive explanation, in contrast, would assume that the observed mating system was ultimately caused by the beneficial consequences it normally induced. In any adaptive analysis, cause and consequence would often be conflated, because the factors that influence the origin or maintenance of an adaptive trait would also be its ultimate causes. Care must be taken to partition the proximate causes of inbreeding and outbreeding into those resulting (1) as side effects of constraints and (2) directly from intrinsic adaptive value.

The three factors likely to be the most important direct causes of any observed mating system are (1) the population's demography, (2) the individual's dispersal decisions and the population dispersal pattern that results, and (3) behavioral preferences for or against particular classes of kin as potential mates. Demography influences the mating system by determining the numbers and classes of kin available as potential mates at any point in time. Dispersal, in contrast, determines the likely dispersion of different classes of kin in space. Together, both control the random component of inbreeding and outbreeding by determining the availability of kin as potential mates, through their control of the size and composition of the potentially randomly mating group. Jointly, they determine the probability that various kinds of kin will come into contact. Once a group of potential mates come into proximity at the same time (i.e., the potential random mating group is constituted), then behavioral preferences would determine the nonrandom component of inbreeding and outbreeding by influencing the acceptance or rejection of different classes of kin as actual rather than potential mates.

The mating system that emerges from the combined influences of these three factors, then, could arise from a variety of nonadaptive constraints as well as because of direct benefits. All three causes are expected to influence the mating system independently, and as determinants of inbreeding or outbreeding, they are not mutually exclusive factors. Demographic factors like birth or death rates are not likely to evolve in response to their adaptive effects on the mating system and so are more likely to act as nonadaptive constraints on inbreeding or outbreeding. Despite this common a priori assumption, however, tracing the causal chains will often be difficult.

For example, if mortality differs between the sexes, then this can favor polygamy (Murray 1984). If polygyny occurs as an adaptive response to a paucity of male breeders due to higher male mortality, then the reduction in N_e and increase in inbreeding that results from the biased sex ratio of parents (Shields 1982) would not in itself be adaptive, but rather might better be considered a nonadaptive side effect of the demography of this population. In contrast, it is at least conceivable that the demographic character of within-family or within-brood sex ratios could evolve owing to its adaptive value in increasing the level or probability of inbreeding (mixed-sex broods) or outbreeding (single-sex broods, which could prevent sibling matings).

Many of the dispersal patterns observed in nature, often dichotomized as philopatric dispersal, which increases the relatedness of potential mates, and wider vagrant or sex-biased dispersal, which result in greater probabilities of outbreeding (Greenwood 1980; Shields 1982), are known to be influenced by numerous nongenetic factors. For example, local crowding, the distribution of critical resources, including potential mates, the risks and energy costs associated with philopatry or dispersal, all can influence dispersal (for review see Shields 1987). If the ecological costs of dispersal were sufficiently high, then it might not be necessary to invoke the benefit of increasing the probability of inbreeding optimally to explain philopatry (e.g., Knowlton and Jackson, chapter 10, this volume). Similarly, if the sexes differ in the kinds or dispersion of critical resources required for reproduction, then it might not be necessary to invoke avoidance of close inbreeding to explain sex-biased dispersal (e.g., Greenwood 1980; Moore and Ali 1984; Shields 1987).

In these cases, demography and dispersal can be thought of as givens, determined by a variety of ecological factors that are independent of their genetic effects. The degree of inbreeding or outbreeding that results, then, would be a side effect, rather than a function, of the observed demographic characters and dispersal behavior. In a proximate sense, the mating system would be caused by ecological factors. This does not mean that the inbreeding or outbreeding that results could not have important consequences for the individuals, populations, or species in question, only that it might be unnecessary to develop an adaptive argument to explain their occurrence.

While it might be difficult to decide whether the random inbreeding or outbreeding that results from demography and dispersal's influences on population structure is an adaption or a mere side effect, it is not impossible (e.g., Shields 1987; Clutton-Brock 1989). In contrast, however, systems of nonrandom mating that result from behavioral avoidance or preference for specific classes of kin as mates within populations *cannot* have originated in response to some constraint on choice. Since the process exemplifies a choice of mate from among all that are available, while demography and dispersal determine who is available, choice necessarily implies an adaptive cause for increased inbreeding or outbreeding. Whenever nonrandom mating on the base of kinship is demonstrated, the resulting mating system can be said to have been caused, at least in part, by increased reproductive success associated with the observed choices.

In the end, then, in investigations of inbreeding and outbreeding, care must be taken to explore all of the potential causes, their relative importance, and whether they are nonadaptive or adaptive in each case. It is likely that the degree of inbreeding and outbreeding in nature will have multiple causes, including both ecological factors, themselves nonadaptive with respect to genetic consequences, and genetic factors, which act to control the amount of inbreeding and outbreeding at adaptive levels. In my opinion, in any species in which heterozygosity contributes significantly to fitness (favoring avoidance of too close inbreeding; Mitton, chapter 2, this volume), or in which epistasis and coadaptation are common contributors to the integration of the genome (favoring avoidance of too distant outbreeding; Shields 1982; Templeton 1986), genetic factors should move to the fore. Since there is evidence for both segregational and recombinational load in many species, it seems possible that an optimal level of inbreeding or outbreeding could be favored in many cases (e.g., Bateson 1978; Shields 1982; Waser, chapter 9, this volume; Lacy, Petric, and Warneke, chapter 15, this volume).

THE CONSEQUENCES OF INBREEDING AND OUTBREEDING

Whether caused by nonadaptive constraints or adaptive responses to local conditions, the levels of inbreeding and outbreeding observed in nature are either known or expected to have numerous and varied consequences for the mating individuals and for the breeding populations and species that result. For example, both the detrimental and optimal inbreeding hypotheses assume that the relatedness of mates will affect individual reproductive success and may affect population persistence. Additional hypotheses assume that the mating system is potentially influential in all aspects of the evolutionary process (e.g., Wright 1978; Shields 1982; Chepko-Sade and Halpin 1987) and should be useful in shaping conservation strategies for threatened species or ecosystems (e.g., Soulé and Wilcox 1980; Soulé 1986). Critical reviews of different aspects of this problem are the focus of much of this book (e.g., any chapter in Part 1, this volume). Rather than simply previewing or repeating what those authors have to say, I would rather focus on what I believe are the critical assumptions and problems associated with measuring the fitness effects of inbreeding or outbreeding.

As everyone knows, inbreeding depression is a fact of life. As Lerner (1954: 22) put it, "There is no particular need to discuss here the details regarding the reduction in fitness and viability occurring under inbreeding. Firstly, the deterioration of populations subjected to continuous consanguineous mating represents a generally known phenomenon." As with all dogma, the "facts" are then reiterated: "It is an almost universal observation that severe inbreeding leads to 'inbreeding depression,' a serious reduction of fitness in its various components" (Mayr 1963: 224), or ". . . inbreeding usually leads to a decrease in size, fertility, vigor, yield, and fitness" (Crow and Kimura 1970: 61), or "But inbreeding lowers individual fitness and imperils group survival by the depression of performance and the loss of genetic adaptability" (E. O. Wilson 1975: 80).

Despite its wide acceptance as fact, almost all of the evidence for inbreeding depression, at least in animals, stems from studies of domestic and laboratory organisms (Lerner 1954; Wright 1977; Falconer 1981) or of populations in artificial environments like zoos (e.g., Ralls, Brugger, and Ballou 1979, reviewed in Lacy, Petric, and Warneke, chapter 15, this volume). The same is true for most of the evidence reflecting on outbreeding depression (e.g., Shields 1982; Templeton et al. 1986; Templeton 1986). While such data, if gathered and analyzed appropriately, are certainly useful for developing hypotheses about field systems, they are not, by themselves, sufficient to confirm their truth under the wholly different circumstances afforded by nature. While everyone might agree that Darwin's (1859) marshaling of evidence about artificial selection in domestic organisms lent weight to his speculations about natural selection in the wild, few if any were convinced of the efficacy of natural selection solely because of the demonstration of artificial selection.

Given the importance of inbreeding and outbreeding to our understanding of evolution, then, it is still surprising how little data there are on their actual effects on reproductive success in natural populations. There are so few data that, in essence, questions about the fitness consequences of inbreeding and outbreeding remain as open today as when they were first broached by Darwin. It is not the case that theory provides simple and straightforward answers to questions about inbreeding or outbreeding depression. It turns out that inbreeding and outbreeding depression are themselves expected to be conditioned by the breeding history of any focal population (e.g., Shields 1982; Lande and Schemske 1985; Schemske and Lande 1985; but see D. Charlesworth and B. Charlesworth 1987). It is theoretically possible that many sexual populations could inbreed regularly without harmful effects. Thus, it would seem useful to critically and even skeptically examine the evidence for inbreeding and outbreeding depression.

That does not appear to be standard operating procedure. So ingrained is the dogma that inbreeding is harmful that scholars arguing that inbreeding is rare, owing to special adaptations that evolved in response to inbreeding depression, find it unnecessary to cite *any* evidence that inbreeding depression in fact does occur in their study species. For example, Ralls, Harvey, and Lyles (1986: 35) assert, "Mating between close relatives increases the proportion of loci at which offspring are homozygous . . . This increase in homozygosity can cause inbreeding depression. Although few data from natural populations are available, studies of captive and experimental animals consistently confirm the ubiquity and magnitude of its effects (Ralls and Ballou 1983)." Apparently, they felt that this analogy was sufficient evidence to support their critical assumption that inbreeding depression controls the avian and mammalian mating systems they then proceed to describe.

Their evidence, however, is much shakier than the conclusions would suggest. For example, *all* of the evidence for inbreeding depression in zoo populations is based on a single and early component of fitness, juvenile mortality (reviewed in Ralls and Ballou 1983 and in Lacy, Petric, and Warneke, chapter 15, this volume). While it is indisputable that in these populations juvenile mortality is higher in the progeny born of closely related pairs, no data are presented on the potential effects of inbreeding on other components of fitness that combine to determine total reproductive success.

Assume that most inbreeding depression results from the unmasking of deleterious recessives and that the observed level of inbreeding, while entailing a cost in terms of increased juvenile mortality, also allows a pair to avoid the costs of recombinational or migrational load associated with wider outbreeding. Under such conditions, it is conceivable that, on average, inbreeding pairs might be more fertile than outbreeding pairs, or that the inbred young that do survive might be more vigorous and fertile than their outbred competitors. Such a trade-off of quantity for quality, like the analogous trade-off between egg number and egg size, could be favored in a variety of conditions. For example, if truncation selection were operating, only pairs that produced progeny of the highest quality could expect final success (Wallace 1968; Shields 1982). Thus, even if outbreeders produce more progeny that live to breed (and even that cannot be known from mortality rates alone!), if the inbreds that did survive were higher in quality than any of the outbreds, then inbreeding could be favored despite the costs of increased juvenile mortality.

If one estimates *all* the components of fitness, it may be the case, even in zoos, that *total* reproductive success of inbreeding pairs will exceed that of outbreeding pairs, despite the increased juvenile mortality associated with inbreeding. No one can know for sure until all of the data are gathered. It is hard to disagree with Darlington (1960: 298), who suggested that in exploring inbreeding and fitness, "Thus we are bound to profit from considering the total reproductive result. Though the parts are arbitrary the whole is a reality: it is the great reality in life."

Following his lead, I would suggest that rather than presenting depression in a single component of fitness and then discussing it as inbreeding or outbreeding depression (e.g., Lacy, Petric, and Warneke, chapter 15, this volume), it might be better to reduce the sample size and concentrate on those few studies where the *total* reproductive success associated with different levels of inbreeding and outbreeding are actually known.

Finally, then, we are led to the question of what evidence exists for either inbreeding or outbreeding depression in wild populations. As all admit, the data on both inbreeding and outbreeding depression are at best sparse, especially for animals (e.g., Shields 1982; Ralls, Harvey, and Lyles, 1986; Waser, chapter 9, this volume; and all of Part 2 in this volume). Inbreeding depression in plants, especially in self-incompatible species, has been documented so frequently that little more need be said about it (for reviews see Grant 1975, and D. Charlesworth and B. Charlesworth 1987), though even here there are too few studies in nature, and fewer still that attempt to examine all of the components of fitness (for review see Waser, chapter 9, this volume).

There is good evidence that fitness can be depressed if mates are too distantly related in a few plant species as well. That evidence is unequivocal in that total reproductive success, rather than one or two fitness components, associated with specific levels of inbreeding and outbreeding was examined (reviewed by Waser, chapter 9, this volume, and Waller, chapter 6, this volume). Similarly, there is one elegant and reasonably complete study by Grosberg (1987, discussed by Knowlton and Jackson, chapter 10, this volume), demonstrating outbreeding depression in a marine invertebrate. Finally, there are numerous cases of presumptive outbreeding depression in other invertebrates, and especially in insects (reviewed in Shields 1982; Partridge 1983). While total fitness is well estimated in all of these studies, kinship usually is estimated indirectly on the basis of an assumption that the distance between two individuals is negatively correlated with their relatedness, so even here there is room for debate.

Finally, there is little direct evidence about outbreeding depression from unmanipulated populations of vertebrates (e.g., Ralls, Harvey, and Lyles 1986; Templeton 1986). There are, however, a few classic cases of artificial transplants that have mixed individuals from different populations and produced hybrids so maladapted that they act as exemplars of intraspecific outbreeding depression (reviewed in Shields 1982; Templeton et al. 1986; Templeton 1986). These have included hybrid ibex from crosses between Austrian and Nubian populations that calved in the coldest month of the year in Switzerland (Greig 1979) and salmon fry from crosses between upstream and downstream populations that cannot make up their minds which way to swim during migration (Brannon 1967). More data are needed to determine the generality and quantitative importance of such outbreeding depression in wild populations.

For a variety of reasons, the question of inbreeding depression has been addressed more frequently than outbreeding depression in wild populations of animals. For example, some data on the relative reproductive success of inbred and noninbred pairs have been reported in many of the studies cited by Ralls, Harvey, and Lyles (1986). These species are presented as putative examples of species in which inbreeding is rare, on the assumption that it is avoided owing to its harmful effects. Given this hypotheses, one would predict that pairs that mated "in error" (i.e., the observed inbreeders) should show a significant fitness depression relative to those not inbreeding closely.

In the one case reporting strong evidence for inbreeding depression in the wild, Greenwood, Harvey, and Perrins (1978) documented sixteen cases of incest with appropriate reproductive data in an English population of the great tit (*Parus major*). They reported that, "We have shown that inbreeding pairs have a lower breeding success, resulting from higher nesting mortality, than normal pairs." Their conclusion was based on the use of an inappropriate statistical test that involved comparing single nest attempts of the inbreds (which resulted in integer values for the reproductive variables in question) to the mean values of otherwise similar outbred nest attempts (which resulted in fractional values for the same variables) using a Wilcoxon matched pair test. This test presumes that the two distributions compared are matched events and not an event and a sample. Usually, the test is used to compare drug efficacy or other treatments in before-and-after studies of the *same* individual. Reanalysis of their data indicates that the differences reported are not statistically significant using most standard tests, though there is a trend in the predicted direction, as six of the inbred pairs had more, and ten had fewer, fledglings than the population averages.

Other than this single case, then, all of the other studies reporting sufficient data are inimical to hypotheses predicting widespread and severe inbreeding depression. For example, studies of populations of the great tit in the Netherlands (van Noordwijk and Scharloo 1981) and the splendid fairy-wren in Australia (Rowley, Russell, and Brooker 1986) reported either greater net reproductive success, despite depression in hatchability, or no differences respectively between inbred and outbred pairs (reviewed in Rowley, Russell, and Brooker, chapter 13, this volume). Similarly, the classic mammalian avoider of inbreeding, Hoogland's (1982) black-tailed prairie dog, shows no evidence of inbreeding depression at any inbreeding intensity (reviewed in Smith, chapter 14, this volume). Similarly, Rood (1987: 97) reported that in the dwarf mongoose, "Inbreeding depression was not apparent in the four related pairs (two father-daughter, one mother-son, and one brother-sister). The mean number of offspring raised to yearling status by these related pairs was 2.25, slightly higher than the mean (1.92) raised by the 55 unrelated pairs for which the number of young raised was known."

In the primates, early claims of inbreeding depression (e.g., Packer 1979) have been challenged (e.g., Moore and Ali 1984; Moore, chapter 17, this volume) and new studies claiming an absence of inbreeding depression have been reported (e.g., Bulger and Hamilton 1988). Finally, in one of the few studies that goes beyond early fitness components in humans (but see Rao and Inbaraj 1977), where inbreeding depression is reported as occurring regularly even at the level of first-cousin pairs, Darlington (1960) reported, "From the same average numbers of about six children per original marriage the inbred groups produce more than twice as many ggc [greatgrandchildren] as the outbred groups (28.6 against 12.8)."

In the vertebrates, in every case with sufficient evidence there is either no inbreeding depression or in a few cases even significant inbreeding enhancement. These data do not support a monolithic hypothesis that assumes that inbreeding depression is a fact of life. At the very least they should force us to additional study, revision of our working hypotheses, or both. We should not be led to our beliefs about inbreeding and outbreeding and our recommendations for managing populations by dogma, but rather should continue to ask the organisms what is going on with inbreeding and outbreeding in nature.

PART TWO

Empirical Perspectives

9

Population Structure, Optimal Outbreeding, and Assortative Mating in Angiosperms

Nickolas M. Waser

"In sexual reproduction of plants, where the male and female gametes must be brought together, not voluntarily as in animals, but through extrinsic means, biochemical elements have, to a great extent, replaced consciousness, so that the union of gametes if facilitated or checked by the stimulation or inhibition of the growth process." (Pandey 1969, 461)

"Plants make use especially of the diploid style as a sieve for sorting the pollen delivered to it by a pollinating agency, over which it can exercise no direct control. . . . The control of the mating system is evidently such as will give a controlled degree of hybridity generally intermediate between homozygosity and that extreme heterozygosity which we see to be disastrous in crosses between species." (Darlington and Mather 1949, 253)

Random mating, or panmixis, is the simplest mating system to envision. Under panmixis an individual has an equal chance of mating with any member of its population, including itself—a feat possible to imagine with angiosperms, most of which are hermaphroditic (cosexual) as individuals or single flowers (Yampolsky and Yampolsky 1922). In an infinite panmictic population the chance of mating with oneself or with any particular related (or unrelated) individual is the inverse of infinity, or zero. Thus there is no mating of relatives, i.e., no inbreeding.

An infinite population and pool of potential mates (*mating pool*) is a good starting point, but the "Wrightian" view that population sizes are constrained (Ehrlich and Raven 1969; Endler 1979; Levin 1981; Shields 1982) is more realistic. A finite mating pool invalidates the conclusion just reached about inbreeding, because even random mating will pair relatives sharing alleles descended from a common ancestor, to yield offspring that are autozygous at the locus in question.

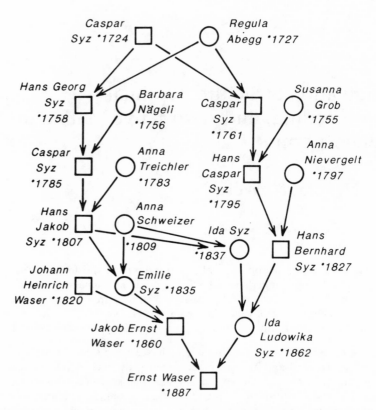

FIGURE 9.1   Part of my pedigree, according to which my paternal grandfather's inbreeding coefficient slightly exceeded 1/16. Arrows indicate the transmission of alleles across generations; years preceeded by asterisks are birth dates.

Consider a human pedigree (figure 9.1). Beginning with the generation of Caspar Syz and Regula Abegg, a reticulate pattern of matings leads to my paternal grandfather, Ernst Waser. His probability of autozygosity (inbreeding coefficient, Wright 1969) was slightly in excess of ¹⁄₁₆, due mostly to the shared descent of his parents from Hans Jakob Syz and Anna Schweizer. A limited mating pool surely played a role: this portion of my ancestors hailed from villages in a single small region, and there was little exchange with other regions. A paucity of potential mates of the proper social class in such a "viscous" population may be one reason why kin matings (e.g., of first cousins) are condoned in many societies.

On the surface, the Syz pedigree approximates those expected in many plant populations. Plants are sessile and often occur in fragmented populations. Even within continuous populations, the mating pool may be restricted by limited seed and pollen dispersal (Levin and Kerster 1974; Endler 1979; Waser and Price 1983; N. M. Waser 1988) due to mechanical constraints on abiotic dispersal and constraints imposed by the foraging strategies of animal pollinators and seed dispersers. Thus most mating possibilities are restricted to a local population, or to a genetic neighborhood (Crawford 1984a) around each individual within a larger population (Levin 1988).

If seed paternity were determined wholly by random pollination in a finite population, inbreeding would be the rule. But in fact there is substantial scope for nonrandom pollination, and for nonrandom processes after pollination that also contribute to final mating patterns. Whenever certain maternal or paternal parents or parental combinations enjoy disproportionate success (Lyons et al. 1989), actual matings will deviate from random expectations based on the mating pool or some other reference population. Net deviations may be in the direction of pollination or postpollination interaction among individuals more similar phenotypically than expected at random, which is termed positive assortment (Roughgarden 1979). In several respects inbreeding can be considered a special case of positive assortment, with the similarity involving alleles shared by descent. Interaction among individuals less similar phenotypically than expected at random is termed negative assortment; outbreeding can be considered a special case in which the dissimilarity is genetic.

In summary, we expect two contributions to realized mating patterns. Forces largely *extrinsic* to plants, such as edaphic requirements and imposed limits on gene dispersal, restrict mating pool size. But the effects of this may be modified by events mainly *intrinsic* to plants that determine which pollinations actually produce offspring.

This chapter discusses the potential for positive and negative assortative mating in plants, focusing mostly on assortment with respect to genetic similarity in natural populations of cosexual angiosperms. I first review stages in angiosperm sexual reproduction and assortative processes at each stage, then consider selection for assortment and the evolution of mate discrimination, and finally touch on evidence from genotypic arrays that assortment has modified the effects of population structure.

MECHANISMS OF ASSORTATIVE MATING

Stages in Reproduction

Angiosperm sexual reproduction begins with the maturation of pollen and ovules. If pollen-bearing stamens are near receptive pistils (female structures) of the same flower or other flowers of a plant, self-pollination may occur with no outside aid. Pollination over longer distances requires intervention by abiotic agents (usually wind) or animal pollinators (insects, birds, mammals). Once deposited on a pistil, a pollen grain must adhere to the stigma (receptive surface of the pistil) and grow a pollen tube through the stigma and down the style to the ovary. On reaching an ovule, one sperm cell in the tube fertilizes the egg cell to form a diploid embryo, and the other fertilizes the associated polar nuclei to form the (usually) triploid nutritive endosperm tissue. Finally, the fertilized ovule must obtain sufficient nutrients to form a seed. I refer to the stages of reproduction just outlined as *pollination, fertilization,* and *seed maturation,* and to events in the latter two stages as *postpollination* events.

Assortative Pollination

When plants within a local area vary in flowering time, pollination necessarily occurs among simultaneously flowering individuals (e.g., Stam 1983; Waser 1983a). Furthermore, pollinators may exhibit flower "constancy" (Waser 1986), enhancing pollen flow among plants of similar size, flower color, or other attributes (Waser 1983a, 1983b; Levin and Watkins 1984). Insofar as phenotypic similarity reflects identity by descent, positive assortative pollination will contribute to inbreeding.

Contributions to negative assortment during pollination seem to be less pervasive. Pollinators may exhibit "win-shift" behavior (Cole et al. 1982), leaving a previously rewarding location when floral resources are depleted, but their movements may be to phenotypically or genotypically similar individuals. Pollinator preference for certain plants (e.g., Stanton et al. 1989) usually causes deviations from random mating, but again, not necessarily in the direction of negative assortment. One contribution to negative assortment imposed by plants themselves is a reduction in self-pollination caused by temporal or spatial separation of male and female structures within flowers, inflorescences, or individuals (Lloyd and Webb 1986; Cruden 1988).

Assortative Fertilization

Opportunities for Nonrandom Fertilization. In most natural situations, pollen reaches a stigma from several donor individuals (Ellstrand 1984; Meagher 1986). But any flower has a finite life, and some pollen may arrive when the stigma is too young or old to be receptive. Likewise, pollen may arrive after a previous load has caused the stigma to become unreceptive temporarily or permanently (Waser and Fugate 1986) or has fertilized all the ovules (Mulcahy, Curtis, and Snow 1983; Epperson and Clegg 1986; Bertin and Sullivan 1988). These phenomena influence the possibilities for multiple paternity within flowers, but their contribution to assortment is less clear.

Fertilization success also may differ among pollen grains arriving simultaneously at a stigma. Some grains may not adhere (Heslop-Harrison 1975; Stead, Roberts, and Dickinson 1979; Knox 1984), germinate a tube (Heslop-Harrison 1975; Ganeshaiah, Uma Shaanker, and Shivashankar 1986), or peneterate the stigma with their tube (Dickinson and Lewis 1973; Heslop-Harrison 1975). Once in the style, tubes may grow at different rates and/or fail to reach the ovary (Ottaviano, Sari-Gorla, and Mulcahy 1980; Mulcahy and Mulcahy 1983; Heslop-Harrison 1983; Knox 1984; Waser et al. 1987; Bertin and Sullivan 1988; Cruzan 1989a, 1989b, 1990; Waser and Price 1991). Tubes penetrating the ovule may fail to fertilize (Schou and Philipp 1983; Knox 1984; Williams et al. 1986; Seavey and Bawa 1986).

Success or failure of male gametophytes (pollen grains and tubes) can be ascribed to a combination of environmental and genetic contributions from each parent and the parental combination (Lyons et al. 1989; Cruzan 1989b). Positive or negative assortment should show up as a parental interaction based on phenotype, and inbreeding or outbreeding as an interaction based on genetic relatedness. Parental interaction is a corollary of assortment, but is not sufficient to demonstrate assortment.

Self-Incompatibility: One Example of Negative Assortment. Complex experimental crosses allowing one to determine maternal, paternal, and interaction contributions to gametophyte success (e.g., Cockerham and Weir 1977b) are performed infrequently, especially in natural populations. But one example of differential success based on parental interaction is easy to recognize. This is *self-incompatibility,* a class of responses in which pollen reaching a stigma of its parent plant, tubes produced by such pollen, or embryos the tubes father (Seavey and Bawa 1986), are rejected. This hinders the closest kind of inbreeding, contributing to negative assortment for shared alleles. Self-incompatibility is widespread in angiosperms (Fryxell 1957; Charlesworth 1985). Rejection of pollen or tubes may depend on a diploid chemical message deposited on pollen by the paternal parent (so-called "sporophytic" incompatibility) or on haploid gene expression in the male gametophyte itself ("gametophytic" incompatibility). Much of what follows refers to gametophytic self-incompatibility.

There is excellent evidence that the maternal parent plays a role in determining the success of self pollen. Complex chemical signaling occurs between pistil and male gametophyte, and there are ultrastructural changes, interpretable as maternal responses (Heslop-Harrison 1983; Knox 1984). These responses, such as the formation of callose (a polysaccharide) that blocks the growth of individual male gameto-phytes, may occur at many points before fertilization (Dumas and Knox 1983). Maternal control also seems likely because pollen tube growth relies beyond some point on nutrients obtained from stylar tissue (Brewbaker 1957; Vasil 1974; Mulcahy and Mulcahy 1982). Furthermore, Saunders and Lord (1989) suggest a maternal role in active transport of pollen tubes toward the ovary.

More specific observations also indicate maternal control. Self pollen may fertilize when alone but not when outcrossed pollen is present ("cryptic self-incompatibility," e.g., Weller and Ornduff 1989). Pollen germination may be prevented until sufficient outcross pollen has accumulated on the stigma (Grove 1983; Murdy and Carter 1988). Tube growth may be halted once all ovules are fertilized (Raff and Knox 1982). Some species produce both outcrossing ("chasmogamous") and selfing ("cleistogamous") flowers on one plant; the former may reject self pollen while the latter accept it (Lord and Eckard 1984). Self-incompatibility may decline when pistils are physiologically altered by exposure to extreme conditions (Stout 1938; Cruzan 1989b.) In vitro growth of self and outcross pollen on agar may be differentially influenced by extracts from pistils (Heslop-Harrison 1983; Malti and Shivanna 1985).

Such observations suggest that self-incompatibility is not "cooperative," with self pollen foregoing an attempt to fertilize. Indeed, a conflict is anticipated theoretically (Waser and Price 1993). Based on coefficients of kinship and assuming the parents are not inbred, a self pollen grain should forego fertilization in favor of outcross pollen only if the fitness of the resulting offspring would be less than one-third that of an outcrossed offspring. In contrast, maternal tissue should favor an outcrossed fertilization if relative fitness would fall below one-half (figure 9.2). There is a conflict of interest when the relative fitness of a selfed offspring would lie between one-half and one-third, so it is unsurprising to see "plant behavior" after self-pollination that suggests pollen-pistil conflict (e.g., Beach and Kress 1980), and phylogenetic patterns that suggest an evolutionary "race" between self pollen and pistil for control over fertilization (Pandey 1969).

Evidence for Graded Assortative Fertilization. Does maternal "choice" of male gametophytes extend beyond the rejection of self discussed above? Many levels of potential inbreeding may occur when pollen reaches a stigma from a finite mating pool, and the region of offspring

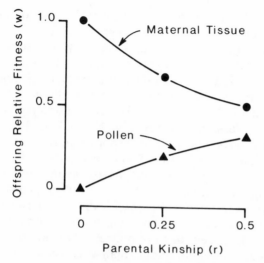

FIGURE 9.2 Conditions favoring divergent prefertilization responses of pollen grains and pistil, when the former represent mating options ranging from selfed ($r = 0.50$), to full sib ($r = 0.25$), to outbred ($r = 0.0$). The "maternal tissue" curve indicates the threshold fitness of an offspring derived from a given class of mating (relative to fitness of an outcrossed offspring) below which it is in the maternal best interest to reject the mating in favor of an outbred one ($r = 0.0$). The "pollen" curve indicates the threshold offspring relative fitness from the genetic perspective of the male gametophyte. Potential for conflict of interest occurs even in a selfed mating, because the genetic perspectives of haploid gametophyte and diploid maternal tissue differ. As kinship of potential mates declines, the two curves diverge. A completely outbred pollen grain should always attempt fertilization, and the pistil should always attempt to block it if there are other outbred grains expected to yield fitter offspring (see Charnov 1979).

fitness values expected to incur a male gametophyte-maternal conflict expands as matings become more outbred (figure 9.2). Furthermore, genetic similarity of potential mates continues to decline even beyond the point where their kinship falls to zero with reference to a given ancestral generation (or equivalently, to some surrounding population of conspecifics). In other words, "outbred" pollen reaching a stigma may be more or less different genetically from the maternal parent, and in the extreme may belong to a distinct species. This variation can be viewed as the result of increasing evolutionary divergence of the gene pools from which pollen and pistil are drawn. Given continuously graded variation in genetic similarity of potential matings available to the maternal parent, selection might favor a correspondingly graded recognition system.

There is some evidence for graded responses. Cruzan (1989b) analyzed reciprocal (diallel) crosses with petunias and found continuous variation in numbers of pollen tubes reaching the ovary. Success depended on parental combination, a corollary of assortative fertilization. In wild radish, J. Nason (personal communication) found that pollen tube growth declined with kinship over a range from outbred to full sib pollinations among plants of known pedigree. Outbred crosses were within a single population, leaving unexplored the additional range of genetic dissimilarity discussed above. Fyfe (1957) and Busbice (1968) reported analogous graded responses in seed set in alfalfa, and Hellman and Moore (1983) found some evidence for them in blueberries. Finally, a few studies (table 9.1) have examined numbers of pollen tubes reaching the ovary as a function of the physical distance between pollen donor and recipient ("outcrossing distance"). Depending on the study, the most successful pollen seems to have been from an intermediate distance, from self, or from nonself (statistical conclusions from these studies are discussed under "Statistical Significance and Power," below).

Could a graded response be under the same genetic control as self-incompatibility? The usual interpretation is that self-incompatibility is controlled by one multiallelic locus (the S gene, Bateman 1952), or a few loci (Lundqvist 1975; Lewis, Verma, and Zuber 1988) at which identity of maternal and pollen alleles (or products of paternal alleles carried on the pollen grain in sporophytic self-incompatibility) causes rejection. A single locus could produce a graded probability of rejection proportional to parental kinship when results are averaged over replicate crosses, since kinship at any autosomal locus should be associated in a genetically structured population with probability of identity at the S locus (Uyenoyama 1988a; chapter 4, this volume). However, a single-locus model of gametophytic self-incompatibility predicts that pollen within any one style either will fully succeed, show a bimodal distribution of success, or fully fail, depending on whether the paternal parent carries 0, 1, or 2 S alleles that match maternal alleles. Thus an index of male success, such as the frequency distribution of pollen tube elongation in a sample of styles at some interval after pollination, should appear trimodal. In contrast, there is some recent documentation of unimodal distributions of pollen tube elongation following selfed or inbred pollinations, suggesting multilocus control (Waser and Price, unpublished; Cruzan 1989b). Indeed, there is evidence to support the view that at least some self-incompatibility systems themselves are polygenic (see Frye 1957; Mulcahy and Mulcahy 1983; Østerbye 1975; Seavey and Bawa 1986; Bowman 1987; Barrett 1988). For example, angiosperms commonly exhibit "pseudoincompatibility," in

TABLE 9.1 Evidence for Assortative Fertilization as a Function of Distance Between Pollen Donor and Recipient (A Correlate of Genetic Similarity)

Species	Distance Range	Outcome[a]	Replicates	Reference[b]
Delphinium nelsonii	1–100 m	10 m: 1 or 100 m ≈ 1.1:1–1.5:1	4	Waser et al. 1987; Waser and Price 1991, and unpublished
Erythronium grandiflorum	0–300 m	3 m: others ≈ 1.3:1–1.9:1	1	Cruzan 1990
Erythronium grandiflorum	0–100 m	self: 100 m = 1.2:1	1	Stratton, Cruzan, and Thomson 1985
Chamaecrista fasciculata	0–100 m	1–100 m: self ≈ 1.2:1	1	Fenster and Sork 1988

[a]Relative success of different distance treatments in delivering pollen tubes to the ovary, expressed as ratios of actual numbers (last two examples) or of numbers predicted controlling for treatment differences in total or germinated pollen loads (first two examples).
[b]Cruzan applied treatments to different stigma lobes within flowers; Waser and Price applied them to different carpels within flowers; others applied them to separate flowers.

which the success of self pollen is reduced but often not to zero, and individuals vary continuously in their response (usually assessed as seed production, which may confound fertilization and seed maturation, but see Cooper and Brink 1940; Griffin, Moran, and Fripp 1987; Hessing 1989; Montalvo 1992). Pseudoincompatibility may be one manifestation of a multilocus system (Mulcahy 1984). Fryxell (1957) cites evidence for self-incompatibility in some 640 angiosperm species, of which about 16% show pseudoincompatibility, and Kenrick (1986) reports in detail on the response in *Acacia*.

At the other extreme of genetic dissimilarity of potential mates are crosses between species. These may evoke an *incongruity* response, a rejection of pollen or embryos reminiscent of incompatibility. There is good evidence that incongruity is a graded multilocus (Pandey 1969) maternal response against dissimilar gametophytes (Martin 1970; Hogenboom 1975; Dumas and Knox 1983; Boyle and Stimart 1986) that intensifies as phylogenetic similarity declines (Sanz 1945; Smith 1968; Coppens d'Eeckenbrugge, Ngendahayo, and Louant 1986). Incongruity enhances positive assortment based on genotype even though all potential crosses are "outbred" as normally defined.

Assortative Seed Maturation

Once double fertilization produces the zygote and endosperm, the embryo matures by sequestering maternal resources in competition with other parental resource needs. Failure to mature takes the form of seed and fruit abortion (Stephenson 1981; Wiens 1984), which often is studied by comparing selfed and outcrossed matings (Crowe 1971; Seavey and Bawa 1986).

Charnov (1979), Westoby and Rice (1982), and Queller (1983) presented inclusive fitness arguments that embryos will attempt to sequester more resources than the parent should supply, and interpreted features of ovules and ovaries in light of this conflict. There is some evidence that maternal responses determine fruit or seed provisioning (Briggs et al. 1987; Haig and Westoby 1988a; but see Uma Shaanker, Ganeshaiah, and Bawa 1988) and that such responses are under polygenic control (Seavey and Bawa 1986; Crowe 1971).

There are two major impediments to empirical studies of assortative seed maturation. First, it is difficult to separate postzygotic from prezygotic events (Lyons et al. 1989). Prezygotic events are sometimes impossible, and usually time-consuming, to observe. Second, it is difficult to distinguish abortion under maternal control from death due to embryo genotype. Fruit or seed set may fail because embryos are inviable in any environment ("embryo lethality") or inferior competitors for maternal resources (Bertin, Barnes, and Guttman 1989; Montalvo 1992). The former is an expression of embryo genotype independent of maternal influence, e.g., of early-acting inbreeding depression. The latter is an interaction of embryo genotype with maternal environment, which makes it hard to distinguish from maternal control, a similar interaction (Lyons et al. 1989).

Some recent studies suggest ways to discriminate embryo lethality from maternal control or differential embryo competitive ability. Stephenson and Windsor (1986) showed that natural fruit abortion in *Lotus corniculatus* (trefoil) resulted in the allocation of resources to remaining fruits containing seeds of superior viability. The aborted fruits were capable of maturing, this excluding embryo lethality (see also Lee and Bazzaz 1982; Casper 1988). Marshall and Ellstrand (1988) found that seed maturation became more selective in *Raphanus sativus* (radish) plants subjected to postfertilization water stress, suggesting maternal control or differential embryo competitive ability, and Bertin (1985) reported a conceptually similar result with *Campsis radicans* (trumpet creeper).

Discriminating maternal control from differential embryo competitive ability is more difficult. Details of embryology can provide evidence for maternal control (Briggs et al. 1987; Haig and Westoby 1988a), as can comparison of embryo growth within fruits and in culture (Nakamura and Stanton 1989). Seavey and Bawa (1986) proposed that a hallmark of maternal control in selfed matings ("late-acting self-incompatibility") would be synchronous embryo failure at an early developmental stage, an outcome that is in the maternal best interest because it minimizes resource allocation to inferior offspring. Conversely, lethality or differential competitive ability would lead to variable timing of embryo failure. If this logic is accepted, it behooves us to characterize the sizes of failed embryos in experiments such as those described above.

Whatever their causes, differential seed and fruit failure certainly can contribute to final patterns of assortative mating when failure depends on level of inbreeding. Also, abortion in *Raphanus* and *Campsis* was affected by the specific parental combination, a corollary of assortative mating, even though parental kinship was not manipulated directly.

SELECTION FOR ASSORTATIVE MATING

Inbreeding Depression and an Intermediate ESS Mating System

The discussion thus far suggests that pollen delivery to stigmas sets the stage for positive assortative mating in angiosperms, but that maternal discrimination may alter the final outcome. What forms of selection should foster the evolution of discrimination?

The most obvious form of selection is inbreeding depression in fitness. It is caused mostly by dominance and overdominance at fitness loci, but "ecological" factors such as resource competition among inbred siblings also might contribute (Price and Waser 1979, 1982; Charnov 1987; Schmitt and Ehrhardt 1990). Because the rejection of matings that are inbred at one locus reduces inbreeding across loci (Uyenoyama 1986; Holsinger 1988), it is favored with sufficient inbreeding depression. This does not mean that rejection systems will evolve automatically (see "The Evolution of Mate Discrimination" below).

The usual view is that there is a threshold level of inbreeding depression that determines whether selection favors rejection or acceptance of self (Kimura 1959). Based on this view, and assuming that the use of pollen in self-fertilization does not cause an exactly corresponding loss of ability to fertilize other plants (incomplete "pollen discounting"), Lande and Schemske (1985) concluded that selfing or complete outbreeding are the only stable mating systems. This implies that nothing more elaborate will evolve than postpollination mechanisms for acceptance or rejection of self matings. However, the conclusion about a simple threshold results from simplifying model assumptions. Uyenoyama (1986, 1988a, 1988b, 1988c) and Holsinger (1988) showed that the past history of inbreeding influences the expected inbreeding depression for a given class of matings, and thus the nature of any threshold. Furthermore, models allowing simultaneous coevolution of mating system modifiers and fitness effects yield conditions under which there is no threshold, and the evolutionarily stable strategy (ESS) is a mixture of selfing and outbreeding (Holsinger 1988; Uyenoyama and Waller 1991a, 1991b; Waller, chapter 6, this volume). Likewise, models that allow classes of inbred matings other than selfing ("biparental inbreeding") yield conditions under which the ESS is mixed selfing and outbreeding or an intermediate degree of outbreeding (Uyenoyama 1986, 1988b, 1988c; Waller, chapter 6, this volume). Thus selection on the mating system does not depend simply on inbreeding depression following selfing, which implies that postpollination mechanisms might evolve for graded acceptance or rejection of matings depending on the level of outbreeding they represent.

Outbreeding Depression and the Intermediate Optimum

If offspring fitness declines under outcrossing beyond some point, i.e., if there is *outbreeding depression* (Müller 1883; Bateson 1978; Price and Waser 1979; Slater and Clements 1981; Templeton 1986), it is even more likely that genetic similarity will influence the optimal mating system. With both inbreeding and outbreeding depression, the possibility of an optimal intermediate similarity is improved (see Ritland and Ganders 1987b). When outbreeding depression alone occurs, the optimal mating system may be selfing ("optimal" implies "most favorable for fitness" after a standard dictionary definition, rather than that intermediate outbreeding does best).

Price and Waser (1979) and Shields (1982) proposed that outbreeding depression might stem either from "genetic" or "ecological" mechanisms (see also Endler 1979). The genetic mechanism relies on intragenomic (i.e., interlocus) coadaptation that differs among neighborhoods, populations, or higher taxonomic units. Crossing different genomes disrupts coadaptation between homologous chromosomes (in the F_1) and between coadapted portions of individual chromosomes (mostly in the F_2, after recombination during F_1 gametogenesis). Thus there is epistasis for fitness in a wide cross (Vrijenhoek and Lerman 1982; Shields 1982; Maynard Smith 1983; Templeton 1986). Genetic drift acts as the driving force by establishing allelic differences at fitness loci, with selection subsequently favoring alleles at other loci that work well with those at the fitness loci (e.g., Slatkin 1975). This version of the genetic mechanism is maximally distinct from the ecological mechanism, which relies on the development of adaptation to local biotic and abiotic conditions, so that wide outcrossing produces F_1 offspring carrying some alleles maladapted to each parental environment. Here selection is the driving force, and the spatial scale for outbreeding depression depends on the scale of heterogeneity in selection regimes, rather than the scale of gene flow alone.

The two mechanisms often are not distinguished, with claims of outbreeding depression taken to imply a genetic mechanism (e.g., Hedrick 1984). This mechanism is feasible if gene pools are sufficiently small and isolated (Templeton 1986). But it is unclear to what extent genetic drift creates the necessary "isolation by distance" (local genetic structure as opposed to random heterogeneity) in continuous populations. Isolation by distance often is presumed to occur in plant populations, based on direct estimates of gene flow and the theoretical predictions of Wright (1951, 1968). But empirical evidence is mixed at best (table 9.2). Even those studies listed in the table as demonstrating isolation by distance usually show patterns in poor quantitative agreement with theoretical predictions (e.g., Campbell and Dooley 1992). In contrast, there is good evidence that local adaptation can develop within continuous populations. In fourteen transplant or similar experiments (table 9.3), the estimated selection coefficient against "foreign" individuals growing a few to hundreds of meters from their natal site averaged about 0.5. This appears to set the stage for substantial outbreeding depression due to an ecological mechanism.

Empirical Examples of Inbreeding and Outbreeding Depression

Inbreeding and outbreeding depression are attributes of the offspring of inbred and outbred matings, not of parents. As already noted, however, it is difficult to distinguish parental from offspring characters; seed and fruit fates may reflect differential fertilization, maternal abortion, and/or early expression of offspring genes. Maternal influences can persist once seeds are shed (Roach and Wulff 1987). These complications should be kept in mind when assessing evidence for inbreeding and outbreeding depression, since much of it involves seed and early seedling characters.

Darwin (1876) was the first to amass evidence for inbreeding depression in plants. After following crossed and self-pollinated lineages through several generations, he concluded (p. 436) that ". . . cross-fertilisation is generally beneficial, and self-fertilisation injurious." Charlesworth and Charlesworth (1987) agreed, based on fifteen species for which offspring fitness components were measured in nature. At the opposite extreme of genetic similarity, one expects outbreeding depression in crosses between species or geographically isolated populations of single species. Confirmation comes in the form of hybrid breakdown and "crossing barriers" over one or more generations (Darlington and Mather 1949; Kruckeberg 1957; Martin 1963; Hughes and Vickery 1974; Banyard and James 1979; Levin 1981).

Is there evidence for fitness effects when the genetic similarity of mates lies between the extremes of selfing and wide outbreeding? I know of twenty-five studies with twenty-two species that provide such evidence (table 9.4). In the absence of pedigree information, which is difficult to obtain at best, all studies rely on outcrossing distance as a correlate of genetic similarity. This is reasonable when the population is structured by drift and/or selection, although the correlation need not be linear (Campbell and Waser 1987; Waser and Price 1989). In

TABLE 9.2 Evidence for and against Isolation by Distance in Continuous Populations of Plants with Short to Moderate Gene Dispersal Distances

Species	Distance Range[a]	Number of Loci	Method	Isolation by Distance	Reference
Ipomopsis aggregata	0–35m	7	Spatial autocorrelation	Yes	Campbell and Dooley 1992
Linanthus parryae	3–95m	1–2	Fixation indices	Yes	Wright 1968
Ipomoea purpurea	10–150m	1	Spatial autocorrelation	Yes	Epperson and Clegg 1986
Impatiens capensis[b]	0–250m	8	Spatial autocorrelation	Perhaps	J. S. Shoemaker and D. M. Waller, unpubl.
Picea abies[c]	50–900m	11	Kinship estimate	Perhaps	Brunel and Rodolphe 1985
Chamaecrista fasciculata[d]	1–100m	7	Genetic distance	Perhaps	Fenster 1988
Pinus albicaulis	1–384m	11	Genetic distance	No	Furnier et al. 1987
Delphinium nelsonii	5–160m	5	Spatial autocorrelation	No	N. M. Waser 1987
Psychotria nervosa	1–30m	2	Spatial autocorrelation	No	Dewey and Heywood 1988
Liatris cylindraceae	4–34m	15	Spatial autocorrelation	No	Sokal and Oden 1978
Plantago lanceolata	2–65m	8	Genetic distance	No	Bos, Harmens, and Vrieling 1986
Silene maritima	50–250m	4	Plot method	No	Baker, Maynard Smith, and Strobeck 1975

Note: Genetic differentiation was measured at putatively neutral or nearly neutral electrophoretic loci in all but two cases. Two studies of flower color are included, but this trait actually may be subject to selection in the species in question (Brown and Clegg 1984, P. Bierzychudek personal communication).

[a] "Distance range" refers to the range of distances studied, not to the range of any isolation by distance. Indeed, spatial autocorrelations in the first two studies listed were positive only over distances of a few meters.

[b] This study is listed as inconclusive because isolation by distance was not detected at all loci or by all tests used.

[c] This study is listed as inconclusive because it involved a managed population not much larger than a single genetic neighborhood.

[d] This study is listed as inconclusive because it tested for isolation by distance with a rank correlation using mean genetic identities across several samples and relatively broad distance classes. This approach can give an impression of pattern even though parametric analysis of raw data shows no pattern (see N. M. Waser 1987 for discussion).

TABLE 9.3 Evidence for Local Adaptation within Plant Populations

Species	Distance	Reciprocal?	Character	Mean Selection against Unadapted Types[a]	Reference
Delphinium nelsonii	50 m	Y	Lifetime F_1 fitness	0.34	Waser and Price 1985
Ipomopsis aggregata	30 m	N	Lifetime F_1 fitness	0.84	Waser and Price 1989
Impatiens pallida	30–60 m	Y	Lifetime F_1 survival	0.35	Schemske 1984
Trifolium repens	10–100 m	Y	Biomass at 12 mo.	≈0.40	Turkington and Harper 1979
Ranunculus repens	< 100m	Y	Ramet production Survival to 18 mo.	0.43	Lovett Doust 1981
Anthoxanthum odoratum	< 10 m	Y	Survival	0.36	Davies and Snaydon 1976
Anthoxanthum odoratum	< 10 m	N	Survival	0.95	Hickey and McNeilly 1975
Anthoxanthum odoratum	< 10 m	N	Growth	0.65	Jain and Bradshaw 1966
Agrostis tenuis	< 10 m	N	Survival	0.30	Hickey and McNeilly 1975
Agrostis tenuis	< 10 m	N	Growth	0.68	Jain and Bradshaw 1966
Agrostis tenuis	< 10 m	N	Growth	0.50	Jain and Bradshaw 1966
Agrostis stolonifera	< 10 m	N	Growth	0.65	Jain and Bradshaw 1966
Plantago lanceolata	< 10 m	N	Survival	0.68	Hickey and McNeilly 1975
Rumex acetosa	< 10 m	N	Survival	0.14	Hickey and McNeilly 1975

Note: From reciprocal and nonreciprocal transplant experiments over short distances, or logically equivalent experiments.
[a]Mean selection coefficients against unadapted types are those given in each study, or were calculated from fitness estimates given in the study.

TABLE 9.4 Evidence for Fitness Effects of outcrossing Distance within Single Plant Populations

Species	Distance Range	Seed Maturation[a]	Rep.	Offspring[b]	Rep.	Reference	Comment
Longest Distance Performs Best							
Erythronium americanum	0–80 m	Shortest loses by ≈ 20%	1	—	—	Harder et al. 1985	% seed set
Amianthum muscaetoxicum	2–60 m	Shortest loses by ≈ 10%	3	—	—	Redmond, Robbins, and Travis 1989	Fruit and seed set
Blanfordia noblis	2–200 m	Shortest loses by ≈ 25%	1	—	—	Zimmerman and Pyke 1988	Seed set
Phlox drummondii	0–200 m	Shortest loses by ≈ 15%	10	—	—	D. A. Levin 1984	Seed abortion
Inga brenesii	0–300 m	Shortest loses by ≈ 70% (≈ 98%)	1	—	—	Koptur 1984	Fruit set
Inga punctata	0–3,000 m	Shortest loses by ≈ 73% (≈ 91%)	1	—	—	Koptur 1984	Fruit set
Clintonia borealis	0–200 m	Shortest loses by (65%)	1	—	—	Galen, Plowright, and Thomson 1985	% seed set
Shortest Distance Performs Best							
Mimulus guttatus	0–500 m	Longest loses by ≈ 10%	4	No clear pattern	4	Ritland and Ganders 1987b	F$_1$ fitness in growth chamber, within-population crosses only
Epacris impressa	0–1,000 m	Longest loses by ≈ 50%	1	—	—	A. Bennett, pers. comm.	Seed set
Carex pachystachya	0–10 m	Longest loses by ≈ 24%	1	Longest loses by ≈ 5%	1	Whitkus 1988, pers. comm.	Seed set, germination

TABLE 9.4 (Continued)

Species	Distance Range	Seed Maturation[a]			Offspring[b]			Reference	Comment
		Short Cross	Wide Cross	Rep.[c]	Short Cross	Wide Cross	Rep.[c]		
Intermediate Distance Performs Best									
Delphinium nelsonii	0–1,000 m	14% (38%)	19%	10	44%	85%	4	Waser, et al. 1987; Waser and Price 1991; N. Waser and M. Price unpubl.	Seed set from crosses with 100–1,000 m treatment, overall fitness to year 7–11
Ipomopsis aggregata	0–100 m	2% (95%)	12%	4	53%	32%	4	Waser and Price 1989	Seed set, lifetime F_1 fitness in field
Ipomopsis aggregata	0–3,000 m	53% (98%)	59%	1	—	—	—	D. Paton pers. comm.	Seed set
Castilleja miniata	0–30 m	≈27% (≈100%)	≈55%	2	—	—	—	Lertzman 1981	Seed set
Calochortus leichtlinii	0–400 m	≈6%	≈8%	1	≈20% (≈22%)	≈35%	1	Holtsford 1984	Seed set, seed mass
Leucocrinum montanum	0–1,000 m	50% (90%)	≈55%	1	—	—	—	W. Schuster pers. comm.	Seed set
Costus allenii	0–300 m	≈5% (≈35%)	≈12%	1	20% (≈30%)	≈10%	1	Schemske and Pautler 1984	Seed set, germination × 5-month biomass

Species	Distance							Reference	Character measured
Picea abies	0–32,000 m	(≈ 83%)	≈ 15%	2	(≈ 25%)	≈ 7%	1	Coles and Fowler 1976	% good seed, size at 3 ½ months
Espeletia schultzii	1–500 m	16%	23%	3	—	—	—	Sobrevila 1988	% filled achenes
Scleranthus annuus	0–100 m	—	—	—	(12%)	19%	1	Svensson 1988	Stamen fertility score in F_1
Scleranthus annuus	0–75 m	—	—	—	(57%)	36%	1	Svensson 1990	% fertile stamens
Impatiens capensis	2–250 m	—	—	—	Intermediate does best		1	McCall, Mitchell-Olds, and Waller 1988	Size at 1 month
Mimulus guttatus	0–4,500 m	(≈ 36%)	26%	1	—	—	—	Waser and Price 1983	Seed set
Fouquieria splendens	0–1,000 m	(96%)	23%	1	—	—	—	Scott 1989	Seed set
Amaryllis belladonna	0–400 m	≈ 45% (98%)	≈ 25%	1	—	—	—	S. D. Johnson pers. comm.	Seed set

Note: Cases in which the longest crosses used performed best show the percentage decrement in the shortest crosses relative to the longest. Cases in which the shortest distance performed best show the percentage decrement in the longest crosses relative to the shortest. Cases in which an intermediate distance performed best show the percentage decrement in the shortest and longest crosses relative to the intermediate.

a Refers to characters of seeds and fruits such as seed set, which may in part be due to postzygotic maternal control of matings.

b Refers to offspring characters such as viability and fecundity.

c Shows the number of individual experiments on which each set of values is based; a "short cross" value in parentheses refers to a selfed cross.

seven studies there is evidence for inbreeding depression, and in three for outbreeding depression in crosses over several hundred meters. Fifteen studies give evidence for both inbreeding and outbreeding depression, i.e., for highest fitness at a relatively short outcrossing distance (presumably an intermediate degree of outbreeding). In seven of these, the conclusion of intermediate optimal outbreeding rests in part or whole on offspring characters, and in one study, of *Ipomopsis aggregata* (scarlet gilia; Waser and Price 1989) it rests on estimates of lifetime offspring fitness in the field.

An important conclusion is that we need more information, especially on offspring fitness components outside immediate maternal control, assessed under natural conditions (Waser and Price 1989). If genotype-environment interactions affect fitness (Shaw 1986), using an unnatural environment—a greenhouse, common garden, or field site that offspring never experience—can be misleading (Mitchell-Olds 1986). Most studies in table 9.4 should be viewed with caution, since they dealt only with seed characters and/or used an artificial environment, and made no attempt to assess the scale of selection heterogeneity in choosing a range of outcrossing distances.

Statistical Significance and Power

Treatment differences in many of the experiments summarized in tables 9.1 and 9.4 were not statistically significant, and authors often concluded there were no biological effects. But lack of confidence in an alternative hypothesis based on sample statistics does not "prove" a null hypothesis. In most studies there probably was little chance of detecting a false null hypothesis (i.e., low power, and a large chance of type II statistical error). It is reasonable to expect demonstration of substantial power before accepting a null hypothesis as a statement about biology, but no study cited includes a power analyses (see Peterman 1990). This, along with the rarity of studies, compromises our ability to conclude much about the frequency and magnitude of inbreeding and outbreeding depression in nature.

Power increases with size of experiments and their repetition. Repetition also has the virtue of allowing one to detect consistency of outcome, which may signal that an effect is biologically important (e.g., to the evolution of reproductive traits) even if it is of small or moderate magnitude (see Nelder 1988). As an example, table 9.5 shows outcrossing effects on seed set from eleven pollination experiments with *Delphinium nelsonii* (Waser and Price 1991). The experiments all tested the same hypothesis that seed set varies with outcrossing distance, reflecting maternal discrimination against overly inbred and outbred matings

TABLE 9.5   Outcrossing Distance Effects on Seed Set in *Delphinium nelsonii*.

Year	Method[a]	0 (self)	1	10	100	1,000
			Outcrossing Distance (m)			
1976	FHP	12.3	13.3	**14.8**	14.4	10.6
1977	FHP	5.3	5.4	**11.9**	6.9	9.1
1977	FHP	8.7	—	**14.8**	—	11.6
1978	FHP	—	10.0	**12.4**	9.5	—
1982	FPT		**12.3**	10.2	7.5	—
1984	FPT	—	8.3	**11.6**	10.2	—
1985	FPT	—	20.2	**22.3**	17.8	—
1986	GHP	—	**11.2**	10.3	10.2	—
1987	GHP	—	5.1	**5.6**	5.1	—
1988	GHP	—	6.2	**7.7**	6.1	—
1990	GHP	—	4.8	**5.6**	4.8	—

Note: Values are maternal per carpel (1987, 1988, and 1990 experiments) or per flower means. The largest mean value in each experiment is shown in boldface. The probability of an intermediate distance (10 m) outperforming longer and shorter distances in 9 of 11 experiments by chance alone is miniscule ($P = .0032$, binomial test). Eliminating the second 1977 experiment that lacks 1- and 100-m treatments, 10 m significantly outperforms 1 m ($\chi^2 = 49.7$, $df = 20$, $P < .001$) and 100 m ($\chi^2 = 44.6$, $df = 20$, $P < .002$), using combined probabilities from t tests with individual experiments (Sokal and Rohlf 1981, 779).
[a]FHP, field hand pollination using pollen donors growing various distances away; GHP, greenhouse hand pollination using donors growing various distances from natal site of each recipient; FPT, field potted transplant moved various distances from natal site and exposed to natural pollination.

that produce inferior offspring (Price and Waser 1979; Waser and Price 1993). Although treatment differences were statistically insignificant in most experiments (the table omits standard deviations to save space), overall analyses ("meta-analyses") show a significant pattern of 10 meters outperforming shorter and longer distances. Persistence with the system also led to discoveries that confirm elements of the original hypothesis. We have documented superior viability and fecundity of offspring from 3-meter and 10-meter crosses under field conditions (Waser et al. 1987; N. Waser and M. Price unpublished) and adaptation to local conditions on a scale of tens of meters (table 9.3, Waser and Price 1985), suggesting an ecological mechanism of outbreeding depression.

THE EVOLUTION OF MATE DISCRIMINATION

Given that intermediate levels of inbreeding and outbreeding can affect offspring fitness, we might expect selection to foster maternal mate discrimination that covers a range of genetic similarity of potential mates and thus transcend the responses against selfed and widely outbred offspring for which there is most evidence. Patterns of prezygotic pollen tube growth (table 9.1) are consistent with this expectation, and there even are three cases of an intermediate "optimal outcrossing distance" in tube growth, which in *Delphinium nelsonii* matches the distance that maximizes offspring viability (Waser et al. 1987; Waser and Price 1991, 1993). However, postpollination assortment sometimes is subtle at best, even when there are fitness consequences. For example, *Ipomopsis aggregata* exhibits a strong outcrossing distance effect on offspring fitness but a weaker effect on seed set (table 9.4; Waser and Price 1989). Is this surprising?

Evolution of discrimination against wide outbreeding is analogous to prezygotic "reinforcement" of reproductive isolation manifested in reduced fitness of "hybrid" offspring. Restricted pollen dispersal in many plant populations should enhance linkage disequilibrium between loci that recognize outbred pollen and those that determine local adaptation or coadaptation, so the situation is at least one of weak parapatry. Initial conditions under parapatry are relatively favorable for reinforcement (i.e., the spread of mating system modifiers), judging from theory (Felsenstein 1981) and experiments (Paterniani 1969; Rice 1985). Conditions are even more favorable if a modifier allele is the same in different subpopulations and its response is simply to reject a male gametophyte too dissimilar from self ("single-allele" substitution, Felsenstein 1981; see also Campbell and Waser 1987). The degree of dissimilarity might be determined by a maternal plant through comparison of gene products from one or more loci that this plant and the male gametophyte (or endosperm or embryo) both express. This scenario assumes that widely outbred pollen arrives with sufficient frequency to cause consistent selection in favor of discriminating against it, which may not always be the case. It also assumes that total fitness does not decline when the mating pool is restricted, as could occur if pollen becomes limiting and the maternal parent has no opportunities to reallocate resources from the flower in question to other fitness-related ends (see Crosby 1970; "Conclusions," below).

The evolution of discrimination against inbred matings has been explored more explicitly. Charlesworth and Charlesworth (1979) showed that conditions for the spread of an allele expressing complete self-incompatibility are stringent, since the allele reduces its own transmission success. Uyenoyama (1988c) suggested that this stringency is reduced if a locus involved in kinship recognition ("antigen" locus) differs from one determining the response of the pistil, and linkage disequilibrium is moderate. However, the evolutionary dynamics remain complex (Uyenoyama 1988a, 1988b, 1988c, this volume), and depend on the strength of the discrimination against self invoked by the antigen allele, the degree of dominance in its expression, the relative costs and allele transmission success associated with different classes of matings (e.g., outbred, sib, selfed), and the statistical association of heterozygosity at an antigen locus and at loci affecting viability ("identity disequilibrium," Weir and Cockerham 1973; Brown 1979).

In summary, the evolution of postpollination influences on mating patterns, in response to offspring fitness differences, is possible but not automatic. Genetic systems discriminating against outbreeding and inbreeding may differ, though it is unclear that this is a theoretical necessity.

ASSORTATIVE MATING AND REALIZED MATING PATTERNS

To what extent do assortative postpollination processes affect final mating patterns? For strong self-incompatibility responses the answer is clear: blockage of self matings enhances outbreeding. A conclusion that strong incongruity reactions against foreign pollen enhance positive assortment for genotype also seems justified. However, the answer is uncertain with regard to continuous variation in genetic similarity between pollen and recipient, or to other kinds of assortment during pollination.

Direct observation of the fates of male gametophytes and ovules representing different average degrees of genetic similarity (tables 9.1, 9.4) indicates that postpollination events have the potential to modify offspring inbreeding and outbreeding. But such studies are rare, partly because few people have persisted in looking for patterns. Studies to date also suffer from limitations including failure to compare performance of gametophytes from mixed pollen loads on a single pistil (but see Cruzan 1990) and to use or mimic natural pollen mixtures.

An alternative to direct observation is to see how genotypic arrays change through the reproductive cycle (Lyons et al. 1989). Pollen genotypic arrays can be specified in experiments, but this may be impossible in nature. At least it is possible to genotype seeds and infer paternity (Meagher 1986; Devlin, Roeder, and Ellstrand 1988). Thus the most common approximation to a complete "genetic demography" (sensu Brown 1979) is to estimate pollen identities from pollen movement, and ask whether they predict seed or adult genotypic distributions. For example, Campbell (1991) found that potential gene movement via pollen is much more restricted in *Ipomopsis aggregata* than realized movement assessed by seed paternity, suggesting reduced postpollination success in inbred crosses (those over a short outcrossing distance). An analagous approach is implicit in Linhart et al. (1981), N. M. Waser (1987, 1988), and Stanton et al. (1989).

CONCLUSIONS

Paradigms to Ponder

If a slight change in perspective is accepted, many botanists will agree with my conclusion that inbreeding patterns in a plant population can be affected by processes that occur after pollination, but may disagree with my terminology and my attempt to paint a picture of assortative mating that goes well beyond recognition of self. In anticipation of this I will mention several paradigms that seem to channel thinking about mating systems. I think there is enough reason for doubt that workers on plant reproduction should be encouraged to entertain alternatives.

At a recent symposium on sexual selection in animals, I approached a distinguished speaker to discuss the evidence for postpollination maternal discrimination. His response, that female choice in plants was preposterous because plants lack a nervous system, is all the more surprising given that zoology only now is extricating itself from the fallacy that higher mental powers are required for female choice. This fallacy can be traced to a debate between Wallace and Darwin (Thornhill 1986), and I doubt it would have persisted had most zoologists been female! Given that plants possess complex physiology (although not neurophysiology), they are no less capable of female choice *in principle* than are ladybird beetles. Furthermore, there is no reason to be surprised by responses that are conditional on circumstances outside the individual flower, as suggested by the experiments of Bertin (1985) and Marshall and Ellstrand (1988).

Confounding the issue is an aversion to the term "choice." But such a term can be given an operational scientific definition different from its lay meaning. If operational definitions are stated, there is no reason to invent new words for biological phenomena such as choice, preference, competition, or infanticide, or to engage in sterile debate about consciousness, motivation, or other implications derived from lay meanings.

Another debate is whether seed production is limited by pollen availability or resources. This dichotomy may be false (Haig and Westoby 1988b; Ackerman and Montalvo 1990), but if not, it still is too simplistic to conclude that maternal choice will occur only under conditions of surplus pollen (e.g., Willson and Burley 1983). Even with insufficient pollen to fertilize all ovules, plants often will have options other than accepting all pollen grains and maturing all embryos. They may be able to reallocate resources from embryo development to additional flowers, enhanced survival, and future reproduction. To be sure, maternal choice seems unlikely to evolve when such options are few or none. Weak outcrossing effects on seed set in *Ipomopsis aggregata*, for example, may reflect the fact that individuals die after one flowering season and are often pollen limited, rather than a genetic constraint on the evolution of physiological machinery for choice (see "The Evolution of Mate Discrimination," above).

Future Goals

Graded postpollination assortment will not be easy to quantify. Careful crossing studies are needed, with large sample sizes and repetition to look for consistency. A promising design is a diallel or factorial cross, in which a range of pollen donors and recipients are paired and pollen mixtures are applied to single pistils (Lyons et al. 1989). Analyses of gametophyte fates, embryo fates, and seed paternities from such crosses hold the potential to determine the contribution of parental interaction that can be ascribed to genetic or phenotypic similarity, although several generations of crossing and elaboration of models appear necessary to deal with the haploid-diploid nature of gametophyte-pistil interaction and triploid genetics of endosperm. It would be ideal to link crossing studies to events in nature using genetic demography of pollen loads and seeds.

Theoretical effort also is needed. Analytical treatments of population genetic structure do not deal with realistically fluctuating and anisotropic gene flow patterns (Slatkin 1985) or with kin-structured dispersal, and treatments of the evolution of inbreeding and outbreeding depression and mate discrimination usually can explore only limited conditions. Given the power of computers, it seems opportune to make more use of numerical methods. As an example, Campbell and Waser (1987) explored the coevolution of population structure and pollen dispersal (and by implication mate discrimination) with a multilocus simulation. However, they did not allow fitness values of matings to evolve. There is much room for improvement.

Implications for Plant Microevolution

At its extreme, a Wrightian view of populations has been taken by some plant ecologists to imply that genetic drift will be a ubiquitous source of local structure, so that it is worthwhile, for example, to characterize small fluctuations in genetic neighborhood size. This does not jibe with electrophoretic surveys, which often show that there is less structure than is predicted by combining direct estimates of gene flow with Wright's models for drift in continuous populations (table 9.2). Slatkin (1985) points out several reasons for expecting little structure at neutral loci. First, the theory may be inadequate. Second, unusual events of extensive gene flow are not easily detected, but might occur sufficiently often to modify drift-generated structure (Campbell and Dooley 1992).

A third possibility involves postpollination assortment. Although discrimination against self matings is liable to be included in direct gene flow estimates, discrimination against other forms of close inbreeding is not. But such discrimination could enhance outbreeding, causing the observed discrepancy between estimated gene flow and gene establishment (sensu Endler 1979). None of this implies that structure cannot develop at loci subject to spatially varying selection (table 9.3, Waser 1987).

Botany Speaks to Zoology

My discussion of postpollination contributions to assortative mating may not seem exotic to many zoologists. To be sure, most animals do not rely for gamete movement on a third party that is capricious (abiotic pollination) or that follows behavioral rules that increase its own fitness (animal pollination, e.g., Waser and Price 1983). Because they rely primarily on behavioral interaction with potential partners before copulating, male animals are less likely to commit their sperm irrevocably than are plants, whose pollen, once it reaches a stigma, has only the options of expiring or attempting fertilization. Beyond this stage, however, there are close parallels, despite the impression given by some theoretical treatments of stages in animal reproduction (e.g., Wade and Arnold 1980). Female animals in many phyla are multiply inseminated, and there are possibilities for differential fertilization, abortion vs. maturation of zygotes, and care of young or infanticide based on sire (e.g., Thornhill and Alcock 1983; Daly and Wilson 1983; Uyenoyama 1988a). Eberhard's (1990) interpretation of male-female conflict of interest in the evolution of sex organs and copulatory behavior in animals is similar to mine of male gametophyte-maternal tissue interaction in plants, although I have focused on genetic similarity of mates rather than on the fundamentally different sexual selection scenarios he discusses. It certainly would behoove botanists and zoologists to discuss such shared issues and their implications.

One major difference between kingdoms may be in the scale and mechanism of outbreeding depression. Viscous population structure tends to foster substantial local genetic differentiation in plants, especially in response to selection (see above and Brown 1979). This capacity is shared by some sessile animals (Alstad and Edmunds 1983; but see Grosberg 1987) but probably by few mobile animals. Thus outbreeding depression in most animals is unlikely to occur on as fine a spatial scale as in plants (although zoologists apparently have not looked on this scale), and is less likely to be due purely to disruption of local adaptation. This view matches evidence (Templeton 1986) that outbreeding depression in vertebrates involves karyotypic differences. However, not all outbreeding depression in plants need follow from an ecological mechanism: the deleterious effects of very wide crosses, e.g., between related species, must involve a large element of disruption of intragenomic coadaptation.

ACKNOWLEDGEMENTS

For ideas and help, thanks to Ellen Bauder, Diane Campbell, Mitch Cruzan, Candi Galen, Diane Marshall, Randy Mitchell, Arlee Montalvo, Mary Price, Linus Svensson, Lisa Rigney, Nancy Thornhill, Ruth Shaw, and Don Waller. Philippe Cohen and Cindy Stead provided a quiet place to think and write. Jürg Waser reconstructed the Syz pedigree. Parts of the work were supported by grants from the National Science Foundation (DEB 8102774, BSR 8313522, BSR 8905808) and the University of California, Riverside Academic Senate.

10

Inbreeding and Outbreeding in Marine Invertebrates

Nancy Knowlton and Jeremy B. C. Jackson

Of the approximately thirty-one phyla of multicellular animals, twenty-five are primarily or exclusively marine or have important marine representatives (Parker 1982). Thus marine invertebrates exhibit an enormous diversity of lifestyles, even when only the major free-living groups are considered. Some, such as corals and sponges, resemble plants, with typically sessile adults, passively dispersed young, and often the capacity for clonal propagation. Others, like most snails and crabs, have freely moving adults with modes of life superficially similar to those of *Drosophila* or mice. Permutations and combinations abound. All, however, live in a medium where floating, and thereby dispersing, is easier relative to comparable situations on land, particularly for preadult stages.

In this chapter, we attempt to address four questions. First, what features of the biology of marine invertebrates influence the potential for inbreeding and outbreeding? We review major theoretical issues in the context of life histories of marine invertebrates, and illustrate important points using a few well-studied model systems. Second, to what extent are natural populations of marine invertebrates in different taxa inbred or outbred? We focus our discussion on twelve major groups belonging to nine phyla. Much of our assessment depends on inference from life history traits related to dispersal and mode of reproduction. Information on the genetic structure of populations and on the extent of inbreeding and outbreeding depression is limited. Third, what characteristics are shared by inbreeding species? We review the taxonomic, morphological, and ecological patterns revealed by our survey. Finally, what processes generate these patterns? Most of the traits involved have important ecological, functional, and genetic consequences that are difficult to disentangle. Consequently, comparisons between sessile marine invertebrates and plants prove particularly instructive. Marine invertebrate zoologists and students of inbreeding in other groups are our intended audience for this first major review of the scattered literature in this area.

GENERAL FEATURES OF LIFE HISTORIES AND MATING SYSTEMS

Inbreeding has been well studied by theoreticians, and a number of models with a variety of parameters have been examined (e.g., Wright 1969: 169–172; Kimura and Ohta 1971: 117–140; Pollak 1987, 1988; Slatkin 1985; D. Charlesworth and B. Charlesworth 1987). Leaving aside much of the detail, inbreeding depends on (1) the extent to which a species is reproductively divided into populations and subpopulations, especially the size of these units and the probability and pattern of migration among them, and (2) the potential for and direction of nonrandom mating within these breeding units.

These two features are sometimes referred to as the random and nonrandom components of inbreeding, respectively (Kimura and Ohta 1971), and similar concepts may be applied to outbreeding. The less technical literature sometimes ignores random (or "background") inbreeding, but both kinds result in a loss of genetic heterozygosity. Nonrandom inbreeding causes a more rapid drop in heterozygosity than does random inbreeding (Futuyma 1986: 123) unless the deviation from randomness in the former case is slight and the population size in the latter case is very small (Pollak 1987, 1988). Levels of random and nonrandom inbreeding can in theory be inferred from Wright's F statistics (F_{IS} and F_{ST} respectively), but to do so without independent evidence of population boundaries and nonrandom mating is problematic.

Below we consider the features of the natural history of marine invertebrates that are most likely to influence inbreeding or outbreeding potential through their effects on population structure and nonrandom mating.

Dispersal

Many models of population structure assume discrete but interconnected breeding units. In such models, routine dispersal distances define the geographic limits of breeding units, while extraordinary dispersal events are responsible for the connections between them (figure 10.1). Connections may occur between all units (island models, figure 10.1C) or be limited to spatially adjacent breeding units (stepping-stone models, figure 10.1D) (Slatkin 1985). Such models can be hierar-

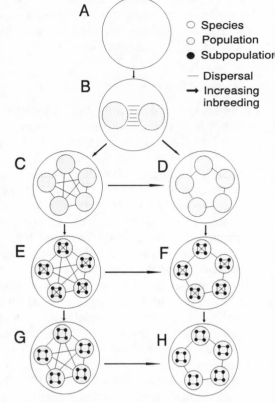

○ Species
○ Population
● Subpopulation

— Dispersal
→ Increasing
inbreeding

FIGURE 10.1   Symbolic representation of the effect of population
structure on random inbreeding.
Illustrated are varying degrees of
population subdivision combined with island and stepping-
stone models of gene flow (see
text) among populations and
subpopulations.

chically organized (Wright 1969; Chakraborty 1980), with subpopulations within populations (figure 10.1E–H). A hierarchy of stepping
stones (figure 10.1H) is probably a good approximation to the more
problematic (Felsenstein 1975; Slatkin 1985) isolation by distance
model. In any model with discrete populations, dispersal influences
both the size of the breeding units, by determining their spatial extent,
and the magnitude and pattern of connections between them. In general, the more limited the dispersal, the greater the potential for inbreeding.

Individual dispersal may be defined as movement from the point of
conception of an individual to the point of conception of its offspring.
In marine invertebrates this movement can take place at various stages
in the life cycle and by various means (table 10.1, figure 10.2). Adults
may be sessile, sedentary, or highly mobile. Mobility of adults may be
active, by walking or swimming, or passive, by floating or rafting

TABLE 10.1 Characters Influencing Dispersal, Their Combined Influence on Overall Dispersal Potential, and Examples

Larval Development	Adult Mobility	Gamete Spawning	Total Dispersal	Examples
Direct/abbreviated	Sessile/sedentary	None	Low	Crustacean symbionts
Direct/abbreviated	Sessile/sedentary	Sperm	Low	Some corals, ascidians
Direct/abbreviated	Sessile/sedentary	Egg + sperm	Low	Some sponges, anemones
Direct/abbreviated	Mobile	None	Moderate/high	Some snails
Direct/abbreviated	Mobile	Sperm	Moderate/high	None?
Direct/abbreviated	Mobile	Egg + sperm	Moderate/high	Some echinoderms
Extended	Sessile/sedentary	None	High	Barnacles
Extended	Sessile/sedentary	Sperm	High	*Membranipora*
Extended	Sessile/sedentary	Egg + sperm	High	Mass-spawning corals
Extended	Mobile	None	High	Many decapod crustaceans
Extended	Mobile	Sperm	High	None?
Extended	Mobile	Egg + sperm	High	Most echinoderms

Note: The combination of direct development and spawned eggs implies that eggs are released prior to fertilization, but that they are nonmobile, and that adult existence begins shortly after hatching. The apparent lack of mobile species with retained eggs and spawned sperm probably reflects past sexual selection favoring males that placed sperm on or in the female.

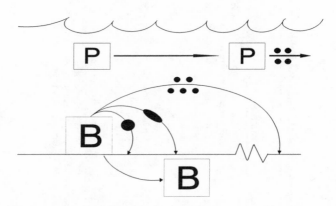

FIGURE 10.2   Mobility in the life cycle of marine invertebrates. P, planktonic (free-float-ing) adult; B, benthic (bottom-dwelling) adult; solid dots, larvae of different sizes; ar-rows, movement.

(Highsmith 1985). Male and female gametes may both be spawned into the water column, females may retain eggs until they are fertilized by spawned sperm, or copulation may occur. The embryo may be brooded by the female, attached to the substratum, or free-floating to varying degrees and for varying periods of time. The larva, if it exists, may float, swim actively, or remain near or on the bottom, with time before metamorphosis ranging from seconds to months. Alternatively, a ju-venile may emerge from the egg case or parent to begin an adultlike existence.

An extended larval phase is generally capable of greater dispersal than a mobile adult phase (Crisp 1978), although exceptions occur in particular taxa. This is because nearly neutrally buoyant larvae dis-perse with little expenditure of energy, while most movement by adults is accomplished by more energetically expensive, active locomotion (species that spend their entire lives drifting in the plankton being an obvious exception).

In contrast, maximum dispersal of ultimately successful gametes is typically much lower than either maximum larval or adult dispersal un-less adults are entirely sessile. Sperm once spawned have relatively limited life spans (e.g., less than 5 minutes in the ascidian *Botryllus schlosseri* [Grosberg 1987]; less than 4 hours in the hydroid *Hydractinia echinata* [Yund 1990]). Sperm spawned in packages (spermatophores) may have longer lifetimes, however. Although not common in spawn-ing marine invertebrates, spermatophores have been reported in sev-eral sessile groups (M. F. Strathmann 1987; R. R. Strathmann 1990). In phoronids they are equipped with sails or filaments (Zimmer 1987) which might aid dispersal. Unfertilized, spawned eggs also have short life spans, although they are often somewhat longer than those reported for sperm (e.g., more than 90 versus less than 20 minutes for eggs and sperm of *Strongylocentrotus droebachiensis* [Pennington 1985], and 4–6 versus 1–2 hours for eggs and sperm of the mussel *Mytilus edulis* [Bayne 1976]).

Short gametic life spans limit matings to relatively nearby individuals, and dilution further narrows the distances that successful gametes travel prior to fertilization. In wave-swept environments, the role of turbulence in limiting typical effective distances between spawning individuals is marked (Denny 1988: 147–151; Denny and Shibata 1989). For example, at distances greater than 20 cm from the sperm source, less than 15% of eggs were fertilized in the sea urchin *S. droebachiensis* (Pennington 1985). Studies of *Hydractinia echinata* (Yund 1990) and the sea urchin *Diadema antillarum* (Levitan 1991) done in calmer waters showed a similar pattern on a slightly greater scale, with low probabilities of fertilization at distances of 7 and 3 m respectively.

Diversity of dispersal modes within a single species' life history can generate complex and hierarchical breeding structures. For example, small species often brood young, which disperse relatively short distances after release (Strathmann and Strathmann 1982), and small species are also more likely to be transported long distances by rafting (Highsmith 1985). This pattern might generate subpopulations defined by larval crawling, connections among subpopulations via occasional storm-mediated longer-distance dispersal of larvae or adults, and rare connections among the populations through rafting. These variable dispersal modes can also create situations where the pattern of connection between breeding units is not the same at all levels of the hierarchy (Stoddart 1988). For the case described above, stepping-stone models are most appropriate for dispersal by larval crawling and storm-mediated dispersal, while island models better describe rafting (e.g., figure 10.1G).

In contrast, other common life histories generate comparatively simple population structures. Some marine invertebrates have larvae that spend months in the plankton and are routinely capable of traveling great distances (Scheltema 1971, 1986a, 1986b). Although the concept of a panmictic planktonic soup is overly simplistic, it nevertheless remains clear that some species with planktonic larvae have much more limited partitioning of the gene pool (as in figure 10.1B) than do species lacking a dispersal phase (Burton and Feldman 1982; Burton 1983; Hedgecock 1986).

Even in the simplest cases, accurate measurements of dispersal patterns are difficult to make. Indirect and direct methods each have their own weaknesses; the former are based on numerous theoretical assumptions, while the latter cannot adequately quantify rare events (Slatkin 1985). Their combined use can be particularly informative, however (e.g., Grosberg 1991, see below).

Effective Population Size

The effective size of a breeding unit, as defined spatially by routine dispersal, is limited by the number of genetically distinct breeding individuals (N) it contains. The effective population size is maximally $2N - 1$ for hermaphrodites or $2N - 2$ for gonochores, but it is generally less because of population fluctuations, unequal contributions of individuals to reproduction, overlapping generations, and biases in sex ratio (Wright 1969; Kimura and Ohta 1971; Crawford 1984b). The smaller the effective population size, the greater the degree of random inbreeding.

The number of genetically distinct individuals in an area necessarily depends on population density, which varies enormously among marine invertebrates. Differences in densities are most critical for species that are highly philopatric at all life stages. In cases where total lifetime dispersal is such that most offspring are found within 1–10 m of the parent, the potential for inbreeding is much higher for low-density species (less than 10 individuals per m²) than for high-density species (more than 1,000 individuals per m²). Once dispersal is widespread, however, most species will have effective population sizes too large for random inbreeding to be important. Among sessile species generally, unitary forms often have much higher densities of genetically distinct individuals than modular or clonal forms (e.g., we crudely estimate densities of 10^0–$10^2/m^2$ for sponges, hydrocorals, and corals; and 10^1–$10^3/m^2$ for hydroids, bryozoans, and colonial ascidians; compared with 10^3–10^6 for solitary ascidians, vermetids, barnacles, and polychaetes).

Prevailing abundances will generate overestimates of effective population size, however, if population bottlenecks have occurred in the past. Population bottlenecks may arise from catastrophic declines in generally abundant species or from chance colonizations by one or a few individuals. Theoretical studies of such events (Nei, Maruyama, and Chakraborty 1975; Chakraborty and Nei 1977) have shown that loss of heterozygosity depends very sensitively on the severity of the bottleneck (e.g., two versus ten individuals) and the rate of recovery as measured by reproductive potential ($r = 0.1$ versus $r = 1.0$). The initial loss occurs comparatively rapidly, the bulk of it within 10 generations (although with larger founding groups and a slower recovery, heterozygosity continues to drop through 100 generations). Most importantly, heterozygosity typically does not begin to recover before 10^5–10^6 generations. This is far more than enough time for populations to return to carrying capacity, even assuming the smaller of the two founding sizes and rates of increase (exponential growth would yield a population of $2e^{100,000}$!). Thus one cannot rule out the effects of random inbreeding simply on the basis of currently large population sizes.

The most dramatic and best-studied catastrophic decline in a marine invertebrate was caused by an epidemic disease in the once very abundant sea urchin *Diadema antillarum*. Between January 1983 and February 1984, the urchin was reduced to between 0.01% and 7% of its original density throughout its neotropical range (Lessios 1988). This catastrophic reduction has produced as yet no measurable genetic changes, however. Persistent low densities may eventually result in loss of genetic variability (Lessios 1988); alternatively, even this bottleneck may not have been sufficiently extreme to produce detectable genetic effects.

Bottlenecks stemming from chance colonizations by one or a very few individuals are probably more common, particularly in groups where routine dispersal is limited. Many large, isolated populations of brooding species may be derived from only a single, accidently transported individual that has the potential to rapidly increase over a relatively short period of time. The best-studied cases involve the snail *Littorina saxatilis*. In South Africa, two populations apparently started by a few introduced individuals had heterozygosities much lower than those of thirteen North Atlantic populations (0.05 versus 0.18 respectively) (Knight, Hughes, and Ward 1987). Reduction in heterozygosity was much less marked, however, for populations on minute, recently emerged islands off the Swedish coast (0.13 on the islets versus 0.16 on the mainland [Janson 1987b]). Janson suggested that this was due to the rapid potential for increase shown by these snails (she estimated the intrinsic rate of increase at $r = 2.0$). However, results in South Africa for the same species (Knight, Hughes, and Ward 1987) and the sensitivity of models to the size of the founding group (Nei, Maruyama, and Chakraborty 1975; Chakraborty and Nei 1977) make it difficult to eliminate slightly larger founding populations as the source of the difference between the two studies. Unfortunately, we entirely lack information on the frequency and severity of bottlenecks for almost all marine invertebrates.

The potential for clonal reproduction also complicates estimates of effective population size, because physiologically separate "individuals" are not necessarily genetically distinct. The two major methods of clonal reproduction are agametic division of the adult body (by fission, budding, etc.) and parthenogenetic production of diploid eggs or larvae (Hughes 1987, 1989). In general, the former is more common among marine invertebrates, and taxa that practice one form tend not to practice the other (Hughes 1989). Clonal division of embryos (polyembryony) and larvae has also been recorded in bryozoans and echinoderms (Ström 1977; Bosch, Rivkin, and Alexander 1989). Clonal reproduction at the expense of sexual reproduction strongly influences effective population size by reducing the number of genetically distinct individuals within the breeding unit.

Another important source of reduction in effective population size is uneven contributions to reproduction. This may occur through differences among genets in size, life span, or ability to acquire mates, whatever their origin. Stoddart's (1984) study of genetic structure of the coral *Pocillopora damicornis* provides a clear example of the first mechanism. On average, each individual genotype was represented by 4.3 colonies, but the two most abundant genotypes of the fourteen detected represented 66% of all colonies in one population, and 95% of the colonies in the largest size category. Wulff (1986) documented similar inequality in abundances of clones in three sponges. In many marine invertebrates, reproductive output is correlated with size (Jackson 1985). Size can vary enormously, particularly when modular construction or cloning results in a body plan with no intrinsic limits on the maximum amount of tissue associated with a single genotype. Clonal organisms also have at most weak tendencies towards senescence (Caswell 1985; Hughes 1989), and survivorship may increase with increasing size, further increasing inequalities in lifetime reproduction between long-lived and short-lived genets. Lack of senescence additionally increases the potential for overlap among generations, which also reduces effective population size (Felsenstein 1971). The potential variation in reproduction is less in unitary (nonmodular) organisms, even when the latter have very steep or exponential size-fecundity relationships, because of their much more limited ranges in body size and life span. In aclonal, behaviorally sophisticated marine invertebrates, there is some evidence for polygyny (e.g., Upton 1987; Christy 1987), but it is not as extreme as that reported for some vertebrates (e.g., Emlen and Oring 1977; Alexander et al. 1979).

Finally, biased sex ratios are not uncommon in species known to reproduce sexually (e.g., male:female ratios of 1:8 for a tetractinomorph sponge [Ayling 1980] and up to 1:20 for an isopod [Carvalho 1989]). Some of these biases may reflect differential rare recruitment and subsequent asexual proliferation of dioecious colonists, however, as reported for starfish (Achituv and Sher 1991) and brittle stars (Mladenov and Emson 1988).

Nonrandom Mating

Nonrandom mating occurs when all members of a breeding unit, as delimited by routine dispersal, are not equally likely to mate. More specifically, nonrandom inbreeding or outbreeding is enhanced when the probability of two individuals mating is a function of their degree of relatedness. The former occurs through higher than expected probability of kin mating and selfing, the latter via behavioral avoidance of kin as mates and genetic or physiological incompatibility systems. Many of the mechanisms that have received much attention in vertebrates (patterns of mate preference and sex-specific dispersal [Bateson 1983; Pusey 1987; Forster Blouin and Blouin 1988]) are largely unstudied in marine invertebrates.

A major component of nonrandom inbreeding for the phyla considered here is the greater probability of fertilization between gametes originating from nearby sources. Copulating partners are necessarily adjacent. In many species females retain eggs until fertilization occurs, and most gametes spawned into the sea probably travel relatively short distances before they are fertilized (see above). This will not lead to inbreeding, however, unless neighbors are close relatives due to cloning, extreme larval philopatry, or kin recognition at settlement.

In contrast, among the best potential candidates for random mating within a population are mass-spawning species, such as epitokous polychaetes (Schroeder and Hermans 1975) and the majority of corals on the Great Barrier Reef. In the latter, synchronous spawning of buoyant egg and sperm packets by the entire population produces enormous egg and embryo slicks visible from the air (Harrison et al. 1984; Oliver and Willis 1987). In typical *Acropora*, which represent a majority of the mass spawners, gametic packets break up when they reach the surface, but a delay in fertilization of 10–20 minutes probably enhances cross-fertilization (Richmond 1990).

Many invertebrates, however, show nonrandom mating patterns such as self-fertilization and sib mating that probably contribute to inbreeding. In the absence of genetic blocks, synchronous hermaphrodites should be particularly susceptible to selfing, especially if they are isolated or large (Maynard Smith 1978: 136), as is frequently true of some clones. Sequential hermaphrodites in which physiological isolation by budding or fragmentation leads to asynchrony in sexual stages may also self-fertilize beyond random expectation, provided the clonemates remain nearby. If clonemates are dispersed, however, only the random component of selfing will increase through reduction in effective population size.

Aggregations of kin in excess of that predicted by dispersal distance alone will increase the probability of sib matings in both hermaphrodites and gonochores. This phenomenon is not well studied in marine invertebrates, but has been reported in two phyla (Keough 1984; Grosberg and Quinn 1986). The bryozoan *Bugula neritina*, for example, is a protandrous hermaphrodite and therefore cannot self-fertilize. However, aggregation of kin at settlement combined with probably synchronized release of eggs and sperm in neighboring colonies (Keough 1984) makes inbreeding probable.

Genetic blocks to self-fertilization, on the other hand, promote nonrandom outbreeding. In the absence of such a block, random mating among N synchronous hermaphrodites would be expected to yield selfed progeny in the proportion $1/N$. Genetically based gametic incompatibility systems are common in plants (D. Charlesworth and B. Charlesworth 1987) but are less well known in marine invertebrates. Heyward and Babcock (1986) provide evidence for gametic incompatibility blocking selfing in the coral *Montipora digitata*, a species which also propagates clonally through fragmentation. The genetics of gametic incompatibility are best understood for ascidians (Oka 1970; Scofield, Schlumpberger, and Weissman 1982). Selfing is probably more commonly prevented by a kind of hermaphroditism which is broadly synchronous, but in which the times of sperm and egg release are slightly offset, thereby preventing fertilization between gametes from the same individual (alternating sequential hermaphroditism of Ghiselin [1974: 109]).

Genetic and temporal systems differ in the extent to which they favor outbreeding. Gametic incompatibility prevents all selfing as well as varying proportions of sib matings, depending on the extent of relationship between the parents. In some ascidians, for example, fertilization is blocked for sperm that share an allele with the maternally derived, diploid egg envelope. Thus 50% of matings between siblings should be blocked, assuming no alleles in common between their parents (Grosberg 1987). In contrast, alternating sequential hermaphroditism will not necessarily prevent sib matings, and will not even block selfing when clonal reproduction produces genetically identical individuals that are physiologically separate and reproductively out of synchrony.

Finally, Hoagland (1978) reported an intriguing comparison of postsettlement behavior among species of *Crepidula* with differing larval dispersal abilities. Species with planktonic larvae have juveniles that are immediately attracted to conspecific adults upon settlement. The resulting clusters function as long-term mating groups. In species in which the females brood the young to metamorphosis, however, juveniles are not immediately attracted to conspecific adults, and breeding attachments are relatively transient. Hoagland (1978) suggested that delayed and transient aggregations in species without planktonic young are the result of selection to avoid inbreeding.

Inbreeding and Outbreeding Depression

Offspring fitness has long been known to vary as a function of the degree of relatedness of the parents. The traditional perspective has two components: inbreeding depression, which occurs when mates are closely related, and outbreeding depression, which occurs at or around the point of interspecific hybridization (figure 10.3). This pattern is undisputed for many species (i.e., Stebbins 1950; Dobzhansky 1951; Wright 1977; D. Charlesworth and B. Charlesworth 1987). Inbreeding depression is thought to result from the unmasking of deleterious recessives and the loss of heterozygote advantage at individual loci (partial dominance and overdominance, respectively) while interspecific outbreeding depression is believed to be due to the combination of incompatible genetic systems (D. Charlesworth and B. Charlesworth 1987).

Others, however, have stressed that in species where inbreeding is common, the theoretical predictions and data are not consistent with the traditional view (e.g., Shields 1982, chapter 8, this volume; Bateson 1983; Schemske and Lande 1985; Waser, chapter 9, this volume). They argue that in the face of persistent inbreeding for whatever reason, the costs of inbreeding are likely to be low or nonexistent (figure 10.3), particularly due to the elimination of unfavorable recessives. At the same time, inbreeding results in linkage disequilibrium and the assemblage of intrinsically well-coordinated gene complexes, while limited dispersal often associated with inbreeding permits the buildup of locally favorable genetic combinations (Endler 1977; Templeton 1986). As a consequence, mating of conspecific individuals from unrelated populations leads to loss of fitness (figure 10.3). The empirical demonstration of intraspecific outbreeding depression (Price and Waser 1979; Willson and Burley 1983: 31, 35) provides particularly compelling evidence for this perspective. Inappropriate experimental design may complicate interpretation of results, however. For example, the mere absence of inbreeding depression may be misleading if the entire area studied represents one inbred population so that no comparisons with truly outbred crosses are made (D. Charlesworth and B. Charlesworth

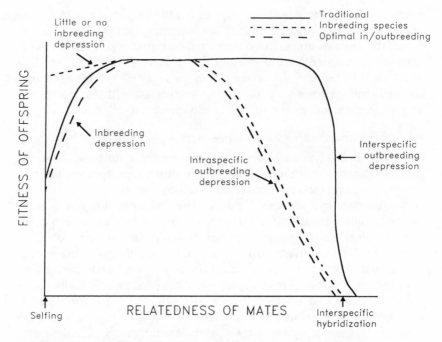

FIGURE 10.3   Proposed relationship between the degree of relatedness between mates and the fitness of their offspring. Three commonly discussed patterns (see text) are illustrated.

1987). Studies of outbreeding that consider only the first generation of crosses may also be deceptive, because initial increases in fitness are likely to be replaced by declines in subsequent generations (Lynch 1991).

In marine invertebrates, studies of the relationship between parental relatedness and offspring fitness are far more limited than in many other groups, due to frequent difficulties with breeding and rearing of offspring. Loss of viability may occur with hybridization of closely related species (e.g., Longwell 1976; Lucas, Nash, and Nishida 1985; Ward, Warwick, and Knight 1986; but see Strathmann 1981), and inbreeding depression has also been reported (Barnes and Crisp 1956; Potswald 1968; Sabbadin 1971; Longwell and Stiles 1973; Longwell 1976; Beckwitt 1982; Beaumont and Budd 1983). Even in species that are almost certainly outbred, however, unambiguous evidence for strong inbreeding depression is not always found (e.g., see Mallet and Haley [1983] for a review of studies on the oyster *Crassostrea virginica*). The evidence for inbreeding depression is particularly scanty for species in which inbreeding is likely (Strathmann, Strathmann, and Emson 1984). Moreover, intraspecific outbreeding depression on scales consistent with dispersal abilities has been reported for the ascidian *Botryllus schlosseri* (Grosberg 1987) and the copepod *Tigriopus californicus* (Burton 1986, 1987). Thus marine invertebrates tend to support the pluralistic view more recently expounded for other taxa.

MODEL SYSTEMS

The diversity of life histories and mating systems reviewed above suggests that marine invertebrates are likely to contain many examples at both ends of the inbreeding-outbreeding continuum. Here we review in greater detail the relevant evidence in four particularly well-studied species. They were chosen because they are representative of several common patterns and because they illustrate the strengths and weaknesses, as discussed by Hedgecock (1982, 1986), of the data available.

The Colonial Ascidian *Botryllus schlosseri*

This widely distributed species often dominates hard substrata in shallow, protected waters. In Eel Pond at Woods Hole, Massachusetts, maximum densities reach 2,000–7,000 colonies per m² (R. Grosberg, personal communication, 1989). Adults are sessile, and although of modular construction, they do not regularly fragment, so that physiologically separate individuals are usually genetically distinct (Grosberg 1987). Some colonies even represent more than one genotype, due to the fusion of adjacent colonies that share a histocompatibility allele. Persistent chimeras usually contain very close relatives, however, due to the high diversity of the histocompatibility system (Grosberg and Quinn 1986) and the long-term instability of fusions between individuals that are not identical at loci governing histocompatibility and resorption (Rinkevich and Weissman 1989). The species is hermaphroditic, but selfing is generally prevented by asynchrony in the release of eggs and sperm within a colony (Sabbadin 1971). All these characters are compatible with outbreeding.

Other aspects of the life history and the data on proximity-dependent mating success strongly suggest, however, that *B. schlosseri* is inbred. Gamete and larval dispersal are generally extremely limited. Fertilization of the eggs takes place in the mother colony. The life span of sperm is less than 5 minutes (Grosberg 1987), and the average distance of sperm dispersal, based on studies using rare electromorphs, is less than 17 cm (Grosberg 1991). Larvae are capable of settlement and metamorphosis immediately upon release. Grosberg (1987) found that over 80% of the larvae settled within 25 cm of their place of birth (although estimates of larval dispersal may be biased due to the use of empty settlement panels when natural substrata are crowded). Other evidence for limited dispersal of larvae comes from spatial patterns in field populations of a number of genetically controlled characters (histocompatibility type [Grosberg 1987]; enzymes and color [Sabbadin 1978]). Larvae also tend to settle next to histocompatible individuals, which are typically siblings, resulting in even tighter associations among close relatives than would be predicted by limited dispersal alone (Grosberg and Quinn 1986). The effect of this behavior on inbreeding is difficult to judge, however, because these individuals may then fuse and form stable chimeras, thus reducing inbreeding by synchronizing the sexual cycle and preventing fertilization between close relatives (Grosberg and Quinn 1986).

The strongest evidence for routine inbreeding in this species comes from Grosberg's (1987) studies of proximity-dependent mating success. He found that the success of fertilization, embryogenesis, and larval metamorphosis all declined significantly with increasing distance between mates. Success rates through larval metamorphosis were 89% for crosses between neighbors, 70% for crosses between individuals separated by 1.5 m, and 58% for crosses between individuals separated by 4.5 m. This pattern is consistent with predictions based on larval dispersal distance. None of the colonies mated were genetically identical, so inbreeding depression with selfing cannot be entirely ruled out from the published data. Indeed, strong inbreeding depression has been reported in other populations of this species (Sabbadin 1971, 1982), and R. Grosberg (personal communication, 1989) has observed sporadic failures of selfed lines in the F_1 generation. Inbreeding depression for matings between close relatives does not occur, however, and intraspecific outbreeding depression over very small spatial scales is striking. These data provide compelling evidence that inbreeding is high in this species.

Grosberg (1991) has also calculated Wright's F statistics for the random and nonrandom components of inbreeding for the Eel Pond population. At three independently segregating loci, F_{IS} and F_{IT} were comparable and ranged from 0.25 to 0.50, while population subdivision (F_{ST}) made a small contribution to the overall F values at spatial scales ranging from 6 m down to 0.5 m. Grosberg's earlier (1987) report of proximity-dependent mating success, which documented intraspecific outbreeding depression over a scale of 3 m, is somewhat puzzling in light of the absence of population subdivision over a 6-m scale.

The absence of differentiation does suggest that gene flow greater than that shown by Grosberg's larval dispersal experiments must occasionally occur, and seasonal or episodic variation in dispersal is to be expected. The calculations of Pollak (1987, 1988) predict the pattern of F statistics found by Grosberg (1991) whenever the probability of matings with close relatives is substantially greater than the reciprocal of the population size. This is because heterozygosity is lost much more rapidly from nonrandom matings than from genetic drift in all but the smallest populations. As a crude approximation for population size, one can use Crawford's (1984a) methods, coupled with Grosberg's (1987) larval dispersal measurements, to estimate a minimum neighborhood area of approximately 0.2 m² (assuming that 86% of larvae settle within 25 cm of their mother, and that this is equivalent to Crawford's formulation of an 86% expectation that a central individual's parents lie within 25 cm). If there were on average 1,000 colonies per m² (yielding a minimum population of 200), then Pollak's criterion would hold when the probability of sib or parent-offspring mating was substantially higher than .005. These conditions seem quite likely for *B. schlosseri.*

The Mussel *Mytilus edulis*

Mytilus edulis and two taxonomically controversial sibling species, *M. galloprovincialis* and *M. trossulus* (Varvio, Koehn, and Väinölä 1988; McDonald and Koehn 1988; Gosling 1984), are the most widely distributed members of the Mytilidae, a family renowned for its dominance of rocky shores (Suchanek 1986). Some of the characters described below come from sources predating the recognition of the species distinctions, but except as noted, the characters are shared by all.

Edulis-like mussels are found in a variety of temperate habitats; they dominate portions of many intertidal zones and may attain very high densities (e.g., nearly 8,000/m²) (Suchanek 1978). Almost all individuals are either males or females for their entire life span (Seed 1976), asexual reproduction is unknown, and there are no reports of strongly biased sex ratios. Eggs and sperm remain viable for several hours (Bayne 1976), and are spawned directly into the water, where fertilization occurs. The larvae spend 3 to 7 weeks in the plankton (Bayne 1976; Strathmann 1987: 325), which is the period of maximum dispersal potential. Levinton and Koehn (1976) estimated a minimum net dispersal distance of nearly 175 km, and Koehn, Newell, and Immermann (1980) documented dispersal by oceanic larvae of over 25 km into Long Island Sound. The juveniles and adults are also capable of some movement (Levinton and Koehn 1976). In sum, due to high dispersal, high densities, and the impossibility of selfing, inbreeding in this species is inconceivable.

Electrophoretic variation in *Mytilus edulis* has been intensively studied, and the data illustrate well the dangers of assessing inbreeding from genetic data alone. Populations from widely separated regions differ electrophoretically (Levinton and Koehn 1976), even when the sibling species are analyzed separately (Varvio, Koehn, and Väinölä 1988). Genetically uniform populations spread over large areas also occur, as in nonestuarine environments south of Cape Cod (Koehn 1983). These patterns are consistent with expectations based on larval dispersal (Slatkin 1981). There are, however, sharp clines over distances as short as a few kilometers (Koehn 1983), and heterozygote deficiencies are also common (Zouros, Romero-Dorey, and Mallet 1988). These latter patterns could in theory stem in part from inbreeding or the Wahlund effect, but careful study has shown that natural selection and the effects of genotype on reproductive seasonality are instead responsible (Koehn 1983; Zouros, Romero-Dorey, and Mallet 1988; Hilbish and Zimmerman 1988).

Unfortunately, breeding studies in this species have not been directed toward the problems of inbreeding and outbreeding depression. There are no published data from inbreeding experiments, nor are there data on the success of matings between conspecifics from distant sites, generally referred to as subspecies (Seed 1976). Matings between *M. edulis* and *M. galloprovincialis* show no evidence of hybrid inferiority (Skibinski 1983; Lubet et al. 1984), however. This suggests that marked outbreeding depression between geographically distant conspecifics is unlikely.

The Snails *Littorina littorea* and *L. saxatilis*

The two preceding examples come from species in different phyla, but widely differing potential for inbreeding may exist within a single genus. The snails *L. littorea* and *L. saxatilis* (sometimes referred to as *L. rudis* [Hannaford Ellis 1983; Janson 1985]) are both common intertidal species found on both sides of the northern Atlantic (Johannesson 1988).

Littorina littorea has a planktonic larval phase that lasts approximately 4 to 6 weeks. Johannesson (1988) suggested that during this period a larva might travel as much as 300 km prior to settlement, and noted (based on Carlton [1982]) that *L. littorea* introduced to the North American east coast spread at a rate of 34 km per year. These data strongly imply outbreeding. The young of *L. saxatilis*, on the other hand, emerge from the mother as crawling juveniles (Janson 1987a), and studies of marked individuals indicate movement of only 1 to 4 m over a 3-month period (Janson 1983). Several studies have shown that densities vary widely depending on habitat, ranging from less than 10 to 3,000 per m² (Hughes and Roberts 1981; Janson 1987b). Thus *L. saxatilis* has the potential for inbreeding due to its limited dispersal, and the effect could be quite marked where densities are low.

The two species have been extensively compared both morphologically and electrophoretically. As expected, shell shape varies more between habitats in *L. saxatilis* than in *L. littorea*. The former shows marked differences between exposed and sheltered shores only 100 m apart, while the latter does not, presumably because gene flow swamps selection when larvae are planktonic (Janson 1987a). Similarly, there is a positive correlation between geographic distance and genetic difference in *L. saxatilis*, but no significant correlation in *L. littorea* (Janson 1987a). In a detailed analysis of microgeographic variation in *L. saxatilis* along 1 km of coast, populations separated by only a few meters sometimes showed significant electrophoretic differentiation (Janson and Ward 1984). Also, the random component of inbreeding (estimated by differences between subpopulations, F_{ST}) was more often and more highly significant than was the component of inbreeding due to nonrandom mating within populations (estimated by F_{IS}). This finding, in contrast with that of Grosberg for the ascidian *B. schlosseri*, is consistent with the greater mobility of snails as adults and the apparent absence of kin recognition and association, making mating with close kin within a population much less likely. Contrary to expectation, however, *L. saxatilis* shows more genetic variability within subpopulations than does *L. littorea* (Janson 1987a). The difference holds true even within Europe, where no founder events have been hypothesized. Unfortunately, no comparisons of the two snails with respect to inbreeding or outbreeding depression have been made.

Taxonomic Patterns

In this section we review the characteristics of the life cycle relevant to inbreeding and outbreeding for the principal groups of marine invertebrates, excluding parasitic forms, and highlight cases within each group where inbreeding is most likely to occur (see table 10.2).

Porifera

Sponges are sessile or passively sedentary after the larval stage (Hyman 1940). More than 90% of all species are in the class Demospongia (Hartman 1982), and the following account is based entirely on that group.

There are two principal groups of demosponges, distinguished both taxonomically and reproductively (Brien 1973; Bergquist 1978). Tetractinomorphs are typically gonochoric (or perhaps sequential hermaphrodites), and spawn relatively small ova that develop upon fertilization into small swimming or crawling larvae. There is evidence that the evolutionary trend in the group has been for precocious fixation and loss of mobility of the larval stage (Borojevic 1970). In contrast, ceractinomorphs are divided approximately equally between simultaneously hermaphroditic and gonochoric species; most incubate comparatively large larvae that swim, or less commonly crawl, on release.

In both groups, free larval life lasts a few hours to days among swimming forms, and from a few to 25 days in exclusively crawling species. Larval aggregation at settlement and postlarval fusion to form chimeras is apparently common, as is agametic clonal reproduction, which occurs by fragmentation, budding, and formation of reduced resting stages (gemmules) and possibly larvae (Fry 1971; Brien 1973; Fell 1974; Bergquist 1978; Wulff 1986).

Exclusively crawling larvae may disperse only a few meters or less from their mother before metamorphosing, even if they crawl for 25 days. Others that roll or bounce along the bottom probably disperse a bit further. Altogether, these entirely benthic larval forms comprise 22% of the sixty-nine species whose movements were compiled by Wapstra and van Soest (1987). A similar percentage of species with crawling larvae has been reported for intertidal sponges in New Zealand (Bergquist and Sinclair 1968; Bergquist, Sinclair, and Hogg 1970). More strikingly, all ten species examined in nearby subtidal sites have exclusively crawling larvae; these include four thinly encrusting ceractinomorphs and six massive tetractinomorphs (Ayling 1980). In addition, budding is more frequent than larval recruitment in several of these species, both intertidal and subtidal. Extensive inbreeding seems very likely among the resulting populations of parents, offspring, and siblings characteristic of such species, but there are no genetic data.

Cnidaria

The diversity of reproductive modes and life histories of cnidarians is exceptionally great. The two most diverse and abundant classes are the Hydrozoa and Anthozoa (Hyman 1940; Dunn 1982). The other two classes (Scyphozoa, Cubozoa) are primarily composed of species whose adults are free-swimming and presumably outbred.

Hydrozoa. Most hydrozoans are apparently gonochoric, although sequential hermaphroditism is hard to rule out in the absence of long-term studies of individuals. Occasional hermaphrodites (1% to 2%) in otherwise gonochoric species are probably chimeras, as has been shown for *Hydractinia* (Hauenschild 1954; Yund and Parker 1989).

Hydroids are the most abundant and diverse hydrozoans (Hyman 1940; Dunn 1982). Many species have both sessile and mobile phases as adults; the former as solitary polyps or, more commonly, vine- or treelike colonies of polyps with interconnecting stolons; the latter as free-swimming medusae, or more rarely as planktonic polyps. Medusae typically produce gametes that fuse and develop into ciliated, mobile planular larvae that settle on the bottom and give rise to the sessile polypoid phase, although various forms of agametic budding from medusae are also known (e.g., Carré and Carré 1990). Polyps in turn agametically bud off new medusae.

There are strong indications that the mobile phases of the life cycle have been increasingly (repeatedly?) suppressed throughout colonial hydroid evolution (Hyman 1940; Mackie 1974). Presumably primitive forms have strongly swimming, frequently large, pelagic medusae. In more derived groups medusae are often smaller, less mobile, or absent (e.g., they swim only near the bottom, creep along the bottom, hold onto the bottom by their tentacles, are never released by the polyp, or are reduced to sessile gonadal structures on the polyp). Larvae also vary in their mobility, ranging from those that swim vigorously, to those that swim and crawl, crawl only, or are varyingly packaged or bound to one another or their parent by mucus (Williams 1965, 1976; Wasserthal and Wasserthal 1973). In addition, most solitary hydroids that lack medusae brood embryos that are ready to settle when they leave their parent (Rees 1957).

Highly derived colonial forms lacking both medusae and swimming larvae are abundant and diverse (Nishihira 1967; Williams 1976), comprising nearly half of the species in many habitats (e.g., Boero 1981). Larvae may disperse only a few centimeters or meters before settlement and metamorphosis (Pyefinch and Downing 1949; Nishihira 1967, 1968; Wasserthal and Wasserthal 1973; Williams 1976). Larvae settle preferentially near one another or on parental tissue, and adjacent postlarvae commonly fuse to form chimeras (Nishihira 1967, 1968; Williams 1976). In such cases inbreeding without selfing is probable, but there are no genetic data. The principal mode of dispersal of such otherwise immobile species is almost certainly by floating or rafting of entire "sessile" colonies on seaweeds, seagrasses, or other floating debris (Jackson 1986).

There are two orders of sessile hydrozoans that construct calcareous skeletons (Hyman 1940). Milliporid hydrocorals have massive encrusting or erect colonies and tiny sexual medusae that shed gametes within a few hours of their release. They are presumably outbred. In contrast, stylasterine hydrocorals form smaller erect or encrusting colonies without a medusoid stage (Ostarello 1973, 1976; Frichtman 1974), and females brood larvae for nearly a year. In Ostarello's study, female colonies more than a meter from the nearest male were frequently unfertilized, so dispersal of sperm must be as short as reported for sea urchins (Pennington 1985). Larvae crawl away from their mother for only a few centimeters before settlement; more than half settled within 10 cm and fully 85% settled within 20 cm of their mothers. Local populations consist of parents and offspring in very close proximity, so that inbreeding must be compounded over many generations, but once again there are no genetic data.

Also included among the hydrozoans are two highly specialized planktonic orders of swimming or floating colonies, the siphonophores (e.g., the Portuguese man-of-war), and the chondrophores (velellids) (Hyman 1940; Dunn 1982). These groups must be routinely outbreeding.

Anthozoa. All anthozoans lack medusae (Hyman 1940). There are two important subclasses (Dunn 1982): the Alcyonaria, which includes soft corals, gorgonians, and pennatulaceans (sea pens); and the Zoantharia, which includes sea anemones, zoanthids, and stony corals.

Soft corals grow as sessile or sedentary encrusting, massive, or erect colonies with spicular skeletons. They are generally gonochoric or sequential hermaphrodites. Many small and some large species of soft corals and encrusting gorgonians brood exclusively crawl-away larvae that settle within a few centimeters of their parents (Gohar 1940; Benayahu and Loya 1983, 1984a, 1984b, 1984c, 1985; Sebens 1983a, 1983b, 1983d). These larvae often aggregate and may fuse to form chimeras (again with the potential to produce rare hermaphrodites in otherwise gonochoric species). These species also characteristically form extensive monospecific stands by fragmentation (Benayahu and Loya 1983, 1985). The combination of extremely short-distance larval dispersal and clonal propagation must result in extensive inbreeding. The same pattern holds for the large erect gorgonian *Capnella gaboensis* (Farrant 1985), but most larger species of soft corals and upright gorgonians spawn small ova that develop on fertilization into swimming larvae (Grigg 1979; Weinberg and Weinberg 1979; Yamazato, Sato, and Yamashiro 1981), and are certainly outbred. There are no appropriate genetic data for any octocoral.

The sea pen *Ptilosarcus gurneyi* forms extensive populations on subtidal sands (Birkeland 1974). Adults are long-lived and sedentary, although some, mostly small, individuals drift into cleared areas. Sexes are separate, and gametes are spawned annually in a neutrally buoyant mucus that is not readily dispersed (Chia and Crawford 1973; Strathmann 1987: 90–91). Development is demersal and juveniles can burrow within 9 days. Settlement occurs in massive sets reflected in abrupt spatial discontinuities in size and age classes, but these commonly overlap presumably parental distributions (Birkeland 1974). Weak inbreeding is therefore possible. The life history of the Atlantic sea pansy, *Renilla*, is similar (Wilson 1883).

Sea anemones are solitary polyps that may be sedentary or mobile, the latter by pedal locomotion or passive drift (Nyholm 1949; Mackie 1974; Wahl 1985a, 1985b). Most species are gonochoric, but sequential and synchronous hermaphrodites are also found (Shick 1991). Clonal propagation by fission, pedal laceration, or ameiotic juveniles is widespread, occurring in 40% of the eighty-five species compiled by Chia (1976). Some species (or species complexes) display gonochorism, hermaphroditism, and parthenogenesis (Schmidt 1967; Rossi 1975; Shick 1991). Reports of brooding and release of crawling larvae or juveniles, limited adult movement, and coordinated spawning between neighbors (Shick 1991) suggest the potential for inbreeding in some species.

All of the best-studied anemones live principally on rocky intertidal shores. Many of these, like *Anthopleura elegantissima* and *Metridium senile*, form extensive monoclonal, unisexual populations (Sebens 1982, 1983c; Hoffman 1986; Francis 1988) so that fertilization may be limited to adjacent clones. However, these species spawn relatively small ova that develop upon fertilization into drifting or swimming larvae (Ford 1964; Jennison 1979) so that genetic exchange between populations is expected and does occur (Hoffmann 1987). In contrast, *Haliplanella lineata* (= *luciae*) is apparently exclusively clonal, at least over large parts of its geographic range (Shick and Lamb 1977; Shick, Hoffmann, and Lamb 1979).

Dense monospecific aggregations may also form by extremely philopatric dispersal of relatively large brooded larvae, as is the case for two intensively studied species of *Actinia* in New Zealand, Australia, and Britain (Black and Johnson 1979; Ottaway 1979; Orr, Thorpe, and Carter 1982; Ayre 1982, 1983a, 1983b, 1984a, 1984b, 1985, 1987, 1988; Brace and Quicke 1985, 1986). However, all larvae that have been examined in these species are apparently clonally produced. Larvae are electrophoretically indistinguishable from their brood parents at four to five polymorphic loci, and gonads of brooders may be male, female, or absent entirely. Apparently normal maturation of ova and sperm occurs in all populations investigated. Sexual reproduction, as yet undocumented, probably involves oviparous spawning and planktonic larval development. This idea is supported by a good fit to Hardy-Weinberg expectation for different genotypes (counting each clone as one individual) at the same locality, or among localities (Ayre 1984b, 1985, 1988).

One well-studied sea anemone, *Epiactis prolifera*, begins life as a female and then develops into a simultaneous hermaphrodite. This species broods large crawl-away larvae that apparently settle in the immediate vicinity of their parents (Dunn 1975a, 1975b, 1977). Electrophoresis indicates that individuals typically self-fertilize with infrequent outcrossing, but there is no genetic evidence for clumping of genetically similar individuals, as would be expected from short-distance larval dispersal (Bucklin, Hedgecock, and Hand 1984). At least three other species of *Epiactis* brood crawl-away larvae (Uchida and Iwata 1954; Fautin and Chia 1986; Strathmann 1987: 94–96).

The natural history of several species strongly suggests inbreeding, although genetic data are absent. The burrowing anemone, *Halcampa duodecimcirrata*, has large crawl-away larvae that develop within sticky gelatinous capsules near their mothers, and females lean toward adjacent spawning males, making fertilization between neighbors highly probable (Nyholm 1949). *Aulactinia incubans* is a small, intertidal hermaphrodite that internally broods its young to the juvenile stage (Dunn, Chia, and Levine 1980). The species is not abundant, but several individuals are typically found close together where it occurs. Isolated clusters of interbreeding kin thus seem likely.

Zoanthids are simultaneously hermaphroditic, morphologically (as opposed to socially) colonial anemones that routinely fragment to form dense clonal populations (Karlson 1981, 1983, 1986, 1988a, 1988b; Fadlallah, Karlson, and Sebens 1984). However, they spawn ova that develop into potentially very long-lived larvae (Scheltema 1971), and are presumably outbred.

Scleractinian corals are typically sessile, solitary or colonial polyps that secrete a calcareous exoskeleton. Most tropical corals are colonial simultaneous hermaphrodites, and spawn ova that develop when fertilized into swimming planular larvae (Fadlallah 1983a; Harrison et al. 1984; Babcock and Heyward 1986; Babcock et al. 1986; Szmant 1986; Richmond and Hunter 1990). Such species are almost certainly outbred, despite demonstrable self-fertilization in some mass-spawning species (Heyward and Babcock 1986; Richmond and Hunter 1990), and the likelihood that some reefs may be colonized primarily by larvae derived from the same reef (Sammarco and Andrews 1988; but see Willis and Oliver 1989).

A substantial minority of tropical scleractinians, however, including perhaps one-quarter of Pacific species and half of those in the Atlantic, brood larvae that crawl or swim and crawl (Fadlallah 1983a; Szmant 1986; Morse et al. 1988). Clutches of brooded larvae are commonly released at widely varying stages of development, with the result that some offspring are competent to settle on release, and do so. This has been shown best for two small faviids, *Favia fragum* (Duerden 1902; Lewis 1974a, 1974b) and *Manicina areolata* (Duerden 1902; Boschma 1929), and two pocilloporids, *Stylophora pistillata* (Loya 1976a, 1976b; Rinkevich and Loya 1979a, 1979b) and *P. damicornis* (Stephenson 1931; Harrigan 1972).

For *P. damicornis*, however, there is a serious problem of interpretation, because all of the hundreds of larvae examined so far are genetically identical to their brood parents at five polymorphic loci (Stoddart 1983, 1984, 1986; Stoddart and Black 1985). The situation seems very similar to that of the anemone *Actinia* (see above), except that sperm in Hawaiian populations of *P. damicornis* have been shown to be reabsorbed just before they would otherwise be released (Martin Chavez 1986). Comparison of gene frequencies among and within local populations suggests that in this case too, dispersal occurs by rare swimming, sexually produced larvae (Stoddart 1984, 1988), but this is speculation. *Pocillopora damicornis* is also commonly rafted for very long distances (perhaps across the entire Pacific!) on pumice and other floating debris (Jokiel 1984, 1990). Bioenergetic estimates and laboratory studies of larvae show that they remain competent to settle after 100 days, allowing them in theory to travel successfully from the central to the eastern Pacific (Richmond 1987).

Two species in the genus *Goniastrea* show apparently related patterns in life history and self-compatibility. Both *G. aspera* and *G. favulus* are common inshore shallow-water species on Australian reefs, although *G. favulus* is generally less common in and less restricted to these disturbed habitats (Babcock 1984). Both are mass-spawning synchronous hermaphrodites with external fertilization and similar investments in sexual reproduction, but *G. aspera* produces more numerous and smaller eggs (Babcock 1984, 1991). This species exhibits the common pattern of releasing buoyant egg/sperm packets that break up and typically mix with gametes from other members of the population prior to fertilization, and the resulting swimming larvae continue to float for at least several days. *Goniastrea aspera* shows some ability to self-fertilize in the laboratory, but self-fertilization is delayed and limited in extent relative to cross-fertilization (Heyward and Babcock 1986). In contrast, *G. favulus* releases negatively buoyant eggs embedded in sticky mucus, and sperm are often released synchronously (Kojis and Quinn 1981; Heyward and Babcock 1986). Spawning occurs during the time of minimum water flow, and fertilized eggs typically remain on or next to the mother colony (Kojis and Quinn 1981). Larvae are at least initially benthic, although they begin to swim after several days in the laboratory (Kojis and Quinn 1981). Consistent with this life history is the finding of high selfing rates in laboratory crosses, with little or no delay in the onset of selfing (Heyward and Babcock 1986; Stoddart, Babcock, and Heyward 1988). Genetic data for one natural population do not show the expected reduction in heterozygotes associated with regular selfing, however, suggesting either that selfed progeny do poorly or that selfing is only common where population densities are not high (Stoddart, Babcock, and Heyward 1988).

Most scleractinians outside the tropics are small, solitary or encrusting, colonial corals. *Balanophyllia* are solitary, gonochoristic brooders that release large crawl-away larvae, which settle less than half a meter from their mothers (Gerrodette 1981; Fadlallah 1983b; Fadlallah and Pearse 1982a) and are almost certainly inbred. In contrast, *Caryophyllia smithi* and *Paracyathus stearnsii* spawn gametes that develop upon fertilization into pelagic larvae that can swim for up to 10 weeks (Lacaze-Duthiers 1897; Fadlallah and Pearse 1982b; Tranter, Nicholson, and Kinchington 1982).

Polychaeta

Polychaetes are the most diverse and abundant worms in the sea (Pettibone 1982) and exhibit a bewildering array of reproductive modes. They include forms ranging from minute to large, and sessile to highly mobile. Fauchald (1983) has outlined three general patterns of polychaete life histories. Annual species are commonly mobile, with moderately large eggs and planktonic larvae. Perennial species are typically sedentary, with moderately large to large eggs and benthic larval or direct (i.e., no free-living larva) development. Lastly, multiannual species (two or more generations per year) are typically sedentary, with relatively small eggs and benthic larval or direct development. This common association of sessile or sedentary adult habit with exclusively benthic development and dispersal suggests the possibility of inbreeding for the latter two categories. Reproductive patterns in polychaetes are not commonly taxonomically constrained, and many closely related forms have very different life histories (Grassle and Grassle 1977, 1978; Blake and Kudenov 1981; McEuen, Wu, and Chia 1983; Wilson 1983, 1991; L. A. Levin 1984b). Moreover, variation is not just a function of egg size; there are also differences in the relative composition of eggs that correlate with developmental patterns and probably reflect differences in the energy requirements of different types of larvae (Eckelbarger 1986).

Most species of polychaetes, particularly those that are large, are gonochoric and spawn small ova (median oocyte volume 300 μ^3, n = 13 species), which develop when fertilized into pelagic swimming larvae that feed or develop on stored yolk (Schroeder and Hermans 1975). Many of these species also produce specialized swimming adults that swarm high in the water column to spawn. On all counts outbreeding is inevitable.

In contrast, a substantial minority of species, particularly small, sessile or sedentary filter feeders, show a different pattern. Approximately one-third of the species surveyed by Wilson (1991) have reproductive characteristics suggesting restricted larval dispersal. Clonal reproduction, sequential and simultaneous hermaphroditism, and potential for self-fertilization are also common. Many of these species brood relatively few, large eggs (median oocyte volume 2,050 μ^3, n = 16 species) that develop into juveniles without an intervening larval stage (Schroeder and Hermans 1975). Moreover, the size, growth rate, and behavior of the larvae depends in many cases on the number of nurse eggs or siblings consumed during development by the individuals that survive to hatch. The greater the number consumed, the more benthic the life history and the lower the probability of dispersal (Blake and Kudenov 1981).

Twelve of seventy-one species listed by Strathmann (1987: chapter 8) and L. A. Levin (1984a) do not swarm, and brood larvae that are benthic or demersal. Of these twelve species, three intertidal tubeworms (*Sabella media, Fabricia limnicola,* and *Axiothella rubrocincta*) brood exclusively benthic larvae (McEuen, Wu, and Chia 1983; Wilson 1983; L. A. Levin 1984a), so that inbreeding may occur despite densities of 20,000/m² or more. The possibility of inbreeding is strengthened for *F. limnicola* by its absence from plankton tows, recruitment experiments, and cleared areas studied over 4 years (L. A. Levin 1984a). *Sabella media* may also self-fertilize, since it is synchronously hermaphroditic, sedentary, solitary, and uncommon, although individuals may be able to collect and store sperm picked up in feeding currents (McEuen, Wu, and Chia 1983). Several intertidal Spionidae also have life histories that could result in inbreeding, including clonal reproduction and brooding of benthic larvae, but these worms are very small, occur in enormous numbers, and actively move or are moved about over short distances (Dauer, Maybury, and Ewing 1981; L. A. Levin 1984a).

The size dependence of brooding is clearly evident among sessile calcareous tubeworms of the families Serpulidae and Spirorbidae. Larger species of genera like *Pomatocerus* and *Hydroides* tend to be sequential hermaphrodites that spawn gametes and lack clonal reproduction (Schroeder and Hermans 1975). Aggregation at settlement (Wisely 1958; Scheltema et al. 1981) is by distantly related larvae that have been dispersed in the water column, just as for gregarious sabellarid tubeworms (Wilson 1968; Pawlik 1988). Alternatively, species of *Salmacina*, *Filograna*, and *Spirorbis*, all of which are very small, include both sequential and simultaneous hermaphrodites, brood larvae in the parental tube, and, except for *Spirorbis*, undergo extensive clonal reproduction (Schroeder and Hermans 1975). All spirorbids brood (Bailey 1969), but larvae of some species, such as *S. borealis*, usually swim briefly upon release, whereas others, like *S. rupestris*, *Pileolaria pseudomilitaris*, and *Janua brasiliensis*, settle and metamorphose immediately if suitable substratum is available (Gee 1963; Beckwitt 1980; Fauchald 1983). Selfing is possible among all spirorbids tested, but results in significant inbreeding depression (Gee and Williams 1965; Potswald 1968; Beckwitt 1982). Nevertheless, inbreeding among adjacent parents, offspring, and siblings appears inevitable for species in which offspring settle near their parents.

Polychaetes that live among sand grains, all of which are relatively small, also show a strong relationship between adult size and larval dispersal. Mean adult length of species with planktonic larvae is 9.8 mm ($n = 7$), while mean length of species with offspring released as juveniles is only 2.1 mm ($n = 8$) (Westheide 1984).

Members of the *Ophryotrocha labronica* species complex are small worms, and include both self-fertile hermaphrodites and gonochores capable of sustaining intense inbreeding for many generations (Åkesson 1977, 1984). In the closely related, less estuarine *O. diadema*, however, selfing does not occur despite gametic self-compatibility. Rather, simultaneous hermaphrodites reciprocally spawn in pairs (Sella 1985). In polychaetes generally, routine selfing does not appear to be common, although Smith (1950) provides strong circumstantial evidence for its occurrence in one burrow-dwelling, estuarine species with direct development.

Mollusca

There are five important classes of mollusks: Polyplacophora (chitons), Gastropoda (snails), Cephalopoda (octopuses, squids, nautiloids), Bivalvia (clams), and Scaphapoda (tusk shells). Clonal reproduction is rare in all groups. Cephalopods are gonochoristic with direct development, but are highly mobile as adults (Arnold and Williams-Arnold

1977; Wells and Wells 1977; Ward 1987); inbreeding is very unlikely. Scaphopods are sedentary burrowers in sediments. They are mostly gonochoristic, with external fertilization and pelagic larvae that settle within 2 to 6 days after fertilization, again making outbreeding inevitable (McFadien-Carter 1979). The other classes all have probable examples of both inbreeding and outbreeding.

Gastropoda. Prosobranchs are the commonest shelled snails in the sea. Many species are highly mobile, but there are also sedentary and cemented forms. Most are gonochoric with sex ratios commonly close to 1:1 or slightly biased toward females (Webber 1977; Hughes 1986). The most primitive reproductive systems occur among the archaeogastropods, which typically broadcast gametes and develop pelagically. Members of all other groups typically copulate, and juveniles develop partially or completely within external egg capsules. Protandric hermaphrodites and brooders are also fairly common. Overall, direct development (no free larval stage) is found in 39% of the 142 species of snails listed by Webber (1977). Clonal reproduction is very rare and occurs only by parthenogenesis.

The likelihood of inbreeding among free-living prosobranchs with entirely direct development depends primarily upon postlarval mobility, behavior, and abundance. For example, some species are highly mobile but aggregate to breed, so the critical data are the size of breeding groups and the faithfulness of snails to their parental group (see Breden [1987] for an analysis of a comparable situation in the toad *Bufo woodhousei*). Sufficient data are available for two intertidal species to suggest that some inbreeding may occur. Approximately 80% of over four thousand surviving marked individuals of *Nucella* (= *Thais*) *lamellosa* returned annually to the same sites and breeding populations over 5 years of observations (Spight 1974), and genetic data are consistent with these observations (Grant and Utter 1988). More strikingly, all marked *Nucella emarginata* moved less than 10 m over 1 year at two sites, and were almost as stationary at a third site (Palmer 1984). Moreover, there are striking morphological, physiological, and genetic differences in local populations of these snails with no regular clinal pattern (Campbell 1978; Crothers 1980, 1981; Palmer 1984; Day and Bayne 1988), just as would be expected if inbreeding occurs. Palmer (1984) did not, however, detect any reduced viability in offspring from matings between individuals from widely distant sites, suggesting that there is little or no intraspecific outbreeding depression.

Vermetids are cemented tubiculous snails that characteristically live in dense aggregations (Hughes 1986); they are dioecious, copulate, and brood embryos for varying periods. Among ten species for which data are available, only two invariably produce planktonic larvae, five hatch crawling juveniles, and three produce planktonic or crawling juveniles, depending on the number of nurse eggs consumed by the developing embryo (Hadfield et al. 1972; Hughes 1978). Because of the sessile habit of adults, inbreeding is possible for species with crawling larvae, but there are no genetic data.

Opisthobranch gastropods have reduced shells or lack shells entirely. Most are simultaneous hermaphrodites with enormously complex reproductive anatomy, and copulation apparently precludes self-fertilization (Beeman 1977). Most species spawn small eggs, which develop quickly into tiny larvae that drift for hours to weeks before settlement. Some are viviparous, including the planktonic pteropods. Clonal reproduction is unknown. Populations of an intertidal nudibrach from sites separated by 3 km showed stable genetic differentiation in patterns suggesting limited gene flow, despite a pelagic larva (Todd, Havenhand, and Thorpe 1988). Nevertheless, outbreeding seems inevitable for all but the ectoparasitic pyramidellids, which usually lack a free larva and develop on their hosts (Hyman 1967).

Polyplacophora. Chitons are sedentary grazers, and most are gonochoric; there are two simultaneous hermaphrodites and one confirmed sequential hermaphrodite (Pearse 1979; Strathmann and Eernisse 1987; Eernisse 1988). Eggs are spawned free or in mucus packages. Fertilization apparently occurs while the eggs are still protected by the female. Most embryos hatch as immature larvae, but in some species are sufficiently developed to settle. Time in the plankton for nonbrooding species is 0 to 19 days with a median of 3.5 days ($n = 11$ species). About thirty species from half the known families brood larvae until they become juveniles; most are smaller than 2 cm as adults but some are much longer. A few other species lay benthic egg masses from which emerge crawling larvae. Most chitons aggregate for spawning. Inbreeding seems likely for species with crawling larvae if adults are stationary (territorial) or return to the same breeding aggregations. Brooders are very patchy in distribution (Eernisse 1988), which suggests limited movement of adults. Where they occur, brooders sometimes reach high densities (e.g., 500 per m²), but these large populations may be the descendants of only a few colonists (Eernisse 1988). Eernisse (1988) found that gonochoristic brooders were less diverse genetically than free-spawners, a pattern consistent with inbreeding caused by limited dispersal. The two simultaneous hermaphrodites, *Lepidochitona fernaldi* and *L. caverna*, are small brooders capable of self-fertilization (Eernisse 1988), and they self in culture even when they have the opportunity to outcross (Eernisse 1984, cited in Strathmann and Eernisse 1987). The preponderance of ovarian tissue in their gonads also suggests that selfing occurs regularly (Charnov 1982: 261), although parthenogenesis (Eernisse 1988) or reciprocal spawnings of animals in pairs (Fischer 1981; Sella 1990) are alternative explanations.

Bivalvia. Most bivalves are sedentary or sessile. Like *Mytilus edulis*, they are usually gonochoric with sex ratios of approximately 1:1 (Sastry 1979; Mackie 1984). Various types of hermaphroditism occur, however, particularly among sessile forms like oysters, shipworms, and giant clams, but also in the mobile scallops and cockles. Most species spawn small ova that are fertilized externally and develop in the plankton through a series of feeding or nonfeeding larval stages. Larval life ranges from 3 to 53 days, with a median of 18 days (n = 22 species). Self-fertilization may occur in some simultaneous hermaphrodites, but at lower rates than cross-fertilization, and parthenogenetic development may be induced in offspring of full-sib crosses of the oyster *Crassostrea virginica* (Longwell 1976). Inbreeding seems extremely unlikely in these spawning species.

Brooding is common among bivalves, especially in small, hermaphroditic species and in high latitudes and the deep sea; larvae are brooded within the mantle, gills, or external egg capsules (Sellmer 1967; Sastry 1979; Mackie 1984). Many brooded embryos are released as advanced pelagic larvae, but more are released as miniature adults (thirteen as larvae versus nineteen as juveniles [Sastry 1979]). Given their sedentary adult existence, inbreeding among close relatives seems possible for brooding bivalves that release benthic juveniles and live in low-energy environments (e.g., Jackson 1968), but there are no appropriate genetic or behavioral data. Most clams in the genus *Lasaea* are minute, intertidal, brooding, simultaneous hermaphrodites. Self-fertilization has been proposed for some species (Beauchamp 1986; Ó Foighil 1987; Tyler-Walters and Crisp 1989).

Crustacea

All crustaceans except barnacles are potentially mobile as adults, although symbionts or crevice dwellers may be quite sedentary (Bruce 1976). Clonal reproduction is uncommon except among Cladocera. The great majority are gonochoric or sequential hermaphrodites and have swimming larvae that pass through one to four stages and many more molts before settlement (Williamson 1982). Simultaneous hermaphroditism is rare except in barnacles. Some barnacles self-fertilize when isolated. They do so only after a delay, however, and such offspring sometimes exhibit reduced viability (Barnes and Crisp 1956; Barnes and Barnes 1958). Barnacles also have long-lived pelagic larvae. Inbreeding is unlikely for all these groups.

Most crustaceans that develop directly into juveniles are small. The most important such group in the ocean is the Peracarida, which includes amphipods, isopods, mysids, tanaids, and cumaceans (Williamson 1982; Highsmith 1985). Small intertidal isopods such as *Sphaeroma serratum* have visible chromatic polymorphisms, and early work on these species revealed striking differences in karyotypes and color patterns along continuous stretches of coast (Bocquet and Lejuez 1974; reviewed in Hedgecock, Tracey, and Nelson 1982). Limited dispersal, highly biased sex ratios, population bottlenecks, and microgeographic genetic differentiation in *Jaera albifrons* strongly suggest inbreeding within subpopulations living in individual crevices (Carvalho 1989). Considerable morphological differences occur among local populations of the isopod *Excirolana braziliensis* on nearby isolated beaches in Panama (Weinberg and Starczak 1988), although these may represent different species (H. A. Lessios, personal communication, 1990).

The harpacticoid copepod *Tigriopus californicus* is abundant in high intertidal and supratidal rock pools (Dethier 1980; Burton and Feldman 1981; Burton 1986). Despite having demersal larvae, adults are freeswimming, and might be expected to disperse readily. Gene flow is extensive among neighboring pools on the same outcrop, but is highly restricted between outcrops separated by stretches of sandy beach. Moreover, F_2 hybrids between distant populations show reduced tolerance to extreme salinities and increased development times compared with offspring from crosses within a single population (Burton 1986, 1987, 1990a, 1990b). Outbreeding depression between widely separated populations does not imply close inbreeding within populations, however. Close inbreeding leads to the disappearance of females within four to five generations in another harpacticoid, *Tisbe reticulata* (Ginsburger-Vogel and Charniaux-Cotton 1982).

In contrast to these forms with direct development, intraspecific geographic variation in heterozygosity and gene frequencies among species with swimming larvae is commonly much lower (Hedgecock, Tracey, and Nelson 1982). Intense selection may result in much local differentiation over very short distances (Hedgecock 1986; Achituv and Mizrahi 1987), however, and the strong swimming ability of crustacean larvae may limit their passive drift to a greater extent than would be predicted by the duration of the larval stage (Burton and Feldman 1982).

Brachiopoda

Brachiopods are bivalved animals that were the dominant shelly fauna of the Paleozoic, but are only locally abundant now. There are two classes, Articulata and Inarticulata. Both groups have separate sexes, but they otherwise differ considerably in aspects of life history important to inbreeding (Jablonski and Lutz 1983; Reed 1987a). Most inarticulates spawn relatively small eggs that develop as planktonic larvae (Paine 1963; Chuang 1977), so that outbreeding is inevitable. In contrast, articulates spawn or brood larger eggs that are competent to settle within about 4 days after fertilization. Larvae are demersal, and in brooding species some percentage may settle just after leaving the mother (Doherty 1979). Thus inbreeding may occur among some brooding species. This possibility is consistent with the frequently clumped distributions of articulates, and the apparently preferential settlement of larvae upon adults (Logan and Noble 1971; Thayer 1977, 1981; Doherty 1979; Noble and Logan 1981). There are no genetic data.

Bryozoa

Bryozoans are sessile or sedentary and colonial. There are two marine classes: Stenolaemata, which includes the cyclostomes, and Gymnolaemata, which includes the ctenostomes and cheilostomes (Ryland 1970, 1982). All bryozoans are hermaphrodites. Some presumably primitive gymnolaemates spawn gametes that develop into planktonic larvae, but the great majority of bryozoans brood short-lived, nonfeeding larvae that settle within a few seconds to hours of release (Ryland 1974; Zimmer and Woollacott 1977; Reed 1987b; Ström 1977). Moreover, many species produce viable larvae when reared in isolation, so self-fertilization or parthenogenesis may be widespread (Maturo 1991).

Brooded larvae are typically positively phototactic on release in the laboratory, swimming upward during part of the brief period before they settle (Cook 1985; Cancino and Hughes 1988). This should prevent inbreeding, but a variety of evidence suggests otherwise. First, recruitment of *Bugula neritina* may fail to occur in favorable habitats less than 100 m away from dense, reproductively active populations, and transplant experiments show that the vacant habitats are indeed suitable (Keough and Chernoff 1987). Second, larvae of *B. neritina* settle preferentially next to siblings (Keough 1984), a phenomenon difficult to understand if larvae were carried more than a few meters from their parents. Third, in the only such experiment done for bryozoans, larvae of the cyclostome *Tubulipora tuba* settled on the same small piece of kelp as their parents at densities ten times greater than on similar pieces of kelp only 1 m away (C. S. McFadden, personal communication, 1989). Moreover, cyclostome larvae multiply clonally in the brood chamber before release (Ström 1977), so the numbers of genotypes present among locally derived recruits should be low. Thus the probability of extensive inbreeding seems high for *T. tuba*.

Echinodermata

There are five extant classes of echinoderms: Echinoidea (sea urchins), Asteroidea (sea stars), Ophiuroidea (brittle stars), Holothuroidea (sea cucumbers), and Crinoidea (sea lilies) (Hyman 1955). All are mobile to varying degrees, and unstalked crinoids can swim. The great majority of all echinoderms are exclusively sexual, gonochoric, and spawn small ova that develop into pelagic larvae, which feed or live on yolk for several days to months before settling (Hyman 1955). For example, of fifty-three common species of echinoids, asteroids, ophiuroids, and holothurians from the American Pacific Northwest, thirty-six spawn pelagic eggs with a median diameter of 236 μm, whereas median egg diameter of the seven brooding species is 850 μm (Strathmann 1987). Outbreeding appears inevitable for the free-spawning species, and this conclusion is supported by genetic data (e.g., the crown-of-thorns sea star [Nishida and Lucas 1988]).

There are cases of clonal fission or autotomy of adults or larvae, hermaphroditism, and brooding scattered through each class (Hyman 1955; Emson and Wilkie 1980; Bosch, Rivkin, and Alexander 1989). The most extensive data are for tropical brittle stars (Emson, Mladenov, and Wilkie 1985; Hendler and Littman 1986; Hendler and Peck 1988; Mladenov and Emson 1988). Of thirty-seven reef-dwelling species from Belize whose reproductive mode is known, twenty-four have planktonic development, nine brood, and four are predominantly clonal species that probably rarely spawn gametes. Small, cryptic habitats, such as branching calcareous algae or the cavities of sponges, support abundant small brooding or clonal species. Clonal species typically build up large populations of sexually immature individuals. Moreover, when individuals with gonads are present, sex ratios are highly biased, presumably due to founder effects (e.g., only males of one species are known from Bermuda). Sexual reproduction, when it occurs, involves free spawning and planktonic development, so outbreeding seems inevitable. In contrast, brooders in the same habitats are hermaphroditic, possibly facultatively parthenogenetic (Hendler 1975), and commonly lack any planktonic stage. Given their cryptic habits and size, adult mobility is probably low, so inbreeding is expected; there are no genetic data.

Sea stars in the genus *Asterina* show a comparable diversity of reproductive modes (Komatsu et al. 1979). Some are gonochores with widely dispersing larvae, and at least one is primarily clonal (Achituv and Sher 1991), while others lack dispersing larvae and are thus potential candidates for inbreeding (Strathmann, Strathmann, and Emson 1984). *Asterina phylactica* is a small simultaneous hermaphrodite that broods its young. Isolated individuals in the laboratory produce viable offspring, and there are no apparent differences in early viability of young from aggregated versus solitary brooders in the field. The abundance of solitary brooding individuals in the field, plus the dominance of ovarian tissue in the gonad (see above) suggest that selfing commonly occurs (Strathmann, Strathmann, and Emson 1984). This species is sometimes found in abundance in large, deep, high-elevation pools on exposed coasts (Emson and Crump [1979] estimate at least ten thousand individuals in one pool), but pools could be initially colonized by a few individuals with generally limited movement between pools. Thus inbreeding may occur even in sizable aggregations. Both *A. phylactica* and the larger *A. gibbosa* (which does not self or brood but also lacks dispersing larvae) showed no enzyme variability at six scorable loci (Bullimore and Crump 1982). *Asterina minor* is another simultaneous hermaphrodite capable of selfing; it lacks dispersing larvae but does not brood and regularly breeds in aggregations whose members are of unknown genetic relatedness (Komatsu et al. 1979).

Inbreeding is also possible for the two small holothurians *Cucumaria lubrica* and *C. pseudocurata*, which occur in dense, stationary aggregations, sometimes exceeding $4,000/m^2$ (Rutherford 1973; Engstrom 1982; McEuen 1987). These species are dioecious, with a sex ratio of 1:1, and females brood eggs externally, with nearly all brooding at peak season. Juveniles are slow to leave their mother and presumably wander very little. Again, local populations are likely to be derived from small numbers of founders, but populations have not been analyzed genetically. Genetic studies of the infaunal brooding holothurian *Leptosynapta clarki* showed no evidence of selfing, close inbreeding, or fine-scale genetic differentiation, however (Hess et al. 1988).

Ascidiacea

Ascidians are usually small and sessile, and include both solitary and colonial forms, often within the same family. Most ascidians are simultaneous hermaphrodites and the few exceptions, including *Botryllus schlosseri*, are sequential hermaphrodites (Millar 1971; Kott 1974; Berrill 1975; Cloney 1987). All species of several large families or genera are self-fertile, but routine self-fertilization is not well documented (Ryland and Bishop 1990).

Colonial ascidians are viviparous, with an average period of embryonic development of 5 to 70 days within the atrium or body of the parent. After release, the large larvae may crawl, drift, or swim for a few minutes to 3 hours before settlement (van Duyl, Bak, and Sybesma 1981; Young 1986; Cloney 1987; Davis and Butler 1989). In *Podoclavella muluccensis* the average larva travels only 2.2 m, and at many sites densities are less than four colonies per square meter (Davis and Butler 1989). Inbreeding seems likely for species with larvae that swim little or not at all, but there are no genetic data except for *B. schlosseri* as noted above.

Most solitary ascidians spawn ova into the sea, where they develop into swimming tadpole larvae within 2 days, and subsequently swim free for a few hours to 6 days before settlement. Inbreeding is unlikely. However, two solitary species show a variety of characteristics that should promote inbreeding. *Corella inflata* is a "fugitive," shallow-water species that can self-fertilize eggs within the atrium; larvae are brooded there until they are ready to settle, often near their parents. There is no evidence of inbreeding depression, in contrast to the closely related, subtidal *C. willmeriana*, which has external fertilization, swimming larvae, and some self-sterility (Lambert 1968; Lambert, Lambert, and Abbott 1981). Many populations of *C. inflata*, however, have genetic characteristics indicating outcrossing, suggesting that self-fertilization occurs naturally only when potential mates are scarce (Cohen 1990). *Molgula pacifica* occurs in patchy aggregations of up to 50/m² (Young et al. 1988). Isolated individuals are completely self-fertile. Eggs are spawned but sticky and adhere to the bottom and to one another; development is entirely within the adhesive coat and without a larval stage. Juveniles hatch 36 hours after fertilization and settle near the parent. Nonbrooding solitary ascidians in the same habitat also build up large populations, but do so by aggregation of presumably unrelated pelagic larvae at settlement.

SUMMARY OF PATTERNS

The preceding review has been necessarily inferential, based as it is almost entirely upon comparative life history traits, particularly the mobility of all phases of the life cycle, and evidence for selfing. The species listed in table 10.2 are those whose life histories are most suggestive of inbreeding. Conclusive demonstration of inbreeding would require direct measurements and experiments (Ralls, Harvey, and Lyles 1986). Such information is unavailable except for *Botryllus schlosseri* (Grosberg and Quinn 1986; Grosberg 1987), and even those data are incomplete. Population genetic surveys are not on their own partic-

TABLE 10.2 Marine Invertebrate Species in Which Inbreeding Is Likely

Taxon	Characters	References
Demospongia[a]		
Stylopus sp. (orange)	*,SH,Fr,V,C	Ayling 1980
Stylopus sp. (pink)	*,SH,V,C	Ayling 1980
Anchinoe sp.	*,SH,V,C	Ayling 1980
Chondropsis sp.	*,SH,Fr,V,C	Ayling 1980
Polymastia robusta	*,O,M,C	Borojevic 1967
Polymastia granulosa	*,O,M,C	Bergquist, Sinclair, and Hogg 1970
Tetilla serica	*,O,M,C	Watanabe 1957, 1960
Tetilla cranium	*,V,C	Borojevic 1970; Bergquist 1978
Raspalia pumila	*,O,M,C,A,F	Lévi 1956; Borojevic 1970
Halichondria moorei	*,G/GH,V,C	Bergquist, Sinclair, and Hogg 1970
Hydroidea[b]		
Sertularia miurensis	*,G,V,B,C,F	Nishihira 1967, 1968
Tubularia larynx	*,G,V,B,C,Exp	Pyefinch and Downing 1949
Clava squamata	*,G,V,B,C,Al	Tardent 1963; Williams 1965, 1976
Plumularia alleni	*,V,B,M,C,Al,F	Williams 1976
Plumularia setacea	*,V,B,C	Williams 1976
Kirchenpaueria pinnata	*,V,B,C,Al	Williams 1976
Nemertesia antenina	*,G,V,B,C,M,Al	Williams 1976; Hughes 1977
Nemertesia ramosa	*,V,B,C,M,Al,F	Williams 1976
Eudendrium rameum	*,V,B,C,M	Wasserthal and Wasserthal 1973
Stylasterina		
Allopora californica	*,G,B,C	Ostarello 1973, 1976
Alcyonacea		
Alcyonium siderium	*,GH,V,B,C,Al,Exp	Sebens 1983 a,b,d
Xenia macrospiculata	*,G,Ma,Fr,V,B,C,M, Al,Exp	Benayahu and Loya 1984a,b,c; 1985
Gorgonacea		
Capnella gaboensis	*,G,Ma,Fr,O,B,M,C	Farrant 1985
Parerythropodium fulvum fulvum	*,G,Fr,O,B,M,C,Al	Benayahu and Loya 1983, 1985
Actinaria		
Epiactis prolifera	#,SH,O,B,C,Self	Dunn 1975a,b, 1977; Bucklin, Hedgecock, and Hand 1984
Epiactis japonica	#,SH,O,B,C	Uchida and Iwata 1954
Epiactis ritteri	#,SH,O,B,C	Strathmann 1987
Halcampa duodecimcirrata	#,G,O,C	Nyholm 1949
Cereus pedunculatus[c]	#,SH,V,C	Rossi 1975
Scleractinia		
Favia fragum	*,SH,V,B,Syp,C/S,A	Duerden 1902; Lewis 1974a,b; Szmant 1986
Manicina areolata	*/#,SH,V,B,C/S,F	Wilson 1888; Duerden 1902; Boschma 1929; Peters 1978

continued

TABLE 10.2 *(Continued)*

Taxon	Characters	References
Siderastrea radians	*/#,G,V,B,C/S,Al,F	Duerden 1902,1904
Stylophora pistillata	*,SH,V,B,Syc,C/S	Loya 1976a,b; Rinkevich and Loya 1979a,b
Goniastrea favulus[d]	*,SH,O,Syp,M,C/S,Self	Kojis and Quinn 1981; Heyward and Babcock 1986
Balanophyllia elegans	*,G,V,B,C	Gerrodette 1981; Fadlallah and Pearse 1982a; Fadlallah 1983b
Balanophyllia regia	*,G,V,B,C	Fadlallah 1983b; Tranter, Nicholson, and Kinchington 1982
Polychaeta		
Axiothella "rubrocincta"[e]	#,O,B,M,C	Strathmann 1987; Wilson 1983
Capitella capitata IIIa	#,B,C	Grassle and Grassle 1976, 1977, 1978
Sabella media	#,SH,O,B,C,Self	McEuen, Wu, and Chia 1983; Strathmann 1987
Fabricia limnicola	#,O,B,C	L. A. Levin 1984a
Fabricia sabella	#,G,B,C	Thorson 1946; Rasmussen 1973
Manayunkia aestuarina	#,SH,B,C	Muus 1967; S. S. Bell 1982
Spirorbis rupestris	*,SH,V,B,C	Gee 1963; Bailey 1969
Pileolaria pseudomilitaris	*,SH,V,B,C,Self,Exp	Beckwitt 1982; Fauchald 1983
Janua brasiliensis	*,SH,V,B,C,Self,Exp	Beckwitt 1982; Fauchald 1983
Prosobranchia		
Nucella emarginata	+,G,V,B,C,Aa,Cop	Palmer 1984
Nucella lapillus	+,G,V,B,C,Aa,Cop	Spight 1974; Crothers 1980, 1981
Littorina saxatillus	+,G,V,B,C,Cop	Janson and Ward 1984; Janson 1987a
Bivalvia		
Lasaea subviridis	#,SH,V,B,C,Self	Ó Foighil 1987
Lasaea rubra	#,SH,V,B,C,Self	Tyler-Walters and Crisp 1989
Polyplacophora		
Lepidochitona thomasi	#,G,O,B,C	Eernisse 1988
Lepidochitona caverna	#,SH,O,B,C/S,Self	Eernisse 1988
Lepidochitona fernaldi	#,SH,O,B,C,Self	Eernisse 1988
Isopoda		
Sphaeroma serratum	+,G,V,B,C,Cop	Bocquet, Lévi, and Teissier 1951; Bocquet, Lejuez, and Teissier 1960; Kaestner 1970
Jaera albifrons	+,G,Fe,V,B,C,Exp,Cop	Carvalho 1989
Excirolana braziliensis	+,G,V,B,C/S,Cop	Weinberg and Starczak 1988
Cyclostomata		
Tubulipora tuba	*,SH,V,B,C,Exp	C. McFadden, pers. comm.; Reed 1987b
Asteroidea		
Asterina phylactica	+,SH,O,B,C,Self	Strathmann, Strathmann, and Emson 1984

TABLE 10.2 (*Continued*)

Taxon	Characters	References
Holothuria		
Cucumaria lubrica	#,G,O,B,C	Engstrom 1982; McEuen 1987
Cucumaria pseudocurata	#,G,O,B,Syp,M,C	Rutherford 1973; McEuen 1987
Ophiuroidea[f]		
Amphioplus abditus	+,O,B,M,C	Hendler 1975
Axiognathus squamatus	+,SH,V,B,C	Hendler 1975; Emson,
		Mladenov, and Wilkie 1985
Ascidiacea		
Trididemnum solidum	*,SH,B,V,C/S	van Duyl, Bak, and Sybesma
		1981
Molgula pacifica	*,SH,O,B,M,C,Self	Young et al. 1988
Botryllus schlosseri	*,GH,V,B,C/	Grosberg and Quinn 1986;
	S?,Al,F,Exp	Grosberg 1987
Corella inflata[g]	*,SH,O,B,C/S?,Self	Lambert 1968; Lambert, Lambert,
		and Abbott 1981
Podoclavella muluccensis	*,B,C/S,Exp	Davis and Butler 1989

Note: This table includes sixty-eight species in which inbreeding is likely based upon life history characteristics. Species that produce larvae asexually and for which sexual larvae have not yet been demonstrated are not included. Symbols: *, sessile adults; #, sedentary adults; +, mobile adults; SH, simultaneous hermaphrodite; G, gonochore; GH, sequential hermaphrodite; Fe, sex ratio female-biased; Ma, sex ratio male-biased; Fr, fragments; P, polyembryony; V, viviparous; O, oviparous; B, broods; Syc, synchronous larval release within colony but not population; Syp, synchronous larval release within population; M, mucus tethering or adhesive packaging of larvae; C, crawling larvae; C/S, crawling or swimming larvae; Al, larvae aggregate with presumed kin; Aa, adults aggregate to copulate with presumed kin; F, larvae fuse with presumed kin; Exp, field experimental evidence for extremely philopatric recruitment; Self, self-fertilizing; Cop, copulates.

[a]Not including six brooding ceractinomorphs of Ayling (1980) for lack of additional information.
[b]All hydroids listed lack free medusae.
[c]Roscoff population.
[d]Genetic data for one population do not show evidence for routine inbreeding (Stoddart, Babcock, and Heyward 1988).
[e]False Bay population.
[f]Not including many brooding species of Emson, Mladenov, and Wilkie (1985), Hendler and Littman (1986), and Hendler and Peck (1988) due to lack of additional information.
[g]Genetic data (Cohen 1990) indicate that inbreeding may only occur in the field when potential mates are rare.

ularly useful for these purposes because such data are subject to widely varying interpretations: equally heterogeneous, fine-scale differentiation of populations may result over a few meters either from natural selection upon widespread recruits or from extreme philopatry and inbreeding (Burton and Feldman 1982; Burton 1983; Hedgecock 1982, 1986). Another important problem is how to evaluate anecdotal information on population densities and the potential for population bottlenecks in species with dispersal patterns suggestive of inbreeding. With all these limitations in mind, a number of patterns are clearly evident.

1. Life histories that almost certainly result in outbreeding are the norm in every major invertebrate phylum in the ocean. Nevertheless, the potential for inbreeding is just as widespread, with likely candidates in all phyla surveyed (see table 10.2). Species with traits conducive to inbreeding are scattered throughout many families and genera that are dominated by species with traits more conducive to outbreeding. Some higher taxa appear to include species with only one set of traits or the other, but we suspect these are the minority in most phyla. This taxonomic pattern is strikingly similar to that found for obligately clonal organisms (Hughes 1989).

2. The taxonomic distribution of species with life history traits conducive to inbreeding suggests that they are secondarily derived from taxa that were primitively exclusively outbred (Hyman 1940; Mackie 1974; Schroeder and Hermans 1975; Webber 1977; Bergquist 1978; Strathmann 1978; Sastry 1979; Boero 1984; Boero and Sarà 1987; McKinney and Jackson 1989; but see Chaffee and Lindberg 1986). The potential for inbreeding has apparently evolved independently many times within each phylum, but there is little evidence for transitions from inbreeding to outbreeding. Again, the same pattern holds for obligately versus facultatively clonal organisms (Hughes 1989).

3. Of the various mechanisms that could lead to inbreeding in marine invertebrates, very low mobility at *all* phases of the life cycle is the most common. Exclusively crawling larvae occur in 84% of the sixty-eight species in table 10.2, and nearly all of the remaining species produce at least some larvae that do not swim before settlement. Moreover, adults are sessile in 59% of the species, sedentary in 28%, and mobile in only 13%. This extreme philopatry must lead to a high incidence of mating among siblings.

4. In striking contrast to the high incidence of philopatry, demonstrated cases of self-fertilization are uncommon. Selfing is found in only thirteen of the sixty-eight species (19%) in table 10.2, despite the high incidence of simultaneous hermaphrodites among those species (54% of the fifty-four species whose mating systems are described). Moreover, experimental confirmations of self-fertilization by isolated individuals commonly result in some form of inbreeding depression (Potswald 1968; Sabbadin 1971; Beckwitt 1982; but see Strathmann, Strathmann, and Emson 1984; Grosberg 1987), even though these same species apparently mate routinely with close relatives. We are unaware of any well-substantiated case of obligate or routine self-fertilization among marine invertebrates, with the exception of the anemone *Epiac-*

tis prolifera (Bucklin, Hedgecock, and Hand 1984) and probably the clams *Lasaea subviridis* and *L. rubra* (Ó Foighil 1987; Tyler-Walters and Crisp 1989). In only three other species does the domination of gonads by ovarian tissue suggest that selfing is common (*Lepidochitona caverna, L. fernaldi* [Eernisse 1988], and *Asterina phylactica* [Strathmann, Strathmann, and Emson 1984]), although parthenogenesis and reciprocal spawning by pairs (see Sella 1990) remain alternative explanations in the absence of other data. We may have underestimated the prevalence of selfing because of inadequate data, but selfing is clearly not the major form of inbreeding in marine invertebrates (Ghiselin 1974: 118). Moreover, genetic data have not supported the hypothesis of regular selfing in several cases where it was predicted from laboratory studies of fertilization and development (Stoddart, Babcock, and Heyward 1988; Cohen 1990).

5. Species with life histories conducive to inbreeding are typically smaller than those unlikely to inbreed, even within the same taxonomic group. The sponges in table 10.2 are thin encrusting or small massive forms rather than large vases, cylinders, or ropes, all of which seem to have dispersed larvae (Bergquist 1978; Ayling 1980). Colonies of potentially inbreeding hydrozoans, alcyonarians, and scleractinian corals are also typically small compared with the species in these groups that dominate most coral reefs and broadcast gametes. Similar patterns occur in most sessile unitary animals. The primary contributor to this pattern is the well-known association between brooding and small size (Strathmann and Strathmann 1982; Strathmann 1990). Simultaneous hermaphroditism is also common among brooders (Strathmann, Strathmann, and Emson 1984), however, and brooding species that self tend to be smaller than closely related brooders with separate sexes (Emson and Crump 1979; Eernisse 1988).

6. Modular and unitary species are equally likely to inbreed (thirty-three versus thirty-five species respectively in table 10.2). However, effective population sizes are generally much lower for modular species, so the chances of inbreeding are much greater. Contrary to previous assertions (Jackson 1985, 1986), chances of strong inbreeding are not closely associated with clonal propagation, which is characteristic of only 7% of the species in table 10.2. Clonal species do have, on average, much shorter periods of larval dispersal than aclonal species (Jackson 1985, 1986; Strathmann 1990), and have other characteristics that reduce effective population size (see above). But the extreme larval philopatry that is probably necessary for strong inbreeding is most common among species that do not commonly undergo budding, fission, or fragmentation (but see below).

7. The environmental distributions of species likely to inbreed are different from those of species that are obviously outbred. Many of the animals in table 10.2 occur principally in intertidal or otherwise highly disturbed environments, including high rocky shores, tide pools, mud flats, reef flats, shallow lagoons, and estuaries (Duerden 1902; Lewis 1974a, 1974b; Bocquet and Lejuez 1974; Beckwitt 1980, 1982; Fauchald 1983; L. A. Levin 1984a; Burton 1986; Emson 1986). This trend can even be seen when comparing potentially inbreeding species with close relatives that inbreed to a lesser extent or not at all (e.g., *Corella inflata* versus *C. willmeriana* [Lambert, Lambert, and Abbott 1981], *Asterina phylactica* versus *A. gibbosa* [Emson and Crump 1979], *Ophryotrocha labronica* versus *O. diadema* [Åkesson 1977; Sella 1985] and the "morphs" of *Cereus pedunculatus* [Rossi 1975] and *Bunodactis verrucosa* [Schmidt 1967]). This environmental pattern is less apparent for many modular or clonal groups, particularly among the sponges, for which it may not apply at all (Bergquist, Sinclair, and Hogg 1970; Ayling 1980). Among reef corals, all the species in table 10.2 are characteristic of shallow subtidal or intertidal habitats, but subtidal brooding species have not been sufficiently studied for comparison. It is possible that the differences in habitat distributions between unitary and modular animals in table 10.2 merely reflect the characteristically greater relative proportions of unitary species in marginal environments (Jackson 1977, 1985) and a relative dearth of comparably detailed work in subtidal habitats (e.g., subtidal, brooding, clonal corals in the genus *Porites*). Nevertheless, Shields' (1982) statement that inbreeding marine invertebrates are characteristic of stable environments is incorrect.

EXPLANATIONS OF PATTERNS

The association between various life history traits and the probability of inbreeding or outbreeding is relatively straightforward. Much more difficult to disentangle are the arrows of causality in these correlations, because characteristics that increase or decrease inbreeding typically have other important selective consequences (D. Charlesworth and B. Charlesworth 1987). This is clearly true for a feature like dispersal, but most characters involved have complex associated costs and benefits. For example, fusion between close kin may prevent close inbreeding when combined with alternating sequential hermaphroditism (Grosberg and Quinn 1986), but may also maximize the benefits of increased size while decreasing the costs of intraspecific parasitism (Buss 1982; Grosberg 1988; Grosberg and Quinn 1988). Added to this is the problem of inadequate data to test alternative hypotheses regarding energetic or demographic trade-offs between alternative developmental and life history patterns (Strathmann 1985).

We consider in turn ecological, functional, and genetic constraints on the breeding systems and life histories of marine invertebrates. We use these terms loosely, realizing that the distinctions are somewhat arbitrary and that the explanations are unlikely to be mutually exclusive. Taxonomic constraints are clearly not an overwhelming factor because of the widespread co-occurrence of alternative life histories among closely related lower taxa within most major phyla. Although taxa certainly differ to some extent in their intrinsic potential for inbreeding (Strathmann 1985), we will not further explore this issue here.

Ecological Constraints

Several ecological arguments predict the observed association of extreme larval and adult philopatry with small adult size. First are a series of models analyzing the relative importance of risk reduction and spreading for large and small individuals (Strathmann 1974; Strathmann and Strathmann 1982). For example, small adults are likely to die sooner and have less total energy available for reproduction than large adults. Thus they cannot produce as many larvae over as long a period of time as can large organisms. Small adults may have the same mean success rate per larva as larger adults producing similar-sized propagules from the same location, but their variance in reproductive success will be necessarily greater. Species that brood typically have less variable recruitment patterns than do species that broadcast (Thorson 1946, 1950), so the hypothesis that brooding reduces variance in reproductive success seems reasonable. The selective advantage of reducing variance is particularly strong when intergenerational variance is involved (Gillespie 1973), as would be likely in small, short-lived species.

Second, the linkage between small size and direct development can be derived from models of the evolution of life cycles in marine invertebrates without considering variance in reproductive success (Roughgarden 1989). Whenever both planktonic habitat quality and benthic habitat quality decline simultaneously, Roughgarden's models predict that adults should become sexually mature at smaller sizes and larvae should develop benthically. The observed correlation between direct development and small adult size would therefore depend on a correlation between the quality of planktonic and benthic habitats. The models also assume that recruitment is limited by intraspecific crowding. We know of no environmental data that broadly support these assumptions.

Third, extremely short-distance dispersal in sessile organisms may permit one of the offspring to grow into space liberated by the death of the parent in situations where space is limiting (e.g., Sebens 1983b, 1986). The probability of parental death per unit of time is often much higher for smaller organisms (Hughes and Jackson 1985; Jackson 1985). Hence inbreeding may be associated with small size because offspring have a higher probability of inheriting parentally controlled space in small, short-lived species than in large, long-lived species (Willson and Burley 1983: 36).

Finally, the link between small size and philopatry may be explained in the context of the other major pattern observed, that of philopatry and highly disturbed habitats. There are three common solutions to the problem of guaranteeing the nearby presence of conspecifics for mating or for other beneficial aspects of group living (Buss 1981; Bertness and Grosholz 1985; Jackson 1985): larval philopatry, clonal propagation, and aggregative larval recruitment. Among serpulids and brittle stars, for example, rapid local buildups of populations can occur by fission, by brooding of sexually produced larvae that crawl a short distance from their parents, and by aggregative larval settlement (Wisely 1958, 1960; Gee 1963; Jackson 1977, 1985; Emson, Mladenov, and Wilkie 1985; Mladenov and Emson 1988). Solitary ascidians cannot propagate clonally, but they also employ aggregative settlement of widely dispersed larvae, or, more rarely, hatch directly as juveniles on substratum adjacent to or on their parents with similar effect (Young et al. 1988).

Which of these three alternatives is favored depends on the ecological and morphological characteristics of the species involved. Aggregative larval recruitment probably overwhelms the effects of either fragmentation or brooding when conditions are favorable. This option, however, is only reliable for species that are abundant, fecund, and widely distributed, because it depends on the regular or synchronized presence of larvae in the water column, and often entails substantial decreases in fecundity and increases in mortality due to excessive crowding. The second alternative, clonal propagation, requires a body plan compatible with fragmentation or budding. This automatically excludes most members of some taxa, for example, mollusks, crustaceans, and solitary ascidians, and is less likely for certain growth forms of modular species. In addition, one of the advantages of producing fragments rather than larvae is that the propagule begins its benthic existence at a relatively large size, and thus avoids the early vulnerable period associated with small size after larval settlement (Jackson 1985). However, if the adult is itself very small, this advantage associated with fragmentation may be lost because the survival of fragments drops sharply with decreasing fragment size (Hughes and Jackson 1985; Knowlton, Lang, and Keller 1989). Thus when adults are themselves small, the production of large, philopatric larvae that are capable of active microhabitat choice (Morse et al. 1988) may be favored over the production of very small fragments. This should be true whether larvae are sexually or asexually produced. Two species of *Goniastrea* illustrate some of these points. Both *G. aspera* and *G. favulus* are massive corals that do not regularly fragment but nevertheless exhibit clumped distributions. The former is more abundant and fecund, and has widely dispersed larvae that apparently aggregate at settlement, while the latter has restricted dispersal of eggs and larvae (Babcock 1984, 1991).

Ecologically, the above three mechanisms may all result in rapid establishment of high-density populations, albeit at potentially very different levels of parental investment in reproduction. The probability of encountering a potential mate will also be increased by each of these mechanisms, except for clonal species that are gonochores or have strict self-incompatibility systems. The evolutionary implications of these mechanisms are entirely different, however. Aggregative recruitment of widely dispersed pelagic larvae leads inevitably to outbreeding, and the production of a few highly philopatric larvae strongly favors extensive sib-sib or parent-offspring matings. Clonally propagating, colonizing species may or may not self-fertilize, depending on their mating system.

Functional Constraints

Strathmann and colleagues (Strathmann and Strathmann 1982; Strathmann, Strathmann, and Emson 1984; Strathmann 1985) argue that large size selects directly against brooding because of changes in fecundity and brood-space relationships with increasing parental size, which secondarily result in increased larval dispersal and outbreeding. Among unitary animals, large masses of eggs may be difficult to ventilate because of surface-to-volume constraints, so brooding should generally be limited to species with small adults. Such physiological constraints may explain why large animals should broadcast (or at least not brood) their eggs, but not why many, but not all, small animals should brood (e.g., contrasting patterns for the small, solitary corals *Balanophyllia*, *Caryophyllia*, and *Paracyathus* discussed previously). Some unitary taxa also fail to show the predicted allometric relationship between body weight and egg volume (H. Hess, personal communication, 1991).

Modular animals provide an important test of the functional argument. Brooding is usually carried out in discrete modules (polyps, zooids) which are typically arranged in two-dimensional arrays (Jackson 1977, 1979), so surface-to-volume constraints should not apply except for extremely large modules (discussed for colonial animals in Strathmann and Strathmann [1982]). Caribbean corals, for example, include abundant species with polyps smaller than 2 mm that brood (*Porites* and *Agaricia* spp.) or broadcast (*Acropora* spp.), and species with polyps 1 cm or larger that also brood (various Mussidae) or broadcast (*Montastrea cavernosa*) (Fadlallah 1983a; Szmant 1986; Morse et al. 1988). Two groups, the faviids and siderastriids, include among them four brooding species (median polyp diameter = 5 mm) and four broadcasters (median = 4 mm). In contrast to this overall similarity in polyp size, *colony* size of these brooding species (*Favia fragum, Manicina areolata, Siderastrea radians*, but not *Diploria labyrinthiformis*) is much smaller than that of their broadcasting relatives (*Montastrea annularis, M. cavernosa, Siderastrea siderea*, and *Diploria strigosa*). Similarly, in *Porites*, all of which have small polyps, species with small adult colony sizes brood, while larger forms are spawners (Szmant 1986; Richmond and Hunter 1990). This correlation of small colony size and brooding within scleractinian families and genera supports an ecological explanation, as discussed above (Babcock 1991). So also does the trivial difference in average polyp size between western Pacific and Caribbean corals (Coates and Jackson 1987), despite the 43-fold difference in the apparent proportions of brooding and broadcasting species in the two regions (Richmond and Hunter 1990).

Genetic Constraints

Many analyses of breeding systems have considered their genetic implications (e.g., Williams 1975; Maynard Smith 1977, 1978; Lloyd 1979; Charlesworth 1980a; D. Charlesworth and B. Charlesworth 1987; Shields 1982, 1987; Lande and Schemske 1985; Schemske and Lande 1985; Campbell 1986; Templeton 1986; Uyenoyama 1986; Holsinger 1988a, 1988b; Hamilton, chapter 18, this volume). Outbreeding is thought to be favored when (1) populations contain large numbers of deleterious recessives or loci that show heterosis, (2) individuals are vulnerable to genetic arms races with parasites and pathogens, and (3) sib-sib competition favors genetic variability. On the other hand, factors favoring inbreeding include (1) the so-called twofold cost of sex, (2) the cost of producing males, (3) coadapted gene complexes, and (4) local adaptations to spatially varying environments.

The patterns we have documented can be plausibly derived from purely genetic considerations. First, selective pressures from parasites might be higher in stable environments, where predictable and adequate abundance of hosts facilitates the persistence of parasites. Long-lived organisms should also be more vulnerable to the effects of arms races with short-lived parasites. The tendency for inbreeding to occur primarily in weedy and small organisms would be a logical consequence of these arguments (Hamilton, chapter 18, this volume). Second, species that colonize new areas with a very small number of propagules would tend to lose deleterious recessives during these population bottlenecks (Lande and Schemske 1985). This would also tend to produce an association between some forms of weediness and inbreeding. Third, large organisms that spawn both eggs and sperm are more likely to suffer from low to moderate levels of accidental selfing. Adults are unable to control the movements of gametes once spawned, and the dispersal of gametes prior to fertilization will be limited relative to the size of the adult. This would result in increased selection for mechanisms to prevent gamete wastage caused by the low viability of selfed offspring. Selfing would be reduced before loss of deleterious recessives occurs, thus preserving outbreeding (Maynard Smith 1978: 136). Finally, large organisms are often more fecund, which could potentially lead to greater intensity of sib-sib competition and thus favor outbreeding (Shields 1982). This last argument is weaker, however, because the extensive larval dispersal characteristic of most outbred marine invertebrates makes sib-sib competition very unlikely (Maynard Smith 1977; Jackson and Coates 1986).

Genetic mechanisms may also strengthen patterns of inbreeding and outbreeding generated by ecological and functional constraints through disruptive selection and positive feedback loops built into the selective processes (Lloyd 1979; Strathmann, Strathmann, and Emson 1984; Waller 1984; Lande and Schemske 1985). Outbred species tend to maintain large numbers of deleterious recessives and heterotic associations that produce substantial inbreeding depression and selective pressures to maintain outbreeding. However, once enforced inbreeding is established through population bottlenecks or low dispersal, then coadapted gene complexes and local adaptation lead to outbreeding depression and increased benefits associated with inbreeding. The models of Lande and Schemske (1985) thus predict that inbreeding and outbreeding are alternative stable strategies. The interaction of these opposing pressures with mating systems can be complex, however, even when only genetic factors are considered (Campbell 1986; Uyenoyama 1986; Holsinger 1988b). Neither can one neglect the costs of the mechanisms required to achieve the genetically optimal level of inbreeding or outbreeding (e.g., dispersal in models by Bengtsson [1978] and Waser, Austad, and Keane [1986]). These and other factors undoubtedly help to break the positive feedback loops leading to higher levels of inbreeding or outbreeding.

Several types of genetic arguments also support the observation that the transition from outbreeding to inbreeding should be more common than the reverse. First, whenever inbreeding occurs through reduced larval dispersal, it may be argued that the loss of a specialized planktonic form is easier to achieve than its creation once lost (Strathmann 1978, 1985). Second, the loss of deleterious recessives during a population bottleneck is a much more rapid process than their accumulation during intervals of large population size (Nei, Maruyama, and Chakraborty 1975). Third, a dioecious or self-incompatible species will lose deleterious recessives during a population bottleneck, but species with nonrandom inbreeding will not reacquire deleterious recessives when large population sizes are reached (Lande and Schemske 1985).

PLANTS VERSUS SESSILE MARINE INVERTEBRATES

In both plants and sessile marine invertebrates, all genetic movement occurs in the gamete (pollen/egg and sperm) and young (seed/larva) stages, or by clonal propagation, suggesting natural parallels between the two groups. The summary of patterns of inbreeding in marine invertebrates was compiled without reference to the extensive literature for plants on inbreeding and outbreeding. When this literature was consulted, we found striking similarities and intriguing differences (see also Strathmann 1990).

Similarities

1.   Outbreeding in plants tends to be the norm, but inbreeding has repeatedly evolved in a variety of taxa. Inbreeding in plants is a secondarily derived character, and the transition from outbreeding to inbreeding appears to be more common than the reverse (Stebbins 1950, 1957; Jain 1976).

2. Plants show a strong relationship between adult size and the probability of inbreeding. Trees are notable in being more likely than herbs to be dioecious or self-incompatible (Bawa 1974; Bawa and Opler 1975) and to have seeds that are carried by animals (Bawa 1980) and thus are probably more widely dispersed (Howe and Westley 1986).

3. Inbred plants are often ecologically weedy in the sense of being characteristic of highly disturbed environments (Stebbins 1950; Baker 1974).

4. Cloning plants do not tend to be strong inbreeders (Stebbins 1950). Indeed, Baker's (1974) alternatives of selfing in annual and clon-

ing in perennial weeds are strikingly parallel to the alternatives we present for marine invertebrates of larval philopatry, clonal propagation, and aggregative settlement by dispersed larvae (the latter not being an option for plants except when animals disperse seeds to distant but spatially restricted sites).

Differences

1. Inbreeding is more extreme in plants than in sessile marine invertebrates. In contrast to marine invertebrates, routine selfing is common in plants; discussions in the plant literature of inbreeding typically refer primarily to selfing (e.g., Jain 1976). Sib and parent-offspring matings are also common in plants (Levin 1986), although comparing the relative frequency of such matings in plants and marine invertebrates is not currently possible. Some plants show both substantial selfing and mating with close relatives (Waller and Knight 1989).

2. Inbreeding in plants is most conspicuously achieved by changes in the mating system, while in sessile marine invertebrates it is the potential for movement, particularly in the larval stages, that is most strikingly affected. In each group there are correlated shifts in mating and dispersal systems, however (see above).

3. Inbreeding is probably more common in plants than in sessile marine invertebrates. Allard (1975) estimated that one-third of all plant species are regularly selfing and thus highly inbred. It seems unlikely that one-third of sessile marine invertebrates are highly inbred, although the scarcity of data makes it difficult to estimate the frequency with any accuracy.

4. At the gamete stage, dispersal is generally more limited for marine invertebrates than for plants. In plants, the transport of pollen by animals and the ability of pollen to move from plant to plant increases the potential for long-distance dispersal in many species. Plant breeders are routinely concerned about unwanted fertilizations from dispersal of pollen over distances of 300 m or more (Levin 1986), while the limited studies of sperm and egg transport prior to fertilization in sessile marine invertebrates (reviewed above) suggest much shorter distances.

5. At the seed or larval stage, marine invertebrates have a much greater potential for dispersal. That larvae can profitably grow while they move, whereas seeds cannot, probably explains why the dustlike seeds of most orchids (Dressler 1981: 12) are the exception while planktotrophic larvae are the rule (Roughgarden 1989; Strathmann 1990). The tail of dispersal distributions is probably longer for marine invertebrates than for plants, even when most recruitment is local, because of the buoyancy provided by water. Dispersing propagules tend to be

expensive in plants because they must attach to or attract the interest of animals, whereas in marine invertebrates, dispersing planktotrophic larvae are cheap to produce compared with those that are immediately ready to settle (Strathmann 1990).

6. At the adult stage, sessile marine invertebrates are more mobile than plants, because marine invertebrates do not need to be attached to acquire nutrients and because water provides a much better medium than air for the rafting and floating of large organisms. Thus marine invertebrate colonists are more likely to arrive in a form that permits sexual reproduction without selfing (small groups of adults or an already fertilized adult) than are plants, which typically arrive as seeds and are often dependent on animals to carry their pollen.

CONCLUSIONS

In this review of inbreeding in marine invertebrates we have found that characteristics suggesting inbreeding are not uncommon, that there is little evidence for taxonomic constraint on the traits affecting the potential for inbreeding, and that there are morphological and environmental correlates of inbreeding in marine invertebrates that parallel those found in plants. Shields' (1982) claim that ecological arguments could not be consistently applied to both plants and marine invertebrates has not been upheld; more detailed review of the ecological characters associated with inbreeding in the latter group has shown that as in plants, inbreeding is commonly associated with unstable habitats.

A variety of ecological, functional, and genetic models make predictions compatible with the patterns observed. Whenever different models make the same general predictions, it is difficult to dissect their relative importance, particularly when derivative hypotheses are not equally amenable to testing. Moreover, many processes may contribute to the patterns, and the importance of various processes may vary among groups. Detailed studies of closely related species or populations with differing degrees of inbreeding offer the best hope of addressing the problem (May 1979). The many complexes of sibling species in marine invertebrates remain largely unexplored from this perspective. Comparative studies should include direct and indirect measures of mating patterns and dispersal, studies of inbreeding and outbreeding depression, and descriptions of ecological and morphological differences. Comparisons of the species listed in table 10.2 with more outbred close relatives (where they exist) is a logical starting place, and we pointed to a few examples above. Conspecific populations differing in inbreeding potential could also be examined.

While awaiting this information, we offer a few tentative conclusions based on the data at hand. In marine invertebrates, the primary characteristics that influence the potential for inbreeding or outbreeding are adult and larval mobility. These features are such fundamental components of ecology and life history that it seems parsimonious to assume that genetic systems initially respond to a level of inbreeding dictated by ecological and functional constraints. This conclusion is supported by the prevalence of inbreeding in plants, with their intrinsic limits on mobility, and the general rarity of close inbreeding in the more mobile vertebrates. Also supporting this interpretation is the fact that both plants and sessile marine invertebrates exhibit various solutions to the problem of colonization that have drastically different implications for inbreeding. In sum, at least on a coarse scale, inbreeding or outbreeding seems more likely to arise due to selection in response to a specific ecological situation rather than as the result of selection for inbreeding or outbreeding per se (see also Grosberg 1987).

Once the basic potential for inbreeding and outbreeding has been determined, however, intrinsic genetic factors appear to have some effect. This is seen most clearly when all members of a group have limited dispersal but only the smallest and weediest species show evidence of selfing. The same process may influence dispersal itself (i.e., general philopatry with selection for extreme philopatry in the smallest, weediest forms), but the data are generally not adequate to detect such subtle differences in dispersal potential.

Note added in proof: Recent evidence suggests that brooding bivalves in the genus *Lasaea* are parthenogenetic rather than self-fertilizing (Ó Foighil and Thiriot-Quiévreux 1991).

ACKNOWLEDGEMENTS

We thank S. Cohen, P. M. Gaffney, R. K. Grosberg, H. Hess, R. N. Hughes, E. G. Leigh, Jr., H. A. Lessios, D. R. Levitan, B. Okamura, R. R. Strathmann, N. W. Thornhill, W. H. Wilson, and P. O. Yund for helpful comments and sharing of unpublished work. We could not have written this chapter without the resources and generous assistance of the staff of the Smithsonian's libraries.

11

Inbreeding and Outbreeding in Fishes, Amphibians, and Reptiles

Bruce Waldman and Jeffrey S. McKinnon

The extent to which individuals mate with their kin is determined proximately by the characteristics of their population structure and by behavioral traits that promote dispersal and mate choice. Asymmetrical patterns of dispersal by sex and kin recognition may effectively act as mechanisms that discourage matings between close relatives. Beyond inbreeding avoidance, the possibility that organisms can achieve an optimal balance between inbreeding and outbreeding has generated much discussion (Shields 1982; Bateson 1983). Supporting data, however, are few or ambiguous, and the area remains controversial (e.g., Bateson 1983; Ralls, Harvey, and Lyles 1986; Shields, chapter 8, this volume). Birds and mammals have been most thoroughly studied (e.g., Pusey 1987; Blouin and Blouin 1988) while other vertebrate groups have been largely ignored. In this chapter, we examine factors that influence the propensity of ectothermic vertebrates—fishes, amphibians, and reptiles—to inbreed and outbreed, and review data on the consequences of those reproductive patterns.

First, we survey studies of population structure to determine how frequently ectothermic vertebrates have the opportunity to mate incestuously (i.e., with relatives as close as cousins). In the absence of such opportunities, incest avoidance behaviors are unlikely to evolve. Second, we ask whether incestuous matings result in inbreeding depression, a decline in offspring fitness. Measurable inbreeding depression should result in selection for incest avoidance mechanisms. Third, we review evidence that fishes, amphibians, and reptiles can recognize their kin. Kin recognition mechanisms might promote inbreeding avoidance or optimal outbreeding, and we examine these possibilities for fishes, amphibians, and reptiles. Finally, we summarize our recent research on the American toad (*Bufo americanus*) to show how molecular methods can be used to assess levels of inbreeding and outbreeding in natural populations.

POPULATION STRUCTURE

Incest can occur only if relatives of opposite sex overlap in time and space when reproductively active. Patterns of dispersal are important determinants of such opportunities, as are other aspects of population structure.

Philopatry and Site Fidelity

In numerous studies, experimentally displaced fishes, amphibians, and reptiles reliably orient toward, and frequently return to, locations from which they were initially collected (e.g., fishes: Kynard 1978; Werner 1979; Horrall 1981; frogs: Brattstrom 1962; Oldham 1967; Dole 1972; Grubb 1973; McVey et al. 1981; Crump 1986; snakes: Hirth 1966; turtles: Nichols 1939; Cagle 1944; Emlen 1969; Ernst 1970; Loncke and Obbard 1977; Chelazzi and Francisci 1979; DeRosa and Taylor 1980; lizards: Weintraub 1970; Krekorian 1977; Strijbosch, van Rooy, and Voesenek 1983; Ellis-Quinn and Simon 1989; crocodilians: Murphy 1981; Webb, Buckworth, and Manolis 1983; Rodda 1984). Some frogs migrate year after year to particular sites to breed (Bogert 1947; Jameson 1957; reviewed in Sinsch 1990), even if the ponds are paved over into parking lots (Heusser 1969). Site fidelity has also been noted in some fishes (e.g., Horrall 1981; L'Abée-Lund and Vøllestad 1985; Crossman 1990).

These findings are often interpreted as evidence of natal philopatry, but in most studies, subjects are initially collected as adults. Thus the failure of juveniles to disperse from natal areas clearly cannot be assumed. This is illustrated by Gill's finding (1978) that adult red-spotted newts (*Notophthalmus viridescens*) were remarkably site-specific, and when displaced, homed with great precision to the pond from which they had been originally captured. Yet each year large numbers of individuals immigrated into each study population, leading Gill (1978) to speculate that opportunities for inbreeding would be limited. Merritt et al. (1984) attribute high levels of genetic similarity between populations of this species to dispersal of the terrestrial eft stage.

Although opportunities for larval dispersal in amphibians may be quite limited, after metamorphosis many appear to demonstrate genuine philopatry. Blair (1953) followed Gulf Coast toads (*Bufo valliceps*) from juvenile stages to sexual maturity and found that very few dispersed. Some newly metamorphosed leopard frogs (*Rana pipiens*) studied by Dole (1971) settled near their natal pond, but others dispersed as far as 5 km away. Red-bellied newts (*Taricha rivularis*) home to specific areas of their natal streams; Twitty (1961) found that only 4 of 73 individuals dispersed from the area of a stream where they had metamorphosed. Of 25,500 marked Fowler's toads (*Bufo woodhousei fowleri*) released at metamorphosis, Breden (1987, 1988) recaptured 37 individuals as breeding adults. The majority bred 2 years later in their natal ponds, and most migrants were found in adjacent ponds. But a minority apparently dispersed over longer distances, at least 2 km away. Juveniles disperse in larger numbers and possibly over longer distances than adults (Breden 1987, 1988). Studies on populations of the wood frog (*Rana sylvatica*) reveal similar evidence of natal philopatry by most individuals, but a minority regularly disperse (Berven and Grudzien 1990). Based on recaptures of 4,000 of 25,000 individuals marked at metamorphosis, fully 75% bred in their natal ponds, and while most of the remaining individuals dispersed to nearby ponds, Berven and Grudzien found some individuals mating up to 2.5 km from their natal sites.

Studies of marked populations of reptiles often reveal that individuals remain in, or return to, specific localities from season to season. In most cases, patterns of juvenile dispersal remain unknown. Despite the tendencies of green iguanas (*Iguana iguana*) to migrate over considerable distances between seasons, females usually return to the same site to nest year after year (Bock, Rand, and Burghardt 1985). Juvenile *Anolis limifrons* lizards disperse on average less than 3 m, and when they mature, their adult home ranges extensively overlap those that they used as juveniles (Andrews and Rand 1983). Yet long-range dispersal by at least some juvenile *Uta stansburiana* lizards is suggested by Spoecker's (1967) observation that while most adults move only short distances, some juveniles travel over much greater distances. Juvenile male leopard lizards (*Crotaphytus wislizeni*) also sometimes disperse over long distances (Parker and Pianka 1976). Although Bull (1987) recaptured 92% of his marked population of *Trachydosaurus rugosus* lizards within 100 m of their original capture site, some individuals apparently migrated over much longer distances. In a 38-year study, Stickel (1989) found that most box turtles (*Terrapene c. carolina*) maintain constant home ranges from year to year, and that individuals usually mate with partners whose ranges overlap with their own (Stickel 1989). Other studies, however, reveal *T. carolina* populations with small numbers of individuals that disperse over long distances (Kiester, Schwartz, and Schwartz 1982). Most snapping turtles (*Chelydra serpentina*) remain in the same home ranges from one year to the next (Obbard and Brooks 1981). Some snakes may return to specific trees or dens to nest each year (e.g., green snakes, *Opheodrys aestivus;* Plummer 1990; also see review by Gregory 1984). Nonetheless, the cryptic appearance and secretive habits of many amphibians and reptiles make precise tracking impractical (see discussion by Macartney, Gregory, and Larsen 1988). Emigration is especially difficult to study, because the disappearance of a study subject may indicate its death.

Marine turtles often forage hundreds or thousands of kilometers from their nesting beaches. Every 2 to 3 years, green turtles (*Chelonia mydas*) return to the same beach to nest, though repeat nesting sites may be up to 1.2 km apart (e.g., Carr and Carr 1972; Carr 1975). Whether this site fixity represents philopatry is unknown, as adults rather than young are tagged. Tagging hatchlings is not a viable option, because few if any would retain their tags through the 30 or more years to sexual maturation (Carr 1967). However, recent molecular surveys of the mitochondrial DNA of green turtle colonies reveal significant differences among colonies, consistent with the hypothesis that females home to natal sites to breed (Meylan, Bowen, and Avise 1990). Other sea turtles show similar degrees of site fidelity (e.g., leatherbacks, *Dermochelys coriacea;* Eckert et al. 1989).

Indirect evidence of natal philopatry is available for several fishes (e.g., Hagen 1967; Aalto and Newsome 1990), but natal homing has been documented best in the salmonids (reviewed by Stabell 1984; Quinn 1984, 1985; Quinn and Tallman 1987). At Waddell Creek, California, for example, 85.1% of coho salmon (*Oncorhynchus kisutch*) marked in the creek as juveniles, prior to dispersal, returned there to spawn (Shapovalov and Taft 1954). Because the population is small (between 37 and 309 breeding females over 6 years), and mortality of eggs and fry varies greatly among nests (T. Quinn, personal communication), incestuous matings are very likely. Homing rates in other salmonids are frequently even higher, commonly 98% or 99% in several *Oncorhynchus* species (Quinn and Tallman 1987). The natal homing of salmonids is often reflected in genetic differences between populations from different drainages and rivers, or even between populations breeding in a single river or lake (e.g., Ryman 1983; Ståhl 1987; Foote, Wood, and Withler 1989; Ferguson and Taggart 1991).

Pelagic fishes, with planktonic eggs and larvae, typically disperse zygotes widely over vast distances (e.g., Barlow 1981; but see Marliave 1986; Shapiro, Hensley, and Appeldoorn 1988). At sexual maturity, individuals thus should rarely encounter relatives, but incestuous matings might be possible if siblings eventually settle at the same location or stay together through maturity (Shapiro 1983). Not surprisingly, available data fail to support this view. Larvae of the coral reef fish *An-*

thias squamipinnis, for example, settle in groups comprising progeny randomly mixed from many matings (Avise and Shapiro 1986).

In contrast, many fishes are territorial, have demersal eggs that may be guarded by parents, and live in relatively localized, fragmented populations with population sizes as small as 100 (FAO/UNEP 1981; Altukhov 1982). Because gene flow between populations is limited, opportunities for close inbreeding will be more common in such species. Population characteristics of fishes such as the coral reef damselfish *Acanthochromis polyacanthus* (Robertson 1973) and the threespine stickleback, *Gasterosteus aculeatus* (e.g., Hagen 1967) suggest that close inbreeding may occur. Rather than dispersing, offspring of the Lake Tanganyika cichlid *Lamprologus brichardi* initially assist their parents both in brood care and territory defense (Taborsky and Limberger 1981). Helpers may remain with the mated pair past reproductive maturity in field populations, and in the laboratory both male and female helpers have been observed to mate with, or at least try to mate with, the resident pair (Taborsky 1985, 1987).

Sexual Asymmetries and Life Histories

Should males and females differ in their propensity to disperse, disperse over different distances, or disperse in different directions, the possibility of incestuous matings would be reduced. Unfortunately, these possibilities have been largely overlooked in studies of ectothermic vertebrates. Indeed, the methods used to assess movement patterns sometimes mask any differences that exist between the sexes. Studies of home range and dispersal often track either males or females, or lump together data from both sexes. For example, the dramatic accounts of site fidelity in green turtles (*Chelonia mydas*) are based on observations only of nesting females (Carr and Carr 1972; Mortimer and Carr 1987). Often, so few individuals are recaptured in studies of dispersal that sample sizes are insufficient to test for sex differences (e.g., Breden 1987).

Asymmetrical dispersal of the two sexes has been reported in a few studies, although overall patterns are not yet apparent in these groups. In a detailed 8-year study of slider turtles (*Pseudemys scripta*), Parker (1984) found that males emigrated more frequently and farther than females. Data on immigration into Parker's study populations also suggest that females show much higher degrees of philopatry than males. In contrast, female painted turtles (*Chrysemys picta*) apparently disperse farther and more frequently than males (Gibbons 1968). Male *Basiliscus basiliscus* lizards disperse more frequently and farther than females, but some young females disperse over short distances (Van

Devender 1982). Male red-cheeked salamanders (*Plethodon jordani*) move greater distances and appear to have more acute homing abilities than females (Madison 1969). Brown (1985) found that adult male mosquitofish (*Gambusia affinis*) dispersed more often than adult females, but at the same location, Robbins, Hartman, and Smith (1987) observed that adult females dispersed more frequently than adult males. Molecular analyses of the population structure of striped bass (*Morone saxatilis*) suggest male dispersal and female philopatry (Chapman 1989, 1990), and similar dimorphisms may be characteristic of other fishes (e.g., walleye, *Stizostedion vitreum;* Ward, Billington, and Hebert 1989). In their asymmetry, the dispersal patterns of these ectotherms thus appear similar to those typical of other vertebrates (Greenwood 1983), and they may promote outbreeding. Whether instances of asymmetrical dispersal represent exceptions in these taxa remains to be seen. Detailed studies of some populations indicate strong natal philopatry both of males and females (e.g., sockeye salmon, *Oncorhynchus nerka*, Foerster 1936; wood frogs, *Rana sylvatica*, Berven and Grudzien 1990; K. Berven, personal communication).

Even when both sexes are philopatric, opportunities for incestuous matings may not exist. Males and females of many species become sexually mature at different ages, and if mortality is high between breeding seasons, individuals are unlikely to encounter their siblings as potential mates. Twitty, Grant, and Anderson (1964) found that male newts (*Taricha rivularis*) homed with a high degree of precision, but females returned to presumed natal areas in large numbers only 2 or 3 years after their initial capture as adults. These results imply that although some males breed every year, females breed only at intervals of 2 to 3 years. Asymmetrical sexual maturation is characteristic of many fishes, amphibians, and reptiles. Typically, males first breed at an earlier age than females (Porter 1972), but in some species, both sexes mature at the same age (e.g., Breden 1988).

The timing of reproduction with respect to dispersal is often a key determinant of the likelihood of incest. For example, males of the viviparous dwarf surf perch (*Micrometrus minimus*) are born sexually mature. Females can mate shortly after birth. Unless offspring disperse immediately following birth (a behavior not noted in this species), opportunities for matings between siblings may be common (Warner and Harlan 1982). In addition, males potentially can mate with their mothers, though observed patterns of size-assortative courtship suggest that such matings are rare (Warner and Harlan 1982).

In species with extended or multiple breeding seasons, related individuals may breed more synchronously than nonrelatives, thus increasing the likelihood of incestuous matings. This possibility has received considerable attention in salmonids. Pink salmon (*Oncorhynchus gorbuscha*) are semelparous and have rigid 2-year life spans, so that even-year and odd-year populations do not interbreed (Bilton and Ricker 1965). Allozyme data suggest that pink salmon stocks differ more between even- and odd-year populations than between locations within years (Beacham, Withler, and Gould 1985). Similar results have been obtained for several species of salmonids with races that spawn in different seasons of the same years (e.g., Kristiansson and McIntyre 1976; Child 1984; Wilmot and Burger 1985). Allele frequencies differ significantly among samples of rainbow trout (*Oncorhynchus mykiss*) taken over the course of a single season (Leary, Allendorf, and Knudsen 1989). Similar findings have been reported for chum salmon (*Oncorhynchus keta*; Altukhov 1981).

Dispersal tendencies and the frequency of matings between close relatives determine, to a large extent, the level of genetic variation within populations. Surveying the literature on reptiles, Gorman, Kim, and Taylor (1977) noted a positive correlation between genetic heterozygosity and movement patterns. Genetically variable species are more mobile and show stronger dispersal tendencies than less variable species. Weak positive correlations between genetic heterozygosity and both dispersal of young and adult mobility also are apparent in comparative studies of amphibians (Nevo and Beiles 1991). Nonterritorial, mobile species presumably have a wider choice of mates than do polygynous, territorial species that move very limited distances during their lifetimes (Gorman, Kim, and Taylor 1977). As individuals disperse, effective population sizes increase, and inbreeding coefficients decrease. Within species, however, if the most heterozygous individuals show the greatest tendencies to disperse, the effective population size of the remaining home population might actually decrease, and the frequency of inbreeding increase.

Effective Population Size

In principle, any factor that tends to decrease effective population size, N_e, is likely ultimately to increase opportunities for incest. Differences in the effective numbers of males and females in a population, due either to differential mortality by sex or to mating systems in which male mating success is strongly skewed, decrease N_e (Lande and Barrowclough 1987). Such mating systems are commonly observed in ectothermic vertebrates (e.g., pupfish, *Cyprinodon* sp.; Kodric-Brown 1977; the tungara frog, *Physalaemus pustulosus*; Ryan 1985). A few dominant males probably obtain the majority of matings in populations of the sailfin molly (*Poecilia latipinna*), which in theory should reduce N_e, and consequently heterozygosity. As expected, heterozygosity is correlated, across populations, with the ratio of dominant males to adult females (Simanek 1978).

N_e is also reduced when variance in family size (i.e., number of an individual's progeny that breed) is large (Falconer 1989), or when populations fluctuate greatly in size or are composed of ephemeral subpopulations (Lande and Barrowclough 1987). Family size typically varies more in fishes, amphibians, and reptiles than in endothermic vertebrates, due to the combined effects of indeterminate growth and a correlation between fecundity and body size (Nelson and Soulé 1987). Large numbers of amphibians breed in ephemeral habitats, and consequently populations tend to be highly inbred with low levels of heterozygosity (e.g., Reh and Seitz 1990). Low levels of heterozygosity in the pool frog (*Rana lessonae*) result from drastic fluctuations in population size caused by reproductive failure in cold years (Sjögren 1991). Over 10 years, population densities of *Anolis limifrons* lizards varied almost sixfold in a tropical rainforest environment, considered by many to be among the most stable habitats (Andrews and Rand 1982); populations fluctuated in size from season to season as well as among years (Andrews, Rand, and Guerro 1983). Similarly, populations of the Mexican topminnow (*Poeciliopsis monacha*) undergo large seasonal fluctuations. When a local extinction occurs, a new population may be founded by a single pregnant female (Vrijenhoek 1985a). Many other fishes possess similar population structures (e.g., mosquitofish, *Gambusia affinis*; Smith, Smith, and Chesser 1983, Smith et al. 1989).

Taking into account these factors, many investigators have computed N_e estimates for reptiles and amphibians. Several methods have been used, and the reliability of some data are unclear. Notwithstanding these limitations, most estimates yield small to moderate effective population sizes. Dispersal of the side-blotched lizard (*Uta stansburiana*) is quite limited, leading to small effective population sizes ($N_e = 14$; Tinkle, 1965, 1967). Rusty lizards (*Sceloporus olivaceus*) disperse somewhat longer distances as juveniles, have larger home ranges, and unlike the monogamous *Uta stansburiana*, they have a polygynous mating system and high levels of immigration (Blair 1960). Estimates of effective population sizes (neighborhood sizes) range from 225 to 270 (Kerster 1964). Gill (1978) estimated effective population sizes of *Notophthalmus viridescens* newts at between 12 and 150 at most breeding sites, although in some ponds, estimates of N_e ranged to 1,100. Immigration was probably sufficient to maintain low levels of inbreeding (Gill 1978; Merritt et al. 1984). Effective population sizes of cricket frogs

(*Acris crepitans*) generally range from 27 to 107, although some populations may be larger (Gray 1984). Breden (1987) estimated N_e values ranging between 38 and 152 for breeding populations of Fowler's toad (*Bufo woodhousei fowleri*), while Berven and Grudzien (1990) computed estimates of from 50 to 178 for the wood frog (*Rana sylvatica*). At the periphery of their range, pool frog (*Rana lessonae*) populations have an N_e of approximately 35 (Sjögren 1991). For populations of the leopard frog (*Rana pipiens*), values range from 2 to 112, with most being in the lower range (Merrell 1968).

Estimates based on levels of allozyme variation often yield higher, possibly inflated, values, reflecting historic levels of genetic variation in founding populations. Taylor and Gorman (1975) computed N_e for a population of *Anolis grahami* to be 641. Easteal's (1985, 1986) estimates of N_e for introduced populations of *Bufo marinus* range from 390 to 460. Estimates may be high because they fail to take into account variation in male reproductive success and in offspring viability (Easteal and Floyd 1986). Even higher estimates of effective population sizes (on the order of 10^3) have been generated, based on biochemical data, for populations of some salamanders (Larson, Wake, and Yanev 1984). Often, though, substantial genetic variation apparent on a microgeographic scale within populations suggests the presence of multiple breeding units together that should be considered separate. Although red-bellied newts (*Taricha rivularis*) are present in large populations, their tendencies to return to portions of streams in which they metamorphosed increase the likelihood that they will encounter close relatives as potential mates (Hedgecock 1978).

The accurate determination of N_e for fish populations has proven especially difficult. For populations of sockeye salmon (*Oncorhynchus nerka*), Altukhov (1981) computed average N_e values of 206. Very approximate values for N_e of 22,900 and 1,800 have been calculated for populations of sea trout (*Cynoscion nebulosus*) and red drum (*Sciaenops ocellatus*) respectively (Ramsey and Wakeman 1987). Absolute population sizes can number from tens to hundreds for a variety of fishes, such as coho salmon (*Oncorhynchus kisutch*; Shapovalov and Taft 1954), the Devil's Hole pupfish (*Cyprinidon diabolis*; Ono, Williams, and Wagner 1983), and the guppy (*Poecilia reticulata*; Haskins et al. 1961). While N_e estimates are unavailable for these species, they will certainly be smaller than the absolute population sizes. Of course, some notable exceptions exist. At one extreme, any estimate of N_e for American eels (*Anguilla rostrata*) would be huge if, as molecular evidence suggests, all individuals in the northern Atlantic Ocean constitute one panmictic population (Williams, Koehn, and Mitton 1973; Koehn and Williams 1978; Avise et al. 1986). At the other extreme, N_e for populations of self-fertilizing hermaphrodites, such as the killifish *Rivulus marmoratus*, is 1 (Vrijenhoek 1985b).

Summary

Relatively few data are available on dispersal and philopatry in ectothermic vertebrates, and reliable estimates of N_e are rare. Yet despite substantial variation, the available data suggest that natal homing and restricted dispersal are common. Furthermore, our knowledge of factors likely to affect N_e, and the few actual estimates of it, indicate that effective population sizes are often small. Thus we can provisionally conclude that opportunities for incestuous matings are probably common in diverse species of fishes, amphibians, and reptiles.

INBREEDING DEPRESSION

Inbreeding depression is a reduction in the fitness of the offspring of inbred matings relative to the progeny of outbred matings. In this section, we review available data on deleterious effects of inbreeding in ectothermic vertebrates, and consider how the inbreeding coefficient F and population history influence the magnitude of those effects.

Experimental Evidence

Few studies of the genetic costs and benefits associated with inbreeding have been conducted in ectothermic vertebrates other than fishes. Among reptiles, experimental studies show evidence of inbreeding depression in the common iguana (*Iguana iguana*) and the western diamondback rattlesnake (*Crotalus atrox*). Banks (1984) noted deleterious effects on a variety of traits of *I. iguana* (see table 11.1) when incestuous matings were allowed. Also, infertility in captive populations of this species has been attributed to three generations of inbreeding (Mendelssohn 1980). Cranial and scale abnormalities were observed in *C. atrox* after a single generation of inbreeding (Murphy et al. 1987).

Indirect evidence points to inbreeding depression in some amphibians. In studies of tiger salamanders (*Ambystoma tigrinum*), individuals expressing heterozygosity at several electrophoretic loci grew faster and had greater metabolic efficiency than homozygous individuals (Pierce and Mitton 1982; Mitton, Carey, and Kocher 1986). In a population of western toads (*Bufo boreas*), individuals heterozygous at enzymatic loci had higher winter survival rates than did homozygous members of their cohorts (Samollow and Soulé 1983). Data linking heterozygosity with growth, survival, and other fitness-associated traits also have been obtained for a variety of fishes (reviewed by Mit-

TABLE 11.1 Effects of Inbreeding on Measures of Reproductive Success

Species	Traits[a]	Inbreeding Coefficients[b]	Reference
Rainbow trout (*Oncorhynchus mykiss*)	Frequency of deformed fry; egg and fry survivorship	0.25	Aulstad and Kittelsen 1971
	Egg and fry survivorship; *fry mass*	0.25	Aulstad, Gjedrem, and Skjervold 1972
	Masses at various ages; fry survivorship; *egg survivorship*	0.125, 0.188, 0.25, 0.375, 0.5	Kincaid 1976a
	Frequency of deformed fry; fry and juvenile survivorship to 84, 147 days; masses at various ages; *egg survivorship*	0.25, 0.375	Kincaid 1976b
	Egg and fry survivorship; recovery rates at two ages from ponds; mass and length at various ages; total mass of egg; feed conversion	0.25, 0.5, 0.594, 0.672	Kincaid 1983
	Egg, alevin, and fry survivorship; adult mass; *juvenile mass*	0.25, 0.375, 0.5	Gjerde, Gunnes, and Gjedrem 1983
	Survivorship to various ages; mass; body condition index at various ages; growth rate for several periods; sexual maturation; *abdominal fat*	0.594	Gjerde 1988
Brook trout (*Salvelinus fontinalis*)	Length, mass at various ages; growth equation parameters	0.25	Cooper 1961
Atlantic salmon (*Salmo salar*)	Recapture frequency	Mainly 0.25	Ryman 1970
Channel catfish (*Ictalurus punctatus*)	Days to egg hatch; length and mass at various ages; *days to "swim up"; egg "hatchability"; total mass of eggs; eggs/kg body mass; individual egg mass; survivorship* to various ages; *frequency of successful spawns*	0.25, 0.375	Bondari and Dunham 1987
	Mass, length at various ages	0.25	Bondari 1981

TABLE 11.1   (*Continued*)

Species	Traits[a]	Inbreeding Coefficients[b]	Reference
Guppy (*Poecilia reticulata*)	Reproductive behaviors: closed courtship displays; total displays; *open displays; gonopodial thrusts*	Approximately 0.5, 1.0	Farr 1983
	Various reproductive behaviors	0.25	Farr and Peters 1984
Carp (*Cyprinus carpio*)	Mass gain	0.25	Moav and Wohlfarth 1968
Zebra danio (*Brachydanio rerio*)	Early egg survivorship; juvenile survivorship; juvenile body length; frequency of deformed fry; *late embryo survivorship*	0.125, 0.25	Mrakovćič and Haley 1979
	Frequency of deformities	Unclear[c]	Piron 1978
Convict cichlid (*Cichlasoma nigrofasciatum*)	Survivorship to 5 months; body length; frequency of deformities	0.25, 0.375, 0.50, 0.594	Winemiller and Taylor 1982
Common iguana (*Iguana iguana*)	Early egg survivorship; late embryo survivorship; juvenile survivorship; *clutch size*	0.25, 0.375, 0.5	Banks 1984
Western diamondback rattlesnake (*Crotalus atrox*)	Cranial and scale abnormalities	0.25	Murphy et al. 1987

[a]Traits showing evidence of inbreeding depression are in roman type; those showing no evidence of inbreeding depression are denoted in italics.
[b]Outbred lines served as controls for each study.
[c]Several generations of inbreeding of a small number of fish were permitted, but coefficients of inbreeding are not known.

ton and Grant 1984; Allendorf and Leary 1986; Zouros and Foltz 1987; Mitton, chapter 2, this volume). For example, mean individual heterozygosity increases with age, from embryo to juvenile stages and from juvenile to adult age classes, in natural populations of *Gambusia holbrooki* (Smith et al. 1989). These studies suggest that heterozygosity is advantageous, at least at the loci examined or at closely linked loci, so the increased homozygosity resulting from inbreeding should lead to lowered fitness.

In fishes, inbreeding depression has been detected in every species for which published data are available (summarized in table 11.1; see

Kirpichnikov 1981 for a review of Soviet studies; see Thorgaard 1983 for a review of inbreeding depression induced through developmental manipulations). The deleterious consequences of close inbreeding are often substantial. The magnitude of inbreeding depression is especially notable given that most studies have been conducted on salmonids, which are derived from a tetraploid ancestor. Few individuals should be completely homozygous at tetraploid loci, and deleterious recessives will be rarely expressed (Waples 1990). Among the traits showing evidence of inbreeding depression are survival and growth rates, frequency of deformities, and female fecundity. For example, Kincaid (1976a) studied over ninety families of rainbow trout (*Oncorhynchus mykiss*) from five different year classes. Comparing the progeny of females mated to their brothers with the outbred offspring of the same females, he observed a 9.8% reduction in fry weight and 16.1% reduction in fry survival at 150 days postfertilization.

Inbreeding also may have detrimental effects on male reproductive behavior, whereas outbreeding proves beneficial. Male progeny from crosses between highly inbred lines of guppies (*Poecilia reticulata*) show higher rates of some courtship displays than either inbred line, while backcrosses show intermediate or parental levels (Farr 1983). Inbreeding over a single generation appears not to affect rates of courtship display, perhaps because more severe inbreeding is necessary to induce negative effects on this trait (Farr and Peters 1984). Display rates of male guppies influence their success in courting females (Farr 1980; Bischoff, Gould, and Rubenstein 1985; but see Houde 1988b), so any decrease in display rates that accompanies inbreeding is likely to reduce fitness. Sexual selection against inbred males may occur in many species (e.g., Maynard Smith 1956; Sharp 1984). Surprisingly, this possibility has been largely ignored in most studies of inbreeding depression.

Although deleterious effects of close inbreeding are evident in a variety of traits (see table 11.1), specific results vary among species—and even among studies of the same species. In *Oncorhynchus mykiss*, for example, Aulstad and Kittelsen (1971), Aulstad, Gjedrem, and Skjervold (1972), and Gjerde, Gunnes, and Gjedrem (1983) report reductions in egg survival after a single generation of brother-sister mating, while in two other studies survival of eggs from sib matings was not different from survival of control eggs (Kincaid 1976a, 1976b). Kincaid (1983) found that one generation of sib mating resulted in a 17.2% reduction in egg survival, while three generations of sib mating resulted in a slight (0.2%) increase. Different histories of the stocks, particularly with respect to previous levels of inbreeding, may explain some, but probably not all, of the variation in results (Kincaid 1983, Gall 1987).

Differences in experimental conditions may also account for some of the variation, especially if some environments select more severely against inbred individuals.

Negative consequences of inbreeding are expected to increase with the inbreeding coefficient, F (Morton, Crow, and Muller 1956; Ralls, Ballou, and Templeton 1988; Falconer 1989). Evidence for further deleterious effects with increasing F beyond initial inbreeding (frequently in sib matings, $F = 0.25$), however, is not strong. Deformed individuals are found most frequently in laboratory populations of zebra danios (*Brachydanio rerio*) with the highest F values (Piron 1978). Convict cichlids (*Cichlasoma nigrofasciatum*) show no evidence of deleterious effects of sib matings until the F_4 and F_5 generations (Winemiller and Taylor 1982). Banks's (1984) study of iguanas also suggests that deleterious effects increase with the level of inbreeding. These studies are all difficult to evaluate, however, as they are based on limited numbers of families, fail to incorporate adequate controls, or both. In *Oncorhynchus mykiss*, the magnitude of deleterious effects sometimes increases with additional inbreeding beyond $F = 0.25$, but negative results also have been obtained (Kincaid 1976a, 1983; Gjerde, Gunnes, and Gjedrem 1983).

Thus, while broadly consistent with the idea that the level of inbreeding depression increases with F, available data suggest that the relationship is not especially strong beyond effects experienced after initial inbreeding. The weakness of the relationship is probably due in part to procedural difficulties. At higher levels of inbreeding, animals are difficult to rear, so sample sizes are often small (Kincaid 1976a). Moreover, selection can eliminate those individuals, or even families, most severely affected by inbreeding depression. Adverse effects associated with further inbreeding are thus reduced, and over the long term deleterious recessive alleles may be eliminated from inbred lines (Aulstad, Gjedrem, and Skjervold 1972; Falconer 1989; see below). If not carefully controlled, changes in density due to differential mortality in more inbred lines also can obscure inbreeding's effects on growth by facilitating the growth of the remaining individuals (Kincaid 1976a). Finally, if a lineage is repeatedly inbred over time, environmental variation can obscure the deleterious effects of inbreeding even when controls are present (Gjerde, Gunnes, and Gjedrem 1983). Problems commonly found in the experimental design and statistical analysis of studies of inbreeding depression are discussed by Lynch (1988a).

Population History

Populations or lineages with a history of close inbreeding may experience fewer effects of inbreeding depression. This decrease is generally attributed to the elimination of deleterious recessive alleles by natural selection (the partial dominance hypothesis; D. Charlesworth and B. Charlesworth 1987; Barrett and Charlesworth 1991; Mitton, chapter 2, this volume). The extent to which deleterious effects are reduced, however, is presently a matter of some debate (Shields 1982; D. Charlesworth and B. Charlesworth 1987).

The expected diminishing effects of inbreeding on fitness have been investigated in ectothermic vertebrates using sophisticated techniques for artificially inducing androgenesis and gynogenesis. Androgenesis produces complete homozygosity (and an inbreeding coefficient of one) in a single generation, as both sets of chromosomes in the offspring are derived from a single sperm cell. Gynogenesis may also produce complete homozygosity, or may simulate a lower level of inbreeding, depending upon the technique used and the species in question (Chourrout 1988). Scheerer et al. (1986) examined the survival and growth of androgenetic juvenile rainbow trout (*Oncorhynchus mykiss*) from two inbred and two outbred populations. Neither survival nor growth varied between the androgenetic *O. mykiss* obtained from inbred and from outbred populations, but high mortality in both groups may have obscured differences (Scheerer et al. 1986; Thorgaard et al. 1990). Survivorship of Mexican axolotls (*Ambystoma mexicanum*) derived from two generations of androgenesis was not greater than that of first-generation androgenetic individuals, despite strong selection against deleterious recessive alleles in the first generation (Gillespie and Armstrong 1981). However, in a more elaborate study including artificial selection among several lines, survivorship of gynogenetically produced homozygous zebra danios (*Brachydanio rerio*) increased with a second generation of gynogenesis (Streisinger et al. 1981). An increase in survival rate and a reduction in the frequency of morphological deformities was also observed after a second generation of gynogenesis in the tilapia (*Oreochromis aureus*; Don and Avtalion 1988).

No attempts have been made to compare the magnitude of inbreeding depression in ectothermic vertebrates between natural populations likely to have experienced different levels of inbreeding over many generations. Vrijenhoek (1985a), however, was unable to produce a "robust" inbred line of the topminnow *Poeciliopsis monacha* from the electrophoretically polymorphic populations of the Rio Fuerte, while such lines have been obtained from isolated populations of *P. lucida* that lack allozyme variation. The robustness of inbred lines of *P. lucida* relative to those of *P. monacha* may be due to historically higher levels of inbreeding in populations of the former species (Angus and Schultz 1983; Vrijenhoek 1985a). Despite their apparent insensitivity to further inbreeding, electrophoretically homozygous populations of topminnows are inferior to more heterozygous populations in a variety of traits, such as fecundity, that are strongly correlated with fitness (Quattro and Vrijenhoek 1989).

Shields (1982) argues that inbreeding depression may be less severe in natural than in laboratory conditions because deleterious effects are expressed mainly in the early stages of life. If sib competition is common at these stages, deaths due to inbreeding depression may coincide with an "excess fecundity" doomed to "inevitable ecological deaths." However, inbreeding depression does not seem restricted to early life stages. In some fishes, as in other organisms (D. Charlesworth and B. Charlesworth 1987), inbreeding leads to a reduction in growth (Cooper 1961; Kincaid 1983; Gjerde, Gunnes, and Gjedrem 1983; Gjerde 1988), lower recapture rates (Ryman 1970), and lessened survival through maturity (Gjerde 1988), as well as less intense male courtship (Farr 1983; but see Farr and Peters 1984) and decreased female fecundity (Kincaid 1983). Results of these studies are summarized in table 11.1.

Available data further show evidence of inbreeding depression in near-natural conditions. Ryman (1970) recaptured fewer inbred juvenile *Salmo salar* than outbred individuals from Swedish rivers into which they had been released. Inbreeding depression is also evidenced by lower recapture rates, growth rates, and total egg masses of *Oncorhynchus mykiss* raised in ponds (Kincaid 1983), reduced growth and survival of *O. mykiss* raised in sea cages (Gjerde, Gunnes, and Gjedrem 1983; Gjerde 1988), and slower growth of carp (*Cyprinus carpio*) raised in ponds (Moav and Wohlfarth 1968). Bondari and Dunham (1987) observed reduced growth in inbred channel catfish (*Ictalurus punctatus*) raised in ponds. In a natural population of the topminnow *Poeciliopsis monacha*, a founder event resulted in inbreeding and drastically reduced heterozygosity; for the following several years, the abundance of *P. monacha* was reduced relative to two competitors, clonally reproducing hybrids between *P. monacha* and *P. lucida* (Vrijenhoek 1989). When a small number of *P. monacha* females were replaced with females from a nearby, more genetically diverse population, however, *P. monacha* returned to its previous population size, which remained stable thereafter. Inbreeding due to the founder event had apparently reduced the *P. monacha* population's ability to compete (Vrijenhoek 1989).

Summary

Incestuous mating usually results in inbreeding depression, manifested by substantial deleterious effects on a variety of traits. These effects are observed not only in laboratory experiments but also in near-natural conditions. Ontogenetically, inbreeding depression is evident from the earliest developmental stages through reproductive maturity. At present there is little firm evidence that a history of inbreeding in a population reduces the deleterious effects of further inbreeding on that population. However, additional studies are needed using natural populations with different histories of inbreeding (e.g., Brewer et al. 1990; Barrett and Charlesworth 1991). Close kin are often accessible as mates for numerous amphibians and reptiles, so investigations of the functional consequences of close inbreeding are warranted. Experiments must be carefully designed, as a variety of factors can confound results.

OUTBREEDING DEPRESSION

Shields (1982) and Bateson (1983) have proposed that philopatry and a preference for mates of intermediate relatedness might be advantageous due to "outbreeding depression." When distant relatives or non-relatives mate, their descendants (either F_1 or F_2s) may suffer reduced viability due to the breakup of the coadapted genomes of the parents, even when both parents are members of the same continuous population. Additionally, "hybrid" offspring may be less well adapted to either of the local environments of their parents than either parent is to its respective home locale (Shields 1987). Outbreeding also entails non-genetic costs, such as those incurred in dispersing and selecting appropriate mates (Partridge 1983; Waser, Austad, and Keane 1986). Our discussion focuses on the more contentious suggestion that outbreeding incurs a genetic burden.

Genomic coadaptation on a broad geographic scale has been documented repeatedly (Endler 1977; Templeton et al. 1986), but within local populations data are scant. Outbreeding depression has yet to be demonstrated in experimental studies of crosses between ectothermic vertebrates from the same population. Hybrid vigor, or heterosis, has been observed in the F_1s of crosses between different strains of several species of fish (e.g., carp, *Cyprinus carpio*, rainbow trout, *Oncorhynchus mykiss*; reviewed by Kirpichnikov 1981; Gjedrem 1985), but negative heterosis has been observed in other instances (e.g., Ferguson, Danzmann, and Allendorf 1985). Outbreeding depression has been reported in the F_2s of crosses between geographic races, for example, in the newt *Triturus cristatus* (Callan and Spurway 1951). To test predictions of optimal outbreeding theory, however, the viability of offspring from within-population crosses of individuals, over a range of relatedness levels, needs to be studied; moreover, F_2s should be examined as well as F_1s (Shields 1982; Templeton et al. 1986).

Local Adaptation

In contrast to fine-scale genomic coadaptation, local adaptation is widely acknowledged as a common and important process (e.g., Templeton 1986). Ample evidence exists for local adaptation on a microgeographic scale in a variety of fishes. Some of the most thorough work has been done on salmonids (reviewed by Ricker 1972; Withler 1982; Quinn 1985; Noakes, Skúlason, and Snorrason 1989). In sockeye salmon (*Oncorhynchus nerka*), numerous characters, including the direction of migration by fry, susceptibility to parasitic infection, and timing of spawning, exhibit local adaptation (Quinn 1985). Pink salmon (*O. gorbuscha*) provide a clear example of the inferior local adaptation of "hybrids." Offspring of interpopulation crosses of *O. gorbuscha* survive as well at sea as pure stock, but return to the stream from which they were released at a lower rate than fish from the native population. Individuals that fail to home probably have no reproductive success (Bams 1976). The relative recovery rate of coho salmon (*O. kisutch*) is negatively correlated with the distance they have been transferred from their native stream (Reisenbichler 1988). These salmonid data are of particular interest because strays do sometimes return to the wrong site and population, where they could be discriminated against as mates (Quinn and Tallman 1987; Quinn, personal communication).

Guppies (*Poecilia reticulata*) show local adaptation over a fine geographic scale. Variation in color patterns among *P. reticulata* populations in Venezuela and Trinidad over distances of less than 5 km is associated with the distribution of predator faunas (Haskins et al. 1961; Endler 1978, 1983). Guppies seem adapted to the predators that they normally encounter in their local environment. Indeed, when the fish are confronted with experimentally altered predator faunas, color patterns shift as selection favors individuals with those patterns best suited for avoiding the introduced predators (Endler 1980). In natural conditions, *P. reticulata* rarely disperse far enough that they would interact with individuals of different color patterns or their predators (Haskins et al. 1961). Thus dispersal distances appear to coincide with the scale of local adaptation.

Amphibian populations are also known to vary in important life history traits, consistent with models of local adaptation. Variation between mountain and lowland populations is particularly pronounced. Larval populations of wood frogs (*Rana sylvatica*) at higher elevations grow and develop more slowly, and metamorphose at larger sizes, than lowland populations do (Berven 1982a, 1982b, 1987). After metamorphosis, mountain populations are characterized by larger adult body size and delayed time of first reproduction as compared with lowland populations. These differences appear to represent adaptations to local environmental conditions, as the need to attain a sufficient size prior to overwintering and shorter juvenile growing seasons in the mountains necessitate metamorphosis at a larger size. Reciprocal transplant experiments of juveniles between populations suggest that local adaptation has both environmental and genetic components (Berven 1982a, 1982b, 1987). Crosses between these populations reveal no evidence of genetic incompatibility; the hybrids consistently grow to an intermediate size (Berven 1982b). Although interbreeding between these populations is not normally expected because the distance separating them is too great, other studies suggest local adaptation over much shorter distances. Comparing cricket frogs (*Acris crepitans*) in closely proximate pond and stream habitats, Gorman and Gaines (1987) found dramatic differences in allele frequencies between populations, and suggest that the differences reflect local adaptation to those very different habitats. Transplantation experiments would be necessary to test this hypothesis.

Heterozygote Deficiencies

Biochemical genetic surveys of natural and seminatural populations of fishes and amphibians frequently reveal significantly fewer heterozygotes than expected under conditions of Hardy-Weinberg equilibrium. Heterozygote deficiencies can arise through several different mechanisms, and these are difficult to resolve (Christiansen et al. 1974). Offspring of outbred matings are likely to be more heterozygous than those of closely related parents (Naylor 1962). If heterozygote deficiencies come about due to selection against heterozygotes, these population characteristics may be symptomatic of outbreeding depression. While such effects might be attributable to negative heterosis, most data reveal instead significant advantages to heterozygosity in traits closely linked with fitness (see previous section on inbreeding depression). More likely, outbreeding disrupts coadapted genomes or otherwise works against local adaptation.

Homozygote excesses also may simply indicate nonrandom proportions of consanguineous matings. To the extent that close kin mate, the proportion of homozygotes in the population should increase. Assortative mating for particular characters, effected by behavioral preferences, might also increase the proportion of homozygous individuals in the population. For example, pairing midas cichlids (*Cichlasoma citrinellum*) show strong positive assortment by color morph in natural populations (McKaye and Barlow 1976). Philopatry, when it leads to genetic substructuring of the population, obviously facilitates assortative mating. Differences in gene frequencies among age classes (Tabachnick 1977) or temporal genotypic variation during breeding seasons (P. J. Smith 1987) also may create homozygote excesses. In many cases, heterozygote deficiencies probably reflect statistical aberrations in sampling procedures. The pooling of several contiguous but genetically distinct breeding populations inflates frequencies of homozygotes (the Wahlund effect; see discussion in Hartl and Clark 1989).

Reports of heterozygote deficiencies repeatedly emerge in fisheries studies (e.g., brown trout, *Salmo trutta*; Ryman, Allendorf, and Stahl 1979; Ferguson 1980; pink salmon, *Oncorhynchus gorbuscha*, Beacham, Withler, and Gould 1985; Pacific cod, *Gadus macrocephalus*, Grant, et al. 1987). In skipjack tuna (*Katsuwonus pelamis*), Fujino and Kang (1968) found that the viability of transferrin homozygotes was superior to that of heterozygotes. Excesses of homozygotes are frequently noted when examining allozyme polymorphisms of marine teleosts, but the bias is greatest during juvenile stages and generally decreases with age. Possibly homozygosity confers advantages early in life but during later ontogenetic periods heterozygotes are at some advantage (Smith and Francis 1984; P. J. Smith 1987). Alternatively, heterozygote deficiencies noted among embryos and juveniles may reflect moderate to high levels of inbreeding, but progeny of those matings may be selected against through later stages of development.

Analyses of the population structure of mosquitofish (*Gambusia affinis*) reveal consistent heterozygote deficiencies in both sexes and all age classes, but especially in adult males (K. L. Brown 1987). Heterozygous males presumably either disperse or are selected against. In apparent contrast, heterozygous *G. affinis* sampled in pond populations were larger in size than homozygous individuals, suggesting that they must either grow faster or live longer (Smith and Chesser 1981; Feder et al. 1984). Populations of red-bellied newts (*Taricha rivularis*) show slight but highly significant deficiencies in proportions of heterozygotes (Hedgecock 1978). Similar results have been obtained on populations of smooth newts (*Triturus vulgaris*; Kalezić and Tucić 1984), red-bellied newts (*Notophthalmus viridescens*; Tabachnick and Underhill 1972; Tabachnick 1977; Merritt et al. 1984), tree frogs (*Hyla* sp.; Case, Haneline, and Smith 1975; Nevo and Yang 1979), common frogs (*Rana temporaria*; Reh and Seitz 1990), and American toads (*Bufo americanus*; Guttman and Wilson 1973). Heterozygote deficiencies also have been found in reptile populations (e.g., slider turtles, *Pseudemys scripta*; Scribner et al. 1986). The consistency with which heterozygote deficiencies are detected, both among species and in numerous populations of some species, suggests that these genotypic biases may have some general significance at the population level.

Before heterozygote deficiencies can be interpreted as evidence of differential viability of homozygotes and heterozygotes, the genetic fine structure of the study population must be carefully evaluated to eliminate other possibilities. Examining variation between mated pairs in albumin and transferrin phenotypes, Christein, Guttman, and Taylor (1979) failed to find evidence of assortative mating in American toads (*Bufo americanus*). Nor was fine-grained genetic segregation apparent in breeding *B. americanus* populations (Christein, Guttman, and Taylor 1979). These findings lend support to the hypothesis that heterozygote deficiencies observed in *B. americanus* populations are attributable to selection against heterozygotes (Guttman and Wilson 1973).

Summary

Few data are available from ectothermic vertebrates on outbreeding depression due to genomic coadaptation over fine geographic scales. Local adaptation, however, clearly occurs over distances such that matings between locally well-adapted residents and poorly adapted dispersers could occur, at least in some fishes. Heterozygote deficiencies are frequently observed in studies of genetic population structure. In most cases these are probably due to pooling of distinct breeding populations, but they may be indicative of outbreeding depression.

KIN RECOGNITION

Even when close relatives are accessible as potential mates, pairings between them may be rare. In principle, kin recognition mechanisms not only permit individuals to avoid mating with close kin, but they can also allow them to strike an optimal balance between inbreeding and outbreeding (Bateson 1983). Kin recognition abilities have been documented in some fishes, numerous amphibians, and at least one reptile. In some birds (Bateson 1982; Burley, Minor, and Strachan 1990) and mammals (Winn and Vestal 1986; Barnard and Fitzsimons 1988), mate choice appears to be influenced by kin recognition abilities. Presently, whether kin recognition influences mate choice in ectothermic vertebrates is a matter of speculation, but available evidence suggests that such discrimination is possible.

Fishes

By means of chemosensory cues, fishes can recognize specific aspects of their natal environment, members of their population, classes of conspecifics, and close kin. Salmonids that home to natal areas within streams may respond to odors characteristic of their early rearing environment (Hasler and Scholz 1983) or to cues associated with their kin or members of their population (Nordeng 1971; Stabell 1984). Homing sockeye salmon (*Oncorhynchus nerka*), tested in Y-maze choice devices, can discriminate water conditioned by members of their own population from that conditioned by individuals from a neighboring lake (Groot, Quinn, and Hara 1986). Coho salmon (*Oncorhynchus kisutch*) show similar abilities to recognize and orient toward members of their own population (Quinn and Tolson 1986). Even more specifically, juvenile *O. kisutch* recognize and orient toward chemosensory cues associated with their siblings in preference to nonsiblings, whether or not they are familiar with these individuals (Quinn and Busack 1985; Quinn and Hara 1986). Juvenile arctic char (*Salvelinus alpinus*) show similar sibling preferences (Olsén 1989).

Whether kin recognition in fishes plays some role in mate choice has not been studied, but kin discrimination has been demonstrated in several other contexts. Fry of threespine sticklebacks (*Gasterosteus aculeatus*) swim toward and associate with their siblings in preference to nonsiblings, regardless of whether they are familiar or unfamiliar (Van Havre and FitzGerald 1988). Van Havre and FitzGerald (1988) suggest that *G. aculeatus* form schools with their siblings in nature. Poeciliid fishes, such as guppies (*Poecilia reticulata*) and mollies (*P. sphenops*), are viviparous, and often cannibalize conspecific fry. Yet in both these species, experimental studies suggest that females cannibalize their own young less than they do the fry of other females (Loekle, Madison, and Christian 1982). Female *G. aculeatus* can distinguish their own eggs from those of other females, and they cannibalize their own eggs less frequently than they do those of other females; males, however, fail to make this discrimination (FitzGerald and Van Havre 1987). In contrast, male pupfish (*Cyprinodon macularius*) discriminate between eggs they have fertilized and those fertilized by other males using chemical cues, but females apparently do not recognize their own eggs (Loiselle 1983). Parental care is provided exclusively by males in this species, and females have limited opportunity to cannibalize their own young.

Female guppies (*Poecilia reticulata*) mate preferentially with newly introduced males or males with "rare" pigmentation patterns in laboratory experiments (Farr 1977, 1980). Farr (1977, 1980) suggests that female preference for rare males evolved to facilitate incest avoidance in small, expanding populations founded by small numbers of individuals (but see Partridge 1988). Evidence of disassortative mating by tailspot pattern is available from natural populations of another poeciliid, the platyfish (*Xiphophorus maculatus*); moreover, tailspot heterozygotes have significantly more offspring than do other females (Borowsky and Kallman 1976). Mosquitofish (*Gambusia affinis*) are not known to recognize kin, but multiple insemination may function to minimize inbreeding. Multiple insemination of population-founding females increases the N_e of subsequent generations (Robbins, Hartman, and Smith 1987).

We know of no direct tests, using ectothermic vertebrates, of the hypothesis that mating preferences exist for individuals of an intermediate level of relatedness within a population (Bateson 1983). Studies have been undertaken, however, of assortative mating by population, usually to investigate speciation and isolating mechanisms. A preference for mates of the same population has been documented in several species, for example the threespine stickleback (*Gasterosteus aculeatus*; Hay and McPhail 1975), sockeye salmon (*Oncorhynchus nerka*; Foote and Larkin 1988), and the guppy (*Poecilia reticulata*; Luyten and Liley 1991). The evolution of different courtship and mating preferences among populations does not require selection against "hybrids," however. Variation might arise through drift, pleiotropic effects of adaptive divergence in other characters, Fisher's runaway process of sexual selection, or some combination of these mechanisms (Fisher 1958; Lande 1981; Templeton 1981; Ryan 1990).

Amphibians

Kin recognition mechanisms of anuran amphibians have been studied in detail (see reviews by Waldman 1991; Blaustein and Waldman, 1992). Larvae of many anurans (e.g., *Bufo americanus, B. boreas, Rana cascadae, R. sylvatica, R. aurora*) recognize and preferentially associate with siblings. In some cases discriminatory responses appear not to be based on social experience. For example, larvae reared in social isolation or just with nonrelatives recognize their siblings on first contact (e.g., Blaustein and O'Hara 1981, 1983; Waldman 1981). Larvae may also demonstrate strong habitat preferences (e.g., Wiens 1972; Punzo 1976; Dunlap and Scatterfield 1982) that vary sufficiently among families to lead to segregation of kin groups (Pfennig 1990). As in fishes, discriminatory behaviors generally are elicited in response to chemosensory cues (Blaustein and O'Hara 1982; Waldman 1985). Indeed, larvae of common frogs (*Rana temporaria*) and wood frogs (*R. sylvatica*) acquire specific preferences for chemical odorants to which they have been exposed prior to hatching, and orient toward those cues through metamorphosis and possibly beyond (Hepper and Waldman 1992). *Rana cascadae* and *R. sylvatica* apparently retain kin recognition abilities through metamorphosis and discriminate their siblings upon metamorphosing

(Blaustein, O'Hara, and Olson 1984; Cornell, Berven, and Gamboa 1989), but kin discrimination is not evident in the responses of froglets tested within the weeks following metamorphosis (Waldman 1989). The acute homing abilities of salamanders have been previously mentioned. In addition, many salamanders can discriminate neighboring from nonneighboring individuals based on their odors. Madison (1975) found that *Plethodon jordani* orient toward the odors of their neighbors, rather than those of strangers, in a laboratory testing device. Prior to the breeding season, males and females show similar preferences for familiar individuals, but during the breeding season, males show no preference, whereas females' preferences reverse: they tend to approach nonneighbors rather than neighbors (Madison 1975). Madison (1975) speculates that neighbors often are close relatives, and that avoidance of neighbors during the breeding season promotes outbreeding.

Work by Jaeger and his colleagues on *Plethodon cinereus* suggests that adults can discriminate between odors of familiar and unfamiliar conspecifics (McGavin 1978; Jaeger and Gergits 1979), possibly through chemical cues present in fecal pellets (Jaeger et al. 1986; Horne and Jaeger 1988). Recently, Walls and Roudebush (1991) found that larval marbled salamanders (*Ambystoma opacum*) were significantly less aggressive toward their siblings than toward nonsiblings in laboratory tests. Discrimination of kin was apparent whether or not opponents were familiar. After metamorphosis, however, discrimination is based more on familiarity than on kinship per se (Walls 1991).

Reptiles

Little is known concerning the kin recognition mechanisms of reptiles, but experimental work suggests that some iguanid lizards can discriminate kin from nonkin. Free-living desert iguanas (*Dipsosaurus dorsalis*) discriminate between neighbors and nonneighbors, presumably through a system of individual recognition (Glinksi and Krekorian 1985). Ferguson (1971) reported that *Sceloporus graciosus* males and *S. undulatus* females distinguished individual mates. Hatchling green iguanas (*Iguana iguana*) form social groups with their siblings in laboratory conditions (Werner et al. 1987). Discrimination is based on fecal and possibly other olfactory cues (Werner et al. 1987).

Summary

In ectothermic vertebrates, kin recognition is well documented among juveniles and in many contexts. Kin recognition during mate choice may facilitate incest avoidance or optimal outbreeding, but this has yet to be demonstrated in any fish, amphibian, or reptile. The frequency of matings between close relatives may be influenced by behavioral patterns and mating preferences other than those typically considered as "kin recognition." Further work is needed, however, to determine whether these are incidental effects or evolved functions.

POPULATION STRUCTURE AND OUTBREEDING IN AMERICAN TOADS

For the exploration of issues raised in this review, American toads (*Bufo americanus*) are in many ways model ectotherms. Presently we are conducting studies on toads to assess levels of inbreeding within populations, the means by which inbreeding avoidance or optimal outbreeding might occur, and the functional consequences of these behavioral tactics. Toads show strong mating site fidelity, yet are genetically highly variable. Thus questions concerning assortative mating and population structure can be readily addressed using molecular methods rather than more tedious mark-and-recapture techniques. Moreover, aspects of toad behavior and ecology can be studied both in natural populations and in laboratory experiments (e.g., see Waldman 1991). A variety of approaches thus are available for determining influences of genetic kinship on population structure and ecology. We present details of some of this work here to illustrate the promise these techniques offer for resolving levels of inbreeding and possible mechanisms of optimal outbreeding.

During five breeding seasons, from 1986 to 1990, we surveyed the genetic structure of breeding populations of *Bufo americanus* at several localities bordering the Estabrook Woods, a 650-acre mixed oak-birch forest in Concord, Massachusetts (Waldman, Rice, and Honeycutt 1992). Although suitable breeding habitats are absent within the reserve, toads are found in large numbers there during the spring, summer, and fall. During late April and May, depending on weather conditions, individuals migrate to breeding pools at the edge of the woods.

Upon arrival, males produce trilled calls, which can easily be heard over distances up to 1 km. These probably serve as advertisement signals for females and may attract other males. At any given time, more males than females are typically visible in a breeding aggregation. Males appear to scramble for females, and indiscriminately clasp any small objects at the water surface, including other males. Such lack of discretion might represent a serious barrier to reproduction were it not for the release calls uttered by clasped males, which signal their unsuitability as potential mates! Females usually arrive at ponds unamplexed, and upon entering the water, they tend to remain submerged and repeatedly approach males. Sometimes females act as if they may be evaluating particular males. Females swim slowly toward males, lifting their heads slightly out of the water, and finally approach one and initiate amplexus (Licht 1976; Howard 1988; Sullivan 1992; Waldman, personal observation). Females sometimes are clasped while swimming, however, and in this case have little opportunity to exercise mate choice (Howard 1988). Pairs occasionally are harassed by unpaired males; the unpaired male struggles with the amplexed male for access to the female. Attempts to dislodge amplexed males are rarely successful (Howard 1988; Waldman, personal observation). Most matings occur within a period of 2 to 5 days at each site, and pairs may remain in amplexus for 24 hours or more before oviposition occurs. Polygyny thus is infrequent (Gatz 1981; Howard 1988).

Breeding ponds were monitored nightly during April and May each year. All amplectant pairs were collected each evening, most after oviposition had been initiated. To assess (1) the frequency of incestuous matings and (2) the extent of genetic differentiation among ponds, we analyzed variation in mitochondrial DNA (mtDNA) extracted from mating pairs. Pure mtDNA was isolated from liver tissue by means of ultracentrifugation in a cesium chloride-ethidium bromide gradient (Densmore, Wright, and Brown 1985, as modified by Honeycutt et al. 1987). The mtDNA was then digested with various enzymes (*Mbo*I, *Taq*I, *Hin*PI, *Msp*I, *Dde*I, *Acc*I) that cleave the circular mitochondrial genome at specific recognition sequences. The fragments produced vary in size and number. Fragments were radioactively labeled (Brown 1980) and separated by electrophoresis on agarose and polyacrylamide gels. The gel patterns are made visible by autoradiography as series of bands corresponding to different sized fragments.

Many aspects of social behavior and population structure can be inferred by examination of restriction fragment length polymorphisms of mtDNA (Harrison 1989). MtDNA is maternally inherited, so siblings, as well as more distant matrilineal relatives, share identical fragment patterns. If individuals differ in mtDNA haplotypes, then barring mutations, they cannot be siblings. The application of mtDNA to studies of population structure would be limited if particular haplotypes predominated, but this is often not the case. In many vertebrates, mtDNA undergoes more rapid evolution than nuclear DNA. Consequently, mitochondrial markers can vary extensively among individuals, both within and between populations. Fishes, frogs, and reptiles have particularly variable mitochondrial genomes (Bermingham, Lamb, and Avise 1986; Avise, Bowen, and Lamb 1989; Waldman and Honeycutt,

unpublished data), sometimes approaching hypervariable minisatellite regions of the nuclear genome of their diversity (used for genetic
"fingerprinting"; Jeffreys 1987).

Genetic Variation among Breeding Localities

Comparisons of the mtDNA haplotypes represented among breeding
individuals in the three ponds reveal significant differences (figure
11.1). Over three-quarters of the 44 individuals present in temporary
pools at Concord Center, south of the woods, belonged to one of three
haplotypes (*A1, B2, B3*), and the remaining individuals comprised five
other haplotypes. Of 18 toads spawning in similarly ephemeral habitat
due east of the woods (Beecher Pond, less than 1 km northeast of Concord Center), six haplotypes were represented. Neither haplotype *B2*
nor *B3*, two of the most common at Concord Center, were present in
Beecher Pond. West of the woods, in Mink Pond (less than 1 km from
Beecher Pond and Concord Center), eight haplotypes were represented among 46 individuals, but again in clearly different proportions
from the other ponds. Haplotype *C1*, for example, was found in Mink
Pond but nowhere else. Haplotype frequencies differed significantly
among the three locations ($\chi^2 = 74.94$, $P < .0001$) but variation among
years within localities was not significant (see Waldman, Rice, and
Honeycutt 1992).

The genetic differentiation evident among the three breeding populations results neither from an inability nor a disinclination of individuals to traverse distances such as those separating the ponds. The common European toad *Bufo bufo* sometimes migrates over 3 km to reach
breeding pools (Heusser 1969; Sinsch 1987), and dispersal patterns of
adult *Bufo americanus* appear quite similar (Blair 1943; Oldham 1966;
Ewert 1969). Maynard (1934) tracked an individual *B. americanus* that
dispersed over 4 km, partly swimming in a river, and Blair (1943) recaptured one subject after it had swum across the rapid currents of the
Illinois River to reach its breeding site. In our study, individuals inhabiting central areas of the woods necessarily had to migrate comparable
distances to reach a suitable breeding pool.

Natal philopatry, facilitated by well-developed homing abilities
(Dole 1971; Sinsch 1987), might lead to genetic variation among breeding populations as demonstrated by our analyses of mtDNA haplotypes. Even small numbers of migrants between populations, however, can prevent particular alleles from becoming fixed within
populations (e.g., see Falconer 1989; Hartl and Clark 1989). Yet considerable levels of overall genetic differentiation should persist, especially
in populations of small to moderate size (Wright 1969; Breden 1987).

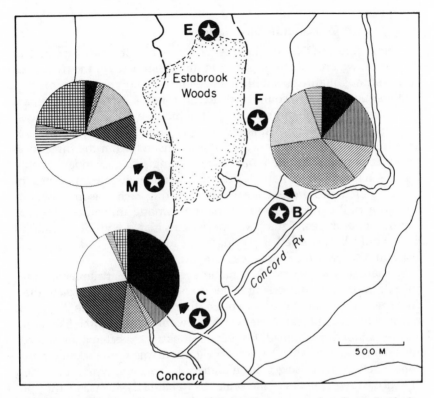

FIGURE 11.1 Distribution of common mtDNA haplotypes among breeding individuals at primary breeding sites surrounding the Estabrook Woods (B, Beecher Pond; C, Concord Center; M, Mink Pond). Advertisement vocalizations of males were recorded at Beecher and Mink Ponds and at two additional localities (E, Evans Pond; F, Freeman Pond). Frequencies of haplotypes (each denoted by a unique pattern) are shown in pie charts. Haplotypes were determined as composite restriction fragment length polymorphisms, based on digests with four restriction enzymes. (From Waldman, Rice, and Honeycutt 1992)

Most of our breeding populations ranged from 25 to 100 individuals, and effective population sizes were considerably smaller. Moreover, breeding success at some localities fluctuated enormously from year to year due to climatic and biotic effects on larval survivorship. These conditions should further amplify levels of genetic differentiation among nearby populations (Crow and Morton 1955), in agreement with our molecular findings. The observed levels of genetic divergence among breeding populations should be accompanied by increased numbers of matings between close kin.

Frequency of Consanguineous Matings

By comparing mtDNA haplotypes of males and females captured in amplexus, we are able to construct minimal estimates of the frequency of incestuous matings. Should mates share haplotypes, they may be siblings, but they also could be related through a more distant female ancestor. Despite the imprecision with which we can determine ancestry, we conclude that incest is exceedingly rare. Of 86 mated pairs collected in the vicinity of Concord, Massachusetts (from the three focal breeding populations and additional breeding sites), members of only 2 pairs had identical haplotypes. Randomly generated pairings of males and females present at each pond during each season, however, lead to a null expectation of 12 matings among individuals bearing identical haplotypes. Thus fewer individuals mated with close relatives than would be expected if pairing were random ($P < .02$, binomial test), but the expected frequencies are low.

Spatial variation in mtDNA haplotypes may be attributable to female philopatry, as dispersing males do not pass mitochondria to their progeny. Our results thus leave open the possibility of male dispersal. Asymmetrical dispersal of the two sexes would decrease the likelihood of close inbreeding. Indeed, few sib matings are expected at our breeding sites in part because haplotype variation was nonrandomly partitioned between the sexes. Individuals of like haplotype more often than not were members of the same sex (table 11.2). Consequently, potential mates were less genetically similar, on average, than members of the population sampled at random. If females are philopatric, sisters probably occur together in breeding aggregations. Brothers might also disperse together. We are currently conducting analyses of variation between breeding sites based on restriction fragment length polymorphisms of nuclear DNA; these will permit us to evaluate more fully the possibility of asymmetrical dispersal by sex.

Outbreeding could be enforced by mechanisms other than dispersal. In some anurans, including some toads, males first breed at a younger age than females do, often a year earlier (e.g., Hemelaar 1981; Gittins 1983). Substantial mortality between breeding seasons would then reduce the likelihood that siblings would mate. In addition, environmental sex determination mechanisms have been reported in amphibians (Bull 1983). If sex determination occurs early in development, siblings exposed to a common environment may be more likely than nonsiblings to be of the same sex. Finally, accounts of possible female choice in toads (Licht 1976; Howard 1988) leave open the prospect that females recognize and avoid mating with their siblings should they come in contact with them.

TABLE 11.2 Sex Ratios of Common Haplotypes in *Bufo americanus* (1987)

Haplotype[a]	Males	Females
S1	2	5
S2	2	6
T1	8	2
T2	2	11
T3	8	0
U2	5	5

Source: Waldman, Rice, and Honeycutt 1992
[a]Composite restriction fragment length polymorphisms (digests with full complement of six restriction enzymes)

Vocalizations as Potential Kin Recognition Cues

Behavioral observations suggest that female toads often have an opportunity to choose their mates. To the extent that the advertisement calls of males are heritable, and thus encode information about the caller's genotype, they serve as potential kinship cues. Mate choice to avoid incest or to optimally outbreed might be possible if females can detect this information and evaluate their genetic relationship to the caller.

To investigate this possibility, we recorded calls of fifteen males in each of four breeding populations around the Estabrook Woods (Waldman, Rice, and Honeycutt 1992). In many anurans, variation in temporal and frequency components of advertisement calls is attributable, in part, to ambient temperature and body size (Sullivan 1992 reviews these effects in *Bufo americanus*). Thus we measured cloacal temperatures of calling males as well as water and air temperatures at the calling sites. We collected calling males and measured them, and then in the laboratory we obtained nuclear DNA "fingerprints" of each. Fingerprinting involves the simultaneous detection of marker alleles at many loci dispersed throughout the nuclear genome, by use of multilocus probes such as those developed by Jeffreys, Wilson, and Thein (1985). Purified genomic DNA, obtained from liver tissue, was digested with *Hin*fl, the fragments were separated by electrophoresis, and then were transferred to nylon membranes by the Southern blot method. We hybridized the DNA with radioactively labeled Jeffreys 33.6 and 33.15 probes, and estimated the genetic similarity of calling males within ponds by the proportion of bands they shared (Wetton et al. 1987). Variation in both the temporal and frequency components of males' calls was analyzed as a function of the callers' genetic similarity using the generalized regression technique proposed by Mantel (1967). The similarity of the temporal—but not the frequency—compo-

TABLE 11.3 Call Dissimilarity as a Function of Genetic Similarity in *Bufo americanus* (Mantel analyses)

Pond	Pulse Duration	Interpulse Interval	Rise Time	Call Duration	Dominant Frequency
Beecher	$P < .005$	$P < .03$	$P < .01$	$P < .05$	ns
Evans	$P < .03$	$P < .05$	$P < .05$	$P < .01$	ns
Freeman	$P < .03$	$P < .03$	ns	ns	ns

Source: Waldman, Rice, and Honeycutt 1992

nents of males' calls was directly related to their genetic similarity as estimated from their fingerprints (table 11.3). Genetically similar individuals (quite likely brothers) had very similar calls, whereas calls of genetically dissimilar individuals were much more variable, even when temperature and size were held constant (figure 11.2). Consistent with the genetic differences observed among localities, temporal properties of advertisement calls differed significantly among ponds, and among individuals within ponds (see Waldman, Rice, and Honeycutt 1992).

These results suggest that females indeed might avoid mating with their brothers, or even choose other "optimally related" males, by listening to their advertisement calls. Whether females actually make this discrimination on hearing these vocalizations, or by perceiving other cues, still has not been established. Moreover, the mechanisms by which females would know what their brothers' calls sound like have yet to be resolved. The recognition of unfamiliar brothers' calls would most likely need to involve comparisons of call features with a genetic template, but the use of such templates in kin recognition systems has never been documented in any organism (Waldman 1987). Alternatively, females might learn their brothers' calls in particular contexts, such as at a common overwintering site (see discussion in Waldman, Rice, and Honeycutt 1992). Further studies, currently under way in our laboratory, should allow us to assess whether females choose between closely related males and nonrelatives based on information encoded in their calls.

CONCLUSION

Opportunities to mate incestuously may be surprisingly common in a wide variety of ectothermic vertebrates. Experimental data, mainly from fishes, indicate substantial levels of inbreeding depression in a variety of traits, in near-natural conditions, and throughout ontogeny from the egg stage through maturation. Few experimental data are

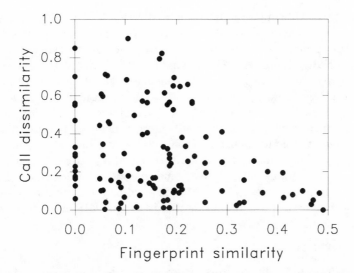

FIGURE 11.2   Call dissimilarity as a function of genetic similarity among calling males. Shown here are analyses of one component of the advertisement call, the interpulse interval, from males collected at Freeman Pond. Similar results are obtained with other temporal call parameters, and at each pond. Identical calls have a dissimilarity value of 0, and increasingly dissimilar calls have higher values. Fingerprint similarity values increase with relatedness (*r*) and inbreeding (*F*) coefficients. (From Waldman, Rice, and Honeycutt 1992)

available on outbreeding depression caused by the breakup of co-adapted genomes. Studies reveal, however, that dispersal and fine-scale local adaptation can lead to outbreeding depression. Heterozygote deficiencies are often observed in natural populations, and may indicate fine-grained spatial separation of genotypes. Alternatively, they may be indicative of outbreeding depression or negative heterosis. Numerous fishes and amphibians, and some reptiles, can recognize their kin, but as of yet, no evidence exists that kin recognition mechanisms play a role in mate choice. Some observed mating patterns and preferences may promote incest avoidance, however, while others may reduce extreme outbreeding.

In our own work with the American toad, *Bufo americanus*, we find evidence of substantial genetic differentiation among breeding ponds, based on mtDNA haplotypes of breeding pairs. These results suggest philopatry, at least by females. Nonetheless our data demonstrate that incestuous matings are rare. Males' advertisement vocalizations may serve as potential cues for kin recognition and female choice. By fingerprinting nuclear DNA, we determined that closely related males have similar calls. Molecular methods offer considerable promise for the resolution of problems of population structure and mating patterns. Mitochondrial genomes of ectothermic vertebrates show extensive variation, making them especially suitable for this work. Moreover, ectothermic vertebrates show high levels of variation in minisatellite nuclear DNA, making possible the use of fingerprinting in studies of genetic population structure. Molecular methods promise to provide accurate measures of both male and female dispersal patterns, and to clarify behavioral mechanisms of mate choice and incest avoidance. Although the determination of absolute levels of relatedness from fingerprinting techniques still poses some theoretical difficulties (Lynch 1988b), the use of these new methods as tools for examining levels of inbreeding in natural populations should greatly facilitate future work.

ACKNOWLEDGEMENTS

We are grateful to the National Science Foundation for support of our work on the ecology and mechanisms of kin discrimination in anuran amphibians (BSR-8717665, BNS-8820043, DIR-8901004, and BSR-9007760 to B. W.). J. S. M. was supported by a postgraduate scholarship from the Natural Sciences and Engineering Research Council of Canada. We thank Tyrone Hayes, John Rice, and Stephanie Seminara for assistance with the molecular work, David Posner for library research, and Felix Breden, Brian Bock, F. Stephen Dobson, Chris Marshall, Gary Meffe, Tom Quinn, Gordon Rodda, Mike Ryan, Christine Tam, and Nancy Wilmsen Thornhill for comments on preliminary versions of this chapter. Fred Allendorf, Brian Bock, Gary Meffe, Tom Quinn, and Gordon Rodda brought numerous references to our attention.

12

Evidence for and Consequences of Inbreeding in the Cooperative Spiders

Susan E. Riechert and Rose Marie Roeloffs

There are approximately 30,000 described species of spiders in the world. Two characteristics of most spiders tend to ensure that they are outbred: their solitary lifestyle and high dispersal capabilities. Most Araneae are not only solitary but maintain territories throughout their lives (see review in Riechert 1982). Individuals defend areas in excess of any web present against intrusion by conspecifics. The size of the area defended is determined by local prey abundances and thermal conditions (time available for foraging). This territoriality leads to a uniformly distributed population.

The aeronautic behavior common to most spider families also leads to outbreeding. Although spiders lack wings, they emit silk threads from spinnerets at the end of the abdomen. The juvenile spiders that exhibit this behavior are lifted off the substrate by air currents that catch these "ballooning" threads. This form of dispersal is a passive one, with individuals often being transported considerable distances.

Given their territorial social structure and high dispersal rates, one might expect that spiders are outbred and that they have developed few mechanisms to avoid inbreeding. However, there are two other kinds of social structures that are represented in the Araneae: communal and cooperative. Communal spiders exhibit reduced aggression and reduced cannibalism. They often form dense aggregations in local areas of high prey abundance. No territories are defended, and adjacent webs generally share the same support threads. However, individual webs are maintained, and there is competition among colony members for web placement within the colony. In the cooperative spider species, there is reduced aggression, and all colony members share in the construction and maintenance of a group web trap. Cooperative behavior has been well documented for sixteen species representing

283

five different families. The population demography of the cooperative spiders is such that they are potentially "naturally" highly inbred. In this chapter we review data that provide insight into the levels of inbreeding present in cooperative spider systems and consider the consequences of inbreeding for them.

THE COOPERATIVE SPIDERS

Five spider families of both cribellate (spinning plate) and ecribellate (spinnerets only) origins have representative species that live in true social groups: the cribellates include *Aebutina binotata* and *Mallos gregalis* (Dictynidae), *Amaurobius socialis* (Amaurobiidae), *Stegodyphus dumicola*, *S. mimosarum*, *S. sarasinorum*, *S. simoni*, and *S. socialis* (Eresidae); the ecribellates include *Agelena consociata* and *A. republicana* (Agelenidae, funnel web spiders), *Achaearanea disparata*, *A. vervoorti*, and *A. wau*, *Anelosimus domingo*, *eximius*, *lorenzo*, and *rupununi* (Theridiidae, comb-footed spiders). All of these species are tropical, and most can be found in the vicinity of the equator (figure 12.1). The cooperative spiders show little aggression toward conspecifics; instead, there is interindividual attraction, which Krafft (1969) has shown to be pheromonal in the case of *Agelena consociata*. These spiders maintain group webs and feed on prey communally. Egg sacs are laid in communal chambers, and there is cooperation in web maintenance, in the rearing of the young, and in capture of large prey items. Altruism is evident in the fact that the individuals that capture prey are often not the ones that feed on it: captured prey is dragged into the nests by the foragers where it is abandoned to others. Food is also regurgitated to any young seeking it. As yet, no division of labor has been attributed to the cooperative spiders, though the larger females are the colony workhorses: males and juveniles are largely inactive. There is also some evidence for differential reproduction among females. As individuals mature they continue to cooperate in the colony's activities and remain there for the duration of their lives. As a consequence, nests persist over many generations, and there is the potential for considerable inbreeding.

Our study system involves the funnel web spider, *Agelena consociata*, an occupant of primary rainforest habitats in equatorial West Africa. We detail *Agelena consociata*'s life history, demographics, and population genetic structure here, referring to data from other social spider species where possible. We will refer to similar data available for the asocial spider *Agelenopsis aperta*, a common spider in the western United States and Mexico. The comparison is an interesting one be-

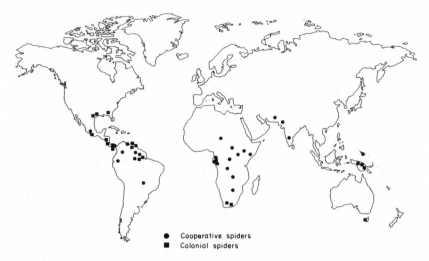

FIGURE 12.1 Distributions of known colonial and cooperative spider species. Most localities are field stations (D. K. Smith, pers. comm)

cause *Agelenopsis* is the New World sister genus to the Old World genus *Agelena*.

Agelena consociata lives in groups of a few to hundreds of individuals. The spiders build their nests in understory trees and shrubs in the rainforest, binding leaves and branches with silk. These nests are often maintained for many generations with overlap among generations. Since *Agelena consociata* breeds continuously, a variety of age classes of spiders are present in the larger nests at all times (table 12.1). The movement of individuals during periods of food shortage and nest fragmentation by falling tree limbs and frequent heavy rains leads to nest budding. Hence groups of related nests are clustered into colonies or local populations (Riechert 1985). Figure 12.2 shows one such colony in our study area at M'Passa, a field station of the Institut de Reserché en Ecologie Tropicale (IRET) in Gabon. As shown in the figure, nests within a colony are often connected by silken bridges, and individuals move freely among these interconnected nests. In a population genetic study of *Agelena consociata*, Roeloffs and Riechert (1988) found the mean genetic distance (Nei 1978) between clustered nests to be 0.001 and between different colonies or clusters of nests in the IRET reserve to be 0.018. The difference between these two means was highly significant ($P < .0001$, t test) and colonies separated by as little as 30 meters were found to be fixed at different alleles. Because *A. consociata* move freely between nests in a cluster, and because electrophoretic evidence sug-

TABLE 12.1　Sex and Age Class Representation in *Agelena consociata*
Colonies by Season

	Proportion of Nests		Relative Representation		
Season	Juveniles and Adults Present	Egg Sacs Present	Adults	(% males)	Juveniles
Major rains (Sept.–Nov.)	1.00	—	0.12 ± 0.03	(3.4 ± 0.89)	0.88 ± 0.03
Minor dry (Dec.–Feb.)	0.98	0.17	0.53 ± 0.01	(4.0 ± 0.24)	0.49 ± 0.01
Minor rains (Mar.–May)	1.00	0.33	0.57 ± 0.01	(4.0 ± 0.24)	0.26 ± 0.03
Major dry (June–Aug.)	1.00	0.93	0.63 ± 0.01	(10.6 ± 0.45)	0.36 ± 0.01

Note: Data from nest collections.

gests that there is little genetic differentiation among clustered nests, we refer to colonies as consisting of clusters of nests. Other workers refer to nests of social spiders as colonies and to nest clusters as local populations. To avoid confusion, in this chapter we refer to the single units in all of the studies as nests and the clusters of nests as colonies.

The facts that (1) nests are occupied by several consecutive generations and (2) the main growth of a colony occurs through budding have led us to consider the extent to which the cooperative spiders might be inbred. It is not possible to directly look for evidence of inbreeding because pedigrees cannot be followed in spiders. However, we and others have collected sufficient data on their mode of dispersal, population demography, and population genetic structure to permit us to infer levels of inbreeding.

EVIDENCE FOR INBREEDING

Dispersal

The population structure of the social spiders is highly viscous, primarily because what little dispersal occurs takes place following mating (table 12.2). Short-distance dispersal of generally less than 10 m occurs in most of the cooperative spider species studied. New nests are formed in the vicinity of the parent nest as a result of accidental fragmentation of the original nest, as is caused by rains and falling tree limbs in the case of *Agelena consociata* (Riechert, Roeloffs, and Echternacht 1986). Spiders of all ages and sexes also may move from the parent nest along silk runways or bridges, establishing new nests in close

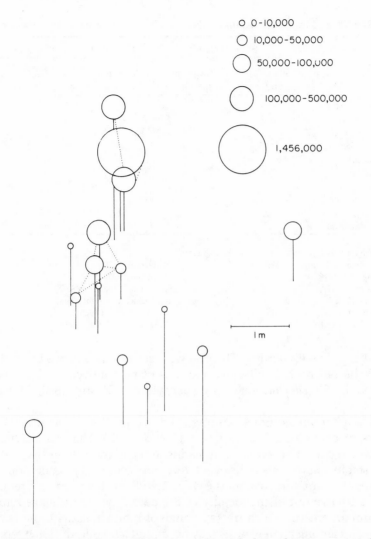

FIGURE 12.2   Example of range of nest sizes and spatial relationships among nests in a colony of the spider *Agelena consociata*. Vertical lines represent nest heights; circles represent nest volumes and numbers of adult females present. Dotted lines show nests that are attached to one another by silk bridges or that share web traps.

proximity to the parent nest. We refer to these processes as colony budding (table 12.2). The impetus for such moves is not known, though Roeloffs and Riechert (1988) did experimentally induce the entire spider population of one nest to move its nest location by cutting off its

TABLE 12.2 Modes of Dispersal Observed in Cooperative Spider Species

| | Short Distance | | Long Distance | | |
| | Silk Bridges | Swarming | Animal | Wind | |
Species	(Budding)	Adult Females	Carriers	Ballooning	Accidental
Achaearanea wau		1			
Achaearanea					
disparata		2,1			
Agelena consociata	3,4		4		
Anelosimus	5				
eximius	6	5			
Stegodyphus					
mimosarum	7,8,9			10	
Stegodyphus					
sarasinorum	11,12				11

Note: Numbers refer to the following references:
1, Lubin and Robinson 1982;   7, Marshall 1898;
2, Darchen 1968;   8, Wickler 1973;
3, Darchen 1978;   9, Seibt and Wickler 1988;
4, Roeloffs & Riechert 1988;   10, Wickler and Seibt 1986;
5, Vollrath 1982;   11, Jambunathan 1905;
6, Christenson 1984;   12, Jacson and Joseph 1973.

food supply with netting. The nest was reestablished above the netting and the old nest. Similar observations of nest movement have been noted in *Stegodyphus* following disruptions (Distant 1898; Marshall 1898).

The phenomenon of swarming is known for the cooperative Theridiids or comb-footed spiders (genus *Anelosimus*). Here adult females leave the parent nest following mating and emigrate along silk bridges to establish new nests. This event does not occur every generation, but is noted in "mature" colonies (Aviles 1986). Furthermore, the females do not move out of the vicinity of the parent nest and hence remain within an area in which free exchange of individuals with the parent nest and among all new nests may occur. In fact, Vollrath (1982) reports that following the establishment of individual nests in *Anelosimus eximius*, many females abandon these to join together, thereby achieving a higher probability of survival.

It is clear from these studies that budding and swarming activities do not lead to gene flow among colonies. They are more mechanisms of colony growth than they are migration. Nor does male dispersal lead to mixing, since no males have ever been observed dispersing in any of these spider social systems. Long-distance dispersal has been observed in the eresids *Stegodyphus mimosarum* and *S. sarasinorum*. It has also been inferred for the funnel web spider *Agelena consociata*. Wickler and Seibt (1986) report observations of ballooning activity in *Stegodyphus mimosarum*. This activity was observed in a previously mated female, though in the laboratory the authors observed both males and females to exhibit the "tiptoe" behavior characteristic of spiders about to balloon. *Stegodyphus* has also been observed to be accidentally carried off by winds (Jambunathan 1905). Thus, storms may be one means by which long-distance dispersal by the social spiders occurs.

Aeronautic behavior is not characteristic of the spider family Agelenidae, and we have obtained no evidence for it in *Agelena consociata*. We never observed "tiptoe" activity, and sticky traps placed in the four compass directions around nests failed to collect any of these spiders in 2 years of trapping for them (i.e., nine nests censused for 3 months [Roeloffs and Riechert 1988] and an additional two nests recently censused for 12 months [unpublished data]). Passive dispersal by vertebrate carriers appears to be the means by which new colonies of *Agelena consociata* are established. Large rifts are frequently observed in the nests, and there is a species of bat that refuges solely in them. We have also observed a monkey eagle passing through one nest. Thus, we experimentally passed cylinders covered with fur and feathers through nests to determine whether spiders might be carried off by this means. The data show that pieces of nest and associated spiders do cling to fur and feathers and that spiders so carried are dispersed considerable distances compared with budding or swarming activity (table 12.3). The data presented in the table represent a comparison of dispersal distances resulting from the dragging of cylinders covered with either fur or feathers through a nest and along a 50 m transect (simulated passive dispersal by animal carriers) with those resulting from the releases of single and multiple spiders at a release point (simulated budding or swarming activity). From these release experiments, we further found that successful colony establishment was significantly correlated with the number of individuals present in the new nests. Individuals forced into moving over the substrate in search of a new nest site suffer a high rate of predation even within the short time period we were able to follow their movements (table 12.4).

Summarizing the dispersal systems characteristic of the social spiders, they are not well developed. Hamilton and May (1977) state that every organism should have a dispersal mechanism. The social spiders seem to avoid dispersal and the gene flow it creates. Perhaps passive dispersal by wind and animal carriers is frequent enough that there has not been selection for a dispersal phase in these spider systems. Investigation of the genetic structure of some of these cooperative spider

TABLE 12.3 Dispersal Distances Achieved by Budding versus Animal Carriers in *Agelena consociata*

Experiment	No. of Individuals/ Nest Formed		Proportion of Nests Formed by Distance from Release Point/Parent Nest						
	Mean	S.E.	1 m	2 m	4 m	8 m	16 m	32 m	>
Simulated swarming or budding	1.18	0.17	0.00	0.17	0.83	0.00	0.00	0.00	0.00
Simulated animal carriers	13.59	0.56	0.10	0.13	0.17	0.10	0.07	0.03	0.40

Source: From Roeloffs and Riechert 1988.

species indicates, however, that passive dispersal is not of sufficient magnitude to lead to outcrossing of their colonies. The population genetic analyses reported below are supported by information collected on cooperative spider responses to foreign individuals. Krafft (1969) mixed individual *Agelena consociata* belonging to different nests and found that these individuals readily accepted one another. Riechert and Roeloffs (unpublished data) found that introduced *A. consociata* are readily accepted into nests, do not differ significantly from natal nest members in the levels and types of activities they exhibit, are treated

TABLE 12.4 Outcome of Experimental Releases (numbers by year)

	1984 May	1987	
		March	August
	32	30	30
Total attacks during observation[a]	5	16	19
Attacks by ants	3	12	18
Attacks by salticid spiders	2	2	0
Attacks by predaceous bugs	0	2	0
Attacks by crickets	0	0	1
Mortality due to attacks	3 (9.4%)	3 (10%)	4 (13.3%)
Injuries due to attacks	2 (6.3%)	1 (3.3%)	0
Nests formed by released individuals	6	2	3

[a]Most individuals were kept in sight for < 1 hour.

by other spiders as if they were natal nestmates, and have the same probability of reproducing as individuals natal to the nest. These observations suggest that no mechanism has evolved to detect intrusion into nests by noncolony members, as might be the case if little colony mixing occurs.

Genetic Structure

As would be expected in a system in which there is little dispersal from the parent nest and where virtually no matings take place between different colonies, cooperative spider systems show high levels of population subdivision (*Achaearanea wau*) [Lubin and Crozier 1985]; *Agelena consociata* [Riechert, Roeloffs, and Echternacht 1986; Roeloffs and Riechert 1988]; *Anelosimus eximius* [Smith 1986]). Between-colony variation is the major contributor to this subdivision. This is exemplified by our work with *A. consociata*. Nests that are separated by as little as 30 m exhibit fixation for different alleles, and nests that are separated by 90 kilometers are no more different than are nests that belong to the same local population but to different colonies (Roeloffs and Riechert 1988).

Whereas the cooperative spiders show some polymorphism, little genetic variation is noted within colonies (table 12.5). Thirty-eight percent of the twenty-two loci examined in *Agelena consociata*, for instance, are polymorphic, but only 5.5% of these loci show multiple alleles within colonies. The levels of polymorphism exhibited in the cooperative spider systems are, in general, an order of magnitude smaller than for asocial spiders and other invertebrates (table 12.5). Even local populations of the territorial spider *Agelenopsis aperta*, which have extremely small population deme sizes (tens of individuals), do not exhibit a similar reduction in polymorphism (table 12.5).

Levels of heterozygosity are also extremely low in cooperative spider systems (table 12.5). While *A. consociata* populations display a fair amount of genetic variability in terms of the number of polymorphic loci, there is a remarkable scarcity of heterozygotes (Mean $\bar{H} = 0.018$, $SE = 0.002$). All of the cooperative species for which data are available have mean heterozygosities of < 0.025, whereas typically outcrossing breeding systems exhibit frequencies of heterozygotes of between 6% and 15% (Selander 1976). Values for the asocial spider *Agelenopsis aperta* show no pattern of either heterozygote deficiency or excess, though small population demes of the species have reduced heterozygosity when compared with large population demes (thousands of individuals) (table 12.5). Other chelicerates and invertebrates, in general, also show the levels of heterozygosity suggested by Selander as indicative of outcrossing breeding systems.

TABLE 12.5 Comparison of Genetic Polymorphism (P) and Heterozygosity ($\bar{H}$) estimates for Spiders with Different Social Structures and Other Invertebrates.

| | P | | |
Taxonomic Group	Population	Within colony	$\bar{H}$
Cooperative Spiders			
Achaearanea wau[a,b]	0.045	0.009	0.005
			0.020
Agelena consociata[c]	0.380	0.055	0.018
Anelosimus eximius[b]			
Panama		0.059	0.017
Surinam		0.03	0.024
Communal Spiders			
Philoponella oweni[b]			0.085
Asocial Spiders			
Agelenopsis aperta[d]			
Small demes	0.540	0.444	0.085
Large demes	0.450	0.443	0.157
Araneus ventricosus[e]	0.330		
Bothriocyrtum[b] *californicum*			0.090
Meta mengei[f]	0.670		
Meta segmentata[f]	0.670		
Nesticus[g]	0.780		
Chelicerates[h]	0.269		0.080
Insects[h]	0.351		0.089
Invertebrates[h]	0.375		0.100

[a]Lubin and Crozier (1985): 30 nests, 22 loci
[b]D. R. Smith (1986): *A. wau*, 22 loci; *A. eximius*, 24 nests, 51 loci; *P. oweni*, 11 loci; *B. californicum*, 11 loci
[c]Riechert and Roeloffs (1988): 67 nests, 22 loci
[d]4 local populations of each deme type, 20–30 individuals representing each, 22 loci
[e]Manchenko (1981)
[f]Pennington (1979): 9 loci
[g]Cesaroni et al. (1981)
[h]Nevo, Beiles, and Ben-Shlomo (1983)

Wright's F_{IS} is frequently used as an indicator of inbreeding, since it measures the reduction of heterozygosity as a result of nonrandom mating within a subpopulation. In the case of the social spiders, F_{IS} measures the amount of inbreeding within nests. Positive values of F_{IS} occur when there are significant influences that increase homozygosity within a population, while negative values occur when conditions create excess heterozygotes. Data are available only for *Agelena consociata* in this instance, and the observed values are positive, indicating that inbreeding does occur (table 12.6). Note, however, the variability

TABLE 12.6 Comparison of Inbreeding Statistic (F_{IS}, Wright 1965) for a Cooperative and an Asocial Spider Species

Locus	Agelena consociata		Agelenopsis aperta	
	1982	1987	Small Demes	Large Demes
GAL	−0.265	1.000		
PEP	1.000			
EST1	−0.151	−0.015	0.278	0.362
EST2	−0.057	0.302		
ACE1		−0.913		
ACE2		0.973		
PGI1		−0.062	−0.014	0.008
IDH1			−0.020	−0.096
PGM1			−0.097	0.055
GOT1			0.191	0.305
GOT2			0.137	−0.014
LDH1			−0.037	−0.019
GAM			−0.135	0.103
MDH2			−0.043	−0.125
FUM2			0.142	0.169
IPO1			0.278	
ODH1			0.203	
HEX1			0.480	
MEAN	0.131	0.164	0.094	0.075

among loci: some values are negative, which would indicate that there is some avoidance of inbreeding as well. It is hard to interpret these results, though comparison with the *A. aperta* system provides some insight. The outcrossed *A. aperta* system also exhibits positive F_{IS} scores, though these inbreeding coefficients are approximately half the magnitude of the estimate for *Agelena consociata*. Further, the scores obtained for small population demes of *A. aperta* are more positive than are the scores obtained for large population demes.

Estimates of relatedness among individuals within colonies are available for *Agelena consociata* (Roeloffs and Riechert 1988; unpublished data). Using the regression method of Pamilo and Crozier (1982), estimates of relatedness within thirty colonies averaged 0.523 (SD = 0.301) for the five polymorphic loci identified for this species. Using the Queller and Goodnight (1989) estimate of relatedness, we obtained values of 0.556 ± 0.200 for thirteen colonies analyzed in 1987 and of 0.85 ± 0.070 for eight groups of full sibs that we reared from egg cases in the laboratory. For comparison we calculated relatedness scores for three local populations of the solitary spider *Agelenopsis aperta* with large population demes (r = 0.014 ± 0.028). These relatedness estimates support the view of social spider colonies as consisting of family groups with limited migration from home nests.

Biased Sex Ratios

The genetic data available to us corroborate the evidence for limited dispersal in cooperative spider systems and suggest that inbreeding regularly occurs and has resulted in loss of genetic variability. All of the cooperative spiders studied to date also exhibit markedly female-biased sex ratios (table 12.7), something that is unusual in diploid systems. There is also evidence that the sex ratio bias noted in the cooperative spiders is a primary one (Vollrath 1982, L. Aviles and Y. D. Lubin, personal communication). These sex ratios are consistent with the view that males are essentially mating with full sibs. Further, courtship in the cooperative spiders is undeveloped relative to that of the asocial species (e.g., Lubin 1986: *Achaearanea wau*; our observations with *Agelena consociata*). One would expect little courtship in a species in which there is no mate competition.

It was Hamilton (1967a) that first proposed the concept of local mate competition to explain female-biased sex ratios in family groups: females should limit the production of sons to those that are necessary to breed their daughters if those sons will be competing with one another for matings. Although this model was specifically developed for systems in which dispersal follows mating each generation, there is no reason why a similar ESS argument might not be made for systems with more limited dispersal (e.g., Frank 1987). Thus, we feel female-biased sex ratios in the cooperative spiders are indicative of a high level of inbreeding. In the remaining sections of this chapter we consider the consequences of inbreeding in a system in which it is a natural consequence of population demography. The cooperative spider system fits this description well.

CONSEQUENCES OF INBREEDING

Classically, the increase in homozygosity associated with inbreeding is thought to have two major consequences: (1) an immediate effect on viability or rigor, which is referred to as inbreeding depression; and (2) a potential deleterious effect in the event of environmental change due to the lack of genetic variability within group members. There is less information available on the consequences of inbreeding for the cooperative spiders than there is information documenting its existence. Nevertheless we do have some data for *Agelena consociata* that pertain to both of these areas. But first, let us consider one potential beneficial effect of inbreeding, fixation of an altruistic allele.

TABLE 12.7 Spider Sex Ratios (Males/Females)

Species	No. of nests	Mean	Range	Reference
Cooperative				
Achaearanea wau	34	0.160	0.031–0.333	Lubin, unpublished data
Achaearanea vervoorti	2	0.051	0.004–0.145	Lubin, unpublished data
Agelena consociata	8	0.224	0.000–0.670	Darchen 1978
	109	0.117	0.000–1.000	Riechert and Roeloffs, unpublished data
	23[a]	0.086	0.000–0.198	Riechert and Roeloffs, unpublished data
Anelosimus eximius	—		0.500–0.220	Overall and Ferreira da Silva 1982
	9	0.150	0.040–0.400	Aviles 1986
	38	0.150	0.040–0.670	Vollrath 1986
Stegodyphus dumicola		0.070		Cobby 1981, as referenced in Seibt and Wickler 1988
	18	0.114		Seibt and Wickler 1988
S. mimosarum	57	0.108	0.023–0.900	Seibt and Wickler 1988
S. sarasinorum	—	0.140	0.380–0.450	Jambunathan 1905
			0.090–1.100	Kullman, St. Nawabi, and Zimmerman 1971
	—	0.290		Bradoo 1976
Colonial				
Anelosimus jucundus	2	0.501	0.492–0.510	Aviles 1986
Solitary				
Achaearanea mundula	4	0.510	0.487–0.545	Lubin, unpublished data

[a]Colonies

Evolution of Altruism

Various authors have offered models that present kin selection as important to the evolution of cooperative (altruistic) behavior (e.g., Haldane 1932; Hamilton 1964a, 1964b; Breden and Wade 1981; Michod, this volume, for review). The conclusion of these models is that alleles associated with altruistic behavior increase in frequency in inbred populations because inbreeding increases the genetic correlation between individuals. Thus, altruists are increasing the fitness of individuals to whom they are most genetically similar. Using an additive genetic model, Breden and Wade (1981) demonstrated that for conditions that would eliminate altruistic alleles in a randomly mating population, increased levels of within-family mating cause the allele to increase to fixation, a runaway process. Theoretically, then, a cooperative population structure can evolve without regard to direct inbreeding costs or benefits (Wade and Breden 1981). Inbreeding may have been imposed upon the social spiders by limits to dispersal, but at the same time it may well have fostered the evolution of the altruistic behavior observed in these spiders.

Inbreeding Depression

Increased homozygosity is known to have deleterious effects on development, survival, and growth rate (Falconer 1981), though some workers believe that if local populations survive the first few generations of inbreeding, major problems with further inbreeding will probably not arise (e.g., Slatis 1975). Since the cooperative spiders are naturally highly inbred, they obviously have survived the initial episodes of inbreeding, and presumably, deleterious genes have been eliminated from their populations (but see D. Charlesworth and B. Charlesworth 1987). Nevertheless, there are some aspects of their life histories that suggest deleterious inbreeding effects. There is the problem of fecundity, for instance. Most researchers that have worked with the cooperative spiders feel that many females in the colonies fail to reproduce at all (e.g., Vollrath 1982; A. Rypstra, personal communication). Roeloffs and Riechert (unpublished data) found a negative correlation between the level of relatedness in colonies of Agelena consociata and the number of egg sacs produced by those colonies (Spearman rho = .442). These results are consistent with sterility associated with inbreeding depression rather than mere reproductive suppression: sterility is a commonly observed consequence of inbreeding in domestic animals (e.g., McPhee, Russel, and Zeller 1931).

Females that do reproduce also put far less of their biomass into reproduction than is typical for spiders. Pain (1964) states that the cooperative spiders produce between ten and thirty eggs, as compared with asocial congeners of the same size, which produce between fifty and one hundred eggs. Seibt and Wickler (1988) suggest that limited food leads to the low egg production of *Stegodyphus mimosarum*. However, low egg production still exists under unlimited food provisioning in the laboratory. While the asocial *Agelenopsis aperta* puts an average of 42% (SE = 2%) of its body mass into eggs, *Agelena consociata* puts only an average of 6% (SE = 1%) of its body mass into egg production. Further, the relationship between female body mass and weight of eggs produced noted for *A. aperta* and other spiders is absent in *Agelena consociata* (figure 12.3).

One might expect this result if there is increased survivorship of the offspring associated with the care juveniles are given in the social spider colonies. However, data collected by Krafft (1966) and Darchen (1965) for *A. consociata* indicate that egg viability is very low as well: 35% of seventy-eight clutches and 4% of one hundred clutches in respective studies. These are very low egg viabilities for spider clutches. Riechert routinely observes 100% hatch success in *A. aperta*, and Palanichamy and Pandian (1983) report hatch rates of 90% of the eggs in the first seven clutches of the multiple-clutch layer *Cyrtophora cicatrosa*. That egg mortality is not generally higher for tropical spiders is suggested by work with the egg production of a range of tropical species by Eberhard (1979). No mention was made of inviability of eggs laid in this fecundity study, despite the fact that all of the egg cases of multiple populations of several spider species were reared to obtain the reproductive rate estimates.

Taken together, all of these fecundity and viability parameters are suggestive of inbreeding depression, though detailed studies of social spiders and their close congeners that are asocial are needed to test for such an effect.

Adjusting to Changing Environments

Perhaps the cooperative spider species are limited to the tropics because of the relatively constant environments it affords. If change were to occur favoring a different behavioral or physiological phenotype, there might be insufficient variability present in the populations to adapt to the new conditions. Disease die-offs are known for *Agelena consociata*. Krafft (1970) reports an epidemic of a protozoan parasite that caused the extinction of local populations, and we have observed similar losses due to fungal attacks on *Agelena consociata*. Note that these extinctions occur when we bring the colonies back into the laboratory

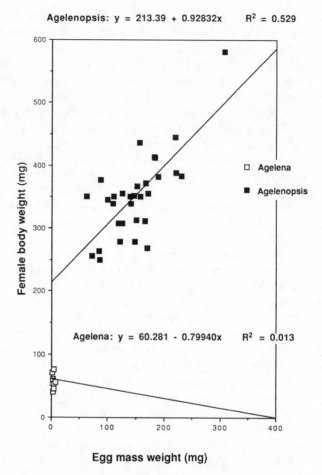

FIGURE 12.3   Comparison of regression relationships between female body weight at oviposition and the mass of eggs produced by asocial *Agelenopsis aperta* versus cooperative *Agelena consociata* spiders fed ad libitum in the laboratory:

where there is opportunity for the spread of infections. *Agelena consociata* apparently has little resistence to disease.

The best example of the failure of cooperative spiders to adjust to changing conditions comes from our monitoring of colonies on the IRET reserve over the last 5 years. *Agelena consociata* suffers its major nest extinctions during two three-month periods of rains, which alternate with dry seasons in Gabon (table 12.8; Riechert 1985). The rains damage the web traps and nests such that these need to be replaced on average 2 out of every 5 days. Prey availabilities are not high enough at

TABLE 12.8   *Agelena consociata* Nests Present at the End of Each Season

Months	Season	No. of Nests
Sept.–Nov.	Major rains	52
Dec.–Feb.	Minor dry	150
Mar.–May	Minor rains	82
June–Aug.	Major dry	144

most sites to permit solitary spiders to rebuild their webs with this frequency. Web trap size, however, does not increase linearly with the number of individuals in a group (figure 12.4). Thus, the larger colonies tend to survive the rains, probably because less energy investment per individual is needed to maintain the web trap (Riechert, Roeloffs, and Echternacht 1986).

We initiated censusing of all colonies within three 900-m² quadrats at the IRET reserve in the spring of 1982. The location of the quadrats and colonies within them for that first census are shown in figure 12.5.

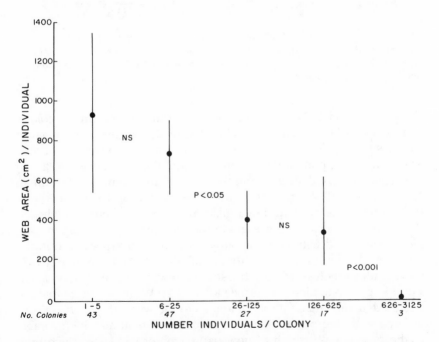

FIGURE 12.4   Silk investment per individual in the web trap and nest as a function of the number of adult female *Agelena consociata* present (medians and 95% confidence intervals). Probability scores represent results of Mann-Whitney test comparisons between adjacent colony sizes.

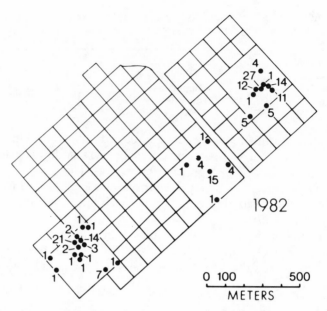

1982

0 100   500
METERS

FIGURE 12.5   Gridded area of M'Passa Reserve, showing location of three 900-m²
quadrats in which the size of colonies of *Agelena consociata* was followed. Points represent colony locations at the initiation of the censusing in 1982. Numbers refer to the
numbers of nests present.

Over the period 1984–1987, we observed a marked decline in colony
and nest numbers (table 12.9). Only ten of the original twenty-seven
colonies were present at the 1987 spring census, a survival rate of only
37% of the local populations. Note also that the extant colonies are of
much smaller size than they and the colonies that went extinct were in
1982 and even 1984. There has, thus, been a general decline of *A. consociata* on the reserve with few replacement colonies being established
(three small new colonies present in the 1987 census). The loss in colonies is correlated with an unusual weather pattern experienced in the
area over the period (table 12.10): rainfall during the major dry season
was significantly high. This is the season in which most of *Agelena consociata's* egg production occurs (see table 12.1). We hypothesize that the
energy demands made by rain damage to the webs of *A. consociata* in-
terfered with egg production. Alternatively, rainfall may be favorable
to the population growth of new pathogens or parasites which then
devastate the spider colonies (see Hamilton, chapter 18, this volume,
on the parasite/pathogen hypothesis).

TABLE 12.9 Growth and Survivorship of *Agelena consociata* Colonies

Quadrat	1982		1984		1987	
	Nests	Adult Females	Nests	Adult Females	Nests	Adult Females
1						
	4	1,047				
	11	4,018				
	1	55	11	222	5	132
	5	9				
	1	13	1	25	5	132
	27	4,301	12	313	4	7
	12	379	1	0	5	207
	1	0	8	211	2	7
	5	753	3	58		
	1	0	1	25		
2						
	4	536				
			1	140		
	1	1				
	1	3				
	15	3,195				
	4	546	1	1		
	1	1				
3						
	1	32	1	79		
					1	33
	1	28	1	149	5	99
	14	1,232	26	523		
	3	26	1	0	1	129
	7	1,041	2	1	1	3
	1	1,096	2	2	1	3
	1	1	4	88		
	21	804	18	416		
	2	190	2	140		
	1	8				
			1	17		
	1		1	39		
	1	39	1	44	1	7
	1	38	1	6		

Note: Censuses over 5-year period in study area grids at M'Passa Reserve in Gabon. Quadrats as shown in figure 12.5. Numbers of nests from absolute counts; numbers of females estimated from nest volumes using equation: # of adult females = 0.0012 nest volume (cm^3) − 4.13. Censuses taken in spring of years shown.

Table 12.10 Precipitation Records for M' Passa Reserve in Gabon

	1950–1983		1984–1987		
Season	Mean	SD	Mean	SD	
Mean Rains	692.2	163.2	479.6	286.5	P < 0.05
Minor Dry	308.5	93.1	221.9	68.0	NS
Major Storms	598.8	118.3	575.0	78.7	NS
Major Dry	97.0	84.4	249.6	176.9	P < 0.008

Note: The table shows unusually high rainfull during major dry seasons between 1984 and 1987, coinciding with high extinction rate of *Agelena consociata* colonies.

If more genetic variability were present in *Agelena* populations, would an adjustment be made to unusual weather patterns and/or to the diseases these may promote? An affirmative answer would be speculative at best. The exciting aspect of this system, however, is that it can lend itself to investigation of the consequences and advantages of inbreeding. Since there is no kin recognition and no apparent population mixing, it is possible to outbreed selected colonies in the field and compare survivorship over the long term. Comparative studies with asocial congeners that will add insight into potential levels of inbreeding depression can also be completed. These are studies we hope to initiate in the near future.

CONCLUSION

The cooperative spiders are inbred because dispersal is costly and limited in the tropical systems these species inhabit. There is little evidence for active dispersal, and most migration that does occur is short-distance, such that there is free exchange among parent and newly created nests. The level of sociality achieved in these diploid systems may, in fact, be dependent on inbreeding in that increased levels of within-family mating can cause an altruistic allele to increase to fixation in a runaway process.

Inbreeding is evidenced by the strong female-biased sex ratios exhibited in the cooperative spiders and in their population genetic structure. There are high levels of population subdivision, with most of the genetic variation occurring among colonies. The levels of genetic polymorphism exhibited in the cooperative spider systems are an order of magnitude smaller than those noted for asocial spiders and other outcrossed invertebrates. Levels of heterozygosity are also extremely low, and relatedness estimates for colony members are high, on the order of full sibs.

Inbreeding is not without its consequences in the social spiders. There is evidence for the reduced fecundity commonly associated with sterility in inbred domestic animals. Egg viability is also exceptionally low when compared with that of the asocial spider species. Finally, colony extinction rates are very high, particularly during periods of unusual weather and with respect to exposure to disease. These observations suggest that there is insufficient variability present to permit adaptation to new conditions, maybe one reason why cooperative spiders are found only in the tropics with their relatively constant environments.

Detailed comparative studies of cooperative spider systems and of close congeners that are asocial are needed to further delineate potential inbreeding depression and the effects of loss of variability. The system is a particularly interesting one from the standpoint of evolutionary biology because the cooperative spiders are diploid, naturally inbred, and are representatives of an order that is known for its highly aggressive and cannibalistic nature.

13

Inbreeding in Birds

Ian Rowley, Eleanor Russell, and Michael Brooker

Interest in the effects of inbreeding among populations of wild animals has greatly increased with the realization that the future survival of many species is highly doubtful and that many current populations are at very low levels. What constitutes the minimum viable population for an endangered species has evoked much debate (Gilpin and Soulé 1986; Ewens et al. 1987; Lande and Barrowclough 1987; Shaffer 1987), and it is now realized that there is no magic figure suitable for all situations, and that each species poses its own problems. There is a demographic concept of minimum viable population size, concerned with the probability of extinction of a population through stochastic demographic forces, and also a genetic concept, based on the rate at which genetic variation in a population is lost. The effects of inbreeding—inbreeding depression, increased homozygosity, the concentration of lethal genes, and the loss of adaptability leading to a lowered capacity to cope with changing environmental situations—are central to the genetic concept of minimum viable population size.

This chapter attempts to review the few data available on inbreeding in natural populations of birds. Departures from random mating leading to inbreeding may be of two types. First, there may be an increased likelihood of close relatives mating, the common idea of inbreeding. Second, the geographic and social structure of a population may inhibit the thorough mixing of genotypes in each generation (Barrowclough and Coats 1985). Long-term studies of marked individuals have provided information on observed pairings from which pedigrees have been calculated, and the incidence of inbred matings has been inferred (table 13.1). We follow Ralls, Harvey, and Lyles (1986) in classifying as

TABLE 13.1 Frequency of Close Inbreeding in Natural Populations of Birds

Species	%	Sample[a] Size	Reference
White-fronted bee-eater	0.0	81	Hegner, Emlen, and Demong 1982
Darwin's medium ground finch	0.1	345	Gibbs and Grant 1989
Pied flycatcher	0.4	276	P. H. Harvey and B. Campbell, unpublished
Florida scrub jay	0.4	280	Woolfenden and Fitzpatrick 1984
Arabian babbler	0.6	>300	Zahavi 1990
Yellow-eyed penguin	0.6	490	Richdale 1957
Purple martin	0.7	140	E. Morton, unpublished
Great tit (Hoge Veluwe)	1.5	460	van Noordwijk and Scharloo 1981
Great tit (Wytham Wood)	1.5	1,000	Greenwood, Harvey, and Perrins 1978
Cliff swallow	1.7	59	P. J. Sikes and K. A. Arnold, unpublished
Great tit (Vlieland)	3.0	834	van Noordwijk and Scharloo 1981
Acorn woodpecker	3.1	225	Koenig, Mumme, and Pitelka 1984
Galápagos mockingbird	7.0	156	Curry and Grant 1990
Mute swan	9.8	184	J. Reese 1980, unpublished
Splendid fairy-wren	25.9	321	I. Rowley unpublished (to 1988)
Song sparrow	4.5	178	Arcese (1989)

Source: Based on table 2 in Ralls, Harvey, and Lyles (1986) with additional unpublished data from R. L. Curry and I. Rowley. Data from sources indicated or from unpublished source quoted by Ralls, Harvey, and Lyles (1986).
Note: Inbreeding between parents and offspring or between full siblings, as determined from observed social pairings.
[a]Sample sizes are number of pair-years (number of pairs × number of years).

close inbreeding matings between parents and their offspring or between full siblings. Shields (chapter 8, this volume) is critical of Ralls et al. for expressing frequency of inbred pairings calculated by different methods. However, the data from Ralls et al. that we have included in table 13.1 are all based on the same measure, namely, the proportion of pair years in which closely related pairs were observed to be mated. We also have used this measure, despite Shields' criticism of it, since it allows better comparisons across species with different mating systems and longevity.

Few studies of bird populations have continued for long enough to provide pedigree information on matings between other than close relatives, so we do not have a direct measure of "background" inbreeding. Coefficients of kinship are usually calculated on the basis that all individuals at the start of a study, and any later immigrants, are completely unrelated, which is probably seldom the case. Indirect evidence of the levels of inbreeding may come from electrophoretic studies and from calculations based on effective populations sizes that estimate inbreeding effects due to geographic structure (Barrowclough 1980; Woolfenden and Fitzpatrick 1984; Barrowclough and Coats 1985; Koenig and Mumme 1987; and see below).

DIRECT EVIDENCE OF INBREEDING: LONG-TERM STUDIES

Electrophoretic studies have shown that observed social pairings do not explain the parentage of all nestlings that are raised and that the distinction made by Wickler and Seibt (1983) between social monogamy and sexual monogamy is clearly an important one. Extra-pair copulations (EPC) take place much more often than had previously been supposed (e.g., indigo bunting [*Passerina cyanea;* Westneat 1987]; zebra finch [*Taeniopygia guttata;* Birkhead, Clarkson, and Zann 1988]; white-crowned sparrow [*Zonotrichia leucophrys;* Sherman and Morton 1988]; splendid fairy-wren [*Malurus splendens;* Brooker et al. 1990]; reviews by Ford 1983; McKinney, Cheng, and Bruggers 1984; Birkhead and Biggins 1987). A significant number of these copulations are effective (Lank et al. 1989). This means that relationships based on observations of pairing alone must be treated with suspicion, genetically speaking. With this qualification, several long-term studies do present direct evidence of inbreeding from observed pairing of related individuals; six are discussed briefly below and a seventh is presented as a detailed case history.

The Great Tit (*Parus Major*)

Long-term studies in Europe have provided pedigree information from marked individuals which can be used to calculate coefficients of inbreeding (F) and to estimate the effects of inbreeding in natural populations of great tits. In the Oxford population discussed by Greenwood, Harvey, and Perrins (1978), 13 out of 885 pairings from 1964 to 1975 (1.6%) were with close relatives (5 mother-son, 7 brother-sister, 1 aunt-nephew), giving $F = 0.0037$. Inbreeding pairs had higher nesting mortality (27.7%) than outbreeders (16.2%), but the stages at which mortality occurred is not known, since eggs that failed to hatch were not distinguished from nestlings that died after hatching.

In the Dutch study discussed by van Noordwijk and Scharloo (1981) and van Noordwijk (1987), two populations were followed, one at a mainland site and the other on an island. For that study, "inbreeding" meant matings between pairs related to the extent of having one great-grandparent in common, or closer. At the mainland site (Hoge Veluwe), the percentage of inbreeding pairs ranged from 1% (for 2,601 clutches where both parents were identified) to 5% (for 460 clutches where both parents had been banded as nestlings in the study area). Coefficients of inbreeding (F) were 0.0011 and 0.0063 for these two samples. For the island (Vlieland) population, there was less immigration into the study area. For 47% of 280 clutches where the male and female were identified, the parents were related with $F = 0.015$. Thus the mainland site was similar to the Oxford population, but the island population, with restricted immigration and dispersal, had a much higher level of inbreeding. The island population provided a large sample of clutches from inbreeding parents for the detection of the effects of inbreeding. Separate effects of the laying female being inbred and of the egg being inbred were detected. The hatching of eggs was reduced by 7.5% for every 10% increase in F. Most of the deaths seemed to occur before or at hatching, and the effect of inbreeding was smaller at the time of fledging. There was also a degree of hatching failure related to the degree of inbreeding of the female herself. Taking both these effects into account, van Noordwijk (1987) calculated the number of grandchildren reaching breeding age that were produced by related and unrelated pairs, and found that related pairs produced twice as many grandchildren; the negative effects of inbreeding at egg hatching were more than compensated for by higher survival after fledging, in both generations, so that inbred individuals were in fact fitter.

The inbreeding coefficients quoted have been calculated on the assumption that actual mating followed observed pairing, which may not in fact be the case if there was any significant level of EPC. In the great tit population of Wytham Woods, near Oxford, Norris and Blakey (1989) used heritability analysis based on the resemblance of offspring tarsus length to that of their parents to estimate the frequency of EPC. Their results suggest that EPC occurs in great tits and may be common in some years; they quoted unpublished data from a study of enzyme polymorphisms that also demonstrated the occurrence of EPC. However, added to the restricted nature of the island population discussed by van Noordwijk and Scharloo (1981) was the fact that the variance in offspring number was such that some families produced a higher number of breeding birds than others, and members of those families had a higher chance of mating with a relative, because more potential mates were likely to be related to them. Even if there is some EPC, the level of inbreeding would still be expected to be high. Van Noordwijk and Scharloo (1981) found no evidence that inbreeding was avoided, and the close approximation between levels of inbreeding and those predicted under a model of random pairing confirms this (van Noordwijk et al. 1985; van Tienderen and van Noordwijk 1988).

The Florida Scrub Jay (*Aphelocoma caerulescens*)

Woolfenden and Fitzpatrick (1984) found a low level of inbreeding in the Florida scrub jay (five cases in 280 group-years from 1970 to 1978). One case involved a helper son who paired briefly with his mother after the death of his father (for one nest only; the female died after incubation). The other four cases involved seven offspring of the one prolific breeder over many years, which not surprisingly encountered each other in the course of pairing and mate replacement in a species with very limited dispersal (males: median 0, mean 0.9 territories away; females: median 2, mean 3.44 territories away). These four pairs were daughter × grandson ($r = 0.25$), granddaughter × son (by different female, $r = 0.125$), and grandson × granddaughter (two cases; $r = 0.125$). In general, however, the pattern of pair establishment and mate replacement observed was such that with the exception mentioned above, jays did not pair with another member of the functional social group in which they were raised. This is a good example of how high variance in reproductive success can lead to inbreeding. Woolfenden and Fitzpatrick observed no indication that EPC was a significant feature of the Florida scrub jay mating system. Courtship is lengthy and ritualized, and no courtship between a female and males other than her observed mate was ever seen.

The Acorn Woodpecker (*Melanerpes formicivorus*)

In the acorn woodpecker of California, a bird generally only became a breeder in its natal group following the disappearance of the breeder or breeders of the opposite sex that were present in the group when the bird was born (Koenig, Mumme, and Pitelka 1984). Ten cases of known or probable inbreeding were recorded in 225 group-years from 1972 to 1982 (4.4%), five resulting from exceptions to the general rule, in which birds bred in their natal group (four father-daughter pairings, one brother-sister). The other five cases resulted from limited dispersal, when close relatives immigrated to the same group at different times and subsequently bred together; two were probably siblings and three relationships were uncertain.

For both jay and woodpecker, these patterns of mate acquisition or replacement have been suggested by the authors (Koenig, Mumme, and Pitelka 1984; Woolfenden and Fitzpatrick 1984) and others (Moore and Ali 1984) as evidence of inbreeding avoidance mechanisms or of incest taboos. However, it is difficult to be certain that the observed patterns of mate replacement do not arise because the potential replacement breeder within the group is competitively inferior to a dispersing individual from outside (see also Shields 1987). The occurrence of matings between sibs outside their natal group suggests that kin recognition is not important. Detailed behavioral observation after experimental removal of individuals is perhaps the only way to resolve the issue.

Darwin's medium ground finch (*Geospiza fortis*)

The long-term study of *Geospiza fortis* on Isla Daphne Major, Galápagos, has provided information for the calculation of levels of inbreeding and the effects of inbreeding on reproductive success (Gibbs and Grant 1989). In the period from 1981 to 1987, 4.6% (27 of 583) of all pairings were between related individuals, with coefficients of kinship ranging from 0.008 to 0.25, and a value of F of 0.00394. Detected matings between related birds increased from 4.6% for all pairs to 24% of 70 pairs for which at least half the grandparents were known. The detected level of close inbreeding (at least $r = 0.25$) over 345 pairs where the parents of both males and females were known was 2.0%. The observed reproductive output of related pairs was not significantly different from the output of unrelated pairs. From computer simulations, in all years, the observed values of inbreeding were extremely close to those expected if pairing between individual birds was random with respect to relatedness, and there was no strong evidence that close kin were avoided while more distant kin were preferred. Gibbs and Grant suggest that population structure is the primary determinant of observed levels of inbreeding, and that there was no evidence of behavioral avoidance of mating with kin.

The Pukeko (*Porphyrio Porphyrio melanotus*)

In the communally breeding pukeko of New Zealand (Craig and Jamieson 1988), groups consisted of one to two breeding females, one to seven breeding males, and up to seven nonbreeding helpers. Each group defended year-round territories, and group females laid in a single nest that was incubated by all individuals that copulated. Dispersal was minimal; most birds remained in their natal territory or at most dispersed to an adjacent one. The extreme philopatry exhibited would greatly increase chances of close inbreeding unless birds that reached reproductive maturity were more likely to disperse when possible parents were present in their group (as suggested for the acorn woodpecker by Koenig, Mumme, and Pitelka 1984). The communal nature of mating and the polygynandrous mating system make it impossible to calculate inbreeding coefficients from proportions of pairs that were related. However, Craig and Jamieson recorded thirty-four occurrences of matings between closely related individuals (daughter-father, son-mother, brother-sister), a measure of when related individuals were reproductively active in the same group. In contrast to this were seven cases of possible inbreeding avoidance (four sons dispersed in the presence of their mothers and three sons had reduced sexual activity in the presence of their mothers), and seventeen examples of likely outbreeding (due to male dispersal or changes in group composition). Thus outbreeding appears less common than inbreeding.

The Song Sparrow (*Melospiza melodia*)

The song sparrow *Melospiza melodia* is resident on Mandarte Island, British Columbia, all year round. A long-term study has continued since 1974 (Smith 1988), and Arcese (1989) reported data on dispersal and levels of inbreeding. Song sparrows are primarily monogamous and live mostly within a territory of up to 1 ha. A median distance of eight territories was traversed in dispersal, and few young birds settled to breed on their natal territory, although in many cases this would have been possible. Of 178 birds of known pedigree settling with their first mate, 8 (4.5%) settled with a close relative, 19 (11%) with a moderately related mate, and 151 (84.5%) with unrelated mates. The rate of inbreeding appeared to decline as population size increased. The proportion of young birds of known lineage that settled with relatives declined from 32.1% in 1982, when the population was low, to 6.1% in 1985, when population size peaked.

The total number of independent young raised by individuals during the study (a measure of lifetime reproductive success) was independent of the relatedness of their first mate. Arcese (1989) suggests that in the Mandarte population of song sparrows, where competition for territories promotes rapid mate replacement and is perhaps the main determinant of reproductive success (Smith 1988), forfeiting a territory when the potential mate is a relative is probably more deleterious than inbreeding.

INDIRECT EVIDENCE FOR INBREEDING

Genetic Structure of Populations: Demographic Studies

Direct information on the level of inbreeding is also provided by data on the genetic structure of populations, aspects of which can be estimated from demographic and electrophoretic data. The various demographic models used depend on whether the populations considered are continuous or in isolated patches. This topic is considered at greater length by Chepko-Sade et al. (1987), Lande and Barrowclough (1987), and Rockwell and Barrowclough (1987), and we follow here the methods outlined therein. The most appropriate general model is Wright's "isolation by distance" model for continuous populations (Wright 1943), which assumes that even in a continuous population, differentiation can develop owing to the finite distances moved during dispersal. In any one generation, genes do not move very far across the range of the species; populations are isolated by distance if not by habitat barriers. Wright suggested that such situations could be modeled by dividing the continuous distributions into a series of "neighborhoods," or demes, each with an effective population size N_e. The actual value of N_e is determined by gene flow, population density, sex ratio, variance in offspring number, fluctuations in population size, and the effect of overlapping generations.

The effective size of these demes and their number in the range of the species are sufficient to calculate F_{ST}, Wright's fixation coefficient, the expected degree of genetic differentiation in the taxon. Values of F_{ST} can vary from zero to one. If gene flow is large and allele frequencies similar across a species range, F_{ST} will be small. If F_{ST} is large, genetic variance among demes is high. F_{ST} is often used as a measure of inbreeding due to finite population size, but Shields (1982: 184–85) emphasizes that this is only so if there is no selection maintaining genetic variation or heterozygosity in the face of inbreeding.

Following Rockwell and Barrowclough (1987), Wright's (1943) equation of neighborhood size N calculates the number of breeding individuals occupying a circle of radius 2σ, where σ is the standard deviation of dispersal distances and ρ is the population density: $N = 4\pi\rho\sigma^2$.

The value of N so calculated can then be adjusted for variance in offspring number, sex ratio, etc., according to the methods outlined by Barrowclough (1980) and Lande and Barrowclough (1987) to give a value for effective population size N_e, from which F_{ST} can be calculated. Estimates of the mean value of F_{ST} in birds from electrophoretic variability (Evans 1987; 23 species), direct measurement of dispersal (Barrowclough 1980; 16 species) or from demographic data (Rockwell and Barrowclough 1987; 3 species) indicate values of the order of 0.1 or less. This suggests that only 10% of the total genetic variation is attributable to locality, and implies substantial gene flow. Since gene flow can retard the rate and effect of inbreeding in local populations, this is a general indication, in the absence of detailed information for many species, that inbreeding is not normally at high levels in bird populations.

There are some significant problems with the estimation of N_e and F_{ST} from direct measurement of dispersal. Barrowclough (1978) showed that there is a sampling bias in dispersal studies that are based on a finite area. The probability of detecting dispersal events is a function of the distance dispersed and the site of origin within the study area; longer dispersals out of the study area from sites near the edge of the study area have less chance of being detected. Consequently the distribution of dispersal distances, if uncorrected for this bias, will reflect a relative excess of short-distance dispersals and a deficit of longer ones. Barrowclough (1978) presents a method of correcting for this sampling bias. Estimates of N_e made for birds by Barrowclough (1980) and by Koenig and Mumme (1987) for the acorn woodpecker are based on corrected distances, but in many cases, dispersal distances are uncorrected (e.g., examples of mammalian dispersal in Chepko-Sade et al. 1987).

Another problem arises with the distribution of dispersal distances. The estimate of neighborhood area generally used is based on the assumption that dispersal distances are distributed normally (Wright 1969), but this is rarely true, especially in any philopatric species. Chepko-Sade and Shields (1987) suggest a method that allows for any form of dispersal distribution. If neighborhood extent is defined as the 85th percentile dispersal distance (calculated in the same way as the 50th percentile, the median), then a circle with this distance as radius would be equivalent to the neighborhood area calculated on the basis of a normal distribution of dispersal distances, and would contain 85% of the potential parents of a central individual. Differences between the two methods depend on how close to normal a distribution is.

The genetic structure of populations, and thus levels of naturally occurring inbreeding, depend in part upon social structure, mating system, and demography. Effective population size is reduced (and chance of inbreeding increased) if: (1) some individuals leave many more offspring than others, either because of the mating system (e.g., polygyny) or variation in reproductive success; (2) generations overlap, so that offspring can mate with their parents; or (3) dispersal is limited. These factors are particularly important in cooperativly breeding species where not only do individuals tend to be long-lived but the offspring of a breeding group frequently remain with the group after they reach sexual maturity. Many long-term studies of marked populations which have provided information on genealogies have been of cooperative breeders: the Florida scrub jay, the acorn woodpecker, and the splendid fairy-wren. These species all show similar characteristics of limited dispersal, high survival, and high variance in reproductive success. Some groups (or families) produce a great number of breeding birds, and the members of these groups have a high chance of mating with relatives, simply because a high proportion of the potential mates available to them are relatives. This high variance in reproductive success occurs not only in cooperative breeders, but in most other species for which long-term studies have been completed (see species discussed in the books on lifetime reproductive success edited by Clutton-Brock [1988] and Newton [1989]).

Long-term studies of cooperative breeders have allowed estimates to be made of their effective population size N_e and the fixation coefficient F_{ST}. The values calculated for N_e are of the same order of magnitude as N_e for birds derived from electrophoretic surveys (on the order of 300; Barrowclough 1983); these values are sufficiently large to indicate that genetic differentiation among subpopulations is unlikely to occur rapidly, and are larger than the population size for which inbreeding effects are likely to be severe. Values of F_{ST} were < 0.1 for both the Florida scrub jay and the acorn woodpecker. The data do not suggest that effective population size for cooperative breeders is particularly small. However, estimates of inbreeding from effective population size are based on the assumption of random mating in a population of finite size. As Lande and Barrowclough (1987) point out, in actual populations, matings between relatives could increase inbreeding considerably beyond that resulting from finite population size, and in cooperative breeders, such matings do occur.

Effective population sizes of < 100 were calculated from long-term demographic data (1975–1991) for two species of Darwin's ground finches on Isla Daphne Major, Galapagos (Grant and Grant 1992). Harmonic mean breeding population sizes were 94 for *Geospiza scandens* and 197 for *G. fortis*. Effective population sizes N_e were much lower, principally as a result of large variance in the production of recruits per parent. Average effective sizes were 38 for *G. scandens* and 60 for *G. fortis*. Despite their relatively small effective population sizes and the likelihood of genetic impoverishment through random drift and inbreeding, they remain genetically variable through gene flow, principally hybridization between *G. scandens*, *G. fortis* and *G. fuliginosa*.

Electrophoretic Studies

Indirect estimates of the level of inbreeding may be obtained by measuring the amount of heterozygosity in a population. If this is lower than expected, or lower than in other populations, then inbreeding may be inferred. Electrophoretic studies of some populations have shown that heterozygosity is such as to suggest that inbreeding is at a low level (Evans 1987). Mean observed heterozygosity from electrophoretic investigations of eighty-six species was calculated by Evans (1987) as 0.044. From levels of heterozygosity, values of F_{ST} can be calculated, reflecting the extent of local differentiation into subpopulations. Evans calculated a value of $F_{ST} = 0.048$ from data for twenty-three species. Few studies have been done on species for which genealogical or demographic information are also available. Johnson and Brown (1980) studied allozyme variation in a population of the cooperatively breeding Australian grey-crowned babbler (*Pomatostomus temporalis*), and found no deficiency of heterozygotes over expected values (average heterozygosity 0.094). However, although they concluded that there was no evidence of close inbreeding, they recognized that while calculated values of the inbreeding coefficient F were not significantly different from zero, the upper confidence limit of F (0.105) was consistent with substantial inbreeding (for a full-sib mating, $F = 0.25$), and concluded that "clearly direct evaluation of inbreeding is preferable to the indirect approach we have used." Similar data were collected for the acorn woodpecker by Mumme et al. (1985). Average heterozygosity was 0.032, again comparable to that of other bird species. The average value of F was 0.172, but the confidence intervals were so large that the data is consistent with interpretations of either no inbreeding or high inbreeding; it should be noted that these estimates were based on only two polymorphic loci.

A CASE HISTORY—THE SPLENDID FAIRY-WREN (*MALURUS SPLENDENS*)

As an example of the difficulties involved in documenting inbreeding in natural populations, we present data from a long-term study of *Malurus splendens*, a small (10 g), highly dimorphic passerine endemic to Australia. A color-banded population has been studied for 16 years and was found to live in stable groups maintaining territories all year round, within which the female builds her nest, lays and incubates her eggs, and raises nestlings that are fed by all group members (Rowley and Russell 1990a).

Relatedness of Social Pairs

Until 1985, we assumed (as do most other ornithologists) that the social bondings we observed were the same as the matings that produced the eggs. When a breeding vacancy occurred, the replacement male or female frequently came from within the social group, and since group members were usually the progeny of earlier years, this led to a large number of apparent parent-offspring matings or to matings between siblings (table 13.2), and this was the situation that we reported in 1986 (Rowley, Russell, and Brooker 1986). Comparison of nesting attempts by related pairs with those of pairs known not to be closely related found no difference in productivity. Since productivity is affected by age and experience, the best comparison is between production by mother-son pairs and that by all experienced females. Leaving out all nests that produced no fledglings because of parasitism by cuckoos, 282 nests of experienced females paired with unrelated males produced 1.7 ± 1.3 fledglings, and 51 nests of experienced females paired with their sons produced 1.8 ± 1.5 fledglings. This figure is less than the value of 2.6 fledglings per nest quoted by Rowley, Russell, and Brooker (1986) because the effect of a severe wildfire in the study area has depressed productivity significantly since 1984 (Rowley and Brooker 1986). Our earlier paper coincided with the review of inbreeding by Ralls, Harvey, and Lyles (1986), which highlighted the exceptional nature of the 21% occurrence of inbreeding in *M. splendens* reported by us then and which is even higher (26%) on current calculation (see table 13.1).

Electrophoretic analysis

Since that time we have sampled the blood of ninety-one nestlings and their "parents" from twenty different groups of wrens and analyzed the allozyme variation by electrophoresis (Brooker et al. 1990). These analyses have shown that more than 60% of the nestlings examined could not have been sired by any male in the social group that raised them. In every case the female that incubated the eggs was genetically compatible with the nestlings sampled, so that neither egg-dumping nor intraspecific brood parasitism need to be considered for this species. Obviously the social bondings were not reflected in fertile matings. Mean observed heterozygosity was 0.051 (average proportion of heterozygous loci per individual), and the average coefficient of inbreeding, F, was calculated to be within the range 0.029 ± 0.068 to 0.063 ± 0.074 ($\pm$ 95% confidence limits). As with other estimates of F, confidence intervals are large, and the lower and upper limits would

TABLE 13.2   Relationship of Individuals in "Inbred" Pairs of *Malurus splendens*

Pair	N
Mother-son	20
Brother-sister[a]	8
Father-daughter	12
Aunt-nephew	1
Uncle-niece	1
Grandfather-granddaughter	1
Half-sibs	1
Assorted[b]	7

Note: N = 51, data for breeding seasons 1973–1987; 321 group-years. Coefficient of inbreeding (F) for the population, calculated from the above data, weighted by the number of nests for each pair (assuming all pairs not known to be related were entirely unrelated): $F = (120 \times 0.25 + 12 \times 0.125 + 2 \times 0.22 + 24 \times 0.375 + 4 \times 0.39 + 2 \times 0.3125 + 3 \times 0.2813)/680 = 0.0646$
[a]Includes individuals with $F > 0.25$; e.g., brother/sister whose father was also the father of their mother.
[b]Includes individuals which were half-sibs via one parent and also connected via the other parent; e.g., mothers were also sisters, etc.

be consistent with either no inbreeding or with significant inbreeding. Such a result must be interpreted with caution, in the light of all available information.

Philandering

In the course of this study we have documented 202 instances in which a male in full breeding plumage intruded into a territory that was not his own (Rowley and Russell 1990b). Such intrusions were quite different from territorial border disputes, which involve several members of the conflicting groups. An intrusion was usually by a solitary male, although occasionally two males from different groups intruded at the same time. In 85 of these 202 instances the intruding male was carrying a flower petal (usually purple or pink), which was only rarely presented to the resident female and only once accepted. On 5 other occasions the female crouched and fluttered her wings, apparently soliciting copulation, in response to the display of the intruding male. In 40 of the 202 cases, the intruding male approached the female and gave a face-fan display, erecting the pale blue feathers of the cheeks laterally. As this is a usual courtship or precopulatory display, and in the absence of conspicuous copulation in this species (39 cases in 16 years) we feel justified in assuming that these intruding males were *philandering*, seeking extra-group copulations. This assumption is further sup-

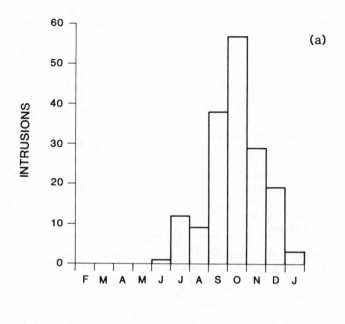

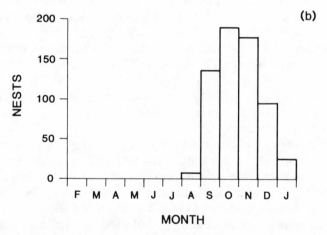

MONTH

FIGURE 13.1   Occurrence and timing of philandering by males in territories of *Malurus splendens*. (a) Frequency per month of intrusions by males. (b) Monthly distrubition of laying date (first egg of each clutch) for 629 nests of *M. splendens*.

ported by the observation that these intrusions occur mainly in the breeding season (Figure 13.1) and often during the fertilization period (sensu Birkhead 1988) of the subject female (table 13.3). A total of 94 of

TABLE 13.3 Timing of 202 Intrusions Relative to the Date of Clutch Initiation by Female Visited

Timing of Intrusion (days from first egg)	Number of Intrusions
June–August	21
>20 days before	r13
11–20 days before	22[a]
Building nest (nest not laid in)	13
1–9 days before	63
0 = day of first egg	10
1–2 days after[b]	8
3–10 days after	14
11–20 days after	14
>21 days after	25
Unknown	7

[a]Includes 6 intrusions when the female was nest-building; some early nests remained empty after completion for 1–2 weeks before the first egg was laid.
[b]The three-egg clutch is completed in 3 days.

202 intrusions (47%) occurred when the female was either building her nest or laying, and 14 (7%) of the intruding males were senior males philandering within 3 days of their "own" female laying her first egg. The majority (64%) of intruders were from an adjoining territory. Most of the known (color-banded) intruders (77%) were what we call "senior" males, the oldest male in a group, the leader in territorial defense, and the primary consort of the breeding female. Some intruders were older helper males; first-year males rarely left their territories except during a boundary dispute.

Most of our earlier observations of philandering were the result of unexpected fortuitous encounters. However, since our interest in this behavior has intensified, watches at nests where females were building or laying have regularly recorded philandering males. The events we saw must, therefore, represent only a fraction of the attempts by sexually mature males to achieve EPC, but there can be little doubt that such EPC are responsible for the effective cuckolding of the putative senior males in most groups, leading to the observed genetic confusion.

Dispersal and Effective Population Size

Our data on dispersal in *M. splendens* suggest that most movements were small (figure 13.2) and that although slightly more females than males dispersed, 40% of birds filling breeding vacancies were in their

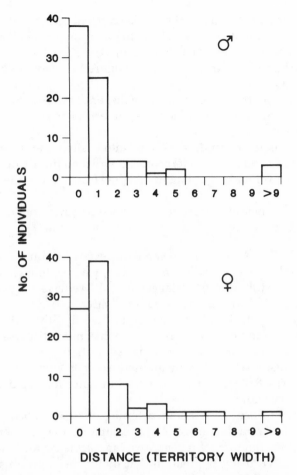

FIGURE 13.2   Dispersal distances between natal and breeding territories for 77 male and 83 female *Malurus splendens*. Distance is measured in territory widths. Dispersal distance of zero indicates that the bird filled a breeding vacancy in its natal territory.

natal territory, 40% were in an adjacent territory, and 2.5% were up to eleven territories away (Russell and Rowley 1992). The standard deviation of dispersal distance is 1.89 (measured in territory widths, taking the mean for 77 males and 83 females that have filled breeding vacancies in their natal territories or elsewhere). The 85th percentile distance for observed dispersal is two territory widths. We have faced the same problems as other researchers who have attempted to measure dispersal, and it is worth commenting on the problems in detail.

1. The perimeter of the study area is searched each breeding season. Considerable areas outside the wren study area form part of a study area for another project; the few long-distance dispersals (> 9 territory widths in figure 13.2) were located in this way.
2. If large numbers of long-distance dispersals were being missed, more dispersal four to eight territories away would be expected.
3. Most unknown immigrants came into territories at the edge of the study area. Central territories (more than three from the edge) never had replacement breeders from outside the study area.
4. Most vacancies for breeders were filled by birds from no more than two territories away.

Undoubtedly we have missed some dispersals, but only a fraction of individuals that disperse will come to occupy a breeding territory, and the skew in distribution of distances is real. We have calculated a revised dispersal distribution based on Waser's (1985) model for a straight-line dispersal with probability t = .28 that the natal territory is vacant (i.e., zero dispersal). This gives a maximum dispersal distance of twelve territory widths and a standard deviation of 2.6, based on 200 individuals dispersing, allowing an increase of 25% on the 160 known dispersers. The 85th percentile dispersal distance for this distribution is five territory widths.

Our present uncertainty about paternity makes it difficult to calculate an effective population size N_e for comparison with other species. Using Wright's equation, $N = 4\pi\sigma^2$, we can calculate a range for neighborhood size, N, using population density (= mean group size; ρ) of 3.26 and the two estimates of the standard deviation of dispersal distance (σ). The first estimate ($SD = 1.9$) is 148 and the second ($SD = 2.6$) is 277. This gives a maximum value of N_e.

Methods of adjustment of N to allow for sex ratio and the effects of overlapping generations depend on a knowledge of which males are breeding and with what success. This is clearly unknown, because a large number of EPC occur, but to get some idea of the size of the effect, we can make certain assumptions. There is no doubt about maternity; all eggs were laid by the designated breeding female. Since most observed intruders were senior males, an assumption of sex ratio of one male per female is not unrealistic. For the calculation of the effect of overlapping generations we have assumed that males and females are equivalent in reproductive success, and have followed the method of

Charlesworth (1980b) based on that of Emigh and Pollak (1979) employed by Barrowclough (1980), Woolfenden and Fitzpatrick (1984), and Koenig and Mumme (1987). Population size may be considered relatively stable for the purpose of these calculations. Because of overlapping generations and other demographic considerations, the effective population size is 0.516 of the neighborhood size. Applying this correction, our estimates yield $N_e = 76$ and 143.

The next step is to adjust for variation in offspring number using the mean and variation in lifetime reproductive success. This is known for females, and we can calculate a similar figure for males assuming that there is no EPC. Since the majority of philandering males that were identified were males already designated as senior "breeding" males in other groups, this may be a reasonable approximation. If EPC increases the variance in male reproductive success, the effective population size will be even further decreased. Following the method used by Koenig and Mumme (1987: 261, equation 3), and using mean and variance of lifetime number of one-year-olds produced of 4.61 ± 13.91 for males and 3.42 ± 14.82 for females, we can calculate a proportional reduction in N of 0.51 for females and 0.70 for males. Assuming equal numbers of males and females in our estimates of N_e (= 76 and 143), we estimate N_m and N_f. Combining these estimates for the two sexes in Wright's (1931) formula,

$$N = \frac{4\, N_m N_f}{N_m + N_f}$$

results in estimated N_e values of 45 and 84. Similarly, we can calculate N_e using two estimates of 85th percentile dispersal distance, two and five territory widths, which gives values for neighborhood size of 41 and 256, and after adjustment in the same way, gives estimates of N_e of 12 and 77. For effective population size of 84, and assuming a universe of 10^6 demes, $F_{ST} = 0.085$.

This very approximate value of N_e (< 100) is small in comparison with other estimates for birds by Barrowclough (1980; 176–7700), Woolfenden and Fitzpatrick (1984; 290–300 for the Florida scrub jay), Barrowclough and Coats (1985; 220 for the spotted owl), Koenig and Mumme (1987; 81–961 for the acorn woodpecker) and Grant and Grant (1989; 75–190 for the large cactus finch *Geospiza conirostris*), but is of the same order of magnitude. However, Grant and Grant (1992) calculated values of N_e for two other species of Darwin's ground finches on Isla Daphne major, Galápagos: 60 for *Geospiza fortis* and 38 for *G. scandens*. Baker (1981) calculated a value of N_e for the white-crowned sparrow of

36, using a slightly different method. Our estimate of F_{ST} of 0.085 is somewhat higher than the estimates by Barrowclough (1980) of 0.04 or less, and the mean of 0.048 from electrophoretic data for twenty-three species calculated by Evans (1987).

In a population such as this, where some matings between close relatives do occur, the level of inbreeding is above that resulting only from the small finite size of a population that is mating at random. In about two-thirds of broods tested by Brooker et al. (1990), at least one young was compatible with a male in the group. Thus we can estimate that although the apparent level of close inbreeding in M. *splendens* is about 26% (at 26% of nests, the breeding female was socially bonded to a closely related male), approximately 5% of offspring were the result of matings between related individuals, which is close to the levels recorded for other species. However, since most intrusions are by neighbors and dispersal distances are small, it is likely that in many EPCs males will mate with a related female.

DISCUSSION

The deleterious effects of inbreeding in small captive populations or remnant populations of endangered species are documented and well known (Ralls and Ballou 1983). The perception of inbreeding as harmful frequently colors our perception of events in natural populations, and leads us to see mechanisms for avoiding inbreeding where they may not in fact exist or where they may not be needed. A balanced view of inbreeding in natural populations must take account of the costs and benefits both of inbreeding and of outbreeding (e.g., Bengtsson 1978; Shields 1982; Bateson 1983).

This chapter and other recent attempts to review inbreeding in natural populations of vertebrates have shown that there are few studies that document actual levels of inbreeding (Greenwood 1987; Hoogland 1992; Smith, chapter 14, this volume; Waldman and McKinnon, chapter 11, this volume). This is largely because such data are difficult to obtain, since large numbers of marked individuals must be followed for many years to obtain the necessary genealogies. After several years, we may know that a mated pair are brother and sister, but it takes many more years of data to know if the brother and sister are more closely related than $F = 0.25$ because of a long history of limited dispersal. Birds which after 10 years we assume to be "unrelated" may share common ancestors at the level of great-grandparent; that information was available for the Dutch study reported by van Noordwijk and Scharloo (1981), but that study was unique because it began in

1912. Indirect biochemical approaches allow the calculation of levels of heterozygosity and values of F or F_{ST}; such data are available for more species than there are genealogical data for, but they are generally not correlated with direct information on inbreeding. It is therefore difficult to know what are "normal" values.

It is clear that demographic variables such as patterns of dispersal, survival, or variance in reproductive success can have a considerable effect on levels of inbreeding (Barrowclough 1980; Lande and Barrowclough 1987). But again, we do not yet know enough to say what is a normal or "expected" level for a certain life history pattern. Patterns of restricted dispersal are characteristic of many vertebrates, with the consequence that close relatives are frequently accessible as mates (Smith, chapter 14, this volume; Waldman and McKinnon, chapter 11, this volume). Some recent reviews of inbreeding in birds (Ralls, Harvey, and Lyles, 1986; Greenwood 1987) have paid little attention to the significance of demographic variables. The demographic factors affecting effective population size have been outlined above and discussed in detail by Chepko-Sade et al. (1987), Lande and Barrowclough (1987), and Rockwell and Barrowclough (1987). In birds, the demographic "climate" most favorable to inbreeding is likely to occur in populations of long-lived resident (nonmigrant) species such as are found in tropical and warmer temperate regions (Fogden 1972; Snow and Lill 1974; Rabenold 1985; Rowley and Russell 1991). In such species, adult survival is high and reproductive rates are frequently low, leading to considerable overlap of generations, with dispersing birds generally not traveling far; consequently effective population sizes are likely to be relatively small, reduced even further by variation in progeny production. Recent complications of studies of lifetime reproductive success suggest that such variation is considerable and generally widespread (see chapters in Clutton-Brock [1988] and Newton [1989]). These are the same regions that appear to favor cooperative breeding (Brown 1987). Disregarding the occurrence of inbreeding between close relatives, the background level of inbreeding could be high simply because of population structure, unless balanced by genetic migration between subpopulations. However, in the context of host-parasite evolution discussed by Seger and Hamilton (1988), it is just these species that should avoid inbreeding, with its effect of reducing recombination.

The likely episodic nature of inbreeding is often not fully appreciated. In theory, the effect of fluctuations in population size in reducing effective population size can be calculated, as discussed by Chepko-Sade et al. (1987) and Lande and Barrowclough (1987). In practice, we can see these effects operating in *M. splendens*, where during a period of population increase, with high productivity and survival, more females remained in groups, frequently breeding as secondary females in their natal group, with the possibility of mating with father or brother. During a period of low productivity after a major fire, the population decreased, female mortality increased, and females were replaced from outside the study area. Thus a period of high probability of inbred matings was followed by a period of immigration by "unrelated" individuals.

Assuming that mating with a close relative could lead to inbreeding depression, it is reasonable to ask whether a certain behavior is the result of selection to reduce the level of inbreeding, but there is as yet no reason to assume that every species has evolved some form of inbreeding avoidance mechanism. In birds, dispersal is frequently female-biased, with females dispersing farther than males (Greenwood 1980), and females that move farther are more likely to avoid inbreeding. Dispersal is most frequently invoked as an inbreeding avoidance mechanism. This view has been criticized by Moore and Ali (1984), and as Dobson and Jones (1985) point out, there is no need for a unitary explanation of dispersal. Alternative hypotheses are possible, for example, that sex differences in dispersal are caused by other factors, and that birds do not in fact avoid inbreeding at all (Greenwood 1987). An experimental study of mate choice in small flocks of zebra finches given a choice between sibs and nonrelatives or between cousins and nonrelatives found no evidence of inbreeding avoidance (Schubert, Ratcliffe, and Boag 1989). In field studies, there was no evidence of inbreeding avoidance in Darwin's medium ground finch (*Geospiza conirostris;* Gibbs and Grant 1989) or the pukeko (*Porphyrio porphyrio;* Craig and Jamieson 1988).

So far, most discussion of inbreeding in natural populations is based on the assumption that observed pairings reflect mating patterns. It is becoming more and more obvious that this is not so, and that in many bird species, levels of EPC are significant. Without more exact determination of parentage by biochemical means, we cannot be sure of levels of inbreeding in natural populations. Mixed mating strategies involving EPC may allow an individual to produce more offspring than it would in a monogamous situation, but they could also have the effect of reducing inbreeding in situations where high survival and limited dispersal increase the chance of pairing with close relatives. The general level of EPC is such that either inbreeding avoidance is rendered unnecessary or EPC is itself either an inbreeding avoidance mechanism or an adaptation for outbreeding. For both male and female, one benefit of EPC is increased genetic diversity of offspring (Westneat, Sherman, and Morton 1990). In a species like *M. splendens*, this includes the possibility of producing both inbred and outbred offspring. The episodic nature of inbreeding due to demographic stochasticity as discussed above may also render inbreeding avoidance unnecessary. In a continuous population, the occasional decline in a local subpopulation may be restored by an influx from outside, thus maintaining levels of heterozygosity.

The costs of inbreeding avoidance must be considered, as well as the benefits. Where the costs of avoiding inbreeding are high, for example, in dispersal or a delay in reproduction, mating with a relative nearby might have some genetic cost, but a considerable benefit to lifetime fitness. This was suggested by Bengtsson (1978) and Waser, Austad, and Keane (1986), and is supported by our data on *M. splendens* (Rowley, Russell, and Brooker 1986; Rowley and Russell 1989), where the most important factor in lifetime reproductive success is the acquisition of a territory as soon as possible. The mate that goes with this territory may be closely related, but the high level of EPC renders that of little significance, especially for the longer-lived male, whose mate is likely to change before his life as a breeder is over.

The possible benefits of inbreeding have been incorporated in the hypothesis of optimal outbreeding as discussed by Bateson (1978, 1983), Price and Waser (1979), and Shields (1982; chapter 8, this volume). Too great a genetic dissimilarity between mates may lead to outbreeding depression, just as too great a similarity leads to inbreeding depression. The hypothesis of optimal outbreeding proposes that there is a level of similarity between mates where fitness is maximized, somewhere between close inbreeding and distant outbreeding.

Implicit in the statement that animals avoid mating with close kin is the necessity for kin recognition. In many cases it is more likely that some behavior such as dispersal (for whatever reason) renders inbreeding unlikely. In polygynous mammals, Clutton-Brock (1989) has shown that where females commonly remain to breed in their natal group, their average age at first conception typically exceeds the average period of residence of adult males in breeding groups. In contrast, where females usually transfer to breed in other groups, the average residence of breeding males typically exceeds the average age of females at first conception. This latter situation applies in *M. splendens*, except that a breeding female is replaced from within the group if a suitable candidate exists. Payne, Payne, and Rowley (1988) demonstrated that *M. splendens* recognized individuals on the basis of social familiarity rather than kinship. Nesting females responded aggressively to the playback of song of close kin from other groups while ignoring the song of unrelated members of their present social group. In the large cactus finch *Geospiza conirostris* on Isla Daphne Major, Galápagos, in spite of the high probability by chance of close relatives mating with each other, no incestuous matings were recorded, and no female mated with a male of the same song subtype as her father (Grant and Grant 1989), suggesting the possibility that females use detailed characteristics of song to avoid mating with close kin, but the sample of females whose father's and mate's song subtypes were known was too small to test this hypothesis.

To date, electrophoretic analysis has shown by parental exclusion that observed pairings and pedigrees are not reliable indicators of genetic relationships. As Greenwood (1987) and Koenig and Mumme (1987) have pointed out, the future use of such technologies as DNA fingerprinting will lead to a new era of precision in the unraveling of true relationships and effective mating strategies in wild populations of birds. Until such time it is not possible to say who copulated effectively with whom and therefore the true genealogies, the frequency of inbreeding, and the true lifetime reproductive success of individuals, especially males, cannot be calculated. Recent work (Quinn et al. 1987; Wetton et al. 1987; Burke et al. 1989; Rabenold et al. 1990) has shown the level of accuracy and confident conclusions that can result from such techniques and their use in the study of mating systems. Their application to long-term studies is eagerly awaited.

Although the detail of effective matings remains obscure, there is an indication that even if inbreeding occurs frequently, it does not appear to decrease significantly either the survival of progeny or the level of individual fitness. Based on observed social pairs, inbred pairs of *Parus major* produced more grandchildren than did unrelated pairs (van Noordwijk 1987). It is the production of fledglings that is depressed by inbreeding; while this is an important component of reproductive success and one that is frequently used as a measure in short-term studies, it is only the starting point of rearing progeny to the age when they can enter the breeding population. Shields (chapter 8, this volume) discusses the need to evaluate the total reproductive success of inbreeding pairs in view of the possibility that despite higher juvenile mortality of the offspring of inbred matings, the inbred young that do survive may be of high quality. Most species of birds have a juvenile mortality of at least 30%—a mortality that is unlikely to be influenced by inbreeding "depression," but which could well be reduced by an increased duration and intensity of parental care devoted to fewer fledglings and by an avoidance of the hazards of dispersal through unknown places in a more sedentary population. In *M. splendens*, the production of fledglings by related pairs was no lower than that of unrelated pairs. Clearly the number of inbred young is much lower than the level of related "social" pairs would suggest, but there is no indication that the level of inbreeding (whatever it is; perhaps about 4% to 5%) leads to reduced fitness in the short term. Long-term effects of increased homozygosity may lead to reduced fitness in the face of changing environmental conditions.

What little data there are on inbreeding in natural populations of birds suggest that the low levels that occur are not such as to lead to any deleterious genetic consequences in the short term. What is perhaps most important to investigate is whether any real inbreeding avoidance mechanisms do actually exist in birds, or whether so-called inbreeding avoidance is merely a consequence of other behavioral or demographic constraints. If these constraints cease to operate, as populations become smaller and more isolated due to environmental change, then the loss of heterozygosity and inbreeding depression may begin to affect species or populations not otherwise considered endangered. On the other hand, true inbreeding avoidance mechanisms operating in small populations may further reduce the effective population size.

CONCLUSIONS

Data on the level of inbreeding in natural populations of birds are few. Studies of observed pairings in marked populations have provided genealogical information and allowed the estimation of the incidence of pairings between closely related individuals, usually less than 5%. Recent evidence of significant levels of effective extra-pair copulations in many species of birds suggests that inferences from observed social pairings of individuals must be treated with caution. The discrepancy between observed pairings and actual mating systems will generally be likely to decrease the apparent levels of inbreeding.

Other estimates of levels of inbreeding provided by observed heterozygosity from electrophoretic studies, and from considerations of genetic population structure, also indicate that inbreeding occurs in natural populations but at levels unlikely to be detrimental.

However, long-term studies of marked individuals, which provide longer pedigrees, indicate that although inbreeding between close relatives may be infrequent, at least 20% of matings may be between more distant kin. Valid inference that inbreeding is avoided or outbreeding preferred in natural populations can be made only if it can be shown that levels of inbreeding are below (or outbreeding, above) what would occur if random mating occurred in the same population (see Shields [chapter 8, this volume] for an easy technique for estimating this). In the few cases where this information is available, no inbreeding avoidance was apparent, despite 20% mating between related individuals (shared grandparent or closer).

Calculations of effective population size for birds, as for other vertebrates, show that many birds live in small, partially isolated and therefore mildly inbred units arising from a variety of social and demographic factors. Before we can appreciate the significance of seeking or avoiding mating with relatives, we need to know the genetic consequences of these different social structures or demographic patterns. Such information is not easy to acquire and little is available.

An understanding of inbreeding in natural populations is necessary for its management in remnant populations. However, such an understanding will not be possible until genealogical data based on known genetic parentage is integrated with information on effective population size derived from studies of demography and social organization.

14

The Natural History of Inbreeding and Outbreeding in Small Mammals

Andrew T. Smith

INTRODUCTION

The phenomena of inbreeding, outbreeding, philopatry, and dispersal all have potential costs and benefits to individuals in natural populations (Shields 1987 and numerous chapters in this volume). Ideally, data from natural populations should give us insight into these phenomena so that we can better understand their evolution and make prescient decisions about population management. In this empirical review I will examine the costs and benefits of these phenomena in small mammals with these objectives in mind. I have introduced philopatry and dispersal at the outset because of the important roles they play in the determination of inbreeding and outbreeding in small mammals. The distinctive biological attributes of small mammals make them ideal subjects for such investigations. However, the results of my survey may be disquieting to some—especially those looking for unambiguous solutions to the problems of whether small mammals preferentially avoid or court inbreeding, or whether they (teleologically) pay any attention to their inbreeding coefficient when deciding to mate or disperse. Although there are some distinctive trends among small mammals, for the present we must be content to appreciate the diversity of approaches they take. And even these (apparent) trends should be considered suspect: very few studies of small mammals directly address inbreeding versus outbreeding and dispersal versus philopatry. I fear that there could be a very real sampling problem when those species that have been studied are compared with the species pool of all small mammals. In this review I will explore the possibilities of adaptive responses of small mammals with regard to inbreeding and out-breeding by presenting detailed examples from a relatively limited sub-set of species. I hope that this approach will focus attention on what is known and, more important, how much more we have to understand.

Definitions

Not only are there difficulties with the data available for such an empir-ical review, but there are formidable problems of term and question definition. For example, what is meant by inbreeding, outbreeding, philopatry, and dispersal? In one sense each of these terms may be con-sidered as a continuous variable, and perhaps that is the best approach for correlating these phenomena with dependent life history variables for single-species investigations. For comparative work, however, it is necessary to construct sharper distinctions to allow compilation of data from diverse studies. Here I treat three levels of genetic similarity among mates: "close inbreeding" (sensu Ralls, Harvey, and Lyles 1986; termed "extreme inbreeding" by Shields 1987; also frequently termed "incest"), which encompasses parent-offspring and sib-sib matings; "moderate inbreeding," mating between individuals related at approx-imately the level of first cousins; and "outbreeding," mating between individuals that are more distantly related than first cousins. The last two categories are necessarily vague because, unlike those in labora-tory pairings, free-living mammals have history on their side, and ani-mals available as mates within a region are likely to share genes by descent. It is for this reason that I do not use the definition of inbreed-ing given by Shields (1987): "the mating of relatives that share greater common ancestry than if they had been drawn at random from an en-tire species." Although this definition properly draws attention to the nonrandom spatial organization of genotypes in nature, it does not al-low for subdivision of matings among closely and distantly related in-dividuals. Shields' definition precludes the description of most natural occurrences of what is generally considered outbreeding. But neither do I concur with those who tout close inbreeding as the only operative definition of this phenomenon (Ralls, Harvey, and Lyles 1986). We lose too much valuable information by ignoring moderate inbreeding, as will be apparent in several of the examples given below (see also Bate-son 1983; Shields 1983).

Definitions of philopatry and dispersal can be equally confusing. Here I treat philopatry as either nondispersal or dispersal that is so restricted that it does not alter the possibility of close inbreeding with nondispersing parents or sibs (normally movement of only one home range from site of origin; see Smith and Ivins 1983a). So defined, philopatry can be viewed as influencing the potential for close inbreeding. This definition is more liberal than that of Waser and Jones (1983), who considered only nondispersal as philopatry, and more restrictive than that of Shields (1987), who allowed movement as great as ten home ranges from the site of origin as philopatry. Because of the nature of the original literature I have had to make value judgments on data for both inbreeding versus outbreeding and philopatry versus dispersal.

Questions

The important questions with regard to inbreeding versus outbreeding and philopatry versus dispersal have been similarly confused. Acrimonious debate has raged over the extent and effect of inbreeding in natural populations, and normally this has been due to differences in the goals, approaches, and definitions used (compare Shields 1982, 1987 and Ralls, Harvey, and Lyles 1986). Here I briefly outline what I consider to be the important questions and how they should be approached.

The most basic questions concern documentation of the occurrence of inbreeding in nature and what this tells us. Shields (1982, 1987) argued that close inbreeding occurs frequently in natural populations of vertebrates. However, in their review of this subject, Ralls, Harvey, and Lyles (1986: 56) determined that close inbreeding rarely occurred in a sample of twenty-eight bird and mammal species. They closed with the statement: "[currently] . . . it is unjustified to conclude that close inbreeding, non-incestuous inbreeding, optimal inbreeding or, indeed, optimal outbreeding prevails in natural populations of birds or mammals." An unfortunate omission from this declaration is that it is equally true that inbreeding avoidance does not prevail in natural populations. I will describe tremendous variability in levels of inbreeding expressed by species of small mammals.

Regardless of the frequency of close inbreeding observed in any population, the global issue revolves around whether documentation of inbreeding in nature speaks directly to the question of inbreeding avoidance. Ralls, Harvey, and Lyles (1986) correctly acknowledge that data on observed frequencies of inbreeding should be matched against models of expected matings to determine accurately whether animals are actively avoiding inbreeding or not; however, they did not pursue this tactic. Instead they argue that the low frequency of observed inbreeding results from inbreeding avoidance. The small mammal data will reveal complexities in the relationship of levels of inbreeding and inbreeding avoidance to other aspects of the life history of a species. The most important of those factors appear to be the high rate of mortality of small mammals and the geometry of the dispersion of natural populations following dispersal (P. M. Waser 1985, 1988). Each of these factors may dramatically reduce the probability of inbreeding independent of inbreeding avoidance. In addition, some animals actively avoid inbreeding, whereas many do not. Clearly, inbreeding avoidance cannot necessarily be predicted by the degree of observed inbreeding.

The next hierarchical question asks whether there are genetic costs or benefits to inbreeding or outbreeding. Shields (1987) summarizes a complex literature on this subject and outlines such potential costs and benefits: inbreeding may unmask deleterious recessive alleles and may result in reduced adaptive variability in progeny, termed "inbreeding depression"; inbreeding may be adaptive due to maintenance of locally adapted gene complexes or intrinsically coadapted gene complexes, termed "optimal inbreeding"; outbreeding may result in the disruption of genetically coadapted complexes of alleles or the production of maladapted hybrid young, termed "outbreeding depression," or outbreeding may enhance viability of progeny and minimize the effects of inbreeding depression. Of these phenomena, inbreeding depression has received the greatest attention. Unfortunately, at present we have few data from inbred matings in natural populations from which to estimate the extent of inbreeding depression. One approach, then, has been to evaluate intensively one of the few good data sets available on inbreeding, that of pedigrees from zoo populations (Ralls, Ballou, and Templeton 1988). These data on forty captive populations belonging to thirty-eight species show an average increase in mortality of 33% in offspring from inbred matings compared with offspring of unrelated matings. Ralls, Ballou, and Templeton (1988: 191) conclude: "The total costs of inbreeding in natural populations are probably considerably higher than our estimates." When extrapolated, these data convincingly demonstrate the potential magnitude of inbreeding depression in natural populations. But as zoo populations do not have the historical precedence of natural population structure, these data must be interpreted with caution. Inbreeding depression is to be expected with recurrent matings between animals with dissimilar genetic backgrounds, typically found in zoo populations. And, theory and data demonstrate that populations with a history of inbreeding express little or no inbreeding depression (Templeton 1987).

A further problem with extrapolations of inbreeding costs from captive to natural populations is that this approach addresses only half of the cost/benefit issue regarding the natural history of inbreeding, that of inbreeding depression. When deciding with whom to mate, animals in the real world must balance the potential deleterious (genetic) effects of inbreeding against other, perhaps more compelling, aspects of their life history. Shields (1987) terms these aspects "somatic" to reflect the ecological stage upon which these decisions are made.

The life history (somatic) trait most often associated with low levels of inbreeding in small mammals is the predominant sex-biased dispersal of males (Greenwood 1980; Dobson 1982). Ralls, Harvey, and Lyles (1986) and Harvey and Ralls (1986) argued that male-biased dispersal has evolved as a mechanism to prevent inbreeding with the more philopatric females. Other authors, most notably Moore and Ali (1984), have countered that juvenile males disperse primarily to reduce competition with resident adult males. The model of Waser, Austad, and Keane (1986) suggests that dispersal is unlikely to have arisen primarily as a mechanism to avoid inbreeding. Dobson and Jones (1985) pointed out the important role of multiple causation in the evolution of mammalian dispersal patterns (see also Gaines and McClenaghan 1980; Stenseth 1983; Johnson 1986). Dobson and Jones (1985) outlined three functional explanations of mammalian dispersal: inbreeding avoidance (IA), competition for mates (CFM), and competition for environmental resources (CFR), and discussed whether each was a necessary and sufficient explanation for observed patterns. This review will examine the data relevant to these three competing hypotheses.

There are additional questions associated with the observation of male-biased dispersal in small mammals: (1) What distribution of dispersal distances is sufficient to characterize a species as exhibiting this trait?; and (2) What are the life history distinctions (here as they relate to inbreeding) differentiating species with different dispersal distribution patterns? In some species all males disperse, and compared with the largely philopatric females, the sex bias in dispersal is clear-cut. But in other studies, only a few males disperse, and most are philopatric and remain associated in space with the majority of philopatric females. For example, muskrats have been characterized as exhibiting male-biased natal dispersal, yet 58% of males (compared with 83% of females) are philopatric (Caley 1987). Only recently has attention focused on philopatric animals and their role in the determination of genetic population structure (Bateson 1983; Shields 1983; Waser and Jones 1983; Chepko-Sade et al. 1987). On the other hand, rare long-distance dispersers can have a tremendous impact on the promotion of an outcrossing population structure (Templeton 1987). That levels of heterozygosity and proportions of polymorphic alleles derived from allozyme studies are lower for mammals than for any other major taxon (Nevo 1978) may be indicative of a trend for genetic structuring due to population viscosity. However, we also know that a variety of historical, demographic, and selective factors can affect heterozygosity levels (Soulé 1976). For now, it appears inappropriate to assume that the stable social groupings frequently observed in small mammals lead to the formation of genetically closed groups until data on their temporal stability or formation are documented (Lidicker and Patton 1987). A primary interface between theoretical and empirical approaches to studies of inbreeding and population structure lies in the resolution of the relative contribution of philopatric and dispersing individuals to the genetic structure of mammal populations.

Attributes of Small Mammals

In many respects small mammals are ideal for studies of the natural history of inbreeding and outbreeding. They are abundant, and they can usually be caught and marked to allow individual identification upon subsequent observation. In general, most other aspects of the natural history of small mammal species (including the mating system and expressions of social dominance), are well known or can be determined. Small mammals (with the exception of bats) are normally confined to a particular geographic region, and their use of space, movements, and potential consorts with related and unrelated conspecifics can be documented.

Our knowledge of the natural history of most species of small mammals also allows us to define clearly the costs and benefits of philopatry and dispersal as they relate to inbreeding (Shields 1987). Compared with most other taxa (especially amphibians, reptiles, and birds, with which they are most often compared), mammals experience relatively high costs of dispersal. Because they are endothermic, mammals require substantially more energy per unit of time during dispersal than do amphibians and reptiles. The need to forage frequently while dispersing across unfamiliar terrain increases predation risks and exposure to potentially agonistic conspecifics. Because they are terrestrial, mammals expend greater energy during locomotion than do birds. Thus mammals, in general, have fewer opportunities to choose among mates or available habitats than birds do. The costs imposed by being endothermic and terrestrial vary depending on the spatial distribution (i.e., patchy versus continuous) of a species' preferred habitat. In summary, because of the generally high cost of dispersal in mammals and our ability to document the social milieu of philopatric animals before they "decide" whether or not to disperse, we are able to put into perspective the magnitude of the somatic costs of dispersal compared with potential genetic costs of inbreeding. These comparisons across species of small mammals are highlighted in this review, but also form a standard for comparison of mammals with other taxa.

The paucity of studies that have capitalized on mammals as ideal subjects for the study of inbreeding attests to the presence of practical difficulties in the documentation of important variables. First, most statements of mating patterns and histories within populations must be evaluated cautiously. The majority of small mammal species are nocturnal and cannot be directly observed, hence descriptions of their life history characteristics rely on indirect (generally trapping) techniques. It is understandable that much of this review concentrates on diurnally active species that may be observed directly, hence the potential for bias in this comparative treatment. Also, even in the most labor-intensive study it is nearly impossible to document the occurrence of all matings. The possibility of multiple paternity (Birdsall and Nash 1973; Hanken and Sherman 1981) and the observation that spatial proximity between breeding-age animals cannot always be used to predict matings (Hoogland 1982; Ågren, Zhou, and Zhong 1989) remain complications.

A second general problem is that despite the general progress that has been made in the area of "motivation" in mammals, we still do not know (and can rarely predict) why an individual animal disperses or decides to remain philopatric. Siblings within litters often have distinct behavioral profiles, and the social interactions leading to dispersal may vary greatly (Bekoff 1977; Lomnicki 1978; Armitage 1986a). Equal levels of observed aggression may cause one animal to disperse while another remains philopatric. Dispersal often occurs when agonistic behavior has not been observed. Our tendency has been to characterize the social environment associated with philopatry or dispersal across populations and to argue parsimoniously as to the adaptedness of such movements with regard to CFR, CFM, and IA. I will be forced to follow this tack here, but we should acknowledge that such conclusions are only approximations of what is undoubtedly a very complex situation.

An associated issue is that it has been frequently documented that mammals can recognize kin at various levels of relatedness (Holmes and Sherman 1983; Hepper 1986). Thus kin recognition may function as a mechanism (termed "opportunity not realized" by Ralls, Harvey, and Lyles 1986) for IA either through mate choice or suppression of reproduction in the presence of kin. Field workers confronted with this problem must determine whether kin recognition exists, what levels of relatedness are encompassed, and whether resulting mate choice (IA or preference for kin) or reproductive suppression is due to kin recognition.

The last critical problem is that we rarely have data sufficient to allow a direct comparison of somatic and genetic costs and benefits resulting from inbreeding versus outbreeding. Inbreeding and outbreeding depression have rarely been documented in natural systems. But in natural systems we should not expect to observe many inbred or outbred pairings in those cases in which genetic costs have prevailed over somatic costs and benefits. Thus we must extrapolate from behavioral, life history, and genetic data to determine whether the mating decisions of individuals may have been determined ultimately by genetic factors. Alternatively, proper experimental pairings of individuals of known degrees of relatedness (combining pedigrees and average levels of genetic homogeneity within subpopulations) should yield genetic data that could be contrasted with somatic constraints. This latter approach is necessary if we are to gain a holistic view of the natural history of inbreeding and outbreeding in mammals.

EXAMPLES FROM NATURAL POPULATIONS

I present a variety of examples drawn from the small mammal literature which speak to the nature of inbreeding and outbreeding. This approach allows me the freedom to explore the nature and quality of data from each species. My goal is to synthesize where we stand and outline the task that lies ahead as we confront the moving target of an increased understanding of inbreeding and outbreeding.

Antechinus stuartii and A. swainsonii

Ralls, Harvey, and Lyles (1986) list three species of well-studied small mammals for which no instances of close inbreeding are documented. Two of these are dasyurid marsupials, *Antechinus stuartii* and *A. swainsonii*. The absence of close inbreeding has been inferred from the highly unusual life history tactics exhibited by these species (Cockburn, Scott, and Scotts 1985). These two species probably come the closest of any mammal to exhibiting semelparity. Males breed as yearlings, showing a sharply seasonal period of reproductive activity that lasts 2 to 3 weeks. Following breeding, and before the birth of their offspring, they die. Females have a single estrous period per year, and most only live long enough to reproduce once (Cockburn, Scott, and Scotts 1985; Lee and Cockburn 1985).

This life history pattern allows a close examination of the mechanisms responsible for the lack of close inbreeding. Cockburn, Scott, and Scotts (1985) argue that close inbreeding is not realized because of the lack of opportunity for females to consort with related males during the breeding season. Dispersal is highly male-biased; most females are philopatric and almost all juvenile males disperse. The proximate cue for male dispersal occurs independent of the number of sibling males in a litter, and dispersal cannot be triggered by adult males because they are not present in the population. Thus CFM is not an adequate explanation for the observed male-biased dispersal. The period of dispersal is marked by an exchange among nests of related males for more distantly related males of the same cohort. As overall and individual nest (home range) density remain similar as a result of this exchange, CFR also appears inadequate to explain the dispersal pattern. Proximately, dispersal appears to result from maternal aggression toward sons, as removal of mothers at the time of weaning led to male philopatry in a majority of cases and to basic disruption of normal demographic processes. Only IA, presumably driven by the potential deleterious affects of inbreeding, can adequately explain this phenomenon (Cockburn, Scott, and Scotts 1985).

A subsequent analysis of one of these species, *A. stuartii*, has determined that it exhibits a lek promiscuity mating system (Lazenby-Cohen and Cockburn 1988). Although it is conceivable that related animals may consort at the lek sites, in reality the long dispersal distances of males preclude their mating with sisters. Only two (less than 1%) of all potential male/female combinations within leks were siblings (A. Cockburn, personal communication).

This example presents one of the strongest arguments that IA may occur in natural populations of mammals and that dispersal may exclusively result from avoidance of inbreeding. However, no data on actual matings are presented, and there was no demonstration that inbreeding depression would result from inbred matings (Cockburn, Scott, and Scotts 1985).

Spermophilus beldingi

The third species of small mammal that is not known to have produced inbred pairings is Belding's ground squirrel (*Spermophilus beldingi*). From a sample of 531 matings of animals of known relatedness, none have involved animals as closely related as mother-son, father-daughter, siblings, cousins, aunt-nephew, or uncle-niece (P. W. Sherman, personal communication; see also Holekamp and Sherman 1989). The primary mechanism for avoidance of inbred matings is lack of opportunity caused by female philopatry and universal dispersal of males (Sherman 1977, 1980; Holekamp 1984b; Sherman and Morton 1984; Holekamp and Sherman 1989). Not only do all juvenile males disperse, but so do most adult males that have mated. The more successful an adult male has been during the breeding season, the farther he disperses between seasons. Juvenile males who have not reached a threshold body mass and adult males who have failed to mate do not disperse (Sherman 1977, 1980; Holekamp 1986). Dispersal in *S. beldingi* appears to be costly even when the habitat over which dispersal occurs is uniform alpine meadow. Dispersing juveniles are heavier than nondispersers of the same age; however, their body mass subsequently drops below that of nondispersing males (Holekamp 1984b). The cost of dispersal appears related to the new social climate in which Belding's ground squirrel finds himself and not to the actual distance moved. Young males frequently negotiate the routes by which they will eventually disperse and then return to their natal burrow (Holekamp 1984b; Holekamp and Sherman 1989).

The timing, magnitude, and cost of male-biased dispersal in *S. beldingi* favors an explanation of IA over those of CFM or CFR. Holekamp (1984b) found no relationship between natal dispersal and food or nest site availability. There is no evidence that conspecific aggression promotes natal dispersal; mothers do not drive sons from their territories. Litter size and gender composition have little influence on dispersal in Belding's ground squirrels, and dispersal does not appear to be density dependent (Holekamp 1984b). Finally, regional density and dispersion of males remain roughly equal following emigration and immigration of males (Sherman 1977; Holekamp 1984b; Sherman and Morton 1984). Only the IA hypothesis is not negated by this suite of observations (Holekamp 1984a; Holekamp and Sherman 1989).

Cynomys ludovicianus

One of the most complete investigations of inbreeding and outbreeding has been conducted by Hoogland on black-tailed prairie dogs, *Cynomys ludovicianus*. This is one of the few studies in which almost all behavioral observations of inbreeding have been confirmed by electrophoretic analysis of blood samples. Close inbreeding occurs infrequently; only 7 of 770 copulations (1%) involved a partner of $r > 0.5$ (Hoogland 1992).

Inbreeding avoidance in *C. ludovicianus* results from two general mechanisms: male dispersal and reluctance of females to copulate with closely related males (Hoogland 1982). Most male juveniles disperse from their natal coterie as yearlings, whereas females are more sedentary; thus mother-son and brother-sister matings are avoided. Adult male prairie dogs that have successfully reproduced within a coterie disperse when their daughters mature. There is no evidence that males obtain more matings in the coteries to which they disperse, thus there appears to be no competitive advantage for adult male dispersers. Females are less likely to come into estrus when their fathers are present in a coterie than when they are absent. Finally, should a female enter estrus when male relatives (father, sons, or brothers) are present in her coterie, she avoids mating with them (Hoogland 1982).

Although instances of close inbreeding are infrequent, prairie dogs frequently mate with more distant relatives (Hoogland 1992). A complete analysis of pedigrees from 14 years has yielded data from 770 copulations. Moderate inbreeding occurred in 198 (26%) of these copulations, and 161 of 557 estrous females (29%) copulated with at least one male relative for which $0.2500 > r \geq 0.0078$. Full and half first- and second-cousin matings were common. These findings are biased downward because in the early years of the study the pedigrees were not complete. In 1988 100% ($N = 44$) of estrous females copulated with at least one male with whom she had a known common ancestor (Hoogland 1992). A thorough analysis between observed and expected frequencies if choice of mates was independent of r showed that prairie dogs neither avoided nor promoted moderate inbreeding (Hoogland 1992).

A detailed analysis using five separate criteria for reproductive success failed to show any significant difference between inbreeding and outbreeding females (Hoogland 1992). Thus, the data from *C. ludovicianus* indicate that avoidance of close inbreeding does not necessarily result from inbreeding depression nor indicate the absence of inbreeding in natural populations. One effect of avoidance of close inbreeding at the population level, however, may be the higher than expected frequency of heterozygotes at polymorphic loci (Foltz and Hoogland 1983).

Marmota flaviventris

Longitudinal studies since 1962 by Armitage and his students have generated pedigrees and behavioral profiles of individual yellow-bellied marmots, *Marmota flaviventris*, that can be used to examine the extent of inbreeding. Close inbreeding occurs infrequently (1.2%, $N = 662$ young; Ralls, Harvey, and Lyles 1986), as determined from proximity and behavior of potential pairs in isolated colonies and verified from genetic analysis (Schwartz and Armitage 1980; Armitage 1986b). The basic life history pattern of marmots is similar to that of other ground-dwelling sciurids: female marmots are generally philopatric, males associate themselves with a colony of females, adult males are territorial and exclude all other males from their home range, and most juvenile males disperse before becoming reproductively active (Armitage 1984; Brody and Armitage 1985). Thus most breeding pairs on a rock outcrop are unrelated. Even with no dispersal, however, the demographic probability of close inbreeding is low (Armitage 1974). Although presence of a father does not inhibit philopatry of daughters, 2-year-olds are less likely to come into estrus when their fathers (Armitage 1984) or adult females, including their mothers, are present (Armitage 1986b). Thus possible avoidance of close inbreeding is confounded with reproductive suppression.

Inbreeding avoidance in *M. flaviventris* appears to be a consequence of male dispersal rather than a cause (Brody and Armitage 1985). Marmots are long-lived, and the probability of a yearling male obtaining residency, and ultimately breeding, on its natal territory is so low that males must emigrate to have a chance of reproductive success (Brody and Armitage 1985). The pattern of this dispersal varies considerably and depends both on the temperament of individuals and the situations they face (Armitage 1982).

Data on inbreeding depression in marmots are limited by small sample size, but litter size and quality of young from litters born of closely related parents cannot be distinguished from that of the litters of outbred pairings (K. B. Armitage, personal communication).

Mus musculus

The house mouse, *Mus domesticus* and *M. musculus*, is the weed of the small mammal world and as such has been studied intensively. It should not be surprising that investigations undertaken throughout its native and introduced range, and in both natural and artificial situations, attest to its incredible flexibility (reviewed in Crowcroft 1966; Bronson 1979; Lidicker and Patton 1987). Of interest here is that many of these studies, although certainly not all, portray the house mouse as living in socially closed groups (Crowcroft 1966; Reimer and Petras 1967; Anderson 1970; Singleton and Hay 1983). The differentiation of these social groups may be expressed in terms of their low effective population sizes (N_e): five to eighty (Petras 1967a, 1967b) in natural populations and as low as four in captive social groups (DeFries and McClearn 1972). Many have been led to assume that these groups are closed genetically (Selander 1970; Singleton and Hay 1983). The genetic analysis of Singleton and Hay (1983) indicated that the social organization of the house mouse has the potential to restrict gene flow. In their longitudinal study they recorded only one instance of gene flow between breeding units. On the other hand, some studies of the temporal population structure of the house mouse have demonstrated that gene flow occurs regularly between groups (Lidicker 1976; Baker 1981).

Movement between social groups is primarily by females, unlike the male-biased dispersal found in most mammals (Selander 1970; Lidicker 1976). In spite of this movement by females, however, females within communal nests are likely to be close relatives (Wilkinson and Baker 1988).

Although the frequency of inbreeding has not been documented in natural populations of house mice, inbreeding may be expected on the basis of their relatively stable closed social groups. Data on the outcome of such inbreeding, however, are mixed. Some reports indicate that house mouse populations brought into the laboratory readily tolerate high levels of inbreeding without expression of deleterious effects (Green 1975). Others have documented various levels of inbreeding depression (Lynch 1977; Connor and Bellucci 1979). Thus in spite of the volumes of research conducted on this form, no clear pattern of the natural history of inbreeding emerges. Perhaps this is to be expected from an animal that is so clearly flexible in all aspects of its biology (Crowcroft 1966; Bronson 1979).

Peromyscus leucopus

A thorough examination of the mating patterns of the white-footed mouse, *Peromyscus leucopus*, indicated that close inbreeding is relatively uncommon (2.2%, N = 135; Wolff, Lundy, and Baccus 1988). *P. leucopus* is nocturnal, hence indirect techniques were required to unravel the mating history: use of space, maternity determined from use of radionuclides (Tamarin, Sheridan, and Levy 1983), and paternity determined from exclusion analysis using electrophoresis. Although dispersal in *P. leucopus* may be characterized as male-biased, animals of all sexes and ages disperse. Only 20% of daughters remain in their natal home ranges. This dispersal pattern leads to local populations where almost all breeding animals are immigrants, thus the potential for close inbreeding is low. The percentage of close inbreeding was determined by examining the relationship of males and females from 135 matings (J. O. Wolff, personal communication). Only 6 of these matings involved cases where potentially related mates were in close proximity. The electrophoresis-radionuclide technique allowed exclusion of 3 of these pregnancies as not resulting from matings between related animals; the possibility that the other 3 matings were incestuous could not be rejected (Wolff, Lundy, and Baccus 1988). These data further indicate that estimates of inbreeding frequency in *Peromyscus* that have been extrapolated from spatial relationships (4%–10% in *P. maniculatus*, based on presumed matings from associations of animals in nest boxes [Howard 1949]; and "highly possible" in *P. leucopus*, based on short dispersal distances [Goundie and Vessey 1986]) should be considered with caution.

Wolff, Lundy, and Baccus (1988) also attempted to determine whether male-biased dispersal in *P. leucopus* results from CFR, CFM, or IA. They conclude (1988: 463) that "the data parsimoniously support inbreeding avoidance as the ultimate causation for sex-biased dispersal. Juvenile males appear to be preprogrammed to disperse independently of density or their social environment." However, the mating system of *P. leucopus* ranges from polygyny and promiscuity to facultative monogamy, thus it is difficult to examine clearly the CFM hypothesis (Wolff, Lundy, and Baccus 1988). And the data do not support the idea that all males are "preprogrammed" to disperse; some do not. Should such a sex bias in dispersal be driven by IA, then the surest way to minimize inbreeding would be for all females to remain philopatric, yet 80% disperse (Wolff, Lundy, and Baccus 1988). Keane (1990a) has also investigated dispersal distances in *P. leucopus*. When he compared the probability of philopatry with a null model he found that females moved only as far as forced to by competition to find a vacant home range. Males appeared to avoid settling in their natal home range, even if the resident adult male had disappeared. Keane (1990a) concluded that although males could be dispersing to avoid inbreeding, other evidence suggested that because females avoided mating with close kin, male dispersal was probably a consequence, not a cause, of inbreeding avoidance (Keane 1990a).

Closely inbred lines of *Peromyscus* show high levels of inbreeding depression (Rasmussen 1968; Hill 1974; Haigh 1983; Keane 1990b), and sibling pairs of deermice, *P. maniculatus*, and cactus mice, *P. eremicus*, show a greater reluctance to mate than nonrelated pairs (Hill 1974; Dewsbury 1982). Recently, Keane (1990b) has investigated the relationship between genetic similarity of mates and their reproductive success over a broad range of degrees of genetic similarity in *P. leucopus*. Nonrelated wild-caught animals from the same population were paired and their reproductive success was measured. Significant inbreeding depression was noted for both full-sibling and half-sibling matings. There was also evidence for a potential outbreeding depression. Matings at the level of cousins (moderate inbreeding) tended to be reproductively superior (Keane 1990b). A series of behavioral choice experiments also showed that estrous females exhibited preferences for cousins over siblings and nonrelatives (Keane 1990b).

Ondatra zibethicus

The primarily monogamous muskrat, *Ondatra zibethicus*, rarely exhibits close inbreeding (Caley 1987), although high levels of moderate inbreeding are likely (M. J. Caley, personal communication). Both males and females are philopatric, although dispersing animals are primarily males. The infrequency of close inbreeding was calculated using data on spatial proximity and overlap of related animals; matings were not observed or confirmed with genetic data or other indirect techniques. The observed pattern of dispersal, demography, and mating system could not be used to discriminate among the CFM, CFR, and IA hypotheses of dispersal.

Microtus pennsylvanicus

The level of close inbreeding in the meadow vole, *Microtus pennsylvanicus*, is high (36.4%, $N = 11$; Pugh and Tamarin 1988). These data were derived from pedigrees of animals in a natural enclosure, and the estimate of inbreeding frequency is conservative because relatedness of parents in the first generation was unknown. Because voles are small and secretive, the pedigrees were established using the electrophoresis-radionuclide technique (Tamarin, Sheridan, and Levy 1983; Sheridan and Tamarin 1986). The method by which paternity was established was robust, because within the population all males, regardless of their spatial orientation, were regarded as potential fathers. Comparison of the observed inbreeding coefficient with the expected inbreeding coefficient (assuming random mating) indicated that the voles were neither courting nor avoiding inbred matings (Pugh and Tamarin 1988). *M. pennsylvanicus* lives in small, ephemeral grassland patches, and the costs of dispersal between patches are apparently high. Should there be a genetic cost to inbreeding (which is unknown), it may benefit individuals to mate with nearby kin rather than forgo reproduction or incur the dispersal costs (Pugh and Tamarin 1988). Vole populations typically fluctuate in density. The population studied by Pugh and Tamarin (1988) was low in density, and they predict that frequency of inbreeding should decline in higher-density populations as more opportunities for outbred matings occur. This prediction, coupled with the wide diversity of social organization and habitat occupancy found in the genus *Microtus*, sets the stage for interesting intra- and interspecific comparisons of the levels of inbreeding in natural populations (see Boonstra et al. 1987).

Dipodomys spectabilis

The frequency of inbreeding or outbreeding is not known for the banner-tailed kangaroo rat, *Dipodomys spectabilis*, but the recent investigation by Jones and Waser of the degree of philopatry versus dispersal and genetic variation in this species is central to our understanding of the natural history of inbreeding. Male and female *D. spectabilis* are philopatric (Jones 1984, 1987; Jones et al. 1988). In normal high-density populations 80% (N = 96) of male and 77% (N = 99) of female juveniles remained within 50 m of their natal sites or points of first capture. Adults are even more sedentary. 89% (N = 70) of males and 92% (N = 72) of females remained similarly philopatric (Jones 1987). Philopatric juveniles gain access to parental resources, large complex burrow systems and food caches, and consequently have higher average survivorship than individuals that successfully disperse (independent of dispersal distance; Jones 1986). It is for this reason that philopatry is more frequent at high densities, when the likelihood of encountering a vacant burrow system is low, whereas dispersal is comparatively more common at low densities, when the habitat is not saturated with animals (Jones et al. 1988). Thus there is high potential for close to moderate inbreeding in banner-tailed kangaroo rats, but matings between kin may vary with density. Electrophoretic data, however, suggest that *D. spectabilis* do not always breed with close neighbors. There was no general pattern of homozygote excess expressed across loci in two populations sampled (Elliott et al. 1989). In addition, there is no evidence for an effect of dispersal distance on reproductive success, as one might expect if optimal outbreeding were occurring (P. M. Waser, personal communication). Understanding the natural history of inbreeding in this system will require direct observations of mating, or indirect genetic or marking techniques allowing the construction of pedigrees.

Meriones unguiculatus

The Mongolian gerbil, *Meriones unguiculatus*, lives in multimale, multifemale age-structured family groups (Gromov 1981; Ågren, Zhou, and Zhong 1989). The largest male of the group is most active in the defense of the family group territory, but offspring help irrespective of gender, age, or status (Ågren 1976; Ågren, Zhou, and Zhong 1989). The temporal demography of these groups is unknown in nature, but Ågren, Zhou, and Zhong (1989) found that when the habitat is saturated, dispersing animals find it difficult to establish themselves in a new territory. Aggressive behavioral interactions with established family group members preclude settlement. A series of studies in large pens and in nature strongly indicate that moderate inbreeding may occur. However, when in estrus, females trespass into neighboring territories and mate preferentially with nongroup members (Ågren 1984a, 1984b, 1984c; Ågren, Zhou, and Zhong 1989). Apparently females thereby minimize the potential somatic costs of dispersal, except for a narrow window of time, and simultaneously minimize the probability of close inbreeding (Ågren 1984a). Long-term observations and the establishment of pedigrees in this species in nature are necessary to support these preliminary data.

Ochotona curzoniae

Another species that occupies family group territories is the black-lipped pika, *Ochotona curzoniae* (Smith et al. 1986; Wang and Smith 1989). Multiple large litters are born in summer to mothers occupying discrete burrow systems on alpine meadow habitat. Behavioral interactions within family groups are primarily amicable, frequently expressed, and cohesive (Smith et al. 1986; Smith and Wang 1991). The Tibetan winters are severe and mortality is high. On average, two of approximately twenty individuals from each family group survive to form the breeding population the next year (Wang and Smith 1988). The average sex ratio is even, but due to stochasticity of survivorship a variety of mating assemblage types are possible (40% monogamy, 34% bigamy, 11% polygyny, 15% polyandry; Wang and Smith 1989). In 1985, 119 copulations were observed, and in each instance the female was on her family group territory. No extra-pair copulations were observed that involved females associated with monogamous males, however, 19% (14/74) of copulations involving females with bigamous mates were initiated by neighbors (Smith and Wang 1991).

Most animals in the population are philopatric. The habitat is saturated and dispersing animals crossing unfamiliar family group territories are met with extreme aggression (Smith and Wang 1991). Limited dispersal occurs during one week of the annual cycle, just prior to the first estrous period of females. At this time most males that have survived the winter leave their natal territories and move to a nearby burrow. Over 50% of males moved next door, and the greatest distance moved in 4 years of observation was four home ranges (Wang, unpublished data). The limited dispersal movements and data on mating patterns indicate that the frequency of moderate inbreeding in *O. curzoniae* may be high. Investigation of this population is ongoing in an attempt to establish pedigrees and more rigorously understand the relationship between mating systems, demography, and inbreeding.

Ochotona princeps

The American pika, *Ochotona princeps*, gives every indication of being a species in which close inbreeding frequently occurs. Movements of pikas within patches of their preferred talus habitat are restricted due to social interactions, and most animals live in proximity to close kin (Smith and Ivins 1983a, 1984; A. T. Smith 1987). Pedigree and mating information are available for six pairings from a 3-year study by Smith and Ivins; 50% of the pairings resulted in close inbreeding (one father-daughter and two mother-son matings; Chepko-Sade et al. 1987). The following aspects of the biology of pikas indicate that these inbred matings probably occur frequently. Adult male and female pikas maintain large individual territories. Once settled on a territory, most adults remain there for life. Pikas are also relatively long-lived; some live to 6 years of age, and annual survivorship is high (Millar and Zwickel 1972; Smith 1978). Thus the occurrence of vacant territories available for settlement by juveniles (or the rare dispersing adult) is infrequent and unpredictable in space and time (Smith and Ivins 1983b, 1984; A. T. Smith 1987). Pikas are facultatively monogamous as a result of the pattern of male and female territories. Male and female pikas are highly aggressive to all conspecifics, but primarily to those of the same sex. Males cannot control sufficient resources (the narrow band of accessible meadow vegetation bordering the talus) to attract multiple females, nor can they control access to multiple females due to their wide dispersion on the talus (Smith and Ivins 1984). Territories normally alternate by sex on the talus, and replacement on vacant territories is almost always by a pika of the same sex as the previous occupant. Distances between the centers of activity of a pair are significantly less than distances to any other conspecific. Almost all amicable interactions (approximately 95%) occur between spatially contiguous males and females, and all instances of mating that have been observed were between animals characterized spatially and behaviorally as a pair (Smith and Ivins 1984; Brandt 1989). However, matings are rarely seen because pikas normally mate while snow blankets the talus. Parturition is timed to occur at the time of average snowmelt in an area, but mating can occur earlier because gestation is not metabolically expensive to mammals (Millar 1972; Smith 1978; Smith and Ivins 1983b). The wide spacing between individuals and the potential difficulty of locating mates on distant territories when the talus is under snow should limit the potential for extra-pair matings, but the frequency of their occurrence is unknown. Finally, frequent inbreeding is likely in this system because pairmates are often closely related. Both male and female juveniles are predominantly philopatric (Krear 1965; Millar 1971; Sharp 1973; Tapper 1973; Smith and Ivins 1983a; Smith and Ivins 1987; Brandt 1985). Smith and Ivins (1983a) found that 65% ($N = 17$) of vacant territories claimed by juveniles were within 50 m (one home range) of their natal burrows.

A second aspect of the biology of pikas that predisposes them to expression of high levels of inbreeding is the isolation of populations on small patches of talus. Pikas experience a high cost of dispersal between habitat patches, and immigration onto a patch is rare. As obligate rock-dwelling mammals, pikas are ill equipped to deal with the terrain they must cross when engaged in interpatch dispersal. Mortality is high due to predation, and pikas are highly sensitive to warm temperatures. Near the lower altitudinal limit of their geographical distribution, where high temperatures commonly occur at the time of dispersal, distances as short as 300 m are effective barriers to dispersal by pikas (Smith 1974a, 1974b, 1980). In addition, the social environment that limits intrapatch dispersal also precludes establishment of immigrants (Tapper 1973; A. T. Smith 1987). In many instances talus patches occupied by pikas are small (frequently two to ten animals), and the probability of nonrandom mating by close relatives is high.

A more complete understanding of the natural history of inbreeding and outbreeding in *O. princeps* will necessitate the generation of pedigrees (with paternity determined using genetic techniques) over several years, and documentation of the genetic effects of intra- and interpatch dispersal on populations. Such an investigation, using both a longitudinal and an experimental approach, is in progress.

Heterohyrax brucei and *Procavia johnstoni*

Two additional species for which a high cost of dispersal has been documented are the bush hyrax, *Heterohyrax brucei*, and the rock hyrax, *Procavia johnstoni*. These species, longitudinally investigated by Hoeck, live on habitat islands (rock outcrops, or kopjes) in the Serengeti. Hyraxes face a multitude of problems when dispersing across the plains to reach another kopje: they are subject to frequent predation, they may not be able to orient in order to locate a target kopje, and they are vulnerable to stress caused by extreme temperatures (Hoeck 1982). Dispersal occurs in all age and sex classes, although it is most pronounced in young males. Recruitment of individuals into the sexually mature age group on a kopje can take place through incorporation of young animals born there or through immigration of animals from other kopjes. Female immigrants are generally recruited when they encounter a kopje, but as most females are philopatric, pedigrees on a kopje are primarily matrilineal. Male immigrants, although they represent the majority of dispersers, are normally expelled from occupied kopjes. Most dispersing males colonize peripheral kopjes without established breeding groups. On the other hand, some male juveniles that failed to disperse eventually became sexually mature on their natal kopje (Hoeck 1982). The tenure of territorial males and breeding females is sufficiently long to allow inbreeding between close kin. Hoeck (1982) observed no such matings, although indirect (spatial and behavioral) evidence pointed to their occurrence.

Populations of hyraxes on many kopjes are significantly isolated from other hyraxes, and lack of immigration (potential gene flow) sets the stage for frequent inbreeding. Hoeck (1982, 1989) has analyzed morphological characteristics and genetic traits (using electrophoresis) of several isolated hyrax populations and concluded that they can persist for at least 16 years under these inbreeding conditions. Some of the morphological traits that were expressed included the alteration of normal pelage patterns to include white patches ranging from a small spot to an almost completely white coat (in *H. brucei*), and white fur behind the nares (in *P. johnstoni;* Hoeck 1982).

Heterocephalus glaber

The naked mole-rat, *Heterocephalus glaber,* is unique among mammals in that it is the only species known to have a colony structure analogous to that of social insects (Jarvis 1981; Sherman, Jarvis, and Alexander 1991). It has now also been determined that one of the highest known rates of close inbreeding also occurs in naked mole-rats. Using the technique of DNA fingerprinting on laboratory colonies established from natural populations, Reeve et al. (1990) have documented extreme homogeneity due to close genetic relationship ($r = 0.81$), itself the result of consanguineous mating. The inbreeding coefficient (F) within colonies of *H. glaber* is 0.62, a figure equivalent to that of a population that was sib-mated for 60 generations of forced inbreeding. There is no evidence of inbreeding depression (Reeve et al. 1990).

Colonies may contain over eighty individuals, but only one female and one to three male mole-rats breed in each colony. The remaining individuals perform various colony tasks. The small animals perform maintenance functions (e.g., food carrying, tunnel clearing) while the largest nonbreeders defend the colony against intrusions by snakes and foreign colonies (Lacey and Sherman 1991). All members of the colony are capable of breeding, but various forms of reproductive suppression inhibit this activity. The environment in which *H. glaber* lives is rigorous. The soil is exceedingly hard, making it difficult to construct burrows, and the primary food sources of the mole-rats (large subterranean tubers, up to 50 kg) are highly dispersed and difficult to find and harvest (Brett 1986). Due to these factors an individual or pair of mole-rats may have a poor chance of surviving or producing young, and successful dispersal appears to be highly unlikely. The communal activities of nonreproductive mole-rats may be associated with the high degree of genetic relatedness within a colony, allowing individuals to ensure the survival and promotion of their own genetic characteristics (Reeve et al. 1990; Sherman, Jarvis, and Alexander 1991).

Chiroptera

Bats (family Chiroptera), comprise almost one-quarter of all species of mammals. However, many aspects of their biology make them very unlike the other small mammals I have discussed. Their capacity for true flight likens them more to birds than to mammals in their ability to disperse; movement over similar distances can be accomplished with less cost than for terrestrial mammals. They also live significantly longer and have lower fecundity rates (generally one pup) than most other small mammals. Maximum longevity for bats commonly exceeds 20 years, and in most species some individuals live more than a decade (Tuttle and Stevenson 1982). Annual survivorship of adult greater spear-nosed bats, *Phyllostomus hastatus* (described below), is over 90%.

McCracken (1987) has reviewed three species of bats for which we have sufficient life history and genetic data to draw conclusions concerning inbreeding and outbreeding. Outbreeding appears to be the norm; dispersal (along with other aspects of a bat's life history) is apparently sufficient to randomize adult population structure and preclude matings among kin. In *P. hastatus*, the breeding unit is a harem composed of one male and approximately eighteen females. All juveniles of both sexes disperse within a year of birth. Young males join bachelor groups, whereas females join with other females to form new stable harems. Genetic analysis has shown that average relatedness among adult females in harems is very low. Instead, most juveniles are related through their common father, and their eventual dispersal minimizes the possibility of inbreeding (McCracken and Bradbury 1981).

In vampire bats, *Desmodus rotundus*, long-lasting stable groups are formed which share common tree roosts. Group size ranges from eight to twelve adult females, and all female offspring are recruited into their natal harems (Wilkinson 1984, 1985). Although this behavior leads to genetic structuring among adult female groups, occasional dispersal by adult females leads to the establishment of multiple matrilines and reduces average relatedness within harems. Probability of inbreeding is further reduced because (1) all young males disperse; (2) the tenure of breeding males is short; (3) all adult males in a roost may succeed in mating, thus lowering any individual male's probability of paternity; and (4) female survivorship to breeding age is low, thus eliminating the probability of father-daughter matings (Wilkinson 1984, 1985).

In the greater white-lined bat, *Saccopteryx biliniata*, most males remain sedentary, and it is likely that those that jointly occupy a roost are related. In contrast, all females disperse, generally over great distances. Only about 16% of marked females were resighted near their natal roost (Bradbury and Vehrencamp 1976). Data on paternity in this system are lacking, and levels of inbreeding have not been determined. However, the genetic analysis of McCracken (1987) provides no evidence for genetic structuring.

SUMMARY AND CONCLUSIONS

This review of the natural history of inbreeding and outbreeding in small mammals indicates a tremendous variety of responses in the relatively few species for which we have some direct or highly indicative indirect evidence of mating patterns. Some species in nature appear to exhibit no close inbreeding, others show various levels of inbreeding. Absence of close inbreeding does not indicate whether or not moderate inbreeding is frequent, nor does it predict whether or not animals are purposely avoiding incestuous matings. Although inbreeding depression is most frequently touted as a driving force for the infrequent observation of close inbreeding, we have seen that these two parameters are not necessarily linked. In several species ecological constraints can readily be coupled to high levels of inbreeding. In others we must rely on inferences from the social context to determine the nature of inbreeding and outbreeding, and these interpretations are generally clouded. Superimposed upon these observations, normally of marked animals in single populations, is the temporal and spatial aspect of genetic structuring. There are as yet no data from investigators working on small mammals that have successfully approached inbreeding from both regional and local perspectives.

I hope that this general discussion of our current state of knowledge of inbreeding and outbreeding in small mammals will be taken as a challenge and will generate excitement in those willing to perform the next generation of studies.

For the present, there are no shibboleths. Extrapolation of these data for the purpose of genetic management of our natural resources must be done with caution. Any management prescription based solely on the assumptions that inbreeding is not found in nature (Harvey and Ralls 1986; Ralls, Harvey, and Lyles 1986) or that all inbreeding is deleterious (Ralls, Ballou, and Templeton 1988), should at this point be regarded as premature. Parsimoniously, we should proceed by attempting to understand the salient life history features of those individual species in need of our protection. If extrapolations from existing data are necessary to enhance our understanding of a species or problem, they should include only data from species experiencing similar ecological and social constraints.

ACKNOWLEDGEMENTS

A review of this nature would be impossible without the data, often laboriously gathered over many years, of many conscientious field biologists. To them this paper is dedicated. This presentation has been aided by spirited correspondance and conversation with of many of these specialists, several of whom offered unpublished data, insightful ideas, and support: Greta Ågren, Kenneth B. Armitage, Ann Eileen Miller Baker, M. Julian Caley, Andrew Cockburn, Hendrik N. Hoeck, John L. Hoogland, Brian Keane, Gary F. McCracken, Paul W. Sherman, Robert H. Tamarin, Wang Xue Gao, Peter M. Waser, and Jerry O. Wolff. I appreciate the careful review of the entire manuscript by F. Stephen Dobson, Mary Peacock, and Harriet Smith.

15

Inbreeding and Outbreeding in Captive Populations of Wild Animal Species

Robert C. Lacy, Ann Petric, and Mark Warneke

That inbreeding, the mating of close relatives, often has deleterious effects has been known for centuries. Darwin (1868) extensively documented inbreeding depression in domesticated stocks in his two-volume book, *The Variation of Animals and Plants under Domestication,* and Wright (1977), Falconer (1981), and others have reviewed data that have accumulated since. With some exceptions (e.g., Shields 1982), the generally (almost universally) deleterious effects of inbreeding have become widely recognized, and almost dogma. Surprisingly, however, the empirical evidence for the effects of inbreeding on vertebrates is quite narrow—primarily restricted to breeding studies of domesticated livestock and laboratory rodents and pedigree studies of human societies. These stocks result from hundreds to thousands of years of intense artificial selection, and it is not clear that they would make good models for inbreeding effects in natural populations. We still do not have a sufficient empirical base for many of the claims that are made about inbreeding depression and its causes, and we certainly have scant data for generalizing about the frequency, magnitude, and correlates of inbreeding depression.

Zoological parks breed a diversity of vertebrate species, and either because of a lack of available animals to pair or because of a lack of concern about the effects of inbreeding, zoos often extensively inbreed their stocks. Therefore, the breeding records of zoos potentially offer a wealth of data on the effects of inbreeding. The use of zoo breeding records to examine the effects of inbreeding has considerable advantages and disadvantages. While published data on inbreeding are accessible for only a few nondomesticated species, as of 1988 international studbooks were being maintained on the pedigrees for eighty-seven species propagated in zoological collections (Olney, Ellis, and

Sommerfelt 1988). These breeding records often extend for 25 to 100 years, encompassing many generations of breeding. In some cases, virtually complete genealogical data exist for more than a thousand specimens (e.g., golden lion tamarin, *Leontopithecus r. rosalia*); such extensive data bases are unparalleled and would be virtually unobtainable from field studies of populations in more natural habitats. Because zoos can provide animals with relatively standard diets, environments, and social groupings, uncontrolled environmental variation can be minimized throughout a pedigreed population, at least relative to the variation that might confound studies of similar magnitude in the field. Even when different zoos follow distinct management procedures, statistical separation of within-zoo from between-zoo variation in data can allow a degree of statistical control over nongenetic variation. Zoo animals rarely die from predation, food stress, or extremes of weather, and this reduced nongenetic mortality relative to natural populations could lead to greater statistical sensitivity in detecting inbreeding depression.

Unfortunately, the use of breeding records on zoo populations to study inbreeding has many disadvantages—often related to the advantages stated above. Although pedigrees can be extensive, they are rarely complete. Few zoos have consistently kept accurate records of all breeding throughout their histories (though many now do keep complete records), and even in seemingly complete data bases stillbirths and neonatal deaths often go unreported. Management practices change over time, and both temporal and between-zoo variations in the captive environment are often correlated with degrees of inbreeding. Although zoos often keep animals in pairs or single-male groups and therefore can be certain of parentage determinations, the parentage of animals born into multimale groups can only be assessed with certainty by molecular genetic analysis. As is the case with field studies, when such molecular studies are undertaken it is often found that parentages assessed by observations of behavior are in error. The ameliorated captive environment, while removing many sources of statistical variation, also results in unnatural, and often unnaturally low, selective pressures on a population. Animals that would be seriously compromised in the wild by genetic abnormalities often live much longer in captivity than they could otherwise. Social interactions and opportunities for mating in captivity are often very different from those in the social systems more typical of wild populations of the species: normally promiscuous animals might be housed as pairs, and naturally solitary species (e.g., orangutans, *Pongo pygmaeus*) are often kept in social groups. Thus, behavioral preferences for or avoidance of inbreeding cannot be assumed to have the opportunity to manifest themselves in captive breeding stocks.

Moreover, it is often not known whether the founding stock of a captive population represents one or a few family groups collected from a local population in the wild, animals collected from divergent geographic regions, a diversity of subspecies (e.g., many captive groups of tigers and lions), or even closely related species (e.g., many captive populations of squirrel monkeys, *Saimiri*, of titi monkeys, *Callicebus*, and of owl monkeys, *Aotus*). Because of the uncertain origins of many captive populations, the effects of inbreeding on zoo populations can be confounded with the effects of outbreeding or even hybridization between formally recognized taxa (Templeton and Read 1984; Templeton 1986). Outbreeding depression, a reduction of fitness when individuals from normally noninterbreeding populations are crossed, results from the disruption of coadapted gene complexes. Coadaptation of genes can occur because multiple genetic loci contribute to a suite of characters that jointly confer adaptation to local environmental conditions (response to extrinsic selection pressures), or because genes evolve in response to other genes under the constraint that genome structure and function must always be compatible (response to factors intrinsic to the genome) (Templeton et al. 1986). A clear example of coadaptation is the constraint on chromosome polymorphisms imposed by the need for proper pairing during meiosis. Crosses between chromosomal races of the owl monkey (*Aotus trivirgatus*) often produce hybrids with reduced or no fertility (de Boer 1982).

Lacking data to indicate otherwise, it must be assumed for genetic analysis that the wild-caught animals used to initiate a captive population are all unrelated. Thus, the first-generation captive-born animals are, by necessity, considered to be noninbred. It is only in the second and later generations of captive breeding that animals known to be inbred can be produced. Founding stocks may differ substantially in their levels of genetic diversity, and inbreeding coefficients calculated from pedigrees can only measure the loss of heterozygosity relative to the founding stocks. If the original wild-caught animals come from diverse small, locally adapted, and perhaps partially inbred populations, the first-generation captive-born descendants will have a complete haploid genome from each of the two parental sources. They are likely to show the hybrid vigor so commonly observed in crosses between genetically divergent strains of domesticated and laboratory animals. In the second- and later-generation descendants, recombination can lead to disruption of coadapted complexes of genes from the parental lineages, resulting in a reduction of fertility or viability. Thus, it is in the second captive generation that we expect to see both the first documentable inbreeding and also the breakdown of coadapted gene complexes. Failure to consider these and other alternative explanations for reduced fitness in the descendant generations is common and can be misleading (Templeton et al. 1986).

Although zoo records are a potential source of extensive data on the effects of inbreeding and outbreeding in numerous species, they have been little utilized to date, and the results must be analyzed with due consideration given to differences between captive and wild populations and certain methodological complications. In this chapter, we will give a brief overview of recently published summaries of inbreeding effects in zoo populations, we will present data from the breeding records of six ungulate species at Brookfield Zoo, and we will present a detailed analysis of the effects of inbreeding and outbreeding on Goeldi's monkey or Callimico. We hope and expect that investigators will mine further the data available in zoo records, so that the tremendous resources invested in captive breeding can be used to maximum benefit for studies of natural history and evolutionary biology, and for the benefit of the animals themselves.

STUDIES OF INBREEDING IN CAPTIVE POPULATIONS

Scattered through the scientific literature are case studies of inbreeding in a variety of species propagated by zoos and private breeders: Przewalski's horse (*Equus przewalskii:* Bouman and Bos 1979; Keverling Buisman and van Weeren 1982), mountain sheep (*Ovis canadensis:* Sausman 1984), gaur, a wild cattle species (*Bos gaurus:* Hintz and Foose 1982), leopards (*Panthera pardus:* Shoemaker 1982), European bison (*Bison bonasus:* Slatis 1960), Dorcas' gazelles (*Gazella dorcas:* Ralls, Brugger, and Glick 1980), white tigers (*Panthera tigris:* Roychoudhury and Sankhala 1979), Speke's gazelle (*Gazella spekei:* Templeton and Read 1983, 1984), rhesus macaques (*Macaca mulatta:* D. G. Smith 1986a, 1986b; Smith et al. 1987), and budgerigars (*Melopsittacus undulatus:* Daniell and Murray 1986). All of these studies reported that inbreeding depressed at least some aspect of fitness in some populations, with infant mortality being the most commonly measured fitness component, but considerable diversity exists among the findings of the studies. For example, Shoemaker (1982) found increased juvenile mortality with inbreeding in one subspecies of leopard, the Persian leopard, but found no such trends in Chinese leopards or Amur leopards. Daniell and Murray (1986) found that inbreeding depressed viability of nestlings in one captive colony of budgerigars, but not in another. Slatis (1960) found higher mortality in inbred European bison, but only among the descendants of a single male that was from a different subspecies than the rest of the remnant world herd of this endangered species. Those bison produced from within-population crosses showed no effect of inbreeding, and it is possible that the "inbreeding depression" reported by Slatis was actually outbreeding depression that appeared in the second, recombinant generation of the between-subspecies crosses. This explanation for apparent inbreeding depression in a captive herd of Speke's gazelle was considered, but then rejected, by Templeton and Read (1984).

Reports of no effect of inbreeding in zoo stocks are very rare, and apparently nonexistent for large data sets, but it is unclear whether that may in part be due to a reporting bias. Kathy Ralls and Jon Ballou of the National Zoological Park have, in a series of papers, summarized data on a number of zoo populations, looking for overall trends (see Ralls and Ballou 1983 for a review). Examining ungulates (Ralls, Brugger, and Ballou 1979; Ballou and Ralls 1982), primates (Ralls and Ballou 1982b), and small mammals (Ralls and Ballou 1982a), they showed in each case that, more often than not, juvenile mortality was higher in the inbred animals. Where sufficient data were available, they have also shown that differences in population density, in management practices among zoos, in birth season and birth order, and between wild-caught and captive-reared dams do not explain the greater average mortality in the inbred offspring (Ballou and Ralls 1982). The sign-tests they used, however, simply demonstrate that inbreeding is more often deleterious than beneficial, and give no indication of how general or how severe inbreeding depression is in zoo stocks.

Recently, Ralls, Ballou, and Templeton (1988) presented a quantitative study of the severity of inbreeding depression in forty populations of thirty-eight species of mammals. Ralls et al. calculated the number of "lethal equivalents" per diploid individual, given by the regression slope of the logarithm of juvenile survival against the inbreeding coefficient. The number of lethal equivalents estimates the number of recessive lethal alleles per individual if the effects of all deleterious recessives were combined into fully lethal genes (Morton, Crow, and Muller 1956). In measuring genetic deaths, it is usually impossible to distinguish between death caused by a single lethal gene and death caused by the combined effects of many sublethal, deleterious genes. A lethal equivalent could be one fully lethal allele, two alleles (at nonlinked loci) each conferring a 50% probability of death, ten alleles causing 10% mortality each, or any combination of deleterious alleles causing on average one death. The cumulative probability of survival through multiple independent risks of mortality would be the product of the individual probabilities (e.g., three 50% lethals result in a 87.5% probability of death); hence, a linear regression of the logarithm of survival probabilities is used to estimate lethal equivalents.

Ralls, Ballou, and Templeton (1988) found lethal equivalents to range from − 1.36 in maned wolves (*Chrysocyon brachyurus*) to 30.32 in Wied's red-nosed rat (*Wiedomys pyrrhorhinos*). (Although no population had significantly less than zero lethal equivalents, a negative value would indicate greater survival in inbred litters.) Much of the scatter among the populations sampled must have been sampling error: for example, two of the populations studied by Ralls et al. (including Wied's red-nosed rat) included only one inbred litter. Because whole litters often die or live as a unit (e.g., the one inbred litter of the red-nosed rats), statistically it is not appropriate to consider each offspring as an independent data point; rather, the number of degrees of freedom in such studies is more likely the number of litters, or perhaps the number of dams. Even accepting that many of the data points presented by Ralls et al. are based on small samples, between-taxa comparisons are possible. Figure 15.1, reproduced from the Ralls et al. paper, displays means, standard deviations, and ranges of lethal equivalents for several mammalian orders. This survey failed to detect clear trends among the mammalian orders, nor did Ralls et al. find trends related to the nature of the source population (captive or wild) from which the pedigreed population was derived.

A recent study of inbreeding in laboratory stocks of eight populations (of two species) of *Peromyscus* mice revealed no relationship between the severity of inbreeding depression and the size and degree of insularity of the natural populations represented by the lab stocks (Brewer et al. 1990). It appears that the limited data on inbreeding of nondomesticated populations of animals do not yet reveal any taxonomic or ecological trends.

HISTORIES OF INBREEDING AMONG UNGULATE POPULATIONS AT BROOKFIELD ZOO

Like many zoos, the Chicago Zoological Park (Brookfield Zoo) maintains much of its hoofstock in herds, generally started from a few individuals and often with little or no input of additional genetic material in subsequent years. Because one or a few males usually monopolize matings, close inbreeding becomes common beginning with the second generation of captive breeding. We present here the histories of several such populations from Brookfield Zoo. The populations chosen

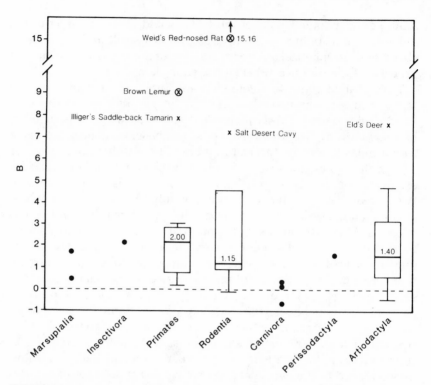

FIGURE 15.1   Box plots of B (regression slopes of log[viability] vs. F) across forty mammalian populations by order. Median effects (middle horizontal lines in the boxes), upper and lower quartiles (ends of the boxes), outlying values (x), and values beyond the outer fences (encircled x) are shown for the distributions in three orders. Results from individual populations in other orders are shown by dots. From Ralls, Ballou, and Templeton (1988); used by permission of Blackwell Scientific Publications, Inc.

for analysis were all those with long-term breeding records (>20 years), large sample sizes, at least thirty, and usually more than one hundred, captive births), and quantifiable evidence of inbreeding from either completely known pedigrees (from which inbreeding coefficients were calculated) or closed herds (in which inbreeding levels must have increased with each generation).

To assess one aspect of inbreeding depression, we examined infant mortality prior to 30 days of age. It may be more appropriate to choose a species-specific criterion, such as one-half the age at which adult size is attained (as was used by Ralls and Ballou 1982a). We noted, however, that most infant mortality occurs within the first few days after birth and that varying the age criterion in the analysis has virtually no effect on the results presented below. We did not attempt to classify the causes of death (as was done, for example, by Ballou and Ralls 1982), nor did we exclude cases of accidental and presumed nongenetic deaths (as was done by Slatis 1960). Often neonatal deaths were noted in the records as being due to a general weakness of the infant, with no specific cause known.

We examined the data by a weighted logit regression model (Snedecor and Cochran 1967: 494), regressing the natural logarithm of the ratio of surviving offspring to dead offspring against the level of inbreeding. In one case, in which the logit model was precluded because many classes of inbreeding had survival rates of 0 or 1, we used a t test to compare the mean level of inbreeding among infants that died to the mean level among those that survived to 30 days of age. (The small sample size correction used by Templeton and Read [1984] and by Ralls, Ballou, and Templeton [1988] to avoid proportions of 0 or 1 produces biased estimates of the effects of inbreeding on survival.) When inbreeding coefficients could be calculated from known pedigrees, we also regressed the natural logarithm of the fraction of young surviving at each level of inbreeding against the inbreeding coefficient, in order to estimate the mean number of lethal equivalents per genome (Morton, Crow, and Muller 1956). When incompleteness of pedigrees prevented calculation of inbreeding coefficients, we more crudely assessed the level of inbreeding as the number of years since the last importation of an unrelated animal into the herd. In some cases, the herds have been maintained as closed breeding stock since the original founding animals were brought to the zoo, and the herds could only have become progressively more inbred (though at an unknown rate). In other cases, breeding males were supplemented or replaced by presumably unrelated animals every few generations, resulting in alternation between periods of inbreeding and outbreeding within the herd.

Banteng *(Bos javanicus)*

Banteng are wild cattle, capable of hybridizing with domesticated cattle. They live in small herds in the dense forests of Indochina, Borneo, Java, and Bali (Macdonald 1984). Little is known of their social dynamics in the wild. Breeding of banteng at Brookfield began with a single pair in 1965, which produced thirty-three descendants in three generations (including offspring of father-daughter, father-granddaughter, and mother-grandson matings, as well as more complex relationships). The logit regression of infant survival against F indicated significantly $(P < .01)$ greater mortality among the increasingly inbred progeny (figure 15.2), with an estimated 2.33 recessive lethal

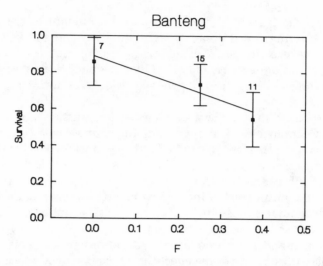

Figure 15.2   Relationship ($P < .01$) between infant survival and inbreeding (F) in banteng (*Bos javanicus*) at the Brookfield Zoo. In this and all subsequent figures, the mean survival to 30 days of age (with standard error based on the expected binomial variance) is given at each level of inbreeding. Sample sizes are given above error bars. The line shows the weighted least-squares linear regression, but (except as noted otherwise below) statistical testing of trends was based on weighted logit regressions.

equivalents per individual in this small herd. A trend toward lower infant survival across the years is nonsignificant, even though inbreeding is correlated with the years since herd formation ($r = 0.66$, $N = 33$).

Wisent *(Bison bonasus)*

The European bison or wisent were once widespread in the forested regions of Europe. The species went extinct in the wild in 1919, but captive and reintroduced wild populations have been propagated from the remnant herds that persisted in a few game preserves through the first few decades of this century. All living wisent descend from just seventeen animals, sixteen of which were of a subspecies (*B. bonasus bonasus*) from the forests of Poland and one of a different subspecies (*B. bonasus caucasius*) from the Caucasus Mountains. The social structure probably consists of small herds, and the wild wisent populations have been fragmented into small remnant populations since the 1700s.

The Brookfield herd of wisent was started in 1956 with a single breeding pair obtained from the Copenhagen Zoo. In 1959 a female full sibling to the breeding male was imported. No animals have been added to the herd subsequently. Inbreeding coefficients can be calculated on all sixty-three descendants of the original trio, if the dominant breeding male is assumed to have been the sire of offspring conceived during a few years in which several males were given access to females. The number of lethal equivalents within the herd is estimated at 1.05, although infant mortality remains low until inbreeding becomes severe ($F > 0.50$) and the trend in figure 15.3 is not statistically significant ($P > .10$).

Although the above estimates of lethal equivalents are based on herds established with very few animals, it is not surprising that the estimated genetic load of the wisent, a bovine species which has been through a severe bottleneck and which was inbred prior to importation to the United States, is just 45% of the load of lethal equivalents estimated for the banteng, a species that has a much broader extant range and is not known to have been forced to inbreed during bottlenecks prior to its captive history.

Siberian Ibex *(Capra ibex sibirica)*

Siberian ibex inhabit the alpine desert regions of Mongolia and northern China. The typical social structure consists of small herds, probably including several males within a dominance hierarchy. The Brookfield herd was established from a single male and four females in 1965 and 1966, and has been maintained with three to fifteen males and five to thirty females. Typically, several of the largest males appear behaviorally dominant, and they limit access to breeding females by subordinate males. From 1966 through 1988 the herd has produced 231 offspring (many of which were sold) without the introduction of any new animals to the herd. The herd has gone through five generations of breeding males and as many as ten generations of females; thus it must be highly inbred. Juvenile mortality has increased significantly through the years (figure 15.4; $P < .01$), unrelated to herd size or other identifiable aspects of management.

Although error bars are often large, there is an indication in figure 15.4 that infant survival may have decreased in a stepwise manner. The sudden shifts in calf mortality coincided with replacement of dominant males by later-generation (and therefore more inbred) males. The founder male was dominant and probably the primary breeder until 1973. He began mating with his daughters in 1969 and was mating with great-granddaughters (as well as with founder females, daughters, and granddaughters) by 1972. An F_1 male was dominant through 1977, and one of his sons was dominant until 1980. Infant survival dropped markedly after 1981, during which time various males, ranging from third- to fifth-generation captive-born, have been socially dominant.

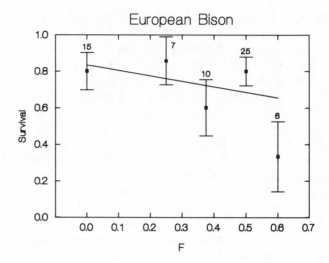

FIGURE 15.3   Relationship ($P > .10$) between infant survival and inbreeding (F) in wisent (*Bison bonasus*) at the Brookfield Zoo.

Sitatunga *(Tragelaphus spekei)*

Sitatunga are spiral-horned antelopes that inhabit swamps of eastern, central, and southern Africa. The Brookfield herd began with a single pair acquired in 1939. Unrelated males were added in 1948, 1951, 1953, and 1964, and a male and two females were added in 1967. Because neither Brookfield nor other zoos had need for additional sitatunga, the sexes have been kept separate for most of the time since 1978, preventing further breeding. During the 40 years (five to ten generations) of alternating inbreeding and outbreeding, 389 sitatunga were born at Brookfield. Infant mortality was consistently very low in the first few years after the introduction of new, unrelated breeding males, followed by highly significant ($P < .01$) increases in infant mortality after the herd had been kept closed long enough for the breeding males to be mating with daughters and granddaughters (figure 15.5). No overall temporal trend (across years of inbreeding and outbreeding) in infant mortality was observed, indicating that the decreased survivorship during years of more inbreeding was not likely a result of changes in husbandry.

Addax *(Addax nasomaculatus)*

Addax are desert antelopes that were once widespread across the Sahara. They are rare and possibly almost extinct in the wild, having

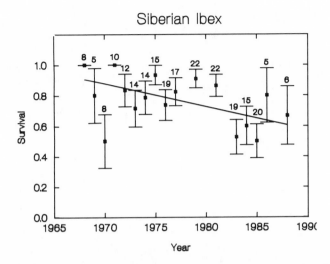

FIGURE 15.4   Relationship ($P < .01$) between infant survival and years since the establishment of a closed herd of Siberian ibex (*Capra ibex sibirica*) at the Brookfield Zoo.

been reduced to small remnant populations by overhunting. Although rarely seen, they apparently live in very tight herds of fewer than twenty individuals (Macdonald 1984). Brookfield began breeding addax in 1935, when a single pair was imported from the Sudan. Additional males were acquired from European zoos in 1955 and 1964. (There is a reasonable possibility that the 1964 import was the offspring of a Brookfield animal that had been sent to zoos in New York, then London, then Paris.) In approximately five generations of breeding, 140 addax have been born in the Brookfield herd. Herd size has been maintained at about one to six males and two to fifteen females. Often only a few addax were born in a year, and in many years all offspring died or all survived; therefore, a logit regression analysis is not possible. A *t* test revealed a significantly greater mean number of years since the last import among those addax that died within 30 days of birth than among those that survived (figure 15.6). As in the sitatunga, there was no overall trend toward increasing infant mortality across years of both inbreeding and outbreeding that could explain the greater mortality during the years of inbreeding.

Collared Peccary (*Tayassu tajacu*)

Collared peccaries inhabit a variety of habitats from the southwestern United States through Central and South America to Argentina. They live in bands of fourteen to fifty animals, consisting of one to three

Sitatunga

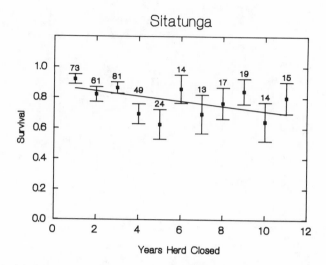

FIGURE 15.5   Relationship ($P < .01$) between infant survival and years since the addition of unrelated breeders to a herd of sitatunga (*Tragelaphus spekei*) at the Brookfield Zoo.

family groups (Macdonald 1984). Brookfield obtained a male and two females in 1965, and an unrelated male replaced the previous breeder in 1979. Typically just one adult male was kept in the herd (and up to several subadult males) with three to fourteen females. Records of parentage were often not kept, precluding calculation of inbreeding coefficients. Longevity can be 10 to 20 years, so many living animals may be offspring of the original trio. At the time breeding was stopped in 1986, 217 peccaries had been produced at the zoo. Across the two decades of breeding, there has been a significant trend toward reduced juvenile mortality, primarily because of high survival since 1980. Even so, juvenile survival decreases, though not significantly, when regressed against the number of years since the introduction of an unrelated breeding male (figure 15.7).

Overall, the above trends add six species to the data base on the effects of inbreeding on juvenile mortality in mammals, with sample sizes generally greater than could be obtained from field studies. In every case, the trend was for increasing neonatal mortality as the herds became increasingly inbred, but the data are too limited for detailed analyses, and we cannot specify the absolute or relative magnitudes of inbreeding depression among these species. We now focus on a species for which the zoo breeding records are sufficient to permit more extensive analysis of inbreeding and outbreeding.

Addax

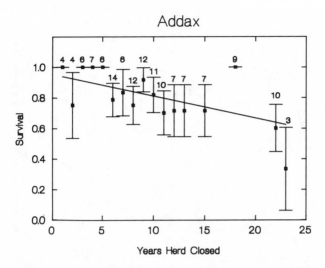

FIGURE 15.6   Relationship between infant survival and years since the addition of unrelated breeders to a herd of addax (*Addax nasomaculatus*) at the Brookfield Zoo. Significance ($P < .05$) of the trend was determined by a t test comparing the mean year of birth of viable offspring versus the mean year of birth of inviable offspring.

CALLIMICO (*CALLIMICO GOELDII*)

Goeldi's monkeys, also known by their generic name, Callimico, are evolutionarily between the Callithricidae and the Cebidae, and are placed in their own family, the Callimiconidae, by many taxonomists (Hershkovitz 1977). They are sparsely distributed in the upper Amazon basin of Brazil, Bolivia, Peru, and Colombia in dense forests. Little is known of their social structure in the wild, but Callimico appear to be monogamous, living in family groups of six to eight, with offspring often staying with the parents while subsequent offspring are reared.

The analyses presented here are based on data presented in the international studbook (Warneke 1988) and more recent data (through studbook number 952) provided by the studbook keeper, Mark Warneke. The captive population of Callimico derives primarily from only a few importations. (Some were illegal shipments that were confiscated at borders by officials and then turned over to zoos.) The pedigrees of most captive-born animals are traceable back to these original imports, paternity is rarely in question (because they are maintained as monogamous pairs), and aborted fetuses, stillbirths, and neonatal deaths have been reliably recorded. Even after eliminating all animals for

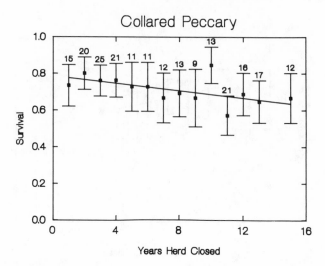

FIGURE 15.7   Relationship ($P > .10$) between infant survival and years since the addition of an unrelated male to a herd of collared peccaries (*Tayassu tajacu*) at the Brookfield Zoo.

which there is question as to one of the captive ancestors, data are available on 790 captive-born Callimico, 111 of which are inbred.

Inbreeding Depression in Callimico

We examined the effect of inbreeding on infant mortality, again looking at mortality before 30 days of age (including abortions and stillbirths). Callimico show marked inbreeding depression (figure 15.8), with a 33% decrease in survival resulting from each 10% increase in inbreeding, and 7.90 ± 0.17 *SE* lethal equivalents per diploid genome estimated from a weighted regression of log[viability] versus F (log and logit regressions significant at $P < .0001$). Separation of the inviable offspring into abortions, stillbirths, and infant deaths revealed twofold greater mortality among inbred than noninbred progeny at each stage of development: of the Callimico pregnancies reported to the studbook keeper, 3% of noninbred and 6% of inbred fetuses were aborted spontaneously, 2% of noninbred and 4% of inbred young were stillborn, and 19% of noninbred and 36% of inbred young died before 30 days of age, leaving 76% of noninbred and 54% of inbred progeny surviving beyond 30 days.

Much of the inbreeding in Callimico has been clustered at a few institutions, so it is possible that the "inbreeding depression" is an artifact of different management procedures at those zoos that inbreed

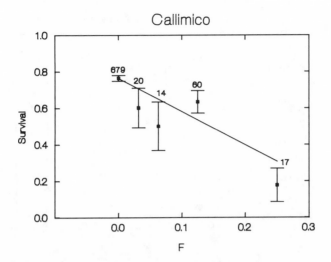

FIGURE 15.8   Relationship ($P < .001$) between infant survival and inbreeding (F) in captive-born Callimico (*Callimico goeldii*). Inbreeding coefficients above 0.00 but below 0.04 were combined in the modal $F = 0.03125$ class; while inbreeding coefficients between 0.04 and 0.08 were combined in the modal $F = 0.0625$ class.

than at those zoos that have few inbred Callimico. Within-zoo analyses of the effects of inbreeding are necessary to eliminate this possible non-genetic cause of the very strong inbreeding depression observed in Callimico. (Another source of bias may be masking an even stronger inbreeding effect: zoos that frequently inbreed their animals may be those with poor record-keeping and therefore may be less likely to report aborted fetuses and stillbirths to the studbook keeper.) Brookfield Zoo has produced 257 Callimico, but almost none have been inbred. The Jersey Wildlife Preservation Trust has produced 106 Callimico, 21 of which have been from half-sib matings. Analysis of the data from Jersey reveals highly significant ($P < .001$) inbreeding depression with 4.86 lethal equivalents. (No other zoo has produced a sufficient number of Callimico to permit statistical analysis.) Thus, we can eliminate inter-zoo differences in management as the sole cause of the inbreeding depression observed, although the somewhat lower estimate of lethal equivalents at Jersey leaves open the possibility that differences in husbandry are in part confounded with inbreeding effects.

Because of the large data set, we can examine the effects of inbreeding separately for males and females. Inbreeding depression in male offspring is much weaker than the overall population trend (figure 15.9a; 3.21 ± 0.20 lethal equivalents), but is still highly significant due

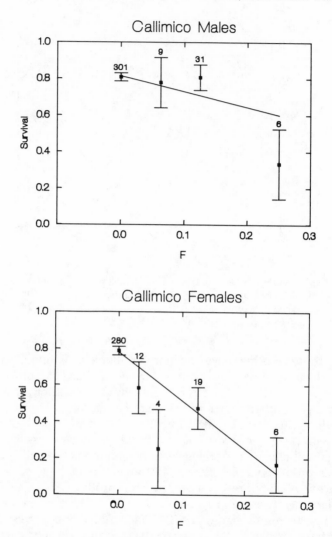

FIGURE 15.9   Relationship between infant survival and inbreeding (F) in (a) male (P < .05) and (b) female (P < .001) captive-born Callimico (*Callimico goeldii*). Lower inbreeding classes pooled as in figure 15.8 for females; inbreeding coefficients above 0.00 and below 0.08 pooled into the 0.0625 class for males.

to the poor survival of males from full-sib and parent-offspring matings (F = 0.25). Inbreeding depression of survival of female infants is quite severe, with 10.87 ± 0.19 lethal equivalents (figure 15.9b). Inbreeding depression in females is significant even at inbreeding levels of 0.0625 and below, that is, matings between first cousins and less closely related animals (chi-square test of proportion of inbred versus noninbred offspring surviving: $P < .01$). This sex-biased trend in inbreeding depression could indicate that heterozygosity is less critical to males, or perhaps that genetically debilitated females are more likely to die than are similarly affected males. (Mortality among noninbred infants is the same, about 20%, in the two sexes, and a 1:1 sex ratio at birth makes it unlikely that most of the unsexed aborted fetuses and stillbirths were males.) Investigations of parental care of inbred infants and of zoo management practices regarding inbred and/or weakened infants might reveal interesting sex biases.

Outbreeding Depression in Callimico

Because the geographic origins of wild-caught animals used to initiate captive colonies are often unknown, the effects of inbreeding and outbreeding can be confounded in captive animals (see above). In small data sets, inbreeding depression and outbreeding depression may be impossible to distinguish. In large, multigenerational data sets, however, the two phenomena can be separated statistically. (Not all second- and later-generation offspring are inbred, and inbreeding is expected to get worse through the generations, while the disruptive effects of recombination generally lessen after the second generation.)

Templeton and Read (1984) proposed a measure, the "hybridity coefficient," for quantifying the disruption of coadapted gene complexes following hybridization. The hybridity coefficient (H) is defined as the fraction of the genome that was subject to recombination between disparate genetic lineages in the parents of an animal. Each wild-caught founder is assumed to represent a unique, coadapted genome. Their captive-born progeny are therefore F_1 "hybrids." Because the F_1 animals have a complete haploid set of chromosomes from each wild-caught parent (perhaps from different geographic populations) they would be expected to have no disruption of coadapted complexes ($H = 0$) and may show hybrid vigor. Because the entire genome of the F_1 animals is subject to recombination between divergent coadapted genomes, the F_2 progeny (whether inbred or not) are assigned hybridity coefficients of one. Backcross progeny between a wild-caught founder and an F_1 hybrid have $H = 0.5$, and could have an inbreeding coefficient of either $F = 0$ (if the founder is not also a parent of the F_1 hybrid) or $F = 0.25$ (if the mating is between a founder and its offspring). Templeton and Read (1984) give the algorithm for calculating the hybridity coefficient of any animal in a pedigree.

Unlike the analysis of hybridity in Speke's gazelle by Templeton and

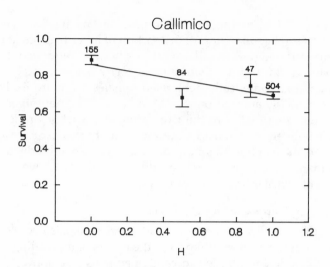

FIGURE 15.10   Relationship ($P < .001$) between infant survival and hybridity (H) in captive-born Callimico (*Callimico goeldii*).

Read, hybridity was found to have a highly significant ($P < .001$ by logit regression analysis) effect on infant survival in Callimico (figure 15.10), with a 3.2% reduction in survival for each 10% increase in H. Table 15.1 shows the distribution of inbreeding and hybridity for the captive Callimico and demonstrates the nonindependence of the two in the data. Some combinations of F and H are impossible in a captive stock: inbred animals must be second- or later-generation or backcross progeny, and therefore cannot have totally intact chromosome complements from wild populations. Multivariate regressions on logit-transformed survival probabilities reveal that both the effect of inbreeding while controlling for hybridity and the effect of hybridity while controlling for inbreeding are significant. The observed inbreeding depression and outbreeding depression are statistically independent, however, with each slope virtually unaffected by the inclusion of the other effect in the regression model. Moreover, analysis of the effect of hybridity among only noninbred Callimico still reveals highly significant outbreeding depression (of unchanged magnitude), and analysis of inbreeding among F_2-generation Callimico ($H = 1$) shows highly significant inbreeding depression. Thus, while Callimico show some of the strongest inbreeding depression reported among mammals, they also show independent and significant, but less severe, outbreeding depression in captive generations subjected to recombination of wild genomes.

TABLE 15.1 Distribution of Inbreeding and Hybridity in Captive Callimico

F	.000	.031	.125	.250	Totals
H					
0.00	155	—	—	—	155
	0.88				0.88
0.50	78	0	1	5	84
	0.71		1.0	0.20	0.68
0.88	31	4	11	1	47
	0.81	0.75	0.55	1.0	0.75
1.0	415	30	48	11	504
	0.73	0.53	0.65	0.09	0.69
Totals	679	34	60	17	790
	0.76	0.56	0.63	0.18	0.73

Note: Distribution by inbreeding coefficient (F, columns) and hybridity coefficient (H, rows), with proportion of infants surviving to 30 days of age (below counts). Dashes indicate combinations of F and H that are impossible in a captive colony. The category $F = 0.031$ includes inbreeding between 0.015 and 0.074, with the majority of values being 0.03125.

We know nothing of the geographic sources of the wild-caught founders of the captive Callimico colonies; thus we cannot speculate on the spatial distances or habitat differences that could lead to coadaptation of the genomes of Callimico. For example, the Brookfield colony was begun with five pairs of Callimico confiscated by the U.S. Fish and Wildlife Service in Miami from an airline flight originating in Bolivia, and may have been a single social unit in the wild or may have come from diverse habitats over a broad area. In an attempt to refine further our measure of outbreeding depression, we calculated "grouped hybridity" coefficients under the assumption that each group of Callimico that were imported together share a coadapted genome uniquely different from the genomes of each other importation. (Ralls and Ballou [personal communication] have independently been using this grouped hybridity coefficient to analyze outbreeding depression in captive stocks.) The grouped hybridity coefficient is an appropriate measure of outbreeding if animals imported together come from a single wild population.

Surprisingly, we found no relationship between infant survival and the grouped hybridity coefficient. A relatively small fraction of the captive population represents second- or later-generation crosses among founder groups (i.e., has grouped hybridity coefficients greater than zero), so the data do not provide a robust test of outbreeding depression at this level. Yet "outbreeding" among founder genomes, but within importation groups, does depress infant viability. As a possible alternative explanation of these data, we note that, by definition, the animals with hybridity coefficients of zero are progeny of two wild-caught founders. It is possible that wild-caught animals are more likely to rear infants in captivity than are captive-born animals. Prior experience could play a role (parents often do not rear their first offspring), but many imported animals were young and likely nulliparous at the time of capture.

Concerned that the effect of hybridity on infant survival could be spurious, we examined the regression of survival against hybridity for those cases in which hybridity was greater than zero. This analysis excludes offspring of wild-caught animals, but still leaves 635 progeny of captive-born Callimico. Among the three levels of hybridity greater than zero, we detected no significant differences in infant survival. Unlike inbreeding depression, which if due to deleterious recessive alleles would show a log-linear response with F, there presently is no theory to predict the shape of outbreeding depression relative to H. It may be that outbreeding depression is a threshold phenomenon, being expressed approximately equally across categories of recombinant descendant generations.

We did not expect to observe outbreeding depression in the Callimico. Unlike most other South American primates, *Callimico goeldii* is a species with no known geographic variation in morphology and no named subspecies. Among other South American primates (e.g., squirrel monkeys, *Saimiri* spp.: Hershkovitz 1984), considerable geographic variation often occurs over a contiguous geographic range. Major rivers frequently divide distributions of taxa that are differentiated chromosomally. Chromosomal polymorphisms have also been noted among captive Callimico. The finding of XY males, males with a translocation of the Y onto an autosome, and males with a translocated piece of the Y (the rest being lost) led Hsu and Hampton (1970: 193) to conclude that it "is possible that a minimum of three different populations exist in South America and the species is diverging." Neither animals of known collecting localities nor the source stocks for the captive populations have yet been karyotyped; thus it is not yet possible to relate the Y-chromosomal polymorphisms observed in captive Callimico either to disjunct natural populations or to the finding of greater mortality in the recombinant second- and later-generation captive-born infants.

In summary, Callimico show evidence of strong inbreeding depression and also of outbreeding depression, with inbreeding depression about tenfold more severe than outbreeding depression in captive environments. The two phenomena are independent conceptually and, in the case of the Callimico, statistically. Inbreeding depression results from poor combinations (homozygosity) of alleles at a genetic locus, while outbreeding depression results from poorly adapted combinations of nonhomologous genes that must interact harmoniously for proper development and adaptation. Because of the constrained structure of matings in captive colonies, inbreeding and outbreeding depression can occur simultaneously in the second and later generations of captive breeding. In wild populations, philopatry could prevent outbreeding depression at the possible cost of inbreeding depression, while dispersal would prevent inbreeding at the risk of outbreeding depression. Unfortunately, it will be difficult to examine the relative effects of inbreeding and outbreeding depression in natural populations without extensive data on pedigrees and fitness components in wild populations that experience both close inbreeding and occasional long-distance dispersal between genetically divergent populations.

CONCLUSION

The breeding records of zoological parks provide data on inbreeding and outbreeding depression for a wide variety of species that would be difficult to study in more natural habitats. Although rigorous experimental controls are usually lacking, it is clear that many captive populations are suffering increased infant mortality as a result of inbreeding. With the notable exception of crosses between chromosomally divergent populations, there has been little documentation of outbreeding depression among zoo stocks. Because the geographic origins of the founders of most zoo stocks are unknown, systematic study of outbreeding depression in zoo populations will be difficult. The few attempts to assess outbreeding depression have focused on verifying that outbreeding depression was not confounding measurement of inbreeding depression. In a South American primate, Callimico, we found evidence that outbreeding depression as well as inbreeding depression affected fitness in second- and later-generation captiveborn animals, and neither effect could be explained as an artifact of the other.

As zoos propagate more species within their animal collections, keep more records on the breeding performance and medical histories of animals, and make the data more widely available through published studbooks and accessible computerized data bases, opportunities for using data on captive populations of wild animal species to examine the effects of inbreeding and outbreeding will multiply. Because zoos are unlikely to conduct well-designed breeding tests (research laboratories are more suited for such experimentation), further research on inbreeding in captive collections will need to focus on collecting sufficient data to allow statistical separation of artifacts of husbandry practices from true genetic effects. The mechanisms of inbreeding and outbreeding depression should be examined more closely within captive propagation programs. Many aspects of fitness should be assessed, in a variety of social contexts. We do not know the relative impact of inbreeding on fertility, growth rates, survival, disease resistance, or social dominance. It is even possible that the observed higher infant mortality among inbred animals could be offset by shorter interbirth intervals in females mating incestuously or by greater reproductive success of the inbred offspring.

Inbreeding depression has been documented in such a variety of species, including most in which data sets are of reasonable size, that it must be a very general phenomenon. Yet the frequency of inbreeding, the mechanisms of inbreeding avoidance, the severity of inbreeding depression, and probably even the causes of inbreeding depression vary tremendously among populations. Given the disparity observed so far among studies of inbreeding, it almost seems as though each population should be considered potentially a special case, or at least a single data point. Experimental studies of inbreeding and outbreeding have focused on too few species to allow generalization. Comparative studies of inbreeding among captive populations at zoological parks have been too poorly controlled to permit valid interspecific comparisons. A marriage of the two approaches is needed, with statistically rigorous analysis of a broader array of taxa in both captive and natural environments, before we can make generalizations about the causes, frequency, and severity of inbreeding and outbreeding depression.

ACKNOWLEDGEMENTS

We thank Pamela Parker, Kathy Ralls, and Jon Ballou for valuable discussions on the analysis of inbreeding and outbreeding in captive populations.

16

Dispersal, Kinship, and Inbreeding in African Lions

Craig Packer and Anne E. Pusey

The extent of inbreeding in natural populations is often estimated by monitoring the dispersal and mating behavior of individually recognized members of a population over several generations. Such long-term studies clearly indicate that close inbreeding (between parents and offspring or siblings) is very rare in birds and mammals (reviewed by Ralls, Harvey, and Lyles 1986). In many species, sex-biased dispersal of individuals from their natal area prevents close relatives from residing together as adults (e.g., Packer 1979, Greenwood 1980, Pusey 1987). Demographic processes such as high mortality, or differential age at maturation and mortality of each sex, have the same effect (Ralls, Harvey, and Lyles 1986; Waldman and McKinnon, chapter 11, this volume). There is also considerable evidence of behavioral avoidance of inbreeding between close relatives in a variety of species (Ralls, Harvey, and Lyles 1986; Pusey and Packer 1987b).

However, it has proved much more difficult to measure from behavioral data the extent to which more moderate levels of inbreeding occur, and this has been the subject of recurrent debate (Ralls, Harvey, and Lyles 1986; Shields, chapter 8, this volume; Smith, chapter 14, this volume). Even where extensive pedigrees of local populations exist, the calculation of degrees of kinship becomes very complex because there are always some individuals breeding in the population whose origins are unknown. Pedigrees based on behavioral data must also be treated with caution because of extra-pair copulation (Rowley et al., this volume), multiple copulation (e.g., Burke et al. 1989; Rabenold et al. 1990) and intraspecific brood parasitism (Emlen and Wrege 1986).

It has recently become possible to measure levels of kinship accurately by molecular genetic techniques. DNA fingerprinting was first developed to estimate parentage (Jeffreys, Wilson, and Thein 1985;

Burke 1989), but it can also be used to estimate more distant degrees of kinship in some circumstances (Lynch 1988b, 1990; Reeve et al. 1990; Westneat 1990; Gilbert et al. 1991; Jones, Lessells, and Krebs 1991). In this chapter we show how DNA fingerprinting can be combined with other techniques to measure inbreeding in natural populations.

The lion populations of the Serengeti and Ngorongoro Crater are among the best-studied vertebrate populations in the world (Schaller 1972; Bertram 1975; Packer et al. 1988; Packer, Pusey, et al. 1991) and we have recently combined long-term genealogical data with DNA fingerprinting (Gilbert et al. 1991; Packer, Gilbert, et al. 1991). In this chapter we review lion dispersal patterns and their effect on the incidence of close inbreeding, we discuss how DNA fingerprinting can be used to measure kinship and therefore to determine relatedness between mates, and we examine the consequences of inbreeding on lion reproductive performance.

LION SOCIAL ORGANIZATION

Lions live in stable social groups ("prides") that typically contain 2–9 adult females (range: 1–18), their dependent young, and a coalition of 2–6 adult males (range 1–9) that has entered the pride from elsewhere (Packer et al. 1988). Prides are territorial and often occupy the same range for generations. Incoming males kill or evict the dependent young of the prior coalition (see reviews by Packer and Pusey 1984; Pusey and Packer 1993). Consequently, females resume sexual receptivity within days, show regular oestrus cycles for about 3 months and mate exclusively with the males of the new coalition by the time they conceive (Bertram 1975; Packer and Pusey 1983). Births tend to be synchronous within a pride (Bertram 1975) and cubs born less than 1 year apart make up a "cohort." DNA fingerprinting analysis confirmed behavioral estimates of maternity for 77 of 78 cubs (Gilbert et al. 1991). The 78th cub belonged to a female pridemate of the assumed mother. The DNA analysis also confirmed that the males of the resident coalition father all cubs born during their tenure. Males generally remain in the same pride for only 2–3 years and typically father only one cohort per pride.

STUDY SITES

African lions in northern Tanzania have been studied continuously since the 1960s (Packer et al. 1988, Packer, Pusey, et al. 1991). Our long-term records include data on over 2,000 individuals in two separate populations: one in the Serengeti National Park and the other in Ngorongoro Crater. These two populations differ greatly in size and extent of isolation. The Serengeti ecosystem covers 25,000 km², the total lion population exceeds 3,000, and males disperse over the entire region (Pusey and Packer 1987a). Over two-thirds of the 118 males that have bred in our 2,000 km² study area originated from other parts of the Serengeti (Packer, Pusey, et al. 1991) (figure 16.1).

In contrast, the 250-km² floor of the Ngorongoro Crater is a small, naturally isolated island of lion habitat. Lions have resided in the Crater for at least a century, but in 1962 an epizootic reduced the population to nine females and one male (Packer, Pusey, et al. 1991). An additional seven males apparently immigrated into the Crater in 1964–65, but there has been no further immigration in the past 25 years. The population had largely recovered by 1969 and has included 25–45 breeding animals since 1970 (figure 16.2). There was considerable variance in the reproductive success of the founders; four of five contemporary prides derive from a single group of four females, and there has been considerable exchange of males among all five prides (Packer, Pusey, et al. 1991). Hence, the Crater population is far smaller than the Serengeti population and has been genetically isolated for five generations.

Lion Dispersal Patterns

Subadults of both sexes are usually forced to emigrate when their natal pride is taken over by a new coalition of males (Hanby and Bygott 1987; Pusey and Packer, 1987a). Females may also disperse when their mothers give birth to a new batch of cubs or to avoid mating with their fathers when they reach sexual maturity. About two-thirds of all female cohorts are recruited into their mothers' prides, while the remainder emigrate and establish new prides nearby (figure 16.3). Dispersing females have significantly lower fitness than females that remain in their natal pride (Pusey and Packer 1987a). Females in our study areas have never transferred successfully from one preexisting pride to another (Packer 1986; Pusey and Packer 1987a).

In contrast, almost all males leave their natal pride before the age of 4 years and undergo a nomadic phase before gaining residence in a new pride (figure 16.3). Large male cohorts often enter new prides intact, whereas cohorts of only 1–2 males may team up with singletons from other prides before gaining residence (Packer and Pusey 1982; Pusey and Packer 1987a). Small cohorts disperse farther from their natal pride than do large cohorts (Pusey and Packer 1987a). Most males

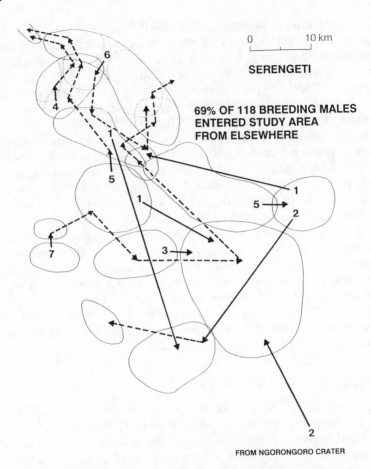

0   10 km

SERENGETI

69% OF 118 BREEDING MALES
ENTERED STUDY AREA
FROM ELSEWHERE

FROM NGORONGORO CRATER

FIGURE 16.1   Size and movements of male coalitions born in the Serengeti study area
that resided in study prides. Approximate pride ranges from 1974–1985 are shown.
Solid arrows indicate male natal dispersal over the same period; dotted arrows indicate
secondary dispersal. Note that most males resident in the study area have entered it
from elsewhere. (Modified from Pusey and Packer 1987a).

in the Serengeti study population have entered the study area from
elsewhere (see figure 16.1). Thus, in contrast to females, males commonly breed far from their natal range and do not appear to suffer such
large losses in fitness from dispersal.

Although much male dispersal occurs as a result of eviction by
usurping coalitions, some features of male dispersal apparently result
from the avoidance of close inbreeding (Pusey and Packer 1987a). First,
young males often disperse when they reach maturity even if their fa-

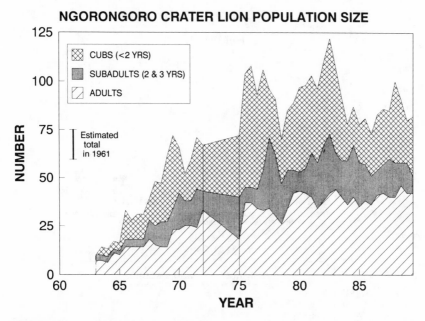

FIGURE 16.2   Population size and composition of the Ngorongoro Crater population. Data for 1961 are based on Fosbrooke's (1963) estimate. Subsequent data give the population size on 1 January and 1 July each year. For each date, the number of individuals in each age class is illustrated by the height of the respective hatched areas. Data for 1973 and 1974 are interpolated. (From Packer, Pusey, et al. 1991).

thers are still resident (although sons will team up with their fathers if they are evicted together), or even in the complete absence of adult males. Second, males almost never return to their natal pride. Third, in the Serengeti, the only two cases in which males have bred with their close relatives occurred when the males joined an offshoot of their natal pride, with whom the males were not as familiar as with their own maternal grouping. This was also true for three of four such cases in the Crater. Fourth, males never reside in the same pride as their mature daughters in the Serengeti, and rarely do so in the Crater. Vigorous coalitions of males sometimes voluntarily abandon prides containing maturing daughters for prides that contain fewer adult females. It should be noted that while the frequency with which males reside with close female relatives in the Crater is higher than in the Serengeti, the opportunity to disperse to prides containing nonrelatives is much more limited in the Crater (see also Packer, Pusey, et al. 1991).

FATE OF SUBADULTS BY 48 MOS

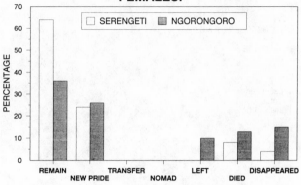

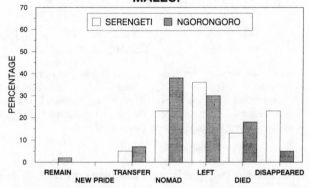

FIGURE 16.3   The fate of subadults by 48 months of age. Remain = individuals still in their natal pride. New pride = females in or adjacent to their natal range, but that no longer associate with their natal pride. Transfer = individuals that have become members of other, preexisting prides. Nomad = individuals whose range includes areas well away from their natal range. Died = individuals whose bodies were found, those last seen in very poor health, or that disappeared singly from their natal pride or cohort. Disappeared = individuals last seen in good health in their natal pride but that disappeared at the same time as at least one other member of their pride. Note that only females are considered to form new prides because males do not show a similar site specificity and male ranges depend on the ranges of the female prides in which they become resident. Data based on 280 subadults that reached 18 months of age between 1970 and 1984. (Reprinted from Pusey and Packer 1987a.)

DNA ANALYSIS

While the demographic data indicate that close inbreeding is typically avoided in lions, it is impossible to estimate the extent to which more moderate inbreeding occurs. It has been suggested that levels of inbreeding might typically be high in populations of large vertebrates because of the limited dispersal of males (Chepko-Sade et al. 1987). Even though Serengeti males typically disperse quite far from their natal pride, there is always the possibility that any two prides are related either by previous pride splits (see below) or by past exchange of males. We have therefore employed DNA fingerprinting to measure precise relatedness between different classes of individuals in the two populations.

Using feline-specific hypervariable probes, DNA fingerprinting analysis was performed on 193 animals from the Serengeti and 23 from the Crater (Gilbert et al. 1991). The DNA was digested with restriction enzymes, and fragments were separated according to molecular weight by agarose gel electrophoresis. DNA fragments carrying different numbers of a hypervariable minisatellite repeat are visualized as bands of varying molecular weight after hybridization with a radioactive minisatellite DNA probe. These bands are inherited in a Mendelian fashion, so close relatives are expected to share more bands by common descent than unrelated individuals do. Percentage of similarity, or "band sharing," between two individuals is defined as $2F_{ab}/ (F_a + F_b) \times 100\%$, where F_{ab} is the number of DNA fragments showing similar molecular weight and intensity carried by both individuals, F_a is the total number of fragments resolved in individual a, and F_b the number resolved in individual b.

Information on matrilineal kinship comes from our long-term records, which were verified by the DNA fingerprinting band-matching analysis of parentage (see above). The accuracy of these data allowed Gilbert et al. to calibrate the extent of minisatellite band sharing against a reliable independent measure of kinship. Average band sharing was calculated for known parents and offspring, known full siblings and half siblings, a variety of more distant matrilineal relatives, and individuals that came from noncontiguous prides in the Serengeti with no known kinship links.

The calibration curve differs between the two populations (Gilbert et al. 1991). In the Serengeti, the mean degree of band sharing is highly correlated with relatedness, but the relationship is not linear (figure 16.4a). The distributions of band sharing for individuals related by 0.125–0.5 overlap sufficiently to prevent classification of a particular pair of animals as siblings or cousins without independent knowledge of parentage. However, the band-sharing data readily distinguish among three classes of relationship: kin related by ≥ 0.125, kin related by 0.02–0.06, and nonrelatives (Gilbert et al. 1991).

The extent of band sharing also declines with decreasing relatedness

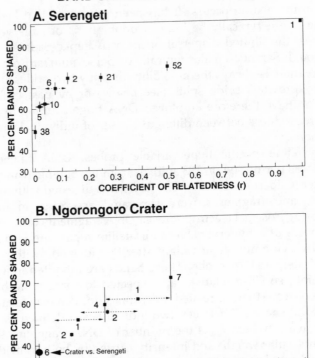

BAND SHARING vs RELATEDNESS

A. Serengeti

B. Ngorongoro Crater

FIGURE 16.4 Calibration curves of minisatellite band sharing vs. coefficient of kinship in (a) the Serengeti population and (b) the Ngorongoro Crater population. Each point is the mean of the average percent similarity (see text) for two restriction enzymes; the vertical line is the standard deviation across the means of all comparisons in the sample. Numbers above each point are the number of comparisons in each relatedness class. For each sample, each individual in a relatedness class was compared at random to only one other individual to assure independent and equivalent weight to each individual's phenotype. Thus, for example, for first-order relatives ($r = 0.5$) in the Serengeti, there were 52 comparisons involving 104 individuals. In the Serengeti, unrelated lions were animals from different prides with no known kinship links. In the Crater, unrelated animals indicate comparison of Crater lions to Serengeti lions (see text). Both sexes were used equivalently in the calibration and there was no sex bias in band sharing. For close relatives, relatedness was known exactly from behavioral observations and pedigrees. For several more distant relatedness categories, only a range of r could be provided. In these cases the mean value of r is plotted and the range is indicated by horizontal arrows. One pair of males of unknown origins are assumed to be identical twins because they share 100% of their bands. No other kinship classifications have been made on the basis of band sharing. The Serengeti curve can only be described by a higher-order statistical analysis involving a piecemeal linear model, whereas the Ngorongoro Crater curve is linear (see Gilbert et al. 1991). All gels were scored by D. Gilbert, who did not know the genealogical relations of most individuals. (Modified from Gilbert et al. 1991.)

in the Crater population, but the relationship is more linear (figure 16.4b). Note that these estimates of r do not consider the history of inbreeding within the population, and that "unrelated" lions here refers to comparisons of Crater lions with those in the ancestral Serengeti population (Gilbert et al. 1991).

Although the relationship between band sharing and relatedness is expected to be linear (Lynch 1988b), the Serengeti calibration curve clearly deviates from linearity: band sharing increases much more rapidly than expected over low values of relatedness and increases only gradually at higher levels of kinship. Similar (though less extreme) nonlinearities have also been detected in other studies (Kuhnlein et al. 1990; Jones, Lessells, and Krebs 1991). One possible explanation for this pattern would be unknown kinship links between individuals assumed to be unrelated (and we had expected the calibration curve to be curvilinear in the inbred Crater population for precisely this reason). However, this cannot be the cause of the nonlinearity in the Serengeti data. First, genealogical records there are so extensive that any inaccuracies would not exceed 1–2% on average. Second, the nonlinearity is apparent within families. In ten cases, two males that were known to be unrelated to each other (and showed band sharing typical of nonrelatives: ca. 50%) each fathered offspring by the same female (to whom they also showed band sharing of only 50%). Each parent showed band sharing of 80% to their own offspring, but the band sharing between the half siblings was 74.5% ± 2.5% (Gilbert et al. 1991). If the relationship between band sharing and kinship was linear, band sharing between these half siblings should only have been about 65%.

The cause of this pattern clearly warrants further study, and data from many other populations are needed to determine how commonly such nonlinearities exist in nature. These results suggest that calibrations of kinship from band sharing should be based on at least three points (see below).

KINSHIP STRUCTURE OF LION POPULATIONS

From the behavioral data, it seemed likely that lion prides show high coefficients of relatedness between most same-sex companions but far lower relatedness between members of different prides or between mating partners (Bertram 1976). These genetic relationships could all be confirmed by the band-sharing analysis.

Kinship between Females

The minisatellite band-sharing data clearly reveal the kinship of females within and between prides. Results are presented separately for the Serengeti and the Crater because of the different relationship between kinship and band sharing in the two populations. In every Serengeti pride, female pridemates show band sharing far in excess of that found between unrelated individuals, and there is very little variation in these scores within each pride (fig. 16.5a). In spite of the more variable relationship between band sharing and kinship in the Crater, female pridemates clearly show higher band sharing than individuals from the most distantly related Crater prides (fig. 16.6a).

Figure 16.7a shows that the initial degree of band sharing between females in Serengeti prides of common ancestry is as high as within each pride, and that band sharing between prides declines through time after the prides have separated. When a cohort of daughters first splits from the parental pride, kinship ties between the two prides are as close as within each pride. But as the mothers die (maximum lifespan of females is about 17 years; Packer et al. 1988) and the separate prides recruit respective sets of daughters fathered by different coalitions of males, the degree of band sharing between prides diminishes. Many neighboring prides in the Serengeti have no known kinship links, and levels of band sharing between unrelated neighbors are as low as those between nonadjacent prides (fig. 16.7). The degree of band sharing between these prides is the same as between other nonrelatives (fig. 16.5a).

In contrast, all prides in the Crater have numerous kinship links. Most contemporary prides are descended from a founding group of four females, and there has been frequent exchange of males between all five Crater prides (Packer, Pusey, et al. 1991). Nevertheless, there is a similar decline in band sharing through time after prides split.

From the Serengeti data, it appears that most female pridemates are at least as closely related as second cousins. Such close kinship between females is consistent with the pattern of male reproductive success within each coalition and with the pattern of female dispersal. Only two males in each pride father most of the offspring (Packer, Gilbert, et al. 1991), which greatly increases the chances that any two females of similar age will be paternal siblings (Bertram 1976). Most new prides are founded by dispersing cohorts of such same-aged females (Pusey and Packer 1987a).

Kinship between Males

Levels of kinship between male coalition partners vary more than between female pridemates. Cohorts of young males remain together after dispersing from their natal pride, but singletons frequently join up with males from other natal prides (Packer and Pusey 1982; Pusey and Packer 1987a). Figure 5b shows that male coalition partners born

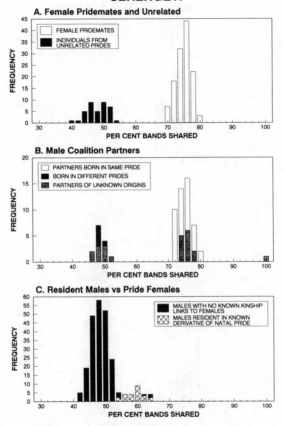

SERENGETI

A. Female Pridemates and Unrelated

B. Male Coalition Partners

C. Resident Males vs Pride Females

FIGURE 16.5   Frequency distributions of minisatellite band sharing within or between different classes of individuals in the Serengeti. For overlapping distributions, the number of pairs showing a particular degree of band sharing is given by the height of the respective hatched region. (Reprinted from Packer, Gilbert, et al. 1991b.) (a) Band sharing between female pridemates compared with band sharing of individuals born in different parts of the park. Data on female pridemates are based on 63 females from 15 prides and include all pairwise combinations of females within each pride (e.g., a pride of 4 females contributes 6 combinations of females). However, each unrelated individual is included in only one combination (76 different individuals in 38 pairs). (b) Band sharing between male coalition partners. Data are based on 45 males in 16 coalitions and include all pairwise combinations within each coalition. Data are plotted separately for partners known to have been born in the same pride, those known to have been born in different prides, and those that had entered the study area from elsewhere. One pair of males of unknown origins shared 100% of their bands and are hence presumed to be identical twins. (c) Band sharing between resident males and pride females. Data based on 44 males in 18 coalitions and 52 females in 15 prides, including all male-female combinations within each pride. Most males had no known kinship links to the females, but one coalition resided in a neighboring pride that derived from the males' natal pride (see text).

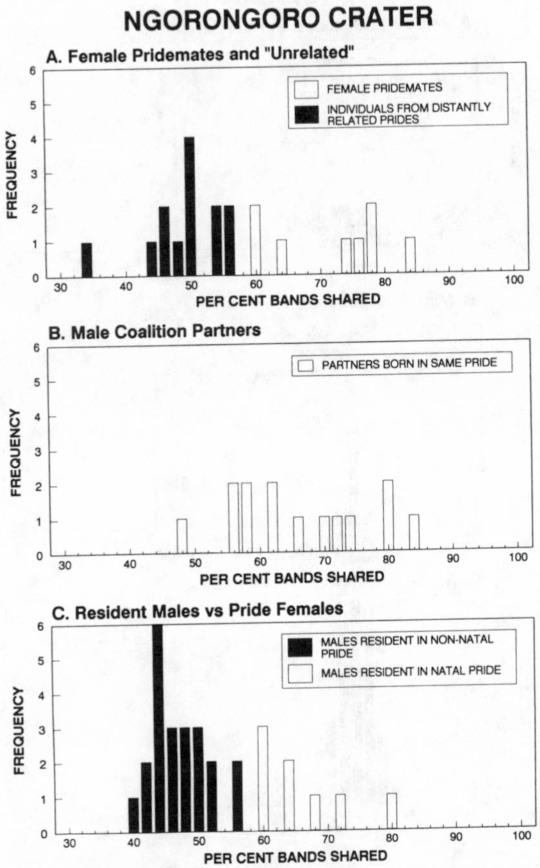

CRAIG PACKER AND ANNE E. PUSEY

NGORONGORO CRATER

FIGURE 16.6 Frequency distributions of band sharing in Ngorongoro Crater, plotted as in figure 1. (Reprinted from Packer, Gilbert, et al. 1991.) (a) Band sharing between female pridemates (8 females in 3 prides) compared with band sharing of 22 individuals from Crater prides with the fewest kinship links. (b) Band sharing between male coalition partners (14 males in 5 coalitions). All partners were known to have been born in the same pride. (c) Band sharing between resident males and pride females (11 males in 4 coalitions and 6 females in 2 prides). All of these males became resident outside their natal pride, but members of one coalition later returned to become resident in their natal pride.

in the same Serengeti pride have the same degree of band sharing as female pridemates, whereas partners born in different prides indeed show band sharing typical of unrelated individuals. Partners in coalitions of unknown origins show a degree of band sharing that clearly belongs to one distribution or the other: a proportion of these partners

INTERPRIDE COMPARISONS

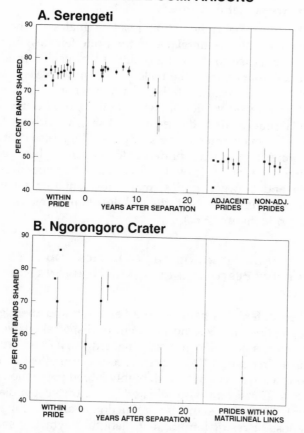

FIGURE 16.7   Degree of band sharing within and between prides. Each point represents a comparison within a pride or between two prides. Where comparisons are made between prides of common ancestry, the data are plotted against the number of years that have elapsed since the prides split apart. Data on adjacent and nonadjacent prides only include prides with no known kinship links (either through males or females) over the past 24 years. The mean and standard deviation across all pairwise combinations of individual females are plotted for each comparison. (Reprinted from Packer, Gilbert, et al. 1991.) (a) Serengeti. All comparisons involve females born in the respective prides. (b) Ngorongoro Crater. The number of between-pride comparisons was enlarged by including males as well as females born in the respective prides.

must have been born in the same pride, whereas the remainder were born in unrelated prides. In our sample from the Crater, the partners in each coalition were born in the same pride, and they show band sharing similar to female pridemates (figure 16.6b).

Kinship between Mating Partners

The band-sharing data show that resident males in the Serengeti prides are almost always unrelated to the pride females (figure 16.5c). One exceptional case concerned a coalition that gained residence in a pride that had split off from the males' natal pride 10 years before the males' births. This coalition fathered no surviving offspring in this pride and abandoned it after one year. Only one other resident male had a similar, intermediate degree of band sharing with the females of his pride. This male entered the study area from elsewhere and his kinship connections to this pride could only have been patrilineal. Even though Crater lions show less clear-cut differences in band sharing within and between prides, males resident in nonnatal prides clearly share fewer bands with the females than do males resident in their natal pride (figure 16.6c).

EFFECTS OF INBREEDING ON THE GENETICS AND REPRODUCTIVE PERFORMANCE OF THE NGORONGORO CRATER LIONS

The marked differences between the Serengeti and the Ngorongoro Crater populations in size and in extent of dispersal enable us to examine the effects of inbreeding on genetic diversity and reproductive performance. In table 16.1, we combine results from the two Tanzanian populations with data from a third, highly inbred population of captive Asiatic lions. Consistent with a history of restricted population size, the Crater lions show a significant reduction in allozyme heterozygosity and a loss of restriction fragment length polymorphism in MHC class I genes, but only a slight decrease in heterozygosity of DNA fingerprint fragments. The lions of the Sakkarbaug Zoo have a history of even more extensive inbreeding and show virtually no genetic variation.

Genetic similarity between the Crater lions and two neighboring populations suggests that the Crater population originated in the Serengeti, and simulations of the breeding history of the Crater lions predict that the average level of heterozygosity in the Crater should have declined by about 10% over the past 20 years (Packer, Pusey, et al. 1991). Thus the rather marked decrease in heterozygosity in allozymes and in the MHC may reflect a long history of repeated bottlenecks in the Crater population. The greater similarity in fingerprint diversity between the two Tanzanian populations is within the range expected from our simulations based on a single bottleneck in 1962, but may also

TABLE 16.1   Genetic and Reproductive Comparisons of Lion Populations

	Serengeti	Crater	Sakkarbaug Zoo
Isozyme polymorphism			
No. heterozygous loci/			
individual	1.47	0.88	0.0
Average heterozygosity (h)	3.3%	2.2%	0.0
(O'Brien et al. 1987;			
Packer, Pusey, et al. 1991)	(n = 79)	(n = 17)	(n = 28)
MHC polymorphism			
Proportion of loci polymorphic	17.0%	5.8%	0.0
Average heterozygosity	21.8%	8.0%	0.0
(Yuhki and O'Brien 1990)	(n = 18)	(n = 15)	(n = 15)
DNA fingerprint			
Average heterozygosity	48.1%	43.5%	2.8%
(Gilbert et al. 1991)	(n = 76)	(n = 22)	(n = 16)
Total sperm abnormalities (%)	24.8 ± 4.0	50.5 ± 6.8	66.2 ± 3.6
(Wildt et al. 1987)	(n = 8)	(n = 9)	(n = 8)
Sperm mobility (%)	89.0 ± 2.1	59.0 ± 8.0	61.0 ± 3.7
(Brown et al. 1991;			
Wildt et al. 1987)	(n = 10)	(n = 6)	(n = 8)

result from the far higher mutation rates in minisatellite regions than in other regions of the chromosome (Jeffreys et al. 1988). Lions in the Crater show a higher variance in band sharing for a given coefficient of kinship (see figure 16.4), which could result from a greater degree of linkage between bands in the Crater lions.

Although the Crater population generally shows considerably lower levels of heterozygosity than the Serengeti population, direct comparisons of fertility and survival between the two populations are complicated by the very different patterns of prey availability in the two areas: the Crater lions have access to a consistently high biomass of prey, whereas the Serengeti lions are regularly subjected to serious food deprivation due to the extensive seasonal migration of their preferred prey (Schaller 1972; Kruuk 1972). Therefore cub mortality is considerably lower in the Crater than in the Serengeti (Packer et al. 1988).

Nevertheless, two lines of evidence suggest that increased levels of inbreeding impair reproductive performance in the Crater lions. First, males in the Crater have a significantly higher proportion of abnormal sperm than Serengeti males do, and the highly inbred Asiatic lions of the Sakkarbaug Zoo have even higher levels of sperm abnormality (table 16.1). In other felids, studies of in vitro fertilization show that sperm from individuals with high levels of sperm abnormality penetrate domestic cat eggs at significantly lower levels (Howard, Bush, and Wildt 1991).

Second, our estimates of average heterozygosity in the Crater over the past 25 years are closely correlated with annual reproductive rates *within* the Crater population (Packer, Pusey, et al. 1991). We used data on allele frequencies of allozymes to estimate the genetic composition of the founding population in the Crater, and then simulated the breeding history of the population using the detailed pedigrees available from the long-term study. These suggest that the overall heterozygosity of the Crater population would have declined by about 10% over the past 20 years, and reproductive performance in the Crater has showed a marked decline over the same period. Multiple regression suggests that productivity in the Crater is significantly correlated with heterozygosity as well as with the proportion of the female population that is subjected to infanticide every biennium (see Packer, Pusey, et al. 1991).

These results are consistent with data presented by Lacy, Petric, and Warneke (chapter 15, this volume) showing that mammalian reproductive performance often declines following the isolation of a small population. Because of the male immigration into the Crater in the mid-1960s, we can also test for evidence on outbreeding depression (on the assumption that the immigrant males were genetically distinct from the surviving Crater animals). Following the method of Templeton and Read (1984), we found no evidence for outbreeding depression in the Crater lions.

DISCUSSION

The demographic data clearly demonstrate that dispersal patterns result in extremely little opportunity for close inbreeding in the Serengeti lions. Although much male dispersal occurs because of male-male competition, several patterns, such as the rarity of return to the natal pride and the abandonment of prides containing maturing daughters, are best explained in terms of the avoidance of close inbreeding. The genetic data indicate that even moderate inbreeding is rare in the Serengeti. Combining data from figure 5c with our long-term records, we can confirm that resident males are typically unrelated to the females in their pride.

Demographic data indicate that levels of inbreeding are higher in the small, isolated Crater population than they are in the Serengeti (Pusey and Packer 1987a; Packer, Gilbert, et al. 1991). But even in the

Crater the incidence of close inbreeding may be rare: males only rarely breed in their natal pride, and the DNA band-sharing data suggest that males are not as closely related to their mating partners when they are resident in a nonnatal pride. The Crater population shows lower levels of heterozygosity and adverse effects of inbreeding on reproductive performance.

Molecular techniques hold considerable promise for resolving controversies over the extent of inbreeding that occurs in natural populations (see Shields, chapter 8, this volume). But it should be emphasized that the relationship between relatedness and DNA band sharing differed between the two populations. Thus the degree of band sharing should be calibrated against an independent measure of kinship (Lynch 1988b; Gilbert et al. 1991). However, it may not be necessary to use genealogical data as extensive as our own: coefficients of kinship could often be interpolated from comparisons of the degree of band sharing between parents and offspring, ($r = 0.5$), half-siblings ($r = 0.25$), and nonrelatives, as has been employed by Westneat (1990) and Jones, Lessells, and Krebs (1991). These three points might often be sufficient to indicate the overall relationship between kinship and band sharing and would also permit testing for nonlinearity in the calibration curve.

Finally, the data in table 1 emphasize the importance of combining numerous genetic techniques for studying the histories of small populations. The population history of the captive Asiatic lions has apparently resulted in the virtual loss of all genetic diversity; but the history of the Crater lions could not have been deduced from the DNA fingerprinting data alone.

ACKNOWLEDGEMENTS

This research was supported by the NSF Program in Population Biology (Grants 8507087 & 8807702), with additional support from the Graduate School of the University of Minnesota and the National Geographic Society. The authors were supported by John Simon Guggenheim Foundation fellowships while writing this chapter.

17

Inbreeding and Outbreeding in Primates: What's Wrong with "The Dispersing Sex"?

Jim Moore

As an order, the Primates differ from most other vertebrates in that the majority of primate species live in stable social groups which usually contain representatives of all age-sex classes (Jolly 1985). This social organization has two consequences that are relevant here. First, dispersal is generally unambiguous; an individual either stays in her/his natal group or leaves it. Among most other mammals, defining and detecting philopatry involves repeated measures of dispersal distance, and sex-biased dispersal may exist only as statistical differences in mean dispersal distance (Waser and Jones 1983; Moore and Ali 1984). To the extent that group structure and mating system are correlated, this semi-closed feature of primate social organization simplifies research on primate inbreeding: if all adult males in a troop were born elsewhere, the possibility of close inbreeding is remote.[1]

Second, the stable social environment either permits or requires elaborated relationships, which are themselves thought to promote the evolution of the apparent altruism, "social intelligence," and increased relative brain size that characterize the order (see Cheney, Seyfarth, and Smuts 1986). Although stable relationships themselves may favor the evolution of such traits for selfish reasons among nonrelatives (Kummer 1978), most authors have emphasized the roles of reciprocal altruism and especially kin selection (e.g., Gouzoules 1984; Walters 1987; P. M. Waser 1988). Thus, our understanding of the evolutionary bases of several theoretically interesting behavior complexes depends

1. The correlation between social structure and mating system may not be as strong as it first appears; specifically, "extragroup" males may father a sizeable proportion of the young in some "unimale" systems (Cords 1987). This factor has yet to be integrated into discussions of primate inbreeding.

to a great extent on a thorough knowledge of breeding and dispersal systems in primates (T. D. Wade 1979; see also Quiatt 1988).

A dramatic example is found in the search for functional explanations of general differences between colobines and cercopithecines (cf. McKenna 1979). Colobines are known for allomothering and generally low rates of intragroup aggression, whereas cercopithecines have a reputation for aggressiveness and competition between matrilines. Gouzoules (1984) argued that because many colobines live in unimale groups, in the absence of female dispersal colobine females would be related patrilineally as well as matrilineally; in comparison, among multimale cercopithecine groups patrilineal relatedness would be less significant. Consequently, average relatedness among females within colobine groups would be higher than in cercopithecines, and relative colobine beneficence could be a result of kin selection. While this logic seems sound and accounts for a puzzling and important behavioral difference, in a review of female dispersal among primates Moore (1984) concluded that intergroup transfer by females was more prevalent among colobines than among cercopithecines, and hence that more colobine troops would contain completely unrelated females. Until we are confident in our knowledge of dispersal patterns and mating systems among these two groups, we cannot successfully interpret the observed behavioral differences (Moore 1984).

Male dispersal is the rule among primates, and in the better-known macaques and baboons females typically remain in their natal troops (Pusey and Packer 1987b). This pattern of male dispersal and female philopatry is common in mammals (Greenwood 1980), and the striking sex bias in dispersal patterns begs for functional interpretation: "Why do virtually all members of one sex disperse in so many different species?" (Packer 1985). As noted by many authors, strongly sex-biased dispersal results in drastically reduced opportunities for close inbreeding; given the demonstrable potential costs of inbreeding depression, the conclusion that sex biases in dispersal function to prevent inbreeding depression is an obvious one.

Coupled with the observation that female dispersion should determine male strategies (Bradbury and Verhencamp 1977), this conclusion makes for a persuasive model of primate social organization: Where female sociality is favored for ecological reasons, opportunities for nepotism make kin preferred partners, thus promoting the evolution of female philopatry and female-bonded social groups (Wrangham 1980). Insofar as inbreeding needs to be avoided, philopatry by one sex necessitates dispersal by the other, and among mammals males are generally "the dispersing sex." The breadth and simplicity of this paradigm make it appealing, and in its general form it has gained wide acceptance. Thus, in discussing inbreeding depression Maynard Smith (1984: 217) writes, "Although in most social mammals it is *the* males which transfer, it is sometimes *the* females" (emphasis added), and Colvin (1986: 131) summarizes mammalian demography: "In primates, as in other mammals, it is males that emigrate, while females remain with their natal group throughout their lives."

The clear split between philopatric females and dispersing males embodied in the phrase "the dispersing sex" is thus central to ideas about the evolution of altruism, the operation of kin selection, and functional explanations of dispersal as a means of avoiding inbreeding.

Unfortunately, the perception that one sex disperses while the other doesn't is simply mistaken, at least as applied to primates; there is no "dispersing sex." There do, however, appear to be taxonomic and ecological patterns to the propensities of males and females of different species to disperse. In this chapter I focus on various aspects of "noise" in these observed patterns of primate dispersal, inbreeding, and outbreeding, in the belief that understanding the noise is likely to hold the key to understanding the pattern. I emphasize that I, at least, do not yet understand that noise. When criticizing conventional wisdom (e.g., that dispersal functions to reduce inbreeding) my point is not that the conventional wisdom is mistaken, but that the data are not yet adequate to accept it as demonstrated.

INBREEDING AND OUTBREEDING DEPRESSION IN PRIMATES

The reproductive consequences of inbreeding have been examined in two field studies of savanna baboons (*Papio cynocephalus*). In 1970, the Beach troop of olive baboons (*P. c. anubis*) at Gombe Stream Research Centre (Tanzania) divided, with daughter "A" troop occupying an adjacent home range. Young adult male Bramble (BRM) subsequently transferred from his natal Beach troop into A troop, thus becoming a postransfer natal male. Packer (1979) compared mortality rates for infants probably sired by BRM with those probably sired by nonnatal males in all three study troops at Gombe, and concluded that inbreeding led to a 40% loss in viability. This conclusion was disputed by Moore and Ali (1984) on the grounds that the infant mortality rate in A troop was unusually high before Bramble joined it; although inbreeding depression probably accounted for some mortality, other (unknown) factors appeared to be involved, and the data were not suitable for a quantitative assessment of inbreeding depression. Further debate (Packer 1985; Moore and Ali 1985) clarified positions, but did little to resolve the question of how costly inbreeding is among these baboons.

In the second study, Bulger and Hamilton (1988) report on infant mortality rates in chacma baboons, *P. c. ursinus,* at Moremi (Botswana). Of thirty-nine adult males who resided in study troops, one was known to be natal and five others were assumed to be natal on the grounds that at the onset of the study they were smaller than any known nonnatal male. Total mortality by 1 month (the cutoff Packer used) for infants probably sired by natal males was no greater than for those probably sired by immigrant males (5%, $N = 20$, versus 7%, $N = 27$). As Bulger and Hamilton point out, these findings do not in any way contribute to the debate over infant mortality at Gombe, because higher rates of inbreeding over time are expected to weed out lethal recessives and thus reduce inbreeding depression. At Gombe few males breed natally and inbreeding depression seems to occur; at Moremi natal breeding is more common and no evidence of inbreeding depression was found. The reasons for the different dispersal patterns have yet to be discovered.

Consorting natal males are known for other baboon and macaque populations (see below), but data pertaining to inbreeding and infant mortality at these sites have not yet been published. Nishida (1966) points out that the incidence of albinism and deformity seems greater within more isolated Japanese macaque troops, and observations of increased birth defect rates following provisioning at several sites are consistent with inbreeding depression, if provisioning affected male dispersal strategies (at Mt. Takasaki rates peaked in 1970, the year after natal male Toku became alpha) (Itani 1975). However, pesticide poisoning also has been implicated in the deformities, and their true cause is not known (Itani 1975).

Using electrophoretic techniques to determine paternity, D. G. Smith (1986b) found increased mortality, decreased fertility, and lower (estimated) birth weights for inbred rhesus monkeys born in three enclosures at the California Regional Primate Center (CRPC); only the birth weight difference was significant. Elsewhere, Smith et al. (1987) indicate that inbred infants were (insignificantly) *heavier* than either hybrid or nonhybrid, noninbred infants by the time of their first weighing; they conclude that the deleterious effects of inbreeding are "probably primarily confined to the prenatal, perinatal, or very early neonatal periods." Variance in first weight was significantly greater among inbred offspring than among hybrids, a finding consistent with other studies, which they attribute to a decline in developmental homeostasis among inbreds.

These findings are important, as Smith's work is by far the most detailed study of actual inbreeding levels and consequences in a nonhuman primate. However, analyses to date have not considered maternal age, and this may call some of the indications of inbreeding depression into question. The three CPRC groups Smith had studied were founded in 1976 (Smith 1982) with presumably unrelated animals, and the earliest that inbreeding could take place would be ca. 1980, when the first cohort born in the cages reached sexual maturity. Thus, while many infants born to young females were not inbred, in the absence of mother-son incest (see below), all inbred offspring were necessarily born to young females. Young, primiparous females may be more likely to have smaller infants, higher infant mortality, and to show longer interbirth intervals (hence apparently lower fecundity) (Case-bolt, Henrickson, and Hind 1985; Small 1984). A further complication is suggested by the possibility of opposing heterozygote versus homo-zygote advantage at different life cycle stages among rhesus monkeys (Smith and Small 1982); females homozygous at the transferrin locus show higher fertility but lower growth rates and possibly higher infant mortality. Until complete reproductive life histories are available and the possible effects of maternal age controlled for, the CPRC data should be considered preliminary.

Despite the inconclusiveness of the above field and quantitative data, there is no question about the potential for inbreeding depression in primates. Ralls and Ballou (1982b) summarize infant mortality rates for inbred and noninbred matings in sixteen captive colonies of four-teen primate species; in only one of the sixteen was infant mortality higher for the noninbred infants. Inbreeding depression is a very real potential problem for maintenance of captive primates, and hence is an important factor in designing strategies for the conservation of endan-gered species (Bercovitch and Ziegler 1989).

The only study that specifically examines the possibility of outbreed-ing depression in a primate is that of Smith et al. (1987). Following the introduction of three rhesus monkeys of Chinese origin into a troop descended from Indian sources, these authors were able to electropho-retically distinguish hybrid from nonhybrid infants, and compare these two groups with a similarly identified group of inbred infants (see above). At ca. 12 months, all three groups of infants weighed less than expected based on standard curves, but this was significant only for the hybrids. There were no significant differences in growth rates or in weights at ca. 11 months among the groups. Finally, variance in weight at ca. 2 months was significantly lower for the hybrids than for the inbreds. Mortality rates were not discussed. Smith and colleagues con-clude that neither inbreeding nor outbreeding depression were de-tected in these monkeys.

Although no inbreeding depression was detected, Smith and colleagues use their findings to argue for the importation of additional rhesus bloodlines from "countries other than India and China" in order to prevent genetic subdivision of domestic stocks. In 1982, controversial attempts by the U.S. State Department to pressure Bangladesh into exporting wild-caught rhesus monkeys to the United States might have decimated that country's already endangered macaque population (IPPL 1982; M. A. R. Khan, personal communication). It would be sadly ironic if the quest for genetic diversity in captive populations were responsible for increasing existing pressure on wild primates.

PROCESSES THAT PREVENT INBREEDING AND OUTBREEDING DEPRESSION

Inbreeding Depression

Given that inbreeding depression potentially exists, it is logical to ask what processes reduce its impact in natural populations. Among primates such processes can be divided into two types: demographic prevention of opportunity through unbiased or biased dispersal patterns, or behavioral avoidance of known or suspected kin (a "taboo"). Most attention has been focused on sex-biased dispersal, reviewed (with markedly different conclusions) by Moore (1984) and Pusey and Packer (1987b). Here I discuss some of the points of ongoing debate over patterns and functions of emigration and immigration.

Dispersal is not the only demographic factor influencing inbreeding: Altmann and Altmann (1979) point out that patterns of mortality and reproductive senescence limit opportunities for inbreeding in most primate populations, and there is a suggestion that patterns of decline in male competitive ability could be influenced by probabilities of paternal incest (see discussion of Clutton-Brock [1989] below). Finally, direct evidence regarding inbreeding and outbreeding in wild primates comes from population genetic surveys of blood protein polymorphism distributions among lineages, troops, and populations (Melnick 1988; Melnick and Pearl 1987).

Unbiased Dispersal. The dispersal in random directions of all young from their natal group will reduce the incidence of inbreeding relative to nondispersal; if young disperse a long way relative to the average group home range, the probability of close relatives finding themselves in the same postdispersal group should be remote (Dobson and Jones 1985; P. M. Waser 1987). Little is known concerning lifetime dispersal distances among primates. Cheney and Seyfarth (1983) found that most natally dispersing vervets went only to the neighboring troop, but they have no data on subsequent transfers; Fukuda (1988) notes observations of male Japanese monkeys who dispersed 60 km; anecdotal accounts of individuals appearing more than 50 km from the nearest conspecific social group are not uncommon (e.g., patas: N. Nicolson, personal communication; grey langurs; D. B. Hrdy, personal communication). Quantitative data on the substantial middle range of distances are lacking.

Without measurements of dispersal distances, it is difficult to be sure dispersal is unbiased in monogamous or solitary primates in which both sexes usually emigrate—average dispersal distance might be the same or different for males and females. If males tend to go farther (as is the case among most mammals; Waser and Jones 1983), functional interpretations need to take into consideration the small expected family size of some primates. A gibbon pair is likely to produce only five or six offspring at intervals of about 3 years (Robbins Leighton 1987), and the chances of randomly dispersing, unpaired, opposite-sex siblings meeting must be remote even in the absence of a sex difference in mean dispersal distance.

Mountain gorillas deserve special mention here. Both males and females usually emigrate; males generally establish new groups (or die), whereas females typically join solitary males or transfer between groups (a pattern similar to that of red howler monkeys). The picture is greatly complicated by the high rate of exceptions, though; males sometimes "inherit" groups from probable fathers, and about 50% of females have at least one infant natally before emigrating—apparently fathered by half-sibs or other relatives other than their fathers, with whom they rarely copulate (Watts 1989, 1991).

Sex-Biased Dispersal. Dispersal by males is almost universal among primates (Pusey and Packer 1987b), and male-biased dispersal is clearly the rule among many species. There is, however, some question as to (1) the prevalence of strongly male-biased dispersal across primate taxa, and (2) the extent and significance of natal breeding by the sex that "normally" disperses in those species that do exhibit a clear sex bias.

I believe Pusey and Packer overestimate the extent of the sex bias, and that this overestimation was caused, ironically enough, by the rigor of their analysis. In their systematic review they included only studies at least 1 year in length, in which at least 6 individuals of one sex were observed to transfer or emigrate, and in which roughly equal numbers of males and females could be identified (Pusey and Packer 1987b: table 21–1, Notes). Because arboreal primates are difficult for humans to observe, identify, and follow, this resulted in a significant bias against arboreal species in their sample.[2] This bias, presumably acting together with historical factors, secondarily resulted in relatively few New World primates being included (three out of twenty-one species in their table 21-1).

Such biases in the sample are likely to influence conclusions, since factors rendering arboreal monkeys difficult for humans to observe also make them difficult for conspecifics to watch, which should affect the behavioral options open to them (Rowell 1988). The bias against arboreal monkeys also biases the sample against more folivorous monkeys, and an association has been predicted between diet and dispersal by females (Wrangham 1980; Moore 1984). Finally, there are indications that female dispersal is relatively common among the New World atelins (McFarland Symington 1988; Rosenberger and Strier 1989; Strier 1990) and howler monkeys (Glander 1980; Moore 1984; Crockett and Eisenberg 1987).

Until more studies of arboreal, folivorous, and New World monkeys have lasted for more years, this sample bias forces a choice between two unsatisfactory strategies: restrict analysis to relatively detailed studies of a taxonomically biased subset of primates (e.g., Pusey and Packer 1987b), or broaden the taxonomic scope by accepting more limited and perhaps sometimes less reliable data sets (e.g., Moore 1984). One can only hope that the interplay of these two strategies ultimately is productive.

Using deliberately liberal criteria, I previously found evidence of "significant" rates of intergroup transfer by females in about 45% of thirty-six primate species for which at least minimal demographic data were available (Moore 1984). As discussed there, the definition of "significant" is partly methodological but, more important, also theoretical. While average levels of within-group relatedness, central to hypotheses concerning inbreeding as well as kin selection, obviously vary with the proportion of individuals who are unrelated immigrants (T. D. Wade 1979), no theoretical or empirical estimates of what constitutes a biologically "significant" proportion are available. How many "exceptional cases" (e.g., immigrant females; natally breeding males) does it take to disconfirm the hypothesis that intragroup affiliation has evolved primarily through kin selection, or that natal male dispersal

---

2. I used Clutton-Brock and Harvey (1977) as a source for arboreal/terrestrial designations and compared the Pusey and Packer sample with theirs. Among sixty-three nonsolitary, nonmonogamous primates, Clutton-Brock and Harvey consider 63% "arboreal"; only 27% of the Pusey and Packer sample are in this category ($\psi^2 = 11.04, p < .001$).

functions to avoid close inbreeding? These questions remain unanswered.

Two recent studies meet the criteria of Pusey and Packer (1987b): Robinson (1988a) observed forty-six immigrations by juvenile and subadult male wedge-capped capuchins (*Cebus olivaceus*); among adults, six male and three female immigrations were reported. Emigration by full size appears close to 100% for males and 0% for females. In 5 years of observation of three troops of *C. capucinus*, six transfers, all by young adult or old males, have been observed (L. Fedigan and C. Chapman, personal communication).

Table 17.1 lists reports of female emigration/transfer not included in Moore (1984), three in colobines and one in howler monkey, consistent with previously noted patterns; male emigration occurs in all four. Evidence for routine female dispersal and perhaps male philopatry among atelins is accumulating; to the extent that their diets resemble those of chimpanzees, they provide further support for Wrangham's (1980) ecological model of female-bonded troops.

Although male dispersal is the norm for most primates, there are occasional references to adult, potentially breeding, natal males in the literature (Pusey and Packer, 1987b). Unless they illustrate a significant theoretical point (e.g., BRM; Packer 1979) such exceptions receive little attention. For example, Cheney et al. (1988: 397) exclude natal male vervets from their analysis of factors affecting male reproductive success on the grounds that "because most natal males were low ranking and seldom attempted to copulate, including natal males . . . might artificially bias results against low-ranking individuals"—despite having noted that natal males' average copulation success was about a third that of nonnatal males', and that one natal male became the dominant breeding male in his troop. One equally well might argue that because many natal males were young and (hence??) low-ranking, the inclusion of young, low-ranking individuals would artificially bias results against natal males.

I emphasize that male dispersal is the rule among most primates. However, exceptions are important for two reasons. First, they can "prove the rule"; at Gombe, both adult natally mating male baboons did so following a troop split, in the half of their natal troop that did not include their mothers. This certainly supports the hypothesis that avoidance of maternal incest is involved in male natal dispersal (though other interpretations are possible; Moore and Ali 1984). Second, there may be enough exceptions to challenge, not prove, the generalization that males in these species normally do not breed natally. As stated above, there are no existing objective criteria for evaluating the biolog-

TABLE 17.1 Recent Reports of Female Emigration or Transfer

Species	Observations
Nasalis larvatus	Male dispersal apparently routine (solitary males and all-male groups common); 2 observed female transfers (1 with infant), 2 observed female immigrations from unknown origins (1 with infant and both possibly accompanied by juveniles), and 2 female/infant pairs briefly joined known groups. Female transfer judged to be "common" during 28 months of study (Bennett and Sebastian 1988; E. Bennett, R. Rajanathan, personal communication)
Presbytis pileata	3 adult female immigrations, 2 juvenile/subadult male emigrations, and one resident adult male emigration "under unclear circumstances" during ca. 15 months (C. Stanford, personal communication)
Ateles paniscus	Circumstantial evidence for 2 subadult females emigrating roughly at sexual maturity, and 3 others disappeared at the same age; 1 adult female disappeared following reproductive failure; 4 males first seen as juveniles associated with presumed mothers remained in same group to at least 6 years old, with 1 seen mating with group females; 2 groups monitored over 4 years. Author concludes male philopatry and female dispersal are the norm (McFarland Symington 1988)
A. geoffroyi	One ca. 3-week-long excursion by a marked female, accompanied by infant and large immature daughter, into a neighboring group's range; no other dispersal events observed in one community over 5 years (C. Chapman, personal communication)
Alouatta caraya	4 adult and 2 subadult females disappeared (2 were seen briefly near their former range) and one adult female joined a pair of adults to form a new reproductive group; at least 4 adult and 4 subadult males disappeared, some during 4 "invasions" by a total of ≥ 7 males; based on regular monitoring of 11 groups over 28 months (D. Rumiz, 1990, personal communication)
Colobus satanas	1 transfer and 2 immigrations by adult females (1 accompanied by an infant) and circumstantial evidence of transfer by 2 males; during 9 months study of 1 troop (M. Harrison, cited in Pusey and Packer 1987b)
C. polykomos	1 apparent immigration by an adult female who was harassed by resident females; 2 adult females and 3 males (juvenile, subadult, and adult) disappeared (1 female probably died) during 17 months (G. Dasilva, personal communication)

ical significance of exceptions to apparent patterns; the reader is cautioned to keep this in mind throughout the following.

Taken by itself, the observation that, say, 98% of males born emigrate neither clarifies why males emigrate nor accurately estimates the proportion of breeding males who breed natally. First, natal emigration may be prompted by the same considerations as breeding dispersal—

TABLE 17.2 Mortality among Immature Male Primates

	% dying during first year	% of those born surviving to sexual maturity	% of 1-year-olds reaching sexual maturity
Cebus olivaceus	21.6	41.9	53.4
Alouatta palliata	45.0	41.7	75.8
A. seniculus	17.0	66.0	79.5
Macaca sinica	39.5	10.4	17.2
Theropithecus gelada	4.5	87.8	91.9
Pan troglodytes	18.8	38.1	46.9

Note: Calculated from Robinson (1988a): table 6. Note that few (if any) males in these species disperse as early as 1 year; also, sexual maturity generally precedes attainment of adult breeding status by a variable length of time.

primarily, access to mates (Packer 1979). Second, many males fail to breed *anywhere* due to high juvenile and subadult mortality (table 17.2) and differential mating success (reviewed by Bercovitch 1986; Smuts 1987). For example, while only 8%–10% of Japanese macaques are still natal by age 8 years, *survivorship* to age 8 may be as low as 16% (Sugiyama 1976). More than 50% of the males likely to become breeders thus would be classed as "atypical" late emigrators.

Table 17.3 lists reports of sexually mature, potentially breeding natal males in species in which females usually remain in their natal group. It is not comprehensive, as not all authors note the natal status of males they study; furthermore, few studies are long enough to demonstrate natal breeding. It appears that while only a small fraction of males ever become potential natal breeders, nearly every long-term study site has had at least one such male. Smith and Smith (1988) point out that without biochemical determination of paternity we simply cannot know what proportion of infants are sired by high-ranking males; the same is clearly true with regard to breeding by (related?) natal males. For Jodhpur langurs this proportion is probably close to 0% (Sommer and Rajpurohit 1989); among some baboons, it may approach 30% (table 17.3). If there are group-living primate species in which 30% of infants have a parent who is a natal member of "the dispersing sex," adaptive explanations for the existing bias need to account for such a high "error rate."

Chimpanzees, one population of red colobus, and perhaps spider monkeys and muriquis are the only primates for which a female bias in dispersal approaches the degree of the male bias seen among the macaques and baboons discussed above. Little of the colobus data has

TABLE 17.3 Observations of Sexually Mature Natal Males in Species in Which Males "Normally" Emigrate

Species	Site	Troop	Observations
Papio cynocephalus (adult ca. 4–7 years) *P. c. cynocephalus*	Amboseli	Alto's	2 known, 2 presumed natal adult males in Alto's group (2 emigrated, then returned when 10 years old); "almost 30% of those reaching adulthood in Alto's group may have spent their entire reproductive careers there . . . [as] . . . fully participating mating adults" (Altmann, Hausfater, and Altmann 1988; J. Altmann, personal communication)
P. c. anubis	Gilgil	PHG	Natal males DV, CL, and HO high-ranking for most of study; together with natal WI they account for 32% of adult male copulations with ejaculation recorded on cycle days −2 and −1 (peak fertility) (Bercovitch, 1986, personal communication; S. Strum, personal communication)
		EBC	Natal males DT, OV, OR, VU, BA occupy top 5 agonistic ranks and rank 1, 2, 6, 9.5, and 12 respectively in consort success during 1983 (Smuts 1985)
	Gombe	A	Following Beach troop fission, natal male BRM transfers into offshoot A troop (Packer 1979
		D	When Beach troop splits again, natal male Sage remains in half that does not contain mother, becomes alpha male (D. A. Collins, personal communication)

continued

TABLE 17.3 *(Continued)*

Species	Site	Troop	Observations
P. c. ursinus	Moremi		Based on consort records, 42% of attributable infants ($N = 47$) and $\geq 20\%$ of all infants ($N = 101$) fathered by 6 high-ranking natal males (Bulger and Hamilton 1988)
P. hamadryas			No males have been recorded breeding outside their natal clan; the majority of females also remain in the natal clan (Sigg et al. 1982)
Theropithecus gelada			Dunbar (1984) estimates that based on observed migration rates, $\approx 70\%$ of harems are held by natal males (p. 174); since all adult males may hold harems (p. 220), these calculations suggest that $\approx 70\%$ of males breed exclusively within the natal band
Macaca mulatta (adult ≤ 4.5 years)	Cayo Santiago	F	Natal males 415 and 580 rank 1 and 2 during 1978 breeding season; m415 lost to tetanus, but m580 most successful male in troop by every measure of RS used (Chapais, 1983)
		A	Ranks 1 and 2 held by natal males, ages 11 and 10; both emigrate in 1970 and become solitary (Missakian, 1973)
		I	Alpha male 9L is natal at age 16 (Hill 1987)
	General		Between 1973 and 1982, 20% of males remain natal to at least 6 years; $\approx 5.5\%$ of 218 males still natal at "8+" years and in 1983 alpha males of 2 troops are

TABLE 17.3 (*Continued*)

Species	Site	Troop	Observations
			natal at ages 12 and 15 (Colvin 1986)
	La Parguera	A	During 1978 mating season, agonistic ranks 1, 2, 5, and 6 held by natal sons of dominant matriline; males 7C and 1V rank 1 and 2 during mating season and 1 and 5 in May 1979 (Tilford 1982) Six-year-old natal male C4 is ranked 1 in 1972 (Drickamer and Vessey 1973)
M. fuscata (sexually mature ca. 4.5; full adult by 8 years)	Ryozen	A	Males *White*, *Black*, and *Dark*, all presumptive natal, all held high ranks (≥ 4) for 1–4 years prior to emigrating at (approx.) ages 12, 10, and 11 during 1969–1973 (Sugiyama 1976). Sugiyama and Ohsawa (1982) reevaluate *Dark*, stating that "he must have immigrated," because no other male has remained natal past 6 years at this site (*White* and *Black* are not mentioned). Roughly half the males who emigrated from the troop remained within/returned to the troop's home range for "some months or years" (Sugiyama and Ohsawa 1982), and such *hanarezaru* males fathered ≥ 2 of 7 infants born in the troop in one year (based on electrophoretic analysis; Sugiyama 1976). (The 2 infants were not fathered by males born in Ryozen A; the point is, some *hanarezaru* do breed.)

continued

TABLE 17.3 *(Continued)*

Species	Site	Troop	Observations
	Toi		7 of 36 males born 1955–1959 still natal at age 8 (N = 6) and 9 (N = 1) years (Sugiyama, 1976)
	Shiga	A	2 of 27 males born 1961–1970 still natal in 1973 at ages 8 and 10 (Sugiyama, 1976). They are presumably males KN and TC, ranked 1 and 2 in 1972–1973 (Enomoto, 1978)
	Katsuyama		5 of 162 males born between 1961 and 1972 still natal at ages 8 (2), 9 (2), and 10 (1) (Sugiyama, 1976)
	Takasakiyama		Males *Monk* and *Bacchus*, ranked 4 and 5 in the original troop (ca. 1956), were evidently natal (Sugiyama 1976: 279)
		A	Natal male *Toku* was ranked 1 between 1/1970 and 1/1973; he remained in his natal troop for 21 years (Sugiyama 1976)
	General		Approximately 8–10% of males remain in their natal troop ≥ 8 years; all have emigrated by age 12 (N = 132) (Sugiyama, 1976: Figure 4)
	Arashiyama	A	Natal alpha *Ao* became solitary in 1968 at age 12 years; in 1972 ranks 4, 8, and 9 were held by natal males BUS-62, KIN-63, and ME-65 (the digits are birth years) (Norikoshi and Koyama, 1975)
		B	Natal males K-65 and K-63 ranked 4 and 5 in 1975, and 5 and 2 in 1978 (Takahata, 1982a). K-63 is alpha male in 1985 (Huffman 1987)

TABLE 17.3 *(Continued)*

Species	Site	Troop	Observations
M. sylvanus			In a multitroop enclosure, roughly 50% of males were still natal at 7 years (adulthood); in the age cohorts that included potentially natal males, more than 75% of sexually mature males in the main study group were natal (Paul and Kuester 1985; *N* = 24–29: Table 1)
Cercopithecus aethiops	Amboseli		"Natal males achieved an average of 9.3% of all copulations" versus 30.2% for nonnatal males; "With the exception of one individual who became the dominant male in his natal group and bred successfully with several females, natal males seldom competed for access to sexually receptive females" (Cheney et al. 1988: 396)

Note: Excludes groups in which actively breeding natal males could not emigrate (e.g., Smith and Smith 1988). Some males are only presumed to be natal; since evidence is inconsistently reported, I simply have accepted the original author's opinions. Natal males who are peripheral or stated to be within 2 years of sexual maturity were not included unless they were specifically noted to be high-ranking during mating periods.

been published (see Pusey and Packer 1987b for a summary). There is no evidence of natal breeding by female *Ateles paniscus* (McFarland Symington 1988), but sample sizes are still quite small. There is some question as to the prevalence of natal breeding among female chimpanzees; although nearly all young adult females leave their natal community for some time, 20% to 40% return and subsequently breed natally. (See Pusey [1987, 1988], Moore [1988], and references therein for discussion.)

Among monogamous (and polyandrous?) species, a subadult might inherit breeding status upon the death of a same-sex parent, thus becoming an incestuous natal breeder. Observations of such natal breeding by gibbons are discussed by Tilson (1981) and Robbins Leighton (1987); McGrew and McLuckie (1986) and Goldizen and Terborgh (1989) discuss the (limited) evidence for inheritance of breeding status in marmosets and tamarins.

Are there any informative intra- or interspecific patterns in the distribution of natal breeding by "the dispersing sex"? Among macaques, males born in dominant matrilines may tend to emigrate later on average and to be more likely to attain high status natally (see Pusey and Packer 1987b); however, not all males who remain natal are from dominant matrilines, and baboons may not show such a tendency at all (J. Altmann, D. Collins, and S. Strum, personal communication). This difference may be because baboon females are less able to agonistically support males (potentially including sons) than are female macaques (Packer and Pusey 1979).

In principle, large troops (provisioned Japanese macaques and rhesus macaques) or subdivided ones (gelada and hamadryas baboons) should make it easier for males to avoid incest without emigration (see also Pusey and Packer 1987b: 259). It is noteworthy that the two species with the lowest reported dispersal rates by both sexes, hamadryas and gelada baboons (Pusey and Packer 1987b), uniquely share a subdivided troop organization, and that the olive baboon population with the highest proportion of reported natal males (Gilgil) is known for the importance of male-female "special relationships" and a lack of correlation between dominance and male reproductive success (Strum 1982; Smuts 1985; Bercovitch 1986), features that are reminiscent of hamadryas baboons. Although Alto's group (yellow baboons) is not unusually large, and although special relationships may not be as prevalent as at Gilgil, Walters (1987) notes that because it had fused with a second baboon troop, "presumably an atypically large number of unrelated adult females were available as mates for these [natal] males." (The case is complicated by the possibility that the two groups may have originated from the fission of a single troop; J. Altmann, personal communication.) Both "natal" male olive baboons at Gombe were in the troop half that did not contain their mothers, and barbary macaques may be a special case (see below); this leaves only the Moremi chacma baboons and single Amboseli vervet "unexplained" in terms of inbreeding avoidance hypotheses. Thus, there are hints that high maternal rank and the size and structure of the group both influence male natal emigration, but the evidence as yet remains inconclusive. The implications of these patterns are discussed below.

Behavioral Avoidance of Inbreeding. If kin can be recognized, and if close inbreeding is deleterious, we should expect to find that close relatives avoid mating with each other (Moore and Ali 1984). Gouzoules (1984) and Walters (1987) review evidence for kin recognition in primates; evidence for some level of recognition of matrilineal kin, based on common association with one's mother, is extensive, but evidence for recognition of paternal or more distant kin is less widespread. As expected, there are a number of reports of avoidance of mother-son incest (see Pusey 1980; Itoigawa, Negayama, and Kondo 1981; Murray and Smith 1983; Moore and Ali 1984; Walters 1987, and references therein.) Behavioral avoidance of incest is not limited to heterosexual interactions; Chapais and Mignault (1991) report avoidance of homosexual "incest" among female Japanese macaques.

Goodall (1986) describes incestuous mating among Gombe chimpanzees. Incest between mother and son and between maternal sibs occurred at rates much lower than expected based on overall mating activity; for most of the combinations no copulations were observed and there was little, if any, sexual interest. There were three exceptions to this pattern of avoidance, two involving Goblin and one, Figan. Goblin was responsible for six of seven attempted maternal copulations, and he was the only male to show no sign of sib-mating inhibition, copulating with sister Gremlin somewhat more than the average male during her first adult cycles (she resisted 30.8% of his attempts, versus fewer than 12% of any other male's). Figan showed little interest in sister Fifi during cycles preceding her first and second offspring, but copulated with her frequently preceding the third; she resisted 41% of his attempts.

All three of these "exceptional relationships" took place ca. 1979–1981, a period during which Goblin (who had risen to high agonistic rank with Figan's help) was actively, aggressively and ultimately successfully challenging Figan for alpha status in the Kasakela community; it was a time of extreme social tension for both males (Goodall 1986: 431–435). There is no direct evidence to connect the two sets of relationships, and no other dominance turnover at Gombe has been associated with incestuous behavior. I note the coincidence here because it bears on a hypothesized neuro/psychological association between sexual and aggressive behavior (see below); I emphasize the anecdotal nature of the possible connection in these data.

Older male chimps show less sexual interest in young females of their own community than in young visitor or immigrant females from neighboring communities, which may represent an adaptation for avoiding father-daughter incest in the absence of known paternity

(Goodall 1986: 469–471). Reluctance on the part of young females to respond to older males seems difficult to interpret as other than a form of incest avoidance, but such behavior is variably expressed (Goodall 1986: 469).

About 30% of male tenures among langurs at Jodhpur are long enough for the single resident male to overlap with sexually mature daughters, but in the only such case with adequate data, an estrous daughter temporarily left her troop and copulated with extragroup males (Sommer and Rajpurohit 1989). Given the relatively well-defined parameters and predictions in unimale systems ("females born $< X$ years into a male's tenure should behave one way, and those born $> X$ years, another"), such avoidance would be a good subject for systematic study.

Using electrophoretic markers to establish paternity of monkeys born within three enclosures, Smith (1982) found no evidence for avoidance of incest among paternal kin: nine of seventeen infants were sired by their mothers' fathers or paternal half-brothers (chance level); this result was confirmed by D. G. Smith (1986b) with a larger sample. Fewer than expected infants were born to matrilineally related parents, leading Smith to conclude that "captive rhesus macaques selectively avoid mating with matrilineal relatives." However, the category "matrilineal relatives" is limited to siblings and sons; more distant relatives are considered nonkin, obscuring any avoidance or preference among less related individuals. Also, though Smith did not distinguish between siblings and sons, kin recognition between mother and son is likely to be easier than between siblings. Based on data in his table 58.6, births to full- and half-sib pairs occurred in close to expected numbers (1 observed versus 2.6 expected, $p > .2$, $N = 16$, binomial test). The observed difference is in the direction expected if matrilineal sibs avoid mating with each other, but a larger sample is needed before concluding that rhesus incest avoidance extends beyond the mother-son relationship (births to mother-son pairs: 0 observed versus 3.3 expected, $p < .05$, $N = 24$). Missakian (1973) observed sibling copulation in 12% of available pairs (5.5% of all copulations); sibling incest avoidance among rhesus is variably expressed, at best.

Barbary macaques provide a very illuminating contrast. In their main study troop, Paul and Kuester (1985) observed no copulations in mother-son or sibling dyads (7.2 and 14.8 mating dyads expected based on mating activity and partner availability); a single uncle-niece mating dyad was close to the expected value of 2.6. Paul and Kuester argue that the difference between Barbary and rhesus macaques is due to their different dispersal patterns: among rhesus, near universal male emigration reduces selection pressure for incest avoidance, and evidence for sibling inhibition is ambiguous; in contrast, the much-reduced male emigration characteristic of Barbary macaques[3] creates abundant demographic opportunity for sibling incest, and behavioral avoidance has been favored by selection.

One hopes these results someday will be confirmed with biochemical paternity determinations and with more field data on both species in their natural habitats; the comparison is an important one for evaluating the role of inbreeding avoidance in the evolution of primate dispersal patterns (see below).

The only other study of inbreeding avoidance to consider mate preferences among distant as well as close relatives is that of Pereira and Weiss (1991), who used DNA fingerprinting to show that ringtailed lemurs in a large enclosure avoided close inbreeding. The authors note that inbreeding was avoided by females, who "repelled sons, matrilineal brothers, and other resident males from attempting to copulate;" ringtailed lemurs are unusual among primates in that females routinely dominate males in agonistic encounters (Richard 1987).

Outbreeding Depression

Though as yet there is no evidence of outbreeding depression in primates, there may be other reasons why behaviors that functioned to avoid outbreeding would be favored (Shields 1982). Outbreeding could be avoided by limited and/or nonrandom dispersal, or by mating preferences for similar/related individuals.

---

3. Pusey and Packer (1987b) and Pusey (1988) state that there is no convincing evidence that Barbary macaque dispersal systems are any different from those of other macaques. They base this conclusion on (1) the fact that field studies have been of short duration and may have missed male transfers, and (2) that Paul and Kuester (1985) found that in a large captive enclosure "most males eventually left their natal group" (Pusey and Packer 1987b). In fact, Paul and Kuester state that (1) more than 50% of the sexually mature males in their main study group were natal; (2) natal and nonnatal males did not differ in mating frequency; (3) only about 50% of males overall emigrated before adulthood but, because adults also transfer, they "assume that most . . . males change groups at least once in their lives;" and (4) they specifically contrast male emigration rates in rhesus and Barbary macaques, noting that although (captive) Barbary macaque males do transfer (contrary to earlier suggestions in the literature), "male mobility was lower than in other macaque species." Mehlman (1986) reports that during a 2-year field study there was no evidence of permanent male transfer or emigration. Instead, he reports twelve cases of males leaving their assumed natal troops for temporary visits to the periphery of other troops, where at least two successfully copulated. Though these findings indicate that Barbary macaques are not as inbred are previously thought, patterns of male dispersal are certainly not typical of other (i.e., rhesus, Japanese) macaques. Unless temporarily visiting peripheral males were responsible for all conceptions, the weight of evidence still indicates relatively reduced gene flow between Barbary macaque troops.

Dispersal. Natal breeding by members of both sexes presumably reduces outbreeding. While certainly not the rule, the extent of this phenomenon is undetermined, so there is little evidence one way or the other as to whether such natal breeding ever *functions* to reduce outbreeding; it seems unlikely to do so.

Troops of Japanese macaques are patchily distributed, forming "local concentrations of troops" (LCTs) with an average of three troops per LCT (Kawanaka 1973). Most male transfer occurs within the LCT, with the result that LCTs are likely to be relatively inbred (Kawanaka 1973; Nozawa et al. 1982). However, as Kawanaka notes, this population structure may be due entirely to habitat disruption by the human population.

Nonrandom dispersal has also been reported for vervet monkeys, in a more natural setting (Cheney and Seyfarth 1983). Neighboring study troops at Amboseli appeared to "exchange" natal males, and females (but not males) of troops that exchanged males showed less aggression toward one another than did members of nonexchanging troops. Cheney and Seyfarth suggest that such nonrandom dispersal benefits natal males in three ways: first, dispersal distance (and therefore physical risk) is minimized by going to an adjacent group; second, the males are able to monitor the adjacent group prior to transfer and so time their move better; and third, they may receive less aggression from familiar, perhaps related, older males who preceded them. (It is not clear how this relates to the elevated intermale aggression observed between "exchanging" troops reported by Cheney [1981]). While such a pattern might increase average relatedness between groups, subsequent male breeding dispersal may reduce inbreeding. Alternatively, Cheney and Seyfarth suggest that nonrandom dispersal gives males "a means of keeping track of close kin that should be avoided as mating partners" (p. 404), thus actually helping them to avoid close inbreeding. Direct genetic measurements will be needed to determine the effects, if any, of nonrandomness in patterns of vervet dispersal.

Mate Preferences. I found no reports of preferences for related males among nonhuman primates, unless one counts a preference for peers among rhesus monkeys at the CPRC colony (Smith and Smith 1988). Smith et al. (1987) claim that Lindburg (1969) reported "preferences for unfamiliar, yet phenotypically similar, mates" among rhesus monkeys. However, I can find no evidence of such a finding in the original paper, unless one simply considers unfamiliar neighbors "phenotypically similar;" Lindburg was not referring to a detectable phenotypically based preference (D. Lindburg, personal communication).

DISCUSSION

The balance between inbreeding and outbreeding may have affected primate social evolution in several ways. Inbreeding within social groups could contribute to speciation rates (Wilson et al. 1975, cited in Templeton 1987); high average relatedness within inbred groups could promote kin-selected altruism (T. D. Wade 1979); or the need to avoid inbreeding could determine dispersal patterns and so exert a powerful influence on social systems and behavior. The bulk of evidence currently available suggests that primate social groups are not routinely inbred, so the speciation rate hypothesis is not supported (Melnick 1988). The distribution of coefficients of relatedness within primate social groups appears to be variable, and as yet there is no convincing evidence that inbreeding, as opposed to matrilineal or patrilineal relatedness within outbred groups, is involved in patterning social behavior; the lack of evidence for close inbreeding itself is of course also relevant here.

The third potential role of inbreeding, that of determining dispersal patterns and hence options regarding the duration and nature of social relationships, is more difficult to evaluate. The central question regarding the role of inbreeding and outbreeding in primate social evolution is, what role does inbreeding avoidance have in the evolution of dispersal strategies? This boils down to a suitably rephrased version of Packer's question: "Why is there such a strong sex bias in dispersal in the majority of the cercopithecines?", noting that the answer will undoubtedly shed light on patterns in other taxa.

Alternative Explanations for Sex-Biased Dispersal

There are several proximate and ultimate hypotheses about sex-biased dispersal to choose among. It is not possible to reject any of them based on available data, and more than one may be correct (Dobson and Jones 1985; Holekamp and Sherman 1989). It is an open question whether the existence of a sufficient proximate explanation for observed patterns obviates ultimate ones (Jamieson 1989; Sherman 1989). I tend to argue against the inbreeding avoidance hypothesis only because it is so widely and prematurely accepted as demonstrated.

Inbreeding Avoidance (IA). Stated simply, the IA hypothesis is that when the costs of inbreeding outweigh the risks of dispersal (factoring in inclusive fitness effects), natal emigration is favored (see Waser, Austad, and Keane 1986). (See Packer [1979], Greenwood [1980], Moore and Ali [1984], and Pusey and Packer [1987b] for further discussion.)

There are two ways for inbreeding avoidance to account for male-biased natal dispersal: either natural selection has favored males who disperse rather than inbreed, or it has favored females who refuse to mate with related males, reducing the number of potential mates in a male's natal group and thus forcing him to disperse to breed (again, a combination theoretically may be possible). Because the cost of inbreeding is generally greater for female primates than for males, theoretical predictions support the second alternative (Moore and Ali 1984; Waser, Austad, and Keane 1986) unless the cost of migration is much lower for males than for females (Packer 1979). There is no evidence that physical risks associated with dispersal are lower for male primates (sexual dimorphism might favor males, but male-biased dispersal occurs in minimally dimorphic species and/or prior to physical maturity). However, Packer (1979) notes that the cost of emigration for males might be devalued by increased numbers of mating opportunities and by aggressive competition from other males in the natal group. Such devaluation may be necessary to invoke given the theoretical prediction that females, not males, should be the first to avoid inbreeding.

Several authors have noted aggressive female resistance during observed or attempted cases of incest (e.g., Goodall 1986; Pereira and Weiss 1991), and Packer (1979) notes that female baboons appeared more careful to avoid inbreeding than did males. Studying captive Japanese macaques, Chapais and Mignault (1991) found that females consistently avoided homosexual "incest," whereas heterosexual pairs were sometimes incestuous. They argue that homosexual incest avoidance among females represents the constraint-free preferred female pattern, and hence that it is males who are responsible for observed heterosexual incest. Of the two forms of the IA hypothesis, female refusal of natal males seems more likely than male avoidance of related females.

Problems with IA. The IA hypothesis predicts that females should be more likely to disperse than males (Moore and Ali 1984; P. M. Waser 1988), the opposite of the observed pattern. Male-biased emigration is consistent with IA if the proximate cost of male emigration is devalued by, for example, intragroup aggressive competition (such that the physical risks of not emigrating approach those associated with emigration); in the extreme case, such devaluation amounts to expulsion and it is unclear whether IA need be invoked.

Ignoring these devaluation effects, IA predicts that the males least likely to breed natally among macaques should be those born to dominant matrilines, since these are likely to have more relatives in the natal troop and perhaps to be more closely related to them (see Moore and Ali 1984). The few data that are available suggest that this is not the case; when there is a pattern, natally mating males tend to be offspring of dominant females (Vessey and Meikle 1987). This should not be regarded as a strong test, given the paucity of data and irregularity of the rank–delayed emigration relationship. (The IA hypothesis makes no prediction about the proportion of sons of dominant mothers who should breed natally.)

Competition and Mate Choice (CMC). The CMC hypothesis suggests that male-male mating competition is adequate for explaining male-biased dispersal among most primates and that female dispersal (when it occurs, regardless of male dispersal) sometimes results from intrasexual competition, but more often involves female choice of mate and/ or range (it is difficult to separate the two in primate field studies). (See Murray [1967], Moore and Ali [1984] and Moore [1984] for further discussion.)

Male-male competition can lead to emigration (natal and breeding) in two ways: direct expulsion, in which the migrant leaves to avoid ongoing, potentially lethal aggression, or through exclusion from mating opportunities. In terms of reproductive success, a subordinate male who is unable to complete a copulation due to mild threats from a dominant male is just as much a victim of intermale competition as if that dominant mercilessly attacked him. Most observations of emigrating males note that they are not preceded by sudden increases in received aggression (Pusey and Packer 1987b). However, subtle changes in intermale relations (e.g., a drop in affiliative greetings) preceded emigration by several months in one study of baboons (Manzolillo 1982). Few studies have gone beyond the absence of dramatic expulsion when discussing the causes of male emigration.

The relation between aggressive expulsion and self-initiated departure has been explored in an innovative experimental study by Mc-Grew and McLuckie (1986) who provided captive cotton-top tamarins (*Saguinus o. oedipus*) with opportunities to "disperse" through ducts into new cages. They found that daughters both received significantly more aggression (five out of fourteen had to be removed to prevent injury or death) and scored highest on several measures of "prospecting" in unfamiliar surroundings. (In this species, only one female breeds in a group, and helper sons may inherit the family territory; the breeding system is similar to that of Florida scrub jays [McGrew and

McLuckie 1986]). The authors expect that in the wild, "tensions leading to emigration would be expressed in milder terms," with avoidance leading to peripheralization and sometimes emigration.

Problems with CMC. The competition hypothesis has been criticized on the grounds that potential immigrants receive as much or more aggression in their new group as they received in their old one (Packer 1985; Pusey and Packer 1987b). This is true for some species; for others, natal emigrants join all-male bands or peripheral male subgroups in which their reception is peaceful (Pusey and Packer, 1987b), and natal males may choose their destination troop so as to minimize the amount of aggression received (Boelkins and Wilson 1972).

Even when males receive more overt aggression upon immigration than they did natally, it is not clear that they are not emigrating to avoid intermale competition. In principle, a natal male is at a twofold disadvantage with respect to resident adult males of his troop. First, as a juvenile, he has learned to be subordinate to them. Bernstein and Ehardt found that subadult rhesus males undergo "avoidance training consequent to the selective aggressive attention of adult males to adolescent males participating in agonistic episodes" that results in subadult males avoiding all conflicts—even to the extent of not aiding matrilineal kin; this "sharp increase in aggression received by males during adolescence is seen as the proximal mechanism inducing the typical ontogenetic change in male rhesus monkey agonistic participation within the social unit" (Bernstein and Ehardt 1986: 224, 225). Inhibition of young/subordinate males by the mere presence of familiar adult dominants has been reported for a number of species (Hayaki 1985; Ruiz de Elvira and Herndon 1986; Estep et al. 1988). The natal male is thus potentially at a psychological disadvantage.

Second, in his natal troop the male is familiar, and other males will have had ample time to evaluate his competitive abilities and experience (cf. Parker 1974). Transfer to a new troop wipes the slate clean, and he may be able to take advantage of residents' uncertainty about his resource-holding power (or fighting ability; W. J. Hamilton III, personal communication) to rise rapidly in rank. Indirect evidence supporting this sort of scenario comes from the observation that the most intense aggression directed against immigrating males may be from other males from their own natal troop who preceded them (Packer 1979); similarly, Cheney (1981: 142) found that "interactions between groups that exchanged males during the study were characterized by higher rates of male-male aggression than interactions between groups where male transfer was not observed." The assessment-related psychology and dynamics of male transfer and mate competition are complex, rarely studied, and almost certainly vary with species and population differences in demography and sexual dimorphism; see Packer and Pusey (1979), Jackson (1988), and especially Smuts (1985: 151–155) and Strum (1982, 1987: 94–116) for discussion. Any conclusions regarding the importance of such factors in male natal emigration would be premature, but so would be their dismissal.

It seems hard to invoke even psychological pressure when natal males emigrate in the absence of adult males: Pusey and Packer (1987b: 257) mention the case of the Ryozen A troop of Japanese macaques, in which young males continued to emigrate even with no adult males in the group. Males of other species have also been reported to emigrate well before sexual maturity, prior to receiving significant aggression from adult males, and this seems to be a problem for CMC.

In fact, this is a problem for both CMC *and* IA. Although natural selection might have favored juvenile males who emigrate before they are aggressed against (Moore and Ali 1984), our suggestion of preemptive emigration does seem, at least superficially, like special pleading. However, by the same token, juvenile emigrants cannot be emigrating due to sexual rejection by related females, since most are not sexually active, and the few who are may find that related females are the only ones who do not reject them (Missakian 1973). Thus, the alternative explanations for emigration by sexually immature males are: avoidance of "anticipated" adult male aggression (anticipated by natural selection, not cognitively), avoidance of anticipated female rejection, or dispersal resulting from selection for innate IA in males rather than in females—which is not in accord with theoretical predictions. None of these explanations seems likely, but the males do leave.

Bet Hedging (BH). Male (and perhaps some female) dispersal may represent a strategy of diversified bet hedging in which individuals act so as to distribute their reproductive success among a number of social units (put their eggs in several baskets) (see Philippi and Seger 1989). If prereproductive mortality rates vary among troops, an individual has the choices of trying to guess the "right" troop and breed only there, risking low reproductive success if he (or she) guesses wrong, or transferring and reproducing in several groups. Due to the various costs of migration it is unlikely that such a disperser would be as successful as one who bred only in the best (local) troop; however, the strategy could be favored if the resulting reduction of variance in mean within-generation fitness resulted in a higher geometric mean fitness over generations (Philippi and Seger 1989).

Although in principle females might play such a strategy, due to their low reproductive potential relative to males I suspect that routine transfer just to hedge bets would rarely be favored among females. Males are a different story.

The group a male breeds in does make a difference. Dittus (1988) reports that between 1975 and 1979, only five of eighteen troops of toque macaques experienced significantly positive population growth; four troops that fissioned had annual growth rates of + 2.3%, + 8.7%, + 13.3%, and − 2.7% in the 5 years preceding fission. In the expanding rhesus macaque population at Cayo Santiago, between 1976 and 1983 Group L grew at only half the rate of Group I (Rawlins and Kessler 1986). For eighteen male tenures longer than 12 months among the langurs (*Presbytis entellus*) of Jodhpur, the rate of births per month was 0.672 ± 0.31 (CV = 46%); this variability was not all due to differences in troop size, since the rate of births per female per month ranged from 0.0052 to 0.0597 (CV = 42%) (Sommer and Rajpurohit 1989). Census data on nineteen troops of red howler monkeys presented by Rudran (1979) demonstrate a mean growth rate of + 1.4% ± 13% (range, − 19% to + 25%) per annum, over 2 years.

Some intergroup variation in reproductive rates is at least partially a consequence of tactical decisions by males themselves (e.g., Robinson 1988b), but some is apparently due to ecological factors (e.g., Rawlins and Kessler 1986; Cheney et al. 1988). Given the complexity of trying to predict future habitat quality as well as second-guessing competing males' predictions (there is no point in everyone trying to join the ecologically most favored group in the area), a bet-hedging strategy of multiple breeding transfers might well be favored.

Problems with BH. Some males remain in a single troop for many years, probably breeding only in that one (see below). If these males are reproductively successful, and if the features that make them so (fighting ability, intelligence, personality, genes) are transferable to other troops, then BH is confronted with the question, why don't they transfer? There is no obvious answer within a purely BH model for male-biased dispersal, but note that there are few such males reported, fewer even than natally mating males.

Psychological (PPR). Several authors have considered the possibility that certain types of relationships are mutually exclusive for proximate psychodynamic reasons; for example, Sade (1968, cited in Murray and Smith 1983) proposed that male dominance is a prerequisite for sex among rhesus monkeys, and since mothers usually remain dominant to their sons, mother-son incest is thereby blocked. This suggestion has not been supported by detailed analysis (Murray and Smith 1983). Surprisingly, there is some evidence to support a second hypothesis relating inbreeding avoidance to aggression and affiliation: If sex and aggression are neurologically linked, strong affiliative bonds may inhibit sexual relationships (Parker 1976).

Among Japanese macaques, unrelated pairs who have affiliative "peculiar-proximate relationships" (PPRs) may avoid mating with each other, as if they were close matrilineal kin, suggesting that a common psychological basis underlies both types of avoidance (Takahata 1982b). At least two interpretations are possible: that PPR avoidance is overgeneralized inbreeding avoidance, or that both kinds of avoidance are based on some underlying psychological principle unrelated to inbreeding. Smuts (1985: 245) points out that PPR avoidance of sex may result from the unusual demography of provisioned troops, arguing that in more "natural" conditions males would probably emigrate after a few years, before incest-avoiding psychological mechanisms are activated. In either case, PPR avoidance seems to be maladaptive at least for the male of the pair, and further investigation of the proximate mechanisms involved in behavioral incest avoidance will be of interest (see Enomoto 1978; Itoigawa, Negayama, and Kondo 1981).

Positive PPR relationships may inhibit sexual behavior; as mentioned earlier, anomalously high rates of incest among Gombe chimps occurred during a period of high tension and aggression between the two males involved. Depending on one's inclinations, this could be viewed as supporting (1) the notion that aggression and striving for dominance promotes pathological behavior, (2) the proposal that male sexuality is strongly linked to domination, or (3) some other equally contentious hypothesis. Until additional relevant observations are available, I believe no firm conclusions can be drawn regarding this coincidence.

A male primate is never likely to have many close female relatives in his natal troop (Altmann and Altmann 1979) and so avoidance of incest need not involve dispersal. However, over several years he could make a number of female "friends" and so a PPR-type mating aversion could in theory play a role in prompting emigration.

Problems for PPR. Two difficulties are obvious. First, natal emigration by immatures is not likely related to "special relationships" restricting mating opportunities, and second, the existence and distribution of the PPR "phenomenon" itself is still in doubt.

Comparative Studies

The strongest evidence in favor of the inbreeding avoidance hypothesis for sex-biased dispersal is the significant negative correlation that Pusey and Packer (1987b; Pusey 1987) found between the proportion of males that are immigrants and the proportion of females who emigrate, across a sample of seventeen species (see figure 17.1). This inverse relationship between male and female dispersal can be interpreted as indicative of a causal relationship connecting male and female dispersal strategies. This relationship is predicted by the inbreeding avoidance hypothesis but not by any of the alternatives.

Other interpretations are possible, however (Moore 1988; but see Pusey 1988). Two issues are involved. The first is the actual correlation: the seventeen species included are a necessarily biased subset of the order (see above) and there is reason to believe that as more populations and additional species are added, they will fall in the region away from the axes (i.e., emigration rates > 0 for both sexes) (above, and Moore 1984, 1988). Whether or not this weakens the correlation remains to be seen.

The second issue is conceptual. Does figure 17.1 illustrate a single relationship in a sample that is homogeneous with respect to the variables, or does it in fact combine two sets of species with entirely different sets of dispersal strategies? Wrangham (1980) suggested that the advantages accruing to females who live among kin should vary in predictable ways according to ecological variables, and that female dispersal should be associated with folivory. This prediction was independently supported by the finding of a significant association between female transfer and morphological adaptations to folivory (Moore 1984). All of the points on the X axis represent species that have diets consistent with Wrangham's predictions for female-bonded primates. It is thus reasonable to consider the sample dichotomously, as female-bonded and non-female-bonded species. According to the inbreeding avoidance hypothesis, the negative correlation found by Pusey and Packer should hold within the non-female-bonded subset, since there is no reason to believe that the severity of inbreeding depression is diet-dependent.

As can be seen in figure 17.1, within the small sample of non-female-bonded primates there is no indication of a correlation between pro-

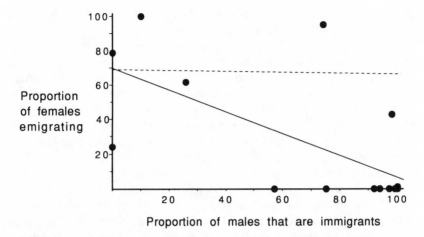

FIGURE 17.1 Relationship between male immigration and female emigration for a sample of seventeen species. Solid line, regression for entire sample; dashed line, regression for species with >0 rate of female emigration. Based on figures, data, and discussion in Pusey and Packer 1987b; Pusey 1987, 1988; and Moore 1988.

portions of male immigrants and the proportion of females who emigrate. Given the small sample size and the methodological problems alluded to above and in Moore (1988), this result is not strong evidence against the inbreeding avoidance hypothesis; it simply shows that more long-term fieldwork is needed before this analysis can be applied with confidence.

If male breeding tenures are shorter than the time it takes for a female to become sexually mature, there is little chance for father-daughter incest. Using a sample of twenty-seven mammals (nineteen primates), Clutton-Brock (1989) has shown that the average age at first conception exceeds average male tenure for all but one of eighteen species with strongly male-biased dispersal, whereas the reverse holds among nine species with female-biased or unbiased dispersal.[4] This result strongly supports the hypothesis that female dispersal in the second group functions to prevent close inbreeding. Among mammals, females are predicted to show a stronger aversion to inbreeding than

4. It should be noted that there is evidence of significant female dispersal for at least three of the species characterized as predominantly male transfer (Campbell's guenon, blue monkey, and purple leaf monkey; Moore 1984), and less well-known populations of red colobus seem to be characterized by unbiased dispersal and relatively short male tenures (e.g., Marsh 1979). Additionally, some of the available sample sizes for calculating average length of male residence are quite small (e.g., Campbell's guenon, $N = 1$).

are males (Waser, Austad, and Keane 1986), and Clutton-Brock's results are consistent with theoretical expectations.

This comparison supports the hypothesis that when male tenure is long, maturing daughters may disperse to avoid incest; Clutton-Brock does not directly address the inverse hypothesis that male tenures may be shortened in order to avoid maturing daughters (Cheney and Seyfarth 1983). Pusey and Packer (1987b) found no evidence in favor of this hypothesis, but it would be supported by a close correlation of male tenure with age of female first conception among species with predominantly male transfer (Moore and Ali 1985). In fact, the two variables are correlated in Clutton-Brock's primate sample (figure 17.2), and the slope (0.85) is close to the predicted value of nearly equal to 1.0. Ironically, however, exclusion of the one species with tenure exceeding age at first conception (*Cebus olivaceus*) reduces the slope to 0.347 (less than 1; $p < .05$) (including nonprimates has little effect on these relationships). A positive correlation is expected on purely allometric grounds: male tenure is presumably a function of the number of prime years a male has, itself related to overall life span, which is closely correlated with female age at first conception (Harvey, Martin, and Clutton-Brock 1987). This problem should be reduced by considering only species in which males attain breeding status in several troops during their prime years; so restricting the set raises the slope from 0.347 to 0.351, still significantly less than 1.

As Clutton-Brock notes, some average male tenures are very close to females' age at first conception. Longer tenures thus may overlap a daughter's productive span (see table 17.4). For some species, even the longest male tenures reported are shorter than the age of female sexual maturity (*C. mitis*); this is clearly consistent with the inbreeding avoidance theory. Others (*C. ascanius, P. entellus*) are more problematic. As illustrated in figure 17.3, it is not the average, or even the range, of tenure lengths that is critical for understanding why males emigrate; it is the distribution of tenure lengths, combined with species (population?)-specific susceptibility to inbreeding depression. As figure 17.3 shows, the data of Drickamer and Vessey (1973) indicate that rhesus males disperse according to an "up or out" rule; tenure indeed. Note that 20% of high-ranking males are potentially co-residing with fertile *grand*daughters. Ultimately, one needs to know the proportion of infants whose fathers might be their grandfathers—a function of breeding success over a realistic distribution of tenure lengths—rather than the relationship between average tenure and age of female sexual maturity.

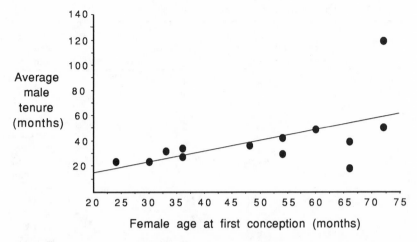

FIGURE 17.2   Relationship between male tenure and female age at sexual maturity. Based on data from Clutton-Brock 1989: table 1; primates with male-biased dispersal.

CONCLUSION

Currently available data are consistent with a scenario in which a mixture of CMC and BH determine male dispersal strategies, and CMC (primarily mate/range choice) and IA determine those of females. When male strategies expose daughters to probable close inbreeding, IA-driven female dispersal occurs; otherwise, inbreeding avoidance has little effect on dispersal patterns. The data are also consistent with a scenario in which migration costs are in an intermediate range and (at least) CMC and IA are both important factors in the evolution of dispersal strategies. At issue is whether dispersal patterns are primary determinants of the severity of inbreeding depression in a species, with some (small?) degree of feedback of IA onto dispersal strategies, or alternatively, whether inbreeding depression is the major determinant of dispersal behavior. This is not yet answerable, though enlargement and refinement of the comparative data base should provide convincing evidence in the future.

Regardless of the direction of causation, dispersal patterns affect levels of inbreeding and hence kinship within social groups; even in the absence of inbreeding, the distribution of relatives among potential interactants is determined by dispersal strategies. We should thus see behavioral consequences. For example, T. D. Wade (1979) has argued that behavioral differences between bonnet and pigtail macaques can

TABLE 17.4 Variability in Duration of Male Tenures

	Age at first conception	Average male tenure	± SD	Maximum tenure	N	Sources
Ceropithecus mitis	66	41.0	22.4	62	9	Clutton-Brock 1989; Cords 1987
c. ascanius	24	22.6	7.8	39	6	Clutton-Brock 1989; Cords 1987
Presbytis entellus	34	26	20.6	>65	39	Sommer and Rajpurohit 1989
Papio c. anubis[a]	54	66	28.0	>108	8	Manzolillo 1982; Melnick and Pearl 1987
Macaca mulatta[b]	45	50	40.3	>120	23	Drickamer and Vessey 1973
		29.2	35.0	>120	50	Drickamer and Vessey 1973
M. mulatta[c]	45	57	30.9	104	14	Hill 1986

Notes: All figures are in months. All studies include censored residencies, so variances and maxima are all underestimated.
[a]Figures based only on nonnatal males present in troop when Manzolillo began her study.
[b]La Parguera; first line includes only top-ranking 50% of nonnatal males; second is total sample of males present in 4 troops in May 1972.
[c]Cayo Santiago Group I; excludes alpha male (natal, age/tenure not given) and a male who emigrated (residence > 107 months).

be explained by differences in dispersal patterns; bonnet macaques are less aggressive and show less discrimination among partners within the group, and Wade argues that this is due to their being inbred relative to pigtails. Similarly, Taub (cited in Mehlman 1986) has suggested that the unusually high levels of infant care shown by male Barbary macaques can be explained by low dispersal rates and consequent inbreeding. Gouzoules (1984) notes that Japanese macaques show elevated care of infants by males relative to rhesus macaques, and one might speculatively suggest a relationship with the outbred nature of rhesus (Melnick 1988) versus the subdivision of Japanese macaques into LCTs (Nozawa et al. 1982).

However, there are important examples of variation in dispersal patterns that do not seem to be reflected in behavior. Female red howler

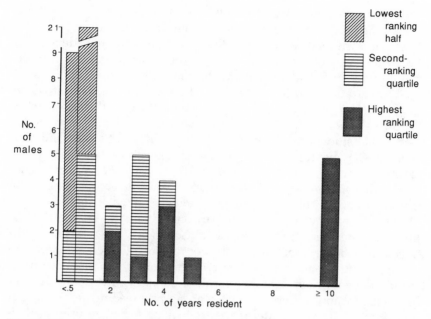

FIGURE 17.3   Distribution of tenure lengths among male rhesus monkeys, La Parguera, in May 1972 (Groups A, C, I, and E). Four years (age at sexual maturity) have
been subtracted from the "tenures" of two natal males. Based on data from Drickamer
and Vessey 1973: table 4.

monkeys emigrate but rarely immigrate and male tenures tend to be
long (Clutton-Brock 1989), a pattern that should result in troops being
single matrilines; mantled howler females transfer, and hence females
in a troop may not be related (Crockett and Eisenberg 1987). Despite
this difference, no obvious behavioral differences between the species
have been reported (Crockett and Eisenberg 1987). At a higher taxonomic level, I have already noted the problems existing with current
interpretations of colobine-cercopithecine differences. (See Moore
[1992] for further examples and discussion.)

The avoidance of inbreeding (or outbreeding) may or may not have
played a role in shaping primate social organization; the evidence is far
from conclusive. The distribution of kin, and perhaps relative inbreeding levels, among social groups of different species is expected to profoundly affect behavior, and there is much evidence that it does so;
however, howlers and colobines illustrate some glaring problems with
our understanding of these effects. "Paired" comparisons such as suggested by the work of Wade, Taub, and McKenna, and broader compar-

ative analyses like those of Pusey and Packer and of Clutton-Brock, should help us to resolve these gaps in our understanding.

As stressed repeatedly above, the existing data are not adequate for confirming any of our specific hypotheses about the roles of inbreeding, outbreeding, and kinship structure in the evolution of primate social organization and behavior. Several things are necessary if we are to profit from suggested comparative studies: First, attention must be paid to standardizing observational methods and vocabulary, so that behavioral studies can be compared. For example, Stewart and Harcourt (1987) operationally define "nonkin [gorillas as] animals with a coefficient of relatedness of less than 0.5" whereas Packer (1979) concludes that an average $r = 0.1$ is sufficiently high to enforce emgiration among baboons. This is problematic enough, but when authors simply discuss "kin versus nonkin" comparisons without specifying the distinction, interpretation is impossible (Gouzoules, 1984). Second, correlations of demography with behavior require attention to differences among populations and species; operating on the assumption that a poorly known group demographically resembles a better-known congener is probably a useful heuristic for captive breeding programs, but can obscure the very patterns we are looking for when used analytically (e.g., Barbary macaques in Pusey and Packer 1987b; see Moore 1988). Third, more attention should be paid to variability in behavioral or demographic parameters; average tenure lengths and proportions of males emigrating are important, but perhaps less informative than the distribution of tenure lengths and which males emigrate. Finally, we need to think about how to distinguish between observations that are important exceptions and those that are just exceptionally noisy.

ACKNOWLEDGMENTS

I am grateful to the following for their very helpful comments on the manuscript (errors and interpretations remain mine alone): W. J. Hamilton III, Anne Pusey, Neal Smith, Karen Strier, and Nancy Thornhill. Fred Bercovitch, Tim Clutton-Brock, Don Melnick, Ken Nozawa, Michael Pereira, David Smith, and Shirley Strum all contributed their expertise to subsets of the manuscript, and Fred Bercovitch, Colin Chapman, Tony Collins, Georgina Dasilva, Bill Hamilton, Catherine Hill, Michael Pereira, Damian Rumiz, Craig Stanford, and David Watts generously provided unpublished data and/or manuscripts (sadly, not all of which I was able to incorporate in a finite-length chapter).

PART THREE

Epilogue

18

Inbreeding in Egypt and in This Book: A Childish Perspective

William D. Hamilton

Sphinx has no more than two kinds of riddle, one relating to the nature of things and the other to the nature of man.—Francis Bacon

The amount of inbreeding in the living world is great if we count species that inbreed (Shields 1982), much less great if we estimate total inbred biomass. This is because most of the world's species are small-bodied and these are the ones most apt to inbreed. Most of the world's biomass is in big organisms like trees and grass. The relation of inbreeding to size (and concomitant longevity) is little discussed in other chapters of this book. I will later devote part of mine to saying why I think such a relation exists but first will illustrate my claim about size and numbers with some examples chosen from among my own earliest experiences of entomology. Although the examples are anecdotes and as such perhaps inappropriate to a serious book about inbreeding, they nevertheless give entry to an ever-surprising world that is strongly linked to our theme.

INBREEDERS ARE SMALL

My mother says that shortly after I was born, ants attacked me in my cradle. The place was Cairo, so I like to guess the ant was "Pharaoh's," *Monomorium pharaonis*, although there is no such particular probability because the species is a worldwide tramp equally at home in any warm town or even a warm London kitchen. There are plenty of other ants in Cairo that could have been interested in a baby. However, I like the idea because, besides being a notorious lover of sweet things, this tiny yellow ant is noted for indiscriminate inbreeding within the nest, including brother with sister. Could this be why it is called "Pharaoh's"—or

429

is it simply that first specimens came from Egypt? The latter is more likely.

To move to my next example, measures that shortly stopped the ants did not, according to my mother, stop another insect which came to eat the buttons on my clothes. Again she can't describe it exactly, but it is very likely to have been the button beetle, *Coccotrypes dactyliperda*, which would have arrived as flying females. The button beetle is even more of an inbreeder. It not only rests its system on a firm foundation of brother-sister incest, like the ant, but continues to other crimes, at least in the case in which a female happens *not* to have been fertilized by her brother. On reaching a button of vegetable ivory (made from seeds of the palm *Phytelephas*) or, probably more commonly in nature, a date stone, such a virgin first excavates a chamber and then, inside, uses parthenogenesis to produce a brood of four or five males. After a wait for the first male to mature she mates with him, eats him except for the hard thorax and head, eats the other sons similarly, and finally proceeds to further enlargement of the chamber and the laying of eggs for her main brood. This consists of about seventy females sired by the first son and about three more males (Buchner 1961). We see at once that small virgin colonizing arthropods such as Pharaoh's ant and the button beetle are much more keen to start breeding somehow, come what may, than are horned scarabs or female academics at the opposite extreme of K-selection, and we see that they act towards this end as if with forethought, paralleling seemingly by ingenious "ROMs" of DNA the subtle but flawed memes of the god-kings and priests of the same land. In reality, the inbreeding achievements of the pharaohs may have been more in the minds of their subjects—how these looked at royal inheritance, accepted social inequality, although of course these remain memes of the imposers—than in substantive homozygous human tissue. An exciting discovery reported by Rowley, Russell, and Brooker (chapter 15, this volume) concerning the reputedly incestuous blue wrens of Australia begins to expose a paternity deception, probably initiated by breeding females in this case, that is surprisingly similar. The birds are small as birds go but nearer to humans than to button beetles or mites on the log scale. The point remains that the pharaohs cannot really have gotten far emulating what some of the tiniest minions of their granaries were doing easily. Their usual distaste for their attempt probably was not much different from what the man in the oasis street, or his camel, or his date palm would feel if asked to breed in the same way—and here we can imagine the palm throwing on the other two its mournful shadows of whole towns of happy button-beetle homes as all three ponder the same theme. This is the contrast:

the big and the small. I admit I do not know much about the tastes and distastes of date palms or what makes the shadow of one mournful, but the evolution of dioecy by palms in general (and the date palm in particular) plus inescapable natural selection to resist such pests as beetles that destroy their seeds suggests a general direction, which I shall return to below. On whether the pharaohs did ever *successfully* inbreed in terms of descendants, a point that seems not quite decided, I will also come back to some rather lightweight thoughts of my own at the end.

I am not sure whether *Locusta migratoria* ever visited me in my cradle but, being a common insect even when not a plague, it is quite likely to have done so. If it did it probably brought me new examples of inbreeding, for it would have had a good chance to carry a certain parasitic mite, *Podapolipus diander.* This mite is named for having two types of male. One occurs in a brood only as a single precocious firstborn (Volkonsky 1940), a son with only one function. As soon as he is born, he shoulders back into his mother's cloaca while she is still feeding and slowly swelling on the haemolymph of the locust. He fertilizes her whether or not she has been fertilized already. After this event she can certainly lay female eggs, for like ants and the button beetle, the mite reproduces by haplodiploidy, the system of reproduction in which haploid males—identical in genotype to their mother's gamete—come from unfertilized ova but normal diploid females appear when ova are fertilized. Later in the brood, different newborn males, uncommon and still very precocious, fertilize their sisters.

But what of the breeding of the locust itself? This animal is admittedly more "weedy" than a camel or date palm, but a locust certainly cannot tolerate inbreeding at the level of the mite. A connected later anecdote may illustrate this. When I worked at Imperial College Field Station we had in the constant-temperature rooms in the basement of Silwood House a colony of *Schistocerca gregaria,* the other main locust of the Middle East, and after a time it became badly infested with mites. The locusts were depressed and sat sluggishly on the floor of the cage. This time for certain the ants that came to swarm over the locusts and cut them up while so immobilized were Pharaoh's, and the name now suited them in a new, rather gruesome way, for they dismantled their huge prey rather like teams quarrying stone for a pyramid. The remembered unpleasant sight reminds me to be less skeptical of my mother's saucers of water she says she placed under the feet of my cradle, for those ants too were probably after not just a baby's spills but meat and serum, as would certainly have been the case had the ant been another incestuous cosmopolitan possibility, the equally tiny *Hy-*

poponera punctatissima, which likes flesh only and has no interest in sugar. To conclude a digression, however, the moral here is that the locusts in the inevitably small colony we kept *appeared to suffer from inbreeding*. This was suggested also by their imperfect wings, poor melanization, etc., and it may have been why, in particular, they were so susceptible to parasitism as discussed above (and again below), and why in general they were so physically incompetent as not to escape the ants, who suffered from inbreeding not at all. *Coccotrypes, Podapolipus,* or an inbreeding hymenopteran parasitoid (but see Crozier 1971) would also certainly not so suffer (Werren, chapter 3, this volume), nor did in fact another middle-of-the-road inbreeder that we had caged in the Silwood basement where it too was doubtless, to me, a returned babyhood companion. This was the small tropical diurnal mosquito, *Aedes aegypti,* the sometime vector of yellow fever. It is definitely not incestuous by habit but is probably nevertheless well accustomed to small-colony bottlenecks brought about by its normal breeding habitat of rain pools in old tires and similar places. Our stock in the basement was as inbred as the locusts, but escaped females often proved to me in my office that they lacked nothing in agile flight or zest for blood.

My fascination for such small animals, begun in my cradle and later accentuated by admiration as I realized their regal contempt for the taboos of humanity, has since turned up a seemingly endless list of other examples (Hamilton 1967, 1978, 1979). Among these, for compounded incestuous sin it is hard to beat a parasitoid of beetles, *Scleroderma immigrans,* whose antlike female readily remates a grandson born of a daughter got from the mother's previous union with a son (Wheeler 1928). At an equal but different extreme—sin aforethought as one might put it, as judged by structure—consider a case that once came to me on a piece of white-rotten wood pulled from a branch. Pearly droplets adhered to the velvety broken surface of mycelium. Under a microscope the drops proved to have heads and legs, and in water on a slide I could see what was happening inside. Each was the enormous body of a female mite distended with eggs and young. Tiny fully formed adults were bursting from eggs as I watched, joining a throng floating or slowly swimming in the maternal fluid. I soon made out there were three kinds, two of them female (both numerous overall, one kind a chelate dispersal morph and the other more plain) and one kind of male. The males were uncommon, one or two to a brood but they were always first to hatch and very active. They spent all their time searching for new sisters in the crowd and copulating with them. All this was happening *within* the yet unruptured mother (Hamilton 1979; for an illustration see Trivers 1985). This mite is in the genus *Pygmephorus.*

As these examples and others mentioned by Werren (chapter 3, this volume) make plain, whether incest is or is not exceptional for large higher organisms—for example, for humans, birds, fishes, trees, and palms—it is certainly far from being so for a huge array of small arthropods. The very size of the array in question may be partly a consequence of the incest, for new species formation is facilitated by inbreeding, as explained by Howard (chapter 7, this volume). Correspondingly, Acarina, which are perhaps the world's greatest practitioners, as already hinted in my examples, happen also to be the world's second most species-rich class. It has been suggested they may even surpass the insects if they are ever equally studied. Among the insects another huge contribution to the faunal list comes from the Hymenoptera Parasitica, whose exceedingly numerous smaller members are likewise very incestuous. Turning to other groups, within lifestyles as social insects and dead tree inhabitants the smaller ants and the male-haploid "bark beetles" of tribe Xyleborini also show up more speciose than noninbred sister groups. Whether this reflects *important* (macro-) evolution as well as species splitting is another question, but the inbreeders are unquestionably highly innovative at times (Hamilton 1978).

Groups and examples like those mentioned should not give the impression that all small arthropods mate close relatives by preference or with impunity. A proposition that really large insects don't practice incest seems firm: no insect the size of the locust mates like the locust's mite *Podapolipus* does, although there are a few fairly large orthopterans and phasmids that are parthenogenetic (Hubbell and Norton 1978). But at the other extreme of size even very small arthropods quite often have adaptations that prevent incest. Examples are the unisexual broods of some cynipid gall wasps, gall midges, sciarid flies, and others. For Acarina, opposite tendencies in animals of rather equal size are well illustrated by the mites of our skin. *Demodex folliculorum*, a wormlike mite that typically lives harmlessly around the roots of our eyelashes, is incestuous and has a very biased sex ratio, whereas our other specifically human associate, which burrows and breeds like a mole, but with more damage, beneath the more open pastures of our skin, *Sarcoptes scabiei*, seems well designed as an outbreeder since it has a 1:1 sex ratio. Later I will show a plausible rationale for this type of contrast based on the degree of damage to the host; but I have none even in prospect yet for another similarly different pair that I find in British woods. There the scolytids *Xyleborus dispar* and *Trypodendron domesticum* make brood tunnels mixed together in the same newly fallen logs of oak or beech. Both are mutualists with "ambrosia" fungi that they carry with them and inoculate into the wood, where later special pseudofruit of the mycelium, presented in smooth lawns spread on the walls of their galleries, feed their larvae. The sizes, lifestyles and even habitat of the two species appear nearly identical, and yet the first insect is incestuous, with its males displaying the modifications that usually accompany such a habit, being few, small, blind, and flightless (Hamilton 1967); while the second is an outbreeder with its males common, large, and similar to the females.

INBREEDING AND REPRODUCTIVE EFFICIENCY

One would think that in a species as thoroughly incestuous as the droplike *Pygmephorus*, having sex at all would have lost all its point, whatever the point might once have been. They must surely, one would think, be so inbred that recombination is ineffective. However, extreme inbreeding does not seem to pass easily to pure female parthenogenesis (thelytoky). Thus although there are certainly thousands of species of incestuous "bark beetles" (for the most part *not* found under the bark but deeper in the wood living on ambrosia, like *X. dispar* above, or else in other obscure plant niches such as the twigs of tea bushes, developing coffee beans, and the date stones and even buttons already mentioned) there is no case yet known to me where such an incestuous species has become parthenogenetic. On the contrary, bark beetles that have adopted parthenogenesis are former outbreeders whose males have well-developed secondary sexual characters (Kirkendall 1983). It is not clear that the same is true of cases of thelytoky in the Hymenoptera Parasitica; here there are groups where both this full type of parthenogenesis (female offspring always) and arrhenotoky (males only from unfertilized eggs following from haplodiploidy, females produced normally through fertilization) seem to be common. Elsewhere in animals, however, a rule that inbreeding and parthenogenesis are not found close together in phylogeny seems to hold fairly well. It certainly holds for plants, in which parthenogens typically appear abruptly in outbred and often self-incompatible species and groups, and arise only rarely among selfers. In chapter 10 (this volume) Knowlton and Jackson show that the same is true of marine invertebrate animals, including such cases as the corals and other "colonials" that have plantlike growth forms. Small animals are the selfers. It seems that when an environment no longer favors outbred sex, species that are adapted to inbreed intensify their practice and reap its efficiency through bias of the sex ratio—that is, they produce fewer of the "wasteful" males. Inbreeding thus provides an adjustable control of their degree of effective sex. Outbreeders, in contrast, are checked in an increasingly unstable position, barred from inbreeding both by habit and by their load of deleterious recessive alleles. Eventually they must take the whole step to parthenogenesis through finding the right mutation, or possibly, as recent evidence suggests, the right executive microbe (Stouthamer, Luck, and Hamilton, 1990). Since species hybridization is so often antecedent to parthenogenesis it may be that inherited microbes gain (or regain) special power over sex when associated with non-coadapted genomes.

How do inbreeders achieve bias of the sex ratio? In some cases the group already has male haploidy and then a mother can produce females at will by control of her stored sperm: a fertilized egg makes a daughter egg and an unfertilized egg makes a son (Werren, chapter 3, this volume). But how to arrive at male haploidy in the first place? The button beetle is probably of fairly recent entry into the mode, representing one among at least three separate entries in Scolytidae, and it hints at an answer using microbial symbionts and inbreeding together. To see how the idea works, consider first the ordinary XY sex determination mechanism. A good reason why in this system the male is usually heterogametic is that having control rest in the female, with her maternally transmitted organelles or symbionts, would too often lead to disastrous female biases in the sex ratio. As explained elsewhere (Hamilton 1979) organelles or symbionts with solely maternal transmission are under strong selection pressure to make a female produce as many daughters as possible, and this can only be done by producing fewer sons—ultimately, if possible, none at all. When females are heterogametic, symbionts are presumably in a strong position to get their way. They can interfere in oogenesis where, for example, they might steer the Y chromosome towards a polar body. In males, on the other hand, almost all kinds of symbionts are doomed and consequently have no interest in the sex ratio: the segregation of the sex chromosomes in a heterogametic male can always proceed unbiased . . . or can it? If the species is an inbreeder, it comes to be in the *kin-selected interest* of the symbionts in the male to bias the sex ratio towards females for the benefit of "cousins," that is, for the benefit of those *related* symbionts that are in the female their bearer is going to mate with. While there is no polar body to which to steer the Y in spermatogenesis, they can attempt to identify the Y-bearing spermatids and kill or disadvantage them. Postfertilization strategies like this by symbionts working against male zygotes in early stages are well known in *Drosophila*. From such rationale comes a possible coevolutionary sequence in which the symbionts in males begin discriminating the Y, perhaps tagging (methylating?) it early in development when it is active and open to recognition, while the host responds by seeking to hide the Y characteristic that is being detected. The host possibly gains a more radical escape from microbial intrigues whenever the real male-determining locus or segment is shifted to another chromosome. Along these lines one may rationalize either of the following courses:

(1) Inactivation of the whole paternally derived set of chromosomes through part of the lifetime of a male, this always being accompanied by rejection of the set during gamete formation. Sometimes staggering onto the printed page under the name parahaplodiploidy, such a condition is surprisingly frequent in small insects.

(2) A sequence in which the sex locus is transferred to an autosome and a previous inactive Y deleted, giving successive constitutions like:

$$\text{Males: } 2n \text{ A } + \text{ XY} \rightarrow 2(n-1) \text{ A } + \text{ X'Y' } + \text{ X}$$
$$\rightarrow 2(n-2) \text{ A } + \text{ X''Y'' } + \text{ X' } + \text{ X} \rightarrow \ldots$$
$$\text{Females: } 2n \text{ A } + \text{ XX} \rightarrow 2(n-1) \text{ A } + \text{ X'X' } + \text{ XX}$$
$$\rightarrow 2(n-2) \text{ A } + \text{ X''X'' } + \text{ X'X' } + \text{ XX} \rightarrow \ldots$$

The latest X^t is the currently effective one determining sex. The sequence ends when all autosomes have been converted to Xs, and male haploidy is attained. The sex of a zygote must now rest entirely on chromosome dosage, and a new group has joined the others to be a touchstone and headache for evolutionary theorists, while, as is most interesting from the point of view of the problem of sex, the group's nonheterozygous males are now ready to contribute their evidence that heterozygote advantage cannot be of great importance for diploid efficiency (Werren, chapter 3, this volume) or for sex, a point to which I return below.

It would take too much space to present the as yet only plausible evidence that exists for such schemes of entry to male haploidy, but two points can be made. Firstly, insects are known whose cytogenetics could illustrate stages on either of the above roads. Multiple sex chromosomes and/or parahaplodiploidy are unusually common in insects known to carry intimate microbial symbionts, as may easily be seen by combining information from the major texts of White (1973) and Buchner (1965). (In the case of the button beetle, indeed, symbionts may have been the focus that led Buchner to the strange life history and sex ratio situation I outlined earlier, although that the tiny companions might in some degree guide the evolution of the life history itself seems not to have occurred to him.) Secondly, it is very interesting, and can also be seen from Buchner's general account, that symbionts in species with ovarial transmission by no means fail in their role or disappear, as we might expect to happen from the hopelessness of their situation, when they find themselves in a male. Sometimes they become housed in well-formed mycetomes near the testes and even more often they are found in close attendance on the events of spermatogenesis. Do they crowd here just because they are useful slaves like mitochondria, providers perhaps of an obscure vitamin, their service entirely at the beck and call of their hosts? Or are they there rather as teats are there on the chest of a male mammal, following blind routines and physiological gradients appropriate to another situation, pointlessly aping the motions of cousins in the female gonad? But, *cousins*—might that be the point? I would guess that neither of the previous explanations is the best; instead I suspect that symbionts are evolved to be present and to act positive roles on behalf of related entities residing in the females their bearer is to meet and mate with, along the lines suggested above.

ECOLOGICAL CORRELATES OF EFFECTIVE ASEX

After sex is effectively abandoned by inbreeding or by the transition to parthenogenesis, it does seem that both types of resulting asex persist best under a certain kind of ecology. Both inbreeders and asexuals tend to be "weedy" inhabitants of extreme and biotically simplified habitats (Glesener and Tilman 1978; Knowlton and Jackson, chapter 10, this volume). Or sometimes they are extreme specialists, incumbents of what might be called crannies rather than niches, outside of the mainstreams of life; or finally, they may be organisms specialized in the sense that they live in the shelter of larger protectors (Law and Lewis 1983). For example, on the last theme, ambrosia beetles mate their brothers while living with fungi that they have planted deep in the wood, far from confamilials in the dizzy arthropodan, fungal, and microbial whirlpool of life in the early decay stages just under the bark. In this more typical under-bark habitat, at least in the temperate zone, the scolytids' approach has been to detect the death of the tree early and get in and out as fast as possible, a policy which however still does not completely save them from the interspecies mayhem of colonizers that follow. Fig wasps (*Blastophaga*) have a "cranny" provided where they are cultivated to inbreed as well as to become, later, outbreeding agents for the trees that they serve, just as the ambrosia beetles become agents for the dispersal and outbreeding of the large fungal invasions which have fed them and which they help to disperse and implant. We note that in both wasps and beetles, rather similar integumental pockets in the adult females have evolved to transport pollen and spores. With these examples in mind, the inbreeding of *Demodex* mites in our eyelash follicles now prompts a thought that perhaps not only do the mites do no harm, they may even do us good, perhaps, for example, protecting our eyes against other infections. The same could be true of our seemingly harmless pinworms, which, as I learned for the first time in this book (Werren, chapter 3, this volume) are also haplodiploid. As for the button beetle, I still remember my surprise in finding that in Britain this scolytid species is much more closely related to outbreeding *Dryocoetes villosus*, a common and rather ordinary inhabitant of the familiar typical diverse community of decaying oak phloem, than it is related to *Xyleborus;* but then what cranny could be quieter than a button to support such a parallel trend? Real weeds—the plants—self-pollinate out in the plowed fields. There they may eventually enter the human following, becoming crops in their own right but remaining inbreeders; but all, both weeds and ex-weeds, grow scattered and short-lived, far apart both from each other and from the dense perennial stands in which their close and fully sexual relatives are found wrestling for the space of the meadow. Cnemidophorus lizards, prickly pears, teddybear chollas burning (perhaps, in Sinai, and unburned brambles of Horeb) perform *their* degenerate sexualities, apomixis and vegetative propagation, out in the Sonoran Desert; *Poeciliopsis* fish theirs (gynogenesis) in the pools of the same desert; *Artemia* shrimp theirs in salt pans often in the same desert again; *Hadenoecus* crickets theirs in caves. Viviparous grasses grow seedlings from their flowers on the moraines of glaciers; oppositely and yet the same, viviparous onions sprout bulbs and plantlets out in the sunny fields of Egypt. Asexual mayflies and chironomids live at the insect limits of depth and temperature in lakes and pools, while asexual ostracods, once more oppositely and yet the same, are in the most shallow and *ephemeral* pools. . . The list goes on. Now it is time to address the question of what makes such extremely various organisms able to evolve an efficiency through reduction of sex that their relatives in denser, more species-rich habitats can't. What is the secret influence connected with size and biotic complexity?

Until we know the answer to this question, as it seems to me, most of the important questions about inbreeding cannot be properly framed, let alone answered. This fact seems to have had too little attention in the rest of this book. I want from here down to give my own view, but as preparation I want first to look at another issue, heterosis, that has been raised in at least three chapters.

OUTBREEDING: HET ADVANTAGE OR HOM COMBINATION

As is made clear by Mitton (chapter 2, this volume), ever since the earliest years of genetics there has been a question about the nature of inbreeding depression. The alternatives most discussed are that it is due to deleterious recessives in mutation-selection balance, or else that it is due to an intrinsic benefit of heterozygosis itself. Mitton presents evidence mostly favoring the latter, but is cautious and admits contrary evidence at some points. Uyenoyama (chapter 4, this volume) assumes intrinsic heterosis as the basis for a theme of eugenic behavior in potential parents. Waser (chapter 9, this volume), attending only to plants, is also cautious but is more on the side of recessive mutations. In the literature in general the problem has been like painting the Forth Bridge, no sooner finished at one end than begun at the other. While not doubting that both of the ideas above are indeed in their way rust-preventive and play a part, I would like now to stir into the paint of this key bridge of biology a third possibility. Not surprisingly, this new idea accords better than either previous one with my own prejudice on the evolution of sex; whether ultimately it will help any better with the issue of inbreeding depression remains to be seen.

It seems possible that some apparent heterozygote advantage, at least in the wild, is actually epistatic homozygote advantage expressed at unseen loci.

Imagine a viscous population for which a study has shown that heterozygosity at various marker loci correlates with components of fitness (as with the evidence for *Pinus ponderosa*, well summarized by Mitton). My point is that the finding does not really show that loci currently heterozygous are on average giving benefit or loci homozygous giving harm. When an occasional distant outcross arises—a pollen grain has blown a long way to claim an ovule and following this a seed from that ovule has grown to be a seedling and then a tree—the zygote so created is indeed highly heterozygous. If the seedling or tree is unusually fit, it must indeed be an indication of heterosis. However we should note that the *next-generation* offspring of such a cross *remain* more heterozygous than the average. Once descendants become inbred again, as the viscosity makes likely, something new has been added: new homozygous combinations are now present, combinations that may never have existed in the locality before. In such inbred descendants, along with the diminishing heterozygosity from the original event, there will be new *multiply homozygous genotypes*. Some will doubtless be bad and likely to die early. Others may be very good; indeed, the good new multiple homozygotes may be the real reward sought in the striving for the distant outcross. The idea has much in common with Uyenoyama's theme, although the selection for incompatibility is based on a less immediate advantage. The main point is that residual heterozygosity may be found correlating with fitness in a way that only creates an illusion of causation: *there might be no over-*

dominance in fitness at any locus at any time and still such correlation could appear.

My approach may seem to betray distrust of heterozygote advantage. It does, and the distrust arises from two sources. One is familiarity with species where perfectly fit and ordinary males cannot possibly be heterotic because they are haploid. It seems to me, for example, that there is hardly any difference either in general evolutionary style or in immediate health and activity between male Hymenoptera and male Diptera. Similarly, there seems to be nothing unhealthy about being a haploid moss (Mishler 1988) or seaweed (De Wreede and Klinger 1988). My other source of distrust is that when I introduce such advantage into a model for evolutionary maintenance of sex the model is always less likely to succeed. The reason for the failure is easy to understand. Heterozygote advantage makes polymorphism very stable but does nothing to make it efficient: in contrast to sexual heterozygotes, which can't breed true, parthenogenetic heterozygotes can, and this immediately favors the switch. Gene duplication followed by selection to homozygosity in opposite ways at the two loci offers a third and perhaps ideal route to the goal of no wasteful segregation. But this is a more difficult step to take, and moreover, it opens up a new kind of inefficiency, that of carrying too much code. I feel that a tenet basic to a whole line of reasoning on the role of heterozygote advantage, that greater diversity of synthesis within an organism is better, needs to be questioned. Again mosses and the competent haploid males of Hymenoptera and several other insect groups seem to say something different. More generally, an admirable belief critique is that of Clarke (1979).

The essence of "Red Queen" models that stabilize sex even against fully efficient parthenogenesis (and that therefore would almost certainly support it against inbreeding, whose efficiency is intermediate) is simply eternal change. This is more likely to occur when selection coefficients that are monotonic across the genotypes switch their slant back and forth without heterozygotes ever becoming the most fit. A role for heterozygote advantage is not completely lost in this kind of alternation, but it is changed in kind. When fitnesses throughout such a sequence are considered together, *geometric* mean fitnesses over the many generations still have to show heterozygote advantage, otherwise one or another allele eventually fixes (Dempster 1955; Frank and Slatkin, 1990); this, however, is a looser advantage than is normally implied.

Now returning in a more experimental mind to the viscous population already discussed, imagine that distant individuals are crossed, and that their progeny are reared at both the parental sites and eventually are scored for their success in life. A greater average success of the F_1 hybrids compared with average success of local controls in this experiment confirms the usual claims: the advantage may be intrinsic heterozygote advantage, local fixation of mildly deleterious mutations, or both together. However, suppose that we inbreed the progeny and continue for one more generation. Now the other effect I have suggested may become perceptible—in addition or instead: Inbred F_2s from the F_1s, similarly reared in two places, may show some individuals *further* to increase their fitness, or else may show fitness increased in some degree even when their parents (F_1s) showed no increase. Heterozygosity would have to be maximal in the F_1, so such further increase in F_2s could not come from that. It could, however, very well be explained by new homozygous combinations. Tests of this point would seem to me well worth trying. As I see it, they must be done in the wild rather than in protected greenhouse, garden, or vivarium conditions so as to expose the broods to the full natural biotic adversity of their habitats. Moreover in such an experiment I would like to see the impact of parasites monitored with particular care, perhaps even experimentally augmented, for reasons that I now discuss.

Outbreeding: Descendant Health

My emphasis on parasites rather than other ecological agencies of stress and selection comes from my belief that parasites are fundamental to the maintenance of sex. If the idea is right then obviously the comparative toll they impose on inbred and outbred individuals is of great interest: the theory says that eventually inbred lines should suffer from the same inflexibility in the face of parasites as parthenogens do and that this will make them die out or else force them back to outbreeding. The reasons to expect parasites to be more effective promoters of sex in hosts than other temporally varying factors of the environment have been given in detail elsewhere (Levin 1975; Hamilton 1982, 1990 and references cited therein) but may here be summarized again as (1) parasite ubiquity, (2) frequency dependence, (3) intimacy, and finally, (4) the combination of (*a*) small size, leading to (*b*) fast turnover, leading to (*c*) fast evolution. I will now raise a few points from such factors that are particularly relevant to inbreeding.

If, instead of heterosis in F_1s, it is the chance to create new combinations for descendants that is the main object of sex, it may be expected that the offspring of wide outcrosses will have adaptations to make best use of the opportunity endowed by their hybridity. For the rapid production of new homozygous combinations, the closer the inbreeding that follows an event of outcrossing the better; and the higher the chance of swift, nonwasteful elimination of bad homozygous combinations, as for example by sibling replacement (Hamilton 1966), the better also. There are few firm data bearing on this, but three suggestive topics may be mentioned. Firstly, we may dimly discern here a possibility for what is happening in some cases of automictic parthenogenesis which, for genes far out on the arms of chromosomes, acts like very close inbreeding after an initial cross. In effect automixis acts like selfing, and parthenogens that have this kind of gametogenesis might be reconsidered from this point of view. The occasional males, if such occur, might in the same spirit be tested for acceptability and competence to inseminate in biotically stressed conditions of the population. (In passing, in fully sexual species such stressed conditions might also, on the same reasoning, be the key to the erratic repeatability in the other kind of 'rare male' study). Unfortunately for the idea about automixis, the best-known exemplars of cyclical parthenogenesis, such as aphids and cladocerans, seem to use apomictic parthenogenesis. Although a process of "endomeiosis" has been claimed as a variant in the clonal reproduction of aphids, its reality is dubious (Blackman 1980). The second possibly relevant set of facts concerns some annual plants that do in effect alternate their generations between emphasis on chasmogamous and on cleistogamous flowers. The phenomenon does not at present favor the idea that the alternation has evolved with the adaptive genetic objectives suggested—other factors seem sufficient explanation (Schnee and Waller 1986). Moreover, the seed numbers are rather too low to provide the level of sib competition I would guess to be ideal (Schnee and Waller 1986). Nevertheless, the existence of such cases is intriguing, and it would seem worth looking for weaker trends in plants that are variable but not dimorphic in their flower type and pollination (e.g., small crucifers). A third possibly relevant set of facts concerns termites, which likewise commonly follow their outbreeding as alates with one or more generations of inbreeding (Myles and Nutting 1988). The alary polymorphism that is deeply embedded in every termite colony's full cycle guides us to think of the phylogenetically diverse set of colonial insects that are usually wingless in their endogamous groups but still produce winged morphs to disperse (Taylor 1981; Hamilton 1978). Winged females in the bursts of dispersers are usually fertilized before they leave; however, it has to be that either circumstances are such as to encourage cofounding of colonies by such females, which is indeed often observed, or that winged males occur along with them, showing that at least some outbreeding is intended, as also observed (e.g. Crespi 1988). Visible polymorphism marks such cases as extremes in the array of tendencies to outbreed or inbreed in colonial insects. Similar phenomena can be expected and are found, though without the visible polymorphism, in the breeding systems of many small mammals (Christian 1961).

The parasite theory of sex predicts that most social insects will not inbreed, and especially will not if they have large colonies which are very apparent and exposed, which means findable by parasites. How species like Pharaoh's ant, *Hypoponera punctatissima*, and *Cardiocondyla* (Stewart, Francoeur and Loiselle 1986; Kinomura and Yamauchi 1987) get away with close inbreeding is not clear, but the answers may be in first part that they are the opposite of "apparent" species and therefore little bothered by parasites, in second part that they are anthropophilic, moving into new man-made habitats where parasites have difficulty in following, and in third and last and perhaps most important part, that they are tramps, and thus like groundsels and dandelions among plants, continually on the move. More specifically in the case of "Pharaoh's," it may well be that in the heated buildings in Scotland where their mating system was first studied and they were found to inbreed they had left even more of their parasites behind than in most other places: if studied in native (?) Egypt, for other example, it might turn out that they do *not* inbreed. Likewise that rather parallel tramp *H. punctatissima*, which today often occupies the same buildings, came at least once to the frontier of Scotland, where in an otherwise inhospitable waste of bracken, heather, and bog, it made its home, as archaeological evidence reveals, in the heated living quarters of the soldiers of Emperor Hadrian, built beside his famous Wall. What the Roman guardians of the Wall thought about their unusual tiny visitor and how much incest it practiced in the recesses of their messy floor are matters unknown. However, I have checked for myself with a colony I obtained from a hospital in London that the workerlike, belligerent, yet utterly lazy males readily mate with sisters when these are available. Such matings, however, are only possible for the dominant male, since fighting excludes all others from the chamber of the royal brood, and in fact all nondominant and defeated males soon die. My observed fighters were brothers or very close, but their adaptations for combat suggest to me that colonies are not always so inbred nor the males so related (Hamilton 1979; Kinomura and Yamauchi 1987; but see also Murray and Gerrard 1984; Murray 1987): perhaps cofounding of nests is not uncommon in big centers for the ant like London even if one has to doubt the event can possibly have been common at Hadrian's Wall. Did the ants come there in some crate of dried, beetle-infested food from Italy? The incident of colonization in such an inhospitable outer environment seems likely to have been unique.

Like a large multicellular body, a large and homeothermic social insect nest containing a closely packed brood offers a good home for parasitic microbes. Providing spread is possible from one larva to another, the nest is almost like the continuous tissues of a large mammalian or avian body, and nurse adults monitoring for disease in this situation (Rothenbuhler 1964) are like lymphocytes. Likewise it is a good living place for larger well-armored predators and for subtle inquilines. I believe that it is the apparency and vulnerability of social insect nests that leads many of the commonest species to be strong outbreeders. They seem to go to great lengths to create diversity of genotypes in their nests (Hamilton 1987; Sherman, Seeley, and Reeve 1988). The many matings of the honeybee queen and the huge simultaneous mating flights of ants and termites are examples, as are the trends to unisexual broods (Nonacs 1986; Boomsma 1988). Pointing the same way is the enormous investment in flying males relative to flightless fertile females shown by true legionary ants. Such ants are particularly abundant and influential in their ecosystems, and doubtless because of this they also support an extraordinary number and variety of inquiline arthropods (Rettenmeyer 1962), not to mention vertebrates like ant birds that follow them to exploit them or their disturbances. One might put it that by daring to become major biotic oppressors in the forest they have ended being also very biotically oppressed.

I hope I have by now shown a reasonable case why large long-lived beings, whether trees, people, or supercolonies, usually retain sex, and for the act itself prefer relative strangers—strangers who should come from as far away, ideally, as their parasites.

OUT OF THE LAND OF EGYPT . . . GENES AND MEMES

Which has been more than enough on that subject, as Herodotus was wont to say when writing his own endlessly digressive first tourist's account of Egypt in the fifth century B.C. Legionary ants have made me think of Africa, where my life and this essay started. Fortunately in Africa, *Dorylus*, harsh biter of explorers and the most formidable genus of the group, does not occur above ground north of the desert, and so definitely was not my attacker. We may leave its swarming armies, populous as the town of London, to the south; thence follow the Nile north and time backward to ancient Egypt and so enter my closing topic. I will now add more detail and new speculation to the discussion of a particular famous historical spate of human inbreeding in that country, and I will suggest what may have spread to the world from the event. And in reality this road to Egypt can start not only south of the Sahara but from almost anywhere—from Texas or indeed from wherever science is practiced, as I now explain.

A pioneer paper already cited on the connection of parasites and sex and outbreeding is Levin (1975). I focus now, unusually, not the content of this paper but on its author's name: the name has an odd connection with Egypt. Perhaps I should apologize at this point for imputing something to a particular name whose origin I don't know about for certain, but Levin is clearly akin to Levins, Levine, Levene, Lewis, etc., and this group as a whole is well known to have Jewish roots. This means, or should mean, that its bearers are more connected in ancestry than most of us are with Egypt. At the very least the root tells that they have *one* more certain connection there, even if a connection across a gap of three thousand years. I thought at first that the names would probably derive from Levi and therefore might claim not just general ancestry through the Israelite period in Egypt but a *male-line connection* to the original Levi, the son of Jacob, who definitely was at least briefly in Egypt according to the Bible story. But surprisingly this root is not considered the most probable (Rottenberg 1977), although there are others, such as Levi itself and Levy, for which it is. Such a direct connection to the original Levi is, however, very definitely claimed for another name swarm typified in Cohen. The point I want to make concerns all Jewish ancestry but specially bears on these particular male lines. It is that even for "Levi" and "Cohen" there is a strong doubt about the claimed pure male line, and it is one which throws an interesting sideways illumination on the whole theme of this book and on other human matters hitherto outside it. The doubt is over the identity of Levi's supposed great-grandsons, Moses and Aaron (Dimont 1962).

Inbreeding was certainly by no means lacking in the Jews of Egypt according to the Bible. Levi himself came of a cousin marriage as well as having consanguinity from links in his parents back through the generation of Terah. Further, although Levi himself outbred, his grandson Amran married an aunt (that is, married Levi's daughter). Amran is the stated father of Moses and Aaron. After Amran, Aaron, but not Moses, again inbred distantly, and it is from Aaron that Cohens and others with related names are supposed to descend. If all this were true, the spread of descendants down from Levi would seem to indicate that the mild level of inbreeding practiced among the patriarchs did the tribes little harm. However, we may ask, how believable are the pedigrees? Dubious as may be the genealogy of the kings of Egypt, with their much stronger inclination to inbreed, or to seem to do so, that of the Jewish patriarchs is more enigmatic still, its claims being merely orally transmitted until some six hundred years later than the events described. In the case of the Jews no mummies were carefully laid down for our assistance.

Scholars agree that the names Moses and Aaron are Egyptian. The tale around them has several confusing and improbable features (Dimont 1962). Difficulties and possible alternative interpretations began to be pointed out in the last century, but a particularly coherent critique and reconstruction is due to Sigmund Freud (1939), himself Jewish and writing his theory with great courage, in old age, under the looming shadow of the Nazis. Freud suggested that Moses was an Egyptian prince or priest: as such he would have *joined* the Israelites, and, as the major point of Freud's argument, would have outbred to them not only his genes (in grandchildren, his wife being foreign even in the story) but his crucial memes as well. Moses' achievement, Freud argued, besides the well-known rescue of the Jews from bondage, was the grafting to their religion of *ideas of a recently suppressed heresy of Egypt which he had kept aflame.* His accepted role in promoting monotheism and iconoclasm in the Hebrew tribes fits very well in time and place with the possibility of his being a missionary or refugee emerging from the counterreformation that convulsed Egypt near to his time. The religion overthrown, of which Moses is suggested to be a surviving adherent, was the sun-disc cult of Pharaoh Akhenaton. It does indeed seem that the presence of the Hebrews in Egypt through the astonishing 17-year revolutionary reign of the heretic king, plus Moses' own name, nature, and described origin, plus the teaching he gave to the Jews as he led them to freedom, is too much to be taken as coincidence. If the thesis is accepted it is likely that his priestly companion and so-called brother, Aaron, was also "grafted" to the Jewish story at the same time. It then follows of course that Aaron was unrelated to the "cousin" Elisheba, and more widely, that the inbreeding of the patriarchs described by the writers of the Pentateuch is more to be seen as a record of what was acceptable and likely than as a record of fact. But that is hardly the point; the other implications of Freud's theory, connecting as it does the history of Jewish thought with that of a unique epoch in the history of Egypt, are by far more dramatic. Its particular relevance to the present chapter is that in the transfer, Akhenaton's undoubtedly revolutionary ideas seem to have become separated into two themes. Or rather three; but the third theme though it came from him was not of his will: this, the idea of overwhelming tragedy, will be mentioned later. Of the other two, the first was his extreme personal indulgence of incest, perhaps the most varied combination ever recorded for one man. His relationships seem to have reflected an extraordinarily close and loving family life, as shown in numerous paintings and reliefs of court events. The lovingness of this life in turn appears to integrate into a unique style and belief that was indeed due to the king himself (Aldred 1988).

The result of the family side of his innovation was disaster. The inability of Akhenaton himself or his sons to produce through their *inbred* unions anything but daughters appeared to the people a confirmation of the withdrawal of the favor of all Egypt's neglected gods, just as the same seemed independently evidenced by the plagues and military setbacks that were affecting the empire from without, coinciding with the monarch's neglect. It was the incestuous, in-turned, family-in-Eden meme of Akhenaton that Moses separated and left behind. Under the laws he gave the Israelites it was still just possible for uncle to marry niece, but other equal and closer unions were prohibited: his tone on sexual behavior generally was stern. Other parts of Akhenaton's beliefs—his iconoclasm, monotheism, and his seemingly almost scientific appreciation that the universe might be interpreted in terms of a single abstract yet life-supporting cause, an attempt, it might be said, at a "unified theory" of his time—Moses forcefully continued. For Akhenaton, the life-supporting Cause was the sun, and perhaps it was partly so also for Moses, a view which may be faintly transmitted to us in the image of the sun-burned bramble by Mount Horeb, withered, punished, and yet ever renewed, like his people, by the light of a Sun. For the Jews after Moses, the Cause was soon to become (perhaps after another priestly fusion soon after the Sinai period, as Freud suggested) the more human-oriented and moralistic creator Jehovah.

Akhenaton bred in succession with his cousin Nefertiti, at least two unrelated wives, his mother Tiye, a daughter by Nefertiti, and perhaps other daughters. Neither of the three close unions produced surviving offspring, and Nefertiti gave six daughters. The two unrelated wives gave at least each a son. These were healthy but died quite young after brief turns as Pharaoh. Akhenaton's legacy to Egypt was an empire diminished, disheartened, and headed for anarchy. His 15-year-old new capital soon lay deserted and defaced, its temples unbuilt to the ground in counterreformatory zeal that began even in the lifetimes of his sons. Apart from these disasters, however, we can hardly doubt that he left another legacy, an outcome that was almost equally guaranteed by the extent of his failure. The annihilation of his religion almost has to have created emigrants. Moses and Aaron, it has been suggested, were two; for hints of others I turn, following Velikovsky (1960), from the Book of Exodus to another myth, that of King Oedipus in Greece.

Velikovsky's first theory, like Freud's last, seems to me more likely to weather time than most other ideas for which these writers are well known. He claimed to give detailed evidence that the story of the heretic king of Egypt reappeared in the legends of Grecian Thebes. The name of Akhenaton's other seemingly more irreverent Paul who brought the story (he called his hero, for example, "King Fat Leg," after the real king's strange physique) has not survived. The story became garbled, barbarized, and mixed with other myths, but still carried along convincing detail from the events of Egypt that are historically known. That genes as well as memes were again implanted is suggested by the extent of the parallel, the very least element of which is the name the Greeks later gave to the Egyptian city, Thebes, linking it to their own. Egyptologists don't seem to like Velikovsky's idea, to judge by his complete noncitation in their books, and it is easy to see how his strong advocacy of a particular version could both cause offense and make him vulnerable. However I cannot see that even the most recent expert accounts of Akhenaton's reign render his general thesis improbable (Redford, 1984; Aldred, 1988).

I have to admit, however, to a bias in my reading of this reinterpretation based on my childhood experiences, as I had better first explain.

Firstly, Velikovsky's theme immediately made sense of something my mother had never properly explained. It was like the "insects" that ate my buttons: impossible!—how could insects have lived on mother of pearl or plastic? I came to understand that by learning of *Phytelephas* and *Coccotrypes:* what Velikovsky unravelled was how the sphinx of the story that my mother read to me had been able to do things in Greece, even in a myth, when *the sphinx itself* had been there so huge and immovable, as she had also often told me, close to my birthplace in Egypt. That confusion soon combined with another, concerning a kind of "sphinx" to Britain itself which likewise was based on an early memory. We once took a picnic to Beacon Hill in Berkshire, and ate it inside the ancient ramparts of the hilltop fort. There in the grass was an iron fence enclosing a mound that was called Tutankhamen's Tomb, and something about this fence and the mound and blue sky made my mother and my great-aunt speak in tones of awe, while inside the fence a sheep was stuck and tried to get out. It was in the time when Mr. Hitler did not let the ships go past, and we were evacuated, staying with my aunt. She kept "chicken," so we had hard-boiled eggs, and there were soft cheese segments wrapped in foil, a great treat. I rolled down a grassy rampart. Only a little later did Pharaoh Tutankhamen, son of Akhenaton, being buried in Berkshire begin to seem odd to me, but then it remained so for a long time. Of course in the end I realized it was not Tutankhamen but Lord Carnarvon, the discoverer of the great tomb, who was buried there. I knew by then that sphinxes were many and could fly and that there was no good reason why the Thebes of the north should not have its own, but even so, granted the mistakes had been childish, Velikovsky still has plenty to be marveled at in his book. Long before reading it, I had thought that although genes and memes don't have to travel together, the bet is that they do whenever transferred detail of culture is great. Moreover, settlement from the civilized to the barbarous is for the most part the more probable, judging by many parallels of recent times. In short, since reading Velikovsky's interpretation it has remained with me strongly that another person or group from Akhenaton's towns did come to Greece. There would be included a person or leader connected nearer to the events than Moses but perhaps more an outsider in other ways—probably not noble or priestly, nor with much interest in the philosophy, perhaps more like a sculptor or a mortician, but above all a storyteller obsessed by a real-life tale. For a parallel of modern times, I think of an ex-hippie of the '60s, perhaps an artist, going native out of the Peace Corps into a tribe of Amazonian Indians and there recounting the Charles Manson affair which he or she happened to have mixed with personally, investing the tale with an atmosphere of Woodstock and disillusion and tragedy. "It could all have been so beautiful." Manson was not leader of an empire nor highly original; how much stronger such an implanted legend might be if he had been!

It is somehow appropriate to the barbarity of the recipients that it was the tragedy of the king's life, not the moral messages and still less the philosophy, that became implanted in Greece; and it is also appropriate to the particular genius of Greece that later it would be the art of Akhenaton that sprung back into life there and flourished. For the other, eastward, migration of the ideas, it seems somehow also appropriate to the spirit of Judaism (at least, as it became molded by the event) that it was more the philosophy and the theology that were accepted.

Much later, we find the artistic naturalism and the philosophy of the strange loved-hated king becoming mixed again in the milieu of still more barbarous peoples and making the foundations of the Science we know. Following such chains, not only do I find "Levis" and "Cohens" of scientific literature now strongly reminding me of their remote and remarkable connection to Egypt, but, when I argue with owners of such names, it even seems to me that I encounter the slants that I am led to expect: they the priestly descendants expecting the universe to be consistent, organized, subject to kingly power, its logic inviolable; for me, the barbarian, no such deep faith existing, logic ready at all times to fail—and if mine, soon doing so. For me it seems the universe only needs to be beautiful, my "science" no more consistent or less tragic than Antigone's story or her sculpted head.

CONCLUSION

Well, to come partly to earth from these stories and to summarize; whether deeply or not, we have to believe in consistency for the purpose of this book and here have a long way to go. From blebs on the side of a locust, and whispering out of a cavity like a caries in an old date stone, listening to the story of their king retold, I have seemed to hear a faint stridulant laughter: "What a fuss!" Well then, for how long can *they* laugh, either? Where, for example, is the button biotype of *Coccotrypes* now that its host is extinct? Do *we* laugh—or did that biotype change back to the diet of date stones? Is inbreeding a long-term failure even for the likes of this beetle, even for those generic empires, each now more than five hundred species strong, of *Xyleborus* of the ambrosia and *Blastophaga* of the figs? Is it disaster for *our* symbionts— say for our lettuce, which is said to be 99% selfed and has been so, perhaps, ever since toddler princess Beketaton was pictured carrying its early ancestor while her father-brother and mother-grandma (as believed by Velikovsky; their identity as Akhenaton and his mother Tiye in the picture is undoubted) walked hand in hand in front? I would think not disaster for lettuce; but worse than for these, what about the three-thousand-year viviparous (and therefore clonal) Egyptian onion? What is the lifetime of such a cultivar—or of an apple? And what of those still smaller helpful enemies, the microbial symbionts within insects—do they escape, not all die, infect outsiders, if their host goes down? How much do they outbreed, if ever? These yet again apart, what other plagues infect button beetles or their symbionts—or Pharaoh's ant—and what plagues may such plagues themselves have? Finally for the Sphinx's second riddle. What was the plague of "Thebes"? It cannot be by inbreeding that the Egyptian king was deformed—he was outbred. Then how came he deformed with traits, such as his strange head, that he passed on? Was he a mutant, was he a would-be starter in a new path of male human physique, as strange as that myrmicine ant that took first stride to be a male *Hypoponera?* Was it hereditary infection? How did his father Amenophis rear his boy, if at all? And how Queen Tiye? Tiye, an empire's correspondent to foreign kings, no cipher consort, strangely and extremely loved, what secret did she have? How does a woman make such a son as she had? Was Akhenaton in infancy, like Oedipus in the legend, like Moses, left out to die?—because he was deformed? It seems to me that notwithstanding all the facts and theories in this book (including mine), we hardly begin to know answers to any of these questions.

CONTRIBUTORS

Michael Brooker
C.S.I.R.O.
Division of Wildlife & Range-
lands Research
Western Australia Laboratory
Locked Bag No. 4.
P.O. Midland, Western Australia
6056

William D. Hamilton
Department of Zoology
South Parks Rd.
Oxford University
Oxford, U.K. OX1 3PS

Daniel J. Howard
Department of Biology
New Mexico State University
Las Cruces, NM 86001

Jeremy B. C. Jackson
Smithsonian Tropical Research
Institute
Apartado 2072
Balboa, Republic of Panama

Nancy Knowlton
Smithsonian Tropical Research
Institute
Apartado 2072
Balboa, Republic of Panama

Robert C. Lacy
Chicago Zoological Park
Brookfield, IL 60513

Jeffrey McKinnon
Biological Laboratories
Harvard University
16 Divinity
Cambridge, MA 02138

Richard E. Michod
Department of Ecology & Evolu-
tionary Biology
The University of Arizona
Tucson, AZ 85721

Jeffry B. Mitton
Department of EPO Biology
The University of Colorado
Boulder, CO 80309-0344

Jim Moore
University of California
Department of Anthropology
C-001
La Jolla, CA 92093

Craig Packer
Department of Ecology, Evolu-
tion and Behavior
The University of Minnesota
Minneapolis, MN 55455

Ann Petric
Chicago Zoological Park
Brookfield, IL 60513

Anne E. Pusey
Department of Ecology, Evolu-
tion and Behavior
The University of Minnesota
Minneapolis, MN 55455

Susan E. Riechert
Department of Zoology
The University of Tennessee
Knoxville, TN 37996-0810

Rose Marie Roeloffs
Department of Zoology
The University of Tennessee
Knoxville, TN 37996-0810

Ian Rowley
C.S.I.R.O.
Division of Wildlife & Range-
lands Research
Western Australia Laboratory
Locked Bag No. 4.
P.O. Midland, Western Australia
6056

Eleanor Russell
C.S.I.R.O.
Division of Wildlife & Range-
lands Research
Western Australia Laboratory
Locked Bag No. 4.
P.O. Midland, Western Australia
6056

William M. Shields
Department of Environmental &
Forest Biology
S.U.N.Y.
Syracuse, NY 13210

Andrew T. Smith
Department of Zoology
Arizona State University
Tempe, AZ 85287-1501

Nancy Wilmsen Thornhill
Evolution and Human Behavior
Program
Rackham Building
University of Michigan
Ann Arbor, MI 48109

Marcy K. Uyenoyama
Department of Zoology
Duke University
Durham, NC 27706

Bruce Waldman
Department of Zoology
University of Canterbury
Private Bag 4800
Christchurch, New Zealand

Donald M. Waller
Department of Botany
University of Wisconsin
Madison, WI 53706

Mark Warneke
Chicago Zoological Park
Brookfield, IL 60513

Nickolas M. Waser
Department of Biology
University of California—
Riverside
Riverside, CA 92521

John H. Werren
Department of Biology
University of Rochester
Rochester, NY 14627

REFERENCES

Aalto, S. K., and G. E. Newsome. 1990. Additional evidence supporting demic behaviour of a yellow perch (*Perca flavescens*) population. *Can. J. Fish. Aquat. Sci.* 47:1959–62.

Abugov, R., and R. E. Michod. 1981. On the relations of family-structured models and inclusive fitness models for kin selection. *J. Theor. Biol.* 88:743–54.

Achituv, Y., and L. Mizrahi. 1987. Allozyme differences between tidal levels in *Tetraclita squamosa* Pilsbry from the Red Sea. *J. Exp. Mar. Biol. Ecol.* 108:181–89.

Achituv, Y., and E. Sher. 1991. Sexual reproduction and fission in the sea star *Asterina burtoni* from the Mediterranean coast of Israel. *Bull. Mar. Sci.* 48(3):670–78.

Ackerman, J. D., and A. M. Montalvo. 1990. Short- and long-term limitations to fruit production in a tropical orchid. *Ecology* 71:263–72.

Adams, J., and J. Neel. 1967. Children of incest. *Pediatrics* 40:55–61.

Adamson, M. L. 1989. Evolutionary biology of the Oxyuridae (Nematoda): Biofacies of a haplodiploid taxon. *Adv. Parasitol.* 28:175–228.

Ågren, G. 1976. Social and territorial behaviour in the Mongolian gerbil (*Meriones unguiculatus*) under semi-natural conditions. *Biol. Behav.* 1:267–85.

———. 1984a. Alternative mating strategies in the Mongolian gerbil. *Behaviour* 91:229–44.

———. 1984b. Incest avoidance and bonding between siblings in gerbils. *Behav. Ecol. Sociobiol.* 14:161–69.

———. 1984c. Pair formation in the Mongolian gerbil. *Anim. Behav.* 32:528–35.

Ågren, G., Q. Zhou, and W. Zhong. 1989. Ecology and social behaviour of Mongolian gerbils, *Meriones unguiculatus*, at Xilinhot, Inner Mongolia, China. *Anim. Behav.* 37:11–27.

Ahearn, J. N. 1989. Interspecific hybrids of *Drosophila heteroneura* and *D. silvestris* I. Courtship success. *Evolution* 43:347–61.

Åkesson, B. 1977. Cross breeding and geographic races: Experiments with the polychaete genus *Ophryotrocha*. *Mikrofauna Meeresboden* 61:11–18.

———. 1984. Speciation in the genus *Ophryotrocha* (Polychaeta, Durvilleidae). In *Polychaete Reproduction*, ed. A. Fischer and H.-D. Pfannenstiel, 299–316. Stuttgart: Gustav Fischer Verlag.

Alcock, J. 1984. *Animal Behavior: An Evolutionary Approach*. Sunderland, Mass.: Sinauer Associates.

Aldred, C. 1988. *Akhenaten, King of Egypt.* London: Thames and Hudson.

Alexander, R. D. 1974. The evolution of social behavior. *Annu. Rev. Ecol. Syst.* 5:325–83.

———. 1979. *Darwinism and Human Affairs.* Seattle: University of Washington Press.

Alexander, R. D., J. L. Hoogland, R. D. Howard, K. M. Noonan, and P. W. Sherman. 1979. Sexual dimorphisms and breeding systems in pinnipeds, ungulates, primates, and humans. In *Evolutionary Biology and Human Social Behavior,* ed. N. A. Chagnon and W. Irons, 402–35. North Scituate, Mass.: Duxbury Press.

Allard, R. W. 1975. The mating system and microevolution. *Genetics* 79 (Suppl.):115–26.

Allen, G. 1965. Random and non-random inbreeding. *Eugenics Q.* 12:181–98.

Allendorf, F. W., and R. F. Leary. 1986. Heterozygosity and fitness in natural populations of animals. In *Conservation Biology: The Science of Scarcity and Diversity,* ed. M. E. Soulé, 57–76. Sunderland, Mass.: Sinauer Associates.

Alstad, D. N., and G. F. Edmunds, Jr. 1983. Selection, outbreeding depression, and the sex ratio in scale insects. *Science* 220:93–95.

Altmann, J., G. Hausfater, and S.A. Altmann. 1988. Determinants of reproductive success in savannah baboons, *Papio cynocephalus.* In *Reproductive Success,* ed. T. H. Clutton-Brock, 403–18. Chicago: University of Chicago Press.

Altmann, S. A., and J. Altmann. 1979. Demographic constraints on behavior and social organization. In *Primate Ecology and Human Origins,* ed. I. S. Bernstein and E. O. Smith, 47–64. New York: Garland Press.

Altukhov, Y. P. 1981. The stock concept from the viewpoint of population genetics. *Can. J. Fish. Aquat. Sci.* 38:1523–38.

———. 1982. Biochemical population genetics and speciation. *Evolution* 36:1168–81.

Anderson, M. A., E. C. Cornish, S.-L. Mau, E. G. Williams, R. Hoggart, A. Atkinson, L. Bonig, B. Grego, R. Simpson, P. J. Roche, J. D. Haley, J. D. Penschow, H. D. Naill, G. W. Tregear, J. P. Coghlan, R. J. Crawford, and A. E. Clarke. 1986. Cloning of cDNA for a stylar glycoprotein associated with expression of self-incompatibility in *Nicotiana alata. Nature* 321:38–44.

Anderson, M. A., G. I. McFadden, R. Bernatzky, A. Atkinson, T. Orpin, H. Dedman, G. Tregear, R. Fernley, and A. E. Clarke. 1989. Sequence variability of three alleles of the self-incompatibility gene of *Nicotiana alata. Plant Cell* 1:483–91.

Anderson, N. O., B. E. Liedl, P. D. Ascher, and S. L. Desborough. 1989. Distinguishing between self-incompatibility and other reproductive barriers in plants using male (MCC) and female (FCC) coefficient of crossability. *Sex. Plant Reprod.* 2:116–26.

Anderson, P. K. 1970. Ecological structure and gene flow in small mammals. In *Variation in Mammalian Populations,* ed. R. J. Berry and H. N. Southern, 299–325. Zoological Society of London Symposium 26. London: Academic Press.

Anderson, W. W. 1968. Further evidence for coadaptation in crosses between geographic populations of *Drosophila pseudoobscura. Genetic Research* 12:317–30.

Andrews, R. M., and A. S. Rand. 1982. Seasonal breeding and long-term population fluctuations in the lizard *Anolis limifrons*. In *The Ecology of a Tropical Forest*, ed. E. G. Leigh, Jr., A. S. Rand, and D. M. Windsor, 405–12. Washington, D.C.: Smithsonian Institution Press.

———. 1983. Limited dispersal of a juvenile *Anolis limifrons. Copeia* 1983:429–34.

Andrews, R. M., A. S. Rand, and S. Guerro. 1983. Seasonal and spatial variation in the annual cycle of a tropical lizard. In *Advances in Herpetology and Evolutionary Biology*, ed. A. G. J. Rhodin and K. Miyata, 441–54. Cambridge, Mass.: Harvard University Museum of Comparative Zoology.

Angus, R. A., and R. J. Schultz. 1983. Meristic variation in homozygous and heterozygous fish. *Copeia* 1983:287–99.

Antonovics, J. 1980. Concepts of resource allocation and partitioning in plants. In *Limits to Action: The Allocation of Individual Behavior*, ed. J. E. R. Staddon, 1–35. New York: Academic Press.

Antonovics, J., and N. C. Ellstrand. 1984. Experimental studies of the evolutionary significance of sexual reproduction. I. A test of the frequency-dependent hypothesis. *Evolution* 38:103–15.

Aoki, K. 1981. Algebra of inclusive fitness. *Evolution* 35:659–63.

Arcese, P. 1989. Intrasexual competition, mating system, and natal dispersal in song sparrows. *Anim. Behav.* 38:958–79.

Armitage, K. B. 1974. Male behaviour and territoriality in the yellow-bellied marmot. *J. Zool.* (Lond.) 172:223–65.

———. 1982. Social dynamics of juvenile marmots: Role of kinship and individual variability. *Behav. Ecol. Sociobiol.* 11:33–36.

———. 1984. Recruitment in yellow-bellied marmot populations: Kinship, philopatry, and individual variability. In *Biology of Ground-dwelling Squirrels*, ed. J. O. Murie and G. R. Michener, 377–403. Lincoln: University of Nebraska Press.

———. 1986a. Individuality, social behavior, and reproductive success in yellow-bellied marmots. *Ecology* 67:1186–93.

———. 1986b. Marmot polygyny revisited: Determinants of male and female reproductive strategies. In *Ecological Aspects of Social Evolution: Birds and Mammals*, ed. D. I. Rubenstein and R. W. Wrangham, 303–31. Princeton, N.J.: Princeton University Press.

Arnold, J. M., and L. D. Williams-Arnold. 1977. Cephalopoda: Decapoda. In *Reproduction of Marine Invertebrates*, Vol. 4, ed. A. C. Giese and J. S. Pearse, 243–90. New York: Academic Press.

Arnold, S. J. 1987. Quantitative genetic models of sexual selection: A review. In *The Evolution of Sex and Its Consequences*, ed. S. C. Stearns, 283–315. Basel: Birkhauser Verlag.

Aulstad, D., T. Gjedrem, and H. Skjervold. 1972. Genetic and environmental sources of variation in length and weight of rainbow trout (*Salmo gairdneri*). *J. Fish. Res. Board Can.* 29:237–41.

Aulstad, D., and A. Kittelsen. 1971. Abnormal body curvatures of rainbow trout (*Salmo gairdneri*) inbred fry. *J. Fish. Res. Board Can.* 28:1918–20.

Avery, P. J. 1984. The population genetics of haplodiploids and x-linked genes. *Genet. Res.* 44:321–41.

Aviles, L. 1986. Sex-ratio bias and possible group selection in the social spider *Anelosimus eximius*. *Am. Nat.* 128:1–12.

Avise, J. C. 1978. Variances and frequency distributions of genetic distance in evolutionary phylads. *Heredity* 40:225–37.

Avise, J. C., B. W. Bowen, and T. Lamb. 1989. DNA fingerprints from hypervariable mitochondrial genotypes. *Mol. Biol. Evol.* 6:258–69.

Avise, J. C., G. S. Helfman, N. C. Saunders, and L. S. Hales. 1986. Mitochondrial DNA differentiation in North Atlantic eels: Population genetic consequences of an unusual life history pattern. *Proc. Natl. Acad. Sci. USA* 83:4350–54.

Avise, J. C., and D. Y. Shapiro. 1986. Evaluating kinship of newly settled juveniles within social groups of the coral reef fish *Anthias squamipinnis*. *Evolution* 40:1051–59.

Avise, J. C., and M. H. Smith. 1977. Gene frequency comparisons between sunfish (Centrarchidae) populations at various stages of evolutionary divergence. *Syst. Zool.* 26:319–35.

Axelrod, R., and W. D. Hamilton. 1981. The evolution of cooperation. *Science* 211:1390–96.

Ayala, F. J., M. L. Tracey, D. Hedgecock, and R. C. Richmond. 1974. Genetic differentiation during the speciation process in *Drosophila*. *Evolution* 28:576–92.

Ayling, A. L. 1980. Patterns of sexuality, asexual reproduction and recruitment in some subtidal marine Demospongiae. *Biol. Bull.* 158(3):271–82.

Ayre, D. J. 1982. Inter-genotype aggression in the solitary sea anemone *Actinia tenebrosa*. *Mar. Biol.* 68:199–205.

———. 1983a. The distribution of the sea anemone *Actinia tenebrosa* Farquhar in southwestern Australia. *West. Aust. Nat.* 15(6):136–40.

———. 1983b. The effects of asexual reproduction and inter-genotypic aggression on the genotypic structure of populations of the sea anemone *Actinia tenebrosa*. *Oecologia* 57:158–65.

———. 1984a. Effects of environment and population density on the sea anemone *Actinia tenebrosa*. *Aust. J. Mar. Freshw. Res.* 35:735–46.

———. 1984b. The effects of sexual and asexual reproduction on geographic variation in the sea anemone *Actinia tenebrosa*. *Oecologia* 62:222–29.

———. 1985. Localized adaptation of clones of the sea anemone *Actinia tenebrosa*. *Evolution* 39(6):1250–60.

———. 1987. The formation of clonal territories in experimental populations of the sea anemone *Actinia tenebrosa*. *Biol. Bull.* 172:178–86.

———. 1988. Evidence for genetic determination of sex in *Actinia tenebrosa*. *J. Exp. Mar. Biol. Ecol.* 116:23–34.

Babcock, R. C. 1984. Reproduction and distribution of two species of *Goniastrea* (Scleractinia) from the Great Barrier Reef Province. *Coral Reefs* 2:187–95.

————. 1991. Comparative demography of three species of scleractinian corals using age-and size-dependent classifications. *Ecol. Monogr.* 61(3): 225–44.

Babcock, R. C., G. D. Bull, P. L. Harrison, A. J. Heyward, J. K. Oliver, C. C. Wallace, and B. L. Willis. 1986. Synchronous spawnings of 105 scleractinian coral species on the Great Barrier Reef. *Mar. Biol.* 90:379–94.

Babcock, R. C., and A. J. Heyward. 1986. Larval development of certain gamete-spawning scleractinian corals. *Coral Reefs* 5:111–16.

Bailey, J. H. 1969. Methods of brood protection as a basis for reclassification of the Spirorbinae (Serpulidae). *Zool. J. Linn. Soc.* 48:387–407.

Baird, P. A., and B. MacGillivray. 1982. Children of incest. *J. Pediatr.* 101:854–58.

Baker, A. E. M. 1981. Gene flow in house mice: Introduction of a new allele into free-living populations. *Evolution* 35:243–58.

Baker, A. J., and A. Moeed. 1987. Rapid genetic differentiation and founder effect in colonizing populations of common mynas (*Acridotheres tristis*). *Evolution* 41:525–38.

Baker, C. M. A., and C. Manwell. 1977. Heterozygosity of the sheep: Polymorphism of "malic enzyme," isocitrate dehydrogenase (Nadp), catalase and esterase. *Aust. J. Biol. Sci.* 30:127–40.

Baker, H. G. 1955. Self-compatibility and establishment after "long-distance" dispersal. *Evolution* 9:347–48.

————. 1974. The evolution of weeds. *Annu. Rev. Ecol. Syst.* 5:1–24.

Baker, J., J. Maynard Smith, and C. Strobeck. 1975. Genetic polymorphism in the bladder campion, *Silene maritima*. *Biochem. Genet.* 13:393–410.

Baker, M. C. 1981. Effective population size in a songbird: Some possible implications. *Heredity* 46:209–18.

Baker, R. J., and J. W. Bickham. 1980. Karyotypic evolution in bats: Evidence of extensive and conservative chromosomal evolution in closely related taxa. *Syst. Zool.* 29:239–53.

Baker, R. J., M. B. Qumsiyeh, and C. S. Hood. 1987. Role of chromosomal banding patterns in understanding mammalian evolution. In *Current Mammalogy*, vol. 1, ed. H. H. Genoways, 67–96. New York: Plenum Publishing Corp.

Ballin, P. J. 1973. Geographic variation in courtship behaviour of the guppy, *Poecilia reticulata*. M.S. thesis, University of British Columbia.

Ballou, J. D., and K. Ralls. 1982. Inbreeding and juvenile mortality in small populations of ungulates: A detailed analysis. *Biol. Conserv.* 24:239–72.

Bams, R. A. 1976. Survival and propensity for homing as affected by presence or absence of locally adapted paternal genes in two transplanted populations of pink salmon (*Oncorhynchus gorbuscha*). *J. Fish. Res. Board Can.* 33:2716–25.

Banks, C. B. 1984. Reproductive history of a colony of captive common iguanas (*Iguana iguana*). *Acta Zool. Pathol. Antverpiensia* 78:101–14.

Banyard, B. J., and S. H. James. 1979. Biosystematic studies in the *Stylidium crassifolium* species complex (Stylidiaceae). *Aust. J. Bot.* 27:27–37.

Barlow, G. W. 1981. Patterns of parental investment, dispersal and size among coral-reef fishes. *Environ. Biol. Fish.* 6:65–85.

Barnard, C. J., and J. Fitzsimons. 1988. Kin recognition and mate choice in mice: The effects of kinship, familiarity and social interaction on intersexual interaction. *Anim. Behav.* 36:1078–90.

Barnes, H., and M. Barnes. 1958. Further observations on self-fertilization in *Chthamalus* sp. *Ecology* 39(3):550.

Barnes, H., and D. J. Crisp. 1956. Evidence of self-fertilization in certain species of barnacles. *J. Mar. Biol. Assoc. UK* 35:631–39.

Barrett, S. C. H. 1988. The evolution, maintenance and loss of self-compatibility systems. In *Plant Reproductive Ecology: Patterns and Strategies,* ed. J. Lovett Doust and L. Lovett Doust, 98–124. New York: Oxford University Press.

Barrett, S. C. H., and D. Charlesworth. 1991. Effects of a change in the level of inbreeding on the genetic load. *Nature* 352:522–24.

Barrowclough, G. F. 1978. Sampling bias in dispersal studies based on finite areas. *Bird-Banding* 49:333–41.

———. 1980. Gene flow, effective population sizes and genetic variance components in birds. *Evolution* 34:789–98.

———. 1983. Biochemical studies of microevolutionary processes. In *Perspectives in Ornithology,* ed. A. H. Brush and G. A. Clark, Jr., 223–62. New York: Cambridge University Press.

Barrowclough, G. F., and S. L. Coats. 1985. The demography and population genetics of owls, with special reference to the conservation of the spotted owl (*Strix occidentalis*). In *Ecology and Management of the Spotted Owl in the Pacific Northwest,* ed. R. J. Guttierez and A. B. Carey, 74–85. Portland, Ore.: U.S. Forest Service.

Bartz, S. H. 1979. Evolution of eusociality in termites. *Proc. Natl. Acad. Sci. USA* 76(11):5764–68.

Bashi, J. 1977. Effects of inbreeding on cognitive performance. *Nature* 266:440–42.

Bateman, A. J. 1952. Self-compatibility systems in angiosperms. I. Theory. *Heredity* 6:285–310.

Bateson, P. P. G. 1978. Sexual imprinting and optimal outbreeding. *Nature* 273:259–60.

———. 1980. Optimal outbreeding and the development of sexual preferences in the Japanese quail. *Z. Tierpsychol.* 53:231–44.

———. 1982. Preferences for cousins in Japanese quail. *Nature* 295:236–37.

———. 1983a. *Mate Choice.* Cambridge: Cambridge University Press.

———. 1983b. Optimal outbreeding. In *Mate Choice,* ed. P. P. G. Bateson, 257–77. Cambridge: Cambridge University Press.

Bawa, K. S. 1974. Breeding systems of tree species of a lowland tropical community. *Evolution* 28(1):85–92.

———. 1980. Evolution of dioecy in flowering plants. *Annu. Rev. Ecol. Syst.* 11:15–39.

Bawa, K. S., and P. A. Opler. 1975. Dioecism in tropical forest trees. *Evolution* 29(1):167–79.

Bayne, B. L. 1976. The biology of mussel larvae. In *Marine Mussels: Their Ecology and Physiology,* ed. B. L. Bayne, 81–120. Cambridge: Cambridge University Press.

Bazzaz, F. A., N. R. Chiariello, P. D. Coley, and L. F. Pitelka. 1987. Allocating resources to reproduction and defense. *BioScience* 37:58–67.

Bazzaz, F. A., and E. G. Reekie. 1985. The meaning and measurement of reproductive effort in plants. In *Studies on Plant Demography: A Festschrift for John L. Harper,* ed. J. White. New York: Academic Press.

Beach, J. H., and W. J. Kress. 1980. Sporophyte versus gametophyte: A note on the origin of self-incompatibility in flowering plants. *Syst. Bot.* 5:1–5.

Beacham, T. D., R. E. Withler, and A. P. Gould. 1985. Biochemical genetic stock identification of pink salmon (*Oncorhynchus gorbuscha*) in southern British Columbia and Puget Sound. *Can. J. Fish. Aquat. Sci.* 42:1474–83.

Beardmore, J. A., and S. A. Shami. 1979. Heterozygosity and the optimum phenotype under stabilizing selection. *Aquilo Ser. Zool.* 20:100–110.

Beauchamp, K. A. 1986. Reproductive ecology of the brooding, hermaphroditic clam *Lasaea subviridis. Mar. Biol.* 93:225–35.

Beaumont, A. R., and M. D. Budd. 1983. Effects of self-fertilization and other factors on the early development of the scallop *Pecten maximus. Mar. Biol.* 76:285–89.

Beckwitt, R. 1980. Genetic structure of *Pileolaria pseudomilitaris* (Polychaeta: Spirorbidae). *Genetics* 96:711–21.

————. 1982. Electrophoretic evidence for self-fertilization in two species of spirorbid polychaetes. *Bull. South. Calif. Acad. Sci.* 81(2):61–68.

Beeman, R. D. 1977. Gastropoda: Opisthobranchia. In *Reproduction of Marine Invertebrates,* vol. 4., ed. A. C. Giese and J. S. Pearse, 115–79. New York: Academic Press.

Beer, A. E., J. F. Quebbeman, Y. Hamazaki, and A. E. Semprini. 1987. Immunotherapy of recurrent spontaneous abortion. In *Immunoregulation and Fetal Survival,* ed. T. J. Gill III, T. G. Wegmann, and E. Nisbet-Brown, 286–99. Oxford: Oxford University Press.

Begon, M. 1977. The effective size of a natural *Drosophila subobscura* population. *Heredity* 38:13–18.

Bekoff, M. 1977. Mammalian dispersal and the ontogeny of individual behavioral phenotypes. *Am. Nat.* 111:715–32.

Bell, G. 1982. *The Masterpiece of Nature: The Evolution and Genetics of Sexuality.* Berkeley: University of California Press.

————. 1988. Uniformity and diversity in the evolution of sex. In *The Evolution of Sex: An Examination of Current Ideas,* ed. R. E. Michod and B. R. Levin, 126–38. Sunderland, Mass.: Sinauer Associates.

Bell, G., and V. Koufopanou. 1986. The cost of reproduction. In *Oxford Surveys in Evolutionary Biology,* vol. 3, ed. R. Dawkins, 83–131. Oxford: Oxford University Press.

Bell, S. S. 1982. On the population biology and meiofaunal characteristics of *Manayunkia aestuarina* (Polychaeta: Sabellidae: Fabricinae) from a South Carolina salt marsh. *Estuar. Coast. Shelf Sci.* 14:215–21.

Benado, W., M. Aguilera, O. A. Reig, and F. J. Ayala. 1979. Biochemical genetics of chromosome forms of Venezuelan spiny rats of the *Proechimys guairae* and *Proechimys trinitatis* superspecies. *Genetica* 50:89–97.

Benayahu, Y., and Y. Loya. 1983. Surface brooding in the Red Sea soft coral *Parerythropodium fulvum fulvum* (Forskål, 1775). *Biol. Bull.* 165:353–69.

————. 1984a. Life history studies on the Red Sea soft coral *Xenia macrospiculata* Gohar, 1940. I. Annual dynamics of gonadal development. *Biol. Bull.* 166:32–43.

————. 1984b. Life history studies on the Red Sea soft coral *Xenia macrospiculata* Gohar, 1940. II. Planulae shedding and post larval development. *Biol. Bull.* 166:44–53.

————. 1984c. Substratum preferences and planulae settling of two Red Sea alcyonaceans: *Xenia macrospiculata* Gohar and *Parerythropodium fulvum fulvum* (Forskål). *J. Exp. Mar. Biol. Ecol.* 83:249–61.

————. 1985. Settlement and recruitment of a soft coral: Why is *Xenia macrospiculata* a successful colonizer? *Bull. Mar. Sci.* 36(1):177–88.

Bengtsson, B. O. 1978. Avoiding inbreeding: At what cost? *J. Theor. Biol.* 73:439–44.

————. 1980. Rates of karyotype evolution in placental mammals. *Hereditas* 92:37–47.

Bengtsson, B. O., and W. F. Bodmer. 1976. On the increase of chromosome mutations under random mating. *Theor. Popul. Biol.* 9:260–81.

Bennet, J. H., and F. E. Binet. 1956. Association between Mendelian factors with mixed selfing and random mating. *Heredity* 10:51–55.

Bennett, E. L., and A. C. Sebastian. 1988. Social organization and ecology of proboscis monkeys (*Nasalis larvatus*) in mixed coastal forest in Sarawak. *Int. J. Primatol.* 9:233–55.

Bercovitch, F. B. 1986. Male rank and reproductive activity in savanna baboons. *Int. J. Primatol.* 7:533–50.

Bercovitch, F. B., and T. E. Ziegler. 1989. Reproductive strategies and primate conservation. *Zoo Biol.* Suppl. 1:163–69.

Berger, J. 1987. Reproductive fates of dispersers in a harem-dwelling ungulate, wild horses. In *Mammalian Dispersal Patterns: The Effects of Social Structure on Population Genetics*, ed. B. D. Chepko-Sade and Z. T. Halpin, 41–54. Chicago: University of Chicago Press.

Berger, J., and C. Cunningham. 1987. Influence of familarity on frequency of inbreeding in wild horses. *Evolution* 4:229–31.

Bergquist, P. R. 1978. *Sponges*. London: Hutchinson.

Bergquist, P. R., and M. E. Sinclair. 1968. The morphology and behaviour of larvae of some intertidal sponges. *N.Z. J. Mar. Freshw. Res.* 2:426–37.

Bergquist, P. R., M. E. Sinclair, and J. J. Hogg. 1970. Adaptation to intertidal existence: Reproductive cycles and larval behaviour in Demospongiae. *Symp. Zool. Soc. Lond.* 25:247–71.

Bermingham, E., T. Lamb, and J. C. Avise. 1986. Size polymorphism and heteroplasmy in the mitochondrial DNA of lower vertebrates. *J. Hered.* 77:249–52.

Bernstein, I. S., and C. L. Ehardt. 1986. Modification of aggression through socialization and the special case of adult and adolescent male rhesus monkeys (*Macaca mulatta*). *Am. J. Primatol.* 10:213–27.

Berrill, N. J. 1975. Chordata: Tunicata. In *Reproduction of Marine Invertebrates*, vol. 2, ed. A. C. Giese and J. S. Pearse, 241–82. New York: Academic Press.

Berry, R. J., M. E. Jakson, and J. Peters. 1978. The house mice of the Faroe Islands: A study in microdifferentiation. *J. Zool.* (Lond.) 185:73–92.

Bertin, R. I. 1985. Nonrandom fruit production in *Campsis radicans:* Between-year consistency and effects of prior pollination. *Am. Nat.* 126:750–59.

Bertin, R. I., C. Barnes, and S. I. Guttman. 1989. Self-sterility and cryptic self-fertility in *Campsis radicans* (Bignoniaceae). *Bot. Gaz.* 150:397–403.

Bertin, R. I., and M. Sullivan. 1988. Pollen interference and cryptic self-fertility in *Campsis radicans. Am. J. Bot.* 75:1140–47.

Bertness, M. D., and E. Grosholz. 1985. Population dynamics of the ribbed mussel, *Geukensia demissa:* The costs and benefits of an aggregated distribution. *Oecologia* 67:192–204.

Bertram, B. C. R. 1975. Social factors influencing reproduction in wild lions. *J. Zool.* (Lond.) 177:463–82.

———. 1976. Kin selection in lions and evolution. In *Growing Points in Ethology*, ed. P. P. G. Bateson and R. A. Hinde, 281–301. Cambridge: Cambridge University Press.

Berven, K. A. 1982a. The genetic basis of altitudinal variation in the wood frog *Rana sylvatica.* I. An experimental analysis of life history traits. *Evolution* 36:962–83.

———. 1982b. The genetic basis of altitudinal variation in the wood frog *Rana sylvatica.* II. An experimental analysis of larval development. *Oecologia* 52:360–69.

———. 1987. The heritable basis of variation in larval development patterns within populations of the wood frog (*Rana sylvatica*). *Evolution* 41:1088–97.

Berven, K. A., and T. A. Grudzien. 1990. Dispersal in the wood frog (*Rana sylvatica*): Implications for genetic population structure. *Evolution* 44:2047–56.

Betzig, L. L. 1986. *Despotism and Differential Reproduction: A Darwinian View of History.* New York: Aldine.

Biemont, C., A. Aouar, and C. Arnault. 1987. Genome reshuffling of the copia element in an inbred line of *Drosophila melanogaster. Nature* 329:742–44.

Biemont, C., and M. Bouletreau. 1980. Hybridization and inbreeding effects on genome coadaptation in a haplo-diploid hymenopteran: *Cothonaspis boulardi* (Eucoilidae). *Experientia* 36:45–46.

Bijlsma, R., R. W. Allard, and A. L. Kahler. 1986. Non-random mating in an open pollinated maize population. *Genetics* 112:669–80.

Bijlsma-Meeles, E., and R. Bijlsma. 1988. The alcohol dehydrogenase poly-morphism in *Drosophila melanogaster:* Fitness measurements and predictions under conditions with no alcohol stress. *Genetics* 120:743–53.

Bilton, H. T., and W. E. Ricker. 1965. Supplementary checks on the scales of pink salmon (*Oncorhynchus gorbuscha*) and chum salmon (*O. keta*). *J. Fish. Res. Board Can.* 22:1477–89.

Birdsall, A. D., and D. Nash. 1973. Occurrence of successful multiple insemi-nation of females in natural populations of deer mice (*Peromyscus manicula-tus*). *Evolution* 27:106–10.

Birkeland, C. 1974. Interactions between a sea pen and seven of its predators. *Ecol. Monogr.* 44(2):211–32.

Birkhead, T. R. 1988. Behavioral aspects of sperm competition in birds. In *Ad-vances in the Study of Behavior,* ed. J. S. Rosenblatt, C. Beer, M.-C. Busnel, and P. J. B. Slater, 35–72. San Diego: Academic Press.

Birkhead, T. R., and J. D. Biggins. 1987. Reproduction synchrony and extra-pair copulation in birds. *Ethology* 74:320–34.

Birkhead, T. R., K. Clarkson, and R. Zann. 1988. Extra-pair courtship, copula-tion and mate-guarding in wild zebra finches, *Taeniopygia guttata. Anim. Be-hav.* 36:1853–55.

Birky, C. W., and J. B. Walsh. 1988. Effects of linkage on rates of molecular evolution. *Proc. Natl. Acad. Sci. USA* 85:6414–18.

Bischoff, R. J., J. L. Gould, and D. I. Rubenstein. 1985. Tail size and female choice in the guppy (*Poecilia reticulata*). *Behav. Ecol. Sociobiol.* 17:253–55.

Bittles, A. H. 1979. Incest re-assessed. *Nature* 280:107.

———. 1980. Inbreeding in human population. *J. Sci. Indust. Res.* 39:768–77.

Bittles, A. H., and E. Makov. 1988. Inbreeding in human populations: An as-sessment of the costs. In *Human Mating Patterns,* ed. C. G. N. Mascie-Taylor and A. J. Boyce, 153–68. Cambridge: Cambridge University Press.

Bixler, R. H. 1982. Hypotheses are like people—some fit, some unfit. *Behav. Brain Sci.* 6:104.

———. 1983. The multiple meanings of "incest." *J. Sex Res.* 19:197–201.

Black, R., and M. S. Johnson. 1979. Asexual viviparity and population genetics of *Actinia tenebrosa. Mar. Biol.* 53:27–31.

Blackman, R. L. 1980. Chromosomes and parthenogenesis in aphids. *Symp. Roy. Entomol. Soc. Lond.* 10:133–48.

Blair, A. P. 1943. Population structure in toads. *Am. Nat.* 77:563–68.

Blair, W. F. 1953. Growth, dispersal and age at sexual maturity of the Mexican toad (*Bufo valliceps* Wiegmann). *Copeia* 1953: 208–12.

———. 1960. *The Rusty Lizard: A Population Study.* Austin: University of Texas Press. (Ed. by D. H. C. Fletcher and C. D. Michener), 287–331. Chichester: John Wiley.

Blake, J. A., and J. D. Kudenov. 1981. Larval development, larval nutrition and growth for two *Boccardia* species (Polychaeta: Spionidae) from Victoria, Aus-tralia. *Mar. Ecol. Prog. Ser.* 6:175–82.

Blaustein, A. R., M. Bekoff, and T. J. Daniels. 1987. Kin recognition in vertebrates (excluding primates): Empirical evidence. In *Kin Recognition in Animals*, ed. D. J. C. Fletcher and C. D. Michener, 359–393. New York: Wiley.

Blaustein, A. R., and R. K. O'Hara. 1981. Genetic control for sibling recognition? *Nature* 290:246–48.

———. 1982. Kin recognition cues in *Rana cascadae* tadpoles. *Behav. Neural Biol.* 36:77–87.

———. 1983. Kin recognition in *Rana cascadae* tadpoles: Effects of rearing with nonsiblings and varying the strength of the stimulus cues. *Behav. Neural Biol.* 39:259–67.

Blaustein, A. R., R. K. O'Hara, and D. H. Olson. 1984. Kin preference behaviour is present after metamorphosis in *Rana cascadae* frogs. *Anim. Behav.* 32:445–50.

Blaustein, A. R., and B. Waldman. 1992. Kin recognition in anuran amphibians. *Anim. Behav.* 44:207–21.

Blouin, S. F., and M. Blouin. 1988. Inbreeding avoidance behaviors. *Trends Ecol. Evol.* 3:230–33.

Bobé, P., G. Chaouat, M. Stanislawski, and N. Kiger. 1986. Immunogenetic studies of spontaneous abortion in mice. II. Antiabortive effects are independent of systemic regulatory mechanisms. *Cell. Immunol.* 98:477–85.

Bock, B. C., A. S. Rand, and G. M. Burghardt. 1985. Seasonal migration and nesting site fidelity in the green iguana. In *Migration: Mechanisms and Adaptive Significance*, ed. M. A. Rankin, 435–43. Port Aransas, Tex.: Marine Science Institute, University of Texas at Austin.

Bocquet, C., and R. Lejuez. 1974. Génétique des populations de *Sphaeroma serratum* (F). XI. Étude des populations de la côte nord de Bretagne (de Brest à Saint-Brieuc). *Cah. Biol. Mar.* 15:169–96.

Bocquet, C., R. Lejuez, and G. Teissier. 1960. Génétique des populations de *Sphaeroma serratum* (F). III. Comparaison des populations mères et des populations filles pour les Sphéromes du Cotentin. *Cah. Biol. Mar.* 1:279–94.

Bocquet, C., C. Lévi, and G. Teissier. 1951. Polychromatisme de *Sphaeroma serratum* (F). *Arch. Zool. Exp. Gen.* 87:245–98.

Bodmer, M., and M. Ashburner. 1984. Conservation and change in the DNA sequences coding for alcohol development in sibling species of *Drosophila*. *Nature* 309:425–29.

Bodmer, W. F., and L. L. Cavalli-Sforza. 1976. *Genetics, Evolution and Man*. San Franciso: W. H. Freeman and Co.

Boelkins, R. C., and A. P. Wilson. 1972. Intergroup social dynamics of the Cayo Santiago rhesus (*Macaca mulatta*) with special reference to changes in group membership by males. *Primates* 13:125–40.

Boero, F. 1981. Systematics and ecology of the hydroid population of two *Posidonia oceanica* meadows. *Pubbl. Staz. Zool. Napoli I: Mar. Ecol.* 2(3):181–97.

———. 1984. The ecology of marine hydroids and effects of environmental factors: A review. *Pubbl. Staz. Zool. Napoli I: Mar. Ecol.* 5(2):93–118.

Boero, F., and M. Sarà. 1987. Motile sexual stages and evolution of Leptomedusae (Cnidaria). *Boll. Zool.* 54:131–39.

Bogert, C. M. 1947. A field study of homing in the Carolina toad. *Am. Mus. Novitates* 1355:1–24.

Bondari, K. 1981. Growth comparison of inbred and random-bred catfish at different temperatures. *Proc. Annu. Conf. SE Assoc. Fish Wild. Agencies* 35:547–53.

Bondari, K., and R. A. Dunham. 1987. Effects of inbreeding on economic traits of channel catfish. *Theor. Appl. Genet.* 74:1–9.

Book, J. A. 1957. Genetical investigations in north Swedish population. The offspring of first-cousin marriage. *Annu. Hum. Genet.* 21:191–223.

Boomsma, J. J. 1988. Empirical analysis of sex allocation in ants: From descriptive surveys to population genetics. In *Population Genetics and Evolution*, ed. G. de Jong, 42–51. Berlin: Springer-Verlag.

Boone, J. W. 1988. Reproductive success among third generation Portugese. In *Human Reproductive Behavior: A Darwinian Perspective*, ed. L. Betzig, M. Borgerhoff-Mulder, and P. Turke. Cambridge: Cambridge University Press.

Boonstra, R., C. J. Krebs, M. S. Gaines, M. L. Johnson, and I. T. M. Craine. 1987. Natal philopatry and breeding systems in voles (*Microtus* spp.). *J. Anim. Ecol.* 56:655–73.

Boorman, S. A., and P. R. Levitt. 1980. *The Genetics of Altruism*. New York: Academic Press.

Borgia, G. 1979. Sexual selection and the evolution of mating systems. In *Sexual Selection and Reproductive Competition in Insects*, ed. M. S. Blum and N. A. Blum, 19–80. New York: Academic Press.

———. 1980. Evolution of haplodiploidy: Models for inbred and outbred systems. *Theor. Popul. Biol.* 17:103–28.

Borojevic, R. 1967. La ponte et le développement de *Polymastia robusta* (Démosponges). *Cah. Biol. Mar.* 8:1–6.

———. 1970. Différenciation cellulaire dans l'embryogenèse et la morphogenèse chez les Spongiaires. *Symp. Zool. Soc. Lond.* 25:467–90.

Borowsky, R., and K. D. Kallman. 1976. Patterns of mating in natural populations of *Xiphophorus* (Pisces: Poeciliidae). I. *X. maculatus* from Belize and Mexico. *Evolution* 30:693–706.

Bos, M., H. Harmens, and K. Vrieling. 1986. Gene flow in *Plantago* I. Gene flow and neighborhood size in *P. lanceolata*. *Heredity* 56:43–54.

Bosch, I., R. B. Rivkin, and S. P. Alexander. 1989. Asexual reproduction by oceanic planktotrophic echinoderm larvae. *Nature* 337:169–70.

Boschma, H. 1929. On the postlarval development of the coral *Maeandra areolata* (L.). *Papers from the Tortugas Laboratory of Carnegie Institution of Washington* 26:129–47.

Bottini, E., F. Gloria-Bottini, P. Lucarelli, A. Polzonetti, F. Santoro, and A. Varveri. 1979. Genetic polymorphisms and intrauterine development: Evidence of decreased heterozygosity in light for dates human newborn babies. *Experientia* 35:1565–67.

Bouman, J. C., and H. Bos. 1979. Two symptoms of inbreeding depression in Przewalski horses living in captivity. In *Genetics and Hereditary Diseases of the*

Przewalski Horse, ed. L. E. M. de Boer, J. Bouman, and I. Bouman, 111–17. Rotterdam: Foundation for the Preservation and Protection of the Przewalski Horse.

Bowman, R. N. 1987. Cryptic self-incompatibility and the breeding system of *Clarkia unguiculata* (Onegraceae). *Am. J. Bot.* 74:471–76.

Boyle, T. H., and D. P. Stimart. 1986. Incompatibility relationships in intra- and interspecific crosses of *Zinnia elegans* Jacq., and *Z. angustifolia* HBK (Compositae). In *Biotechnology and Ecology of Pollen*, ed. D. L. Mulcahy, G. B. Mulcahy, and E. Ottaviano, 265–70. New York: Springer.

Boyse, E. A., G. K. Beauchamp, and K. Yamazaki. 1987. The genetics of body scent. *Trends Genet.* 3:97–102.

Brace, R. C., and D. L. J. Quicke. 1985. Further analysis of individual spacing within aggregations of the anemone, *Actinia equina*. *J. Mar. Biol. Assoc. UK* 65:35–53.

———. 1986. Dynamics of colonization by the beadlet anemone, *Actinia equina*. *J. Mar. Biol. Assoc. UK* 66:21–47.

Bradbury, J. W., and S. L. Vehrencamp. 1976. Social organization and foraging in emballonurid bats. I. Field studies. *Behav. Ecol. Sociobiol.* 1:337–82.

Bradbury, J. W., and S. T. Verhencamp. 1977. Social organization and foraging in emballonurid bats. III. Mating systems. *Behav. Ecol. Sociobiol.* 2:1–17.

Bradoo, B. L. 1976. The sexual biology and morphology of the reproductive organs of *Stegodyphus sarasinorum* Karsch (Araneae: Eresidae). *Entomol. Monthly Mag.* 111:239–47.

Brandt, C. A. 1985. The evolution of sexual differences in natal dispersal: A test of Greenwood's hypothesis. *Contrib. Mar. Sci.* (Suppl.) 27:386–96.

———. 1989. Mate choice and reproductive success of pikas. *Anim. Behav.* 37:118–32.

Brannon, E. L. 1967. Genetic control of migrating behavior of newly hatched sockeye salmon fry. *International Pacific Salmon Commission Progress Report*, no. 16, 31 pp.

Brattstrom, B. H. 1962. Homing in the giant toad *Bufo marinus*. *Herpetologica* 18:176–80.

Breden, F. 1987. The effect of post-metamorphic dispersal on the population genetic structure of Fowler's toad, *Bufo woodhousei fowleri*. *Copeia* 1987:386–95.

———. 1988. Natural history and ecology of Fowler's toad, *Bufo woodhousei fowleri* (Amphibia: Bufonidae), in the Indiana Dunes National Lakeshore. *Fieldiana, Zool.* 49:1–16.

Breden, F., and G. Stoner. 1987. Male predation risk determines female preference in the Trinidad guppy. *Nature* 329:831–33.

———. 1988. Sexual selection in guppies defended. *Nature* 334:201–2.

Breden, F., and M. J. Wade. 1981. Inbreeding and evolution by kin selection. *Ethol. Sociobiol.* 2:3–16.

Breden, F., and M. J. Wade. 1991. "Runaway" social evolution: Reinforcing selection for inbreeding and altruism. *J. Theor. Biol.* 153:323–47.

Brett, R. 1986. The ecology and behavior of the naked mole-rat (*Heterocephalus glaber* Ruppell) (Rodentia: Bathyergidae). D.Ph. thesis, University of London.

Brewbaker, J. L. 1957. Pollen cytology and self-incompatibility systems in plants. *J. Hered.* 48:271–77.

Brewer, B. A., R. C. Lacy, M. L. Foster, and G. Alaks. 1990. Inbreeding depression in insular and central populations of *Peromyscus* mice. *J. Hered.* 81:257–66.

Brien, P. 1973. Morphologie et reproduction. In *Traité de Zoologie*, vol. 3, ed. P.-P. Grassé, 133–461. Paris: Masson.

Briggs, C. L., M. Westoby, P. M. Selkirk, and R. J. Oldfield. 1987. Embryology of early abortion due to limited maternal resources in *Pisum sativum* L. *Ann. Bot.* 59:611–19.

Bristow, A. 1978. *The Sex Life of Plants.* New York: Holt, Rinehart and Winston.

Brncic, D. 1954. Heterosis and the integration of the genotype in geographic populations of *Drosophila pseudoobscura*. *Genetics* 39:77–88.

———. 1961. Integration of the genotype in geographic populations of *Drosophila pavani*. *Evolution* 15:92–97.

Brody, A. K., and K. B. Armitage. 1985. The effects of adult removal on dispersal of yearling yellow-bellied marmots. *Can. J. Zool.* 63:2560–64.

Bronson, F. H. 1979. The reproductive ecology of the house mouse. *Q. Rev. Biol.* 54:265–99.

Brooker, M. G., I. Rowley, M. Adams, and P. R. Baverstock. 1990. Promiscuity: An inbreeding avoidance mechanism in a socially manogamous species. *Behav. Ecol. Sociobiol.* 26:191–99.

Brotschol, J. V., J. H. Roberds, and G. Namkoong. 1986. Allozyme variation among North Carolina populations of *Liriodendron tulipifera*. *Silvae Genet.* 35:131–38.

Brown, A. H. D. 1979. Enzyme polymorphism in plant populations. *Theor. Popul. Biol.* 15:1–42.

———. 1990. Genetic characterization of plant mating systems. In *Plant Population Genetics, Breeding, and Genetic Resources*, ed. A. H. D. Brown, M. T. Clegg, A. L. Kahler, and B. S. Weir, 145–62. Sunderland, Mass: Sinauer Associates.

Brown, B. A., and M. T. Clegg. 1984. Influence of flower color polymorphism on genetic transmission in a natural population of the common morning glory, *Ipomoea purpurea*. *Evolution* 38:796–803.

Brown, J. L. 1987. *Helping and Communal Breeding in Birds: Ecology and Evolution.* Princeton, N.J.: Princeton University Press.

Brown, J. L., M. Bush, C. Packer, A. E. Pusey, S. L. Monfort, S. J. O'Brien, D. L. Janssen, and D. E. Wildt. 1991. Developmental changes in pituitary-gonadal function in free-ranging lions (*Panthera leo leo*) of the Serengeti Plains and Ngorongoro Crater. *J. Reprod. Fert.* 91:29–40.

Brown, J. S., M. J. Sanderson, and R. E. Michod. 1982. Evolution of social behavior by reciprocation. *J. Theor. Biol.* 99:319–39.

Brown, K. L. 1985. Demographic and genetic characteristics of dispersal in the mosquitofish, *Gambusia affinis* (Pisces: Poeciliidae). *Copeia* 1985: 597–612.

———. 1987. Colonization by mosquitofish (*Gambushia affinis*) of a Great Plains river basin. *Copeia* 1987: 336–51.

Brown, W. L. 1957. Centrifugal speciation. *Q. Rev. Biol.* 32:247–77.

Brown, W. M. 1980. Polymorphism in mitochondrial DNA of humans as revealed by restriction endonuclease analysis. *Proc. Natl. Acad. Sci. USA* 77:3605–9.

Bruce, A. J. 1976. Shrimps and prawns of coral reefs, with special reference to commensalism. In *Biology and Geology of Coral Reefs*, vol. 3: *Biology 2*, ed. O. A. Jones and R. Endean, 37–94. New York: Academic Press.

Bruere, A. N. 1974. The segregation patterns and fertility of sheep heterozygous and homozygous for three different Robertsonian translocations. *J. Reprod. Fert.* 41:453–64.

Bruere, A. N., I. S. Scott, and L. M. Henderson. 1981. Aneuploid spermatocyte frequency in domestic sheep heterozygous for three Robertsonian translocations. *J. Reprod. Fert.* 63:61–66.

Brunel, D., and F. Rodolphe. 1985. Genetic neighbourhood structure in a population of *Picea abies* L. *Theor. Appl. Genet.* 71:101–10.

Brussard, P. F. 1984. Geographic patterns and environmental gradients: The central-marginal model in *Drosophila* revisited. *Annu. Rev. Ecol. Syst.* 15:25–64.

Bryant, E. H., S. A. McCommas, and L. M. Combs. 1986. The effect of an experimental bottleneck upon quantitative genetic variation in the housefly. *Genetics* 114:1191–1211.

Bryant, E. H., and L. M. Meffert. 1988. Effect of an experimental bottleneck on morphological integration in the housefly. *Evolution* 42:698–707.

———. 1990. Multivariate phenotypic differentiation among bottleneck lines of the housefly. *Evolution* 44:660–68.

Bryant, E. H., H. van Dijk, and W. van Delden. 1981. Genetic variability of the face fly, *Musca autumnalis* De Geer, in relation to a population bottleneck. *Evolution* 35:872–81.

Buchner, P. 1961. Endosymbiosestudien an Ipiden. I. Die gattung *Coccotrypes*. *Z. Morph. Okol. Tiere* 50:1–80.

———. 1965. *Endosymbiosis of Animals with Plant-Like Micro-organisms*. New York: Wiley Interscience.

Bucklin, A., D. Hedgecock, and C. Hand. 1984. Genetic evidence of self-fertilization in the sea anemone *Epiactis prolifera*. *Mar. Biol.* 84:175–82.

Budge, E. A. T. 1902. *A History of Egypt*, vols. 7 and 8. London: Oxford University Press.

Bulger, J., and W. J. Hamilton III. 1988. Inbreeding and reproductive success in a natural Chacma baboon, *Papio cynocephalus ursinus*, population. *Anim. Behav.* 36:574–78.

Bull, C. M. 1987. A population study of the viviparous Australian lizard, *Trachydosaurus rugosus* (Scincidae). *Copeia* 1987: 749–57.

Bull, J. J. 1983. *Evolution of Sex Determining Mechanisms*. Menlo Park, Calif.: Benjamin/Cummings.

Bullimore, B., and R. G. Crump. 1982. Enzyme electrophoresis and taxonomy of two species of *Asterina* (Asteroidea). In *Echinoderms: Proceedings of the International Conference, Tampa Bay*, ed. J. M. Lawrence, 185–88. Rotterdam: A. A. Balkema.

Bulmer, M. G. 1985. *The Mathematical Theory of Quantitative Genetics*. New York: Oxford University Press.

Burke, T. 1989. DNA fingerprinting and other methods for the study of mating success. *Trends Ecol. Evol.* 4:139–44.

Burke, T., N. B. Davies, M. W. Bruford, and B. J. Hatchwell. 1989. Parental care and mating behaviour of polyandrous dunnocks *Prunella modularis* related to paternity by DNA fingerprinting. *Nature* 338:249–51.

Burley, N., C. Minor, and C. Strachan. 1990. Social preference of zebra finches for siblings, cousins and non-kin. *Anim. Behav.* 39:775–84.

Burton, R. S. 1983. Protein polymorphisms and genetic differentiation of marine invertebrate populations. *Mar. Biol. Lett.* 4:193–206.

———. 1986. Evolutionary consequences of restricted gene flow among natural populations of the copepod, *Tigriopus californicus*. *Bull. Mar. Sci.* 39(2): 526–35.

———. 1987. Differentiation and integration of the genome in populations of the marine copepod *Tigriopus californicus*. *Evolution* 41(3): 504–13.

———. 1990a. Hybrid breakdown in developmental time in the copepod *Tigriopus californicus*. *Evolution* 44(7): 1814–22.

———. 1990b. Hybrid breakdown in physiological response: A mechanistic approach. *Evolution* 44(7): 1806–13.

Burton, R. S., and M. W. Feldman. 1981. Population genetics of *Tigriopus californicus*. II. Differentiation among neighboring populations. *Evolution* 35(6): 1192–1205.

———. 1982. Population genetics of coastal and estuarine invertebrates: Does larval behavior influence population structure? In *Estuarine Comparisons*, ed. V. S. Kennedy, 537–51. New York: Academic Press.

Busbice, T. H. 1968. Effects of inbreeding on fertility in *Medicago sativa* L. *Crop Sci.* 8:231–34.

Bush, G. L. 1982. What do we really know about speciation? In *Perspectives on Evolution*, ed. R. Milkman, 119–28. Sunderland, Mass.: Sinauer Associates.

Bush, G. L., S. M. Case, A. C. Wilson, and J. L. Patton. 1977. Rapid speciation and chromosomal evolution in animals. *Proc. Natl. Acad. Sci. USA* 74:3942–46.

Bush, G. L., and D. L. Howard. 1986. Allopatric and non-allopatric speciation: Assumptions and evidence. In *Evolutionary Processes and Theory*, ed. S. Karlin and E. Nevo, 411–38. Orlando: Academic Press.

Bush, R. M., P. E. Smouse, and F. T. Ledig. 1987. The fitness consequences of multiple-locus heterozygosity: The relationship between heterozygosity and growth rate in pitch pine (*Pinus rigida* Mill.). *Evolution* 41:787–98.

Buss, D. M. 1987. Sex differences in human mate selection criteria: An evolutionary perspective. In *Sociobiology and Psychology: Ideas, Issues and Applications*, ed. C. Crawford, C. Krebs, and M. Smith, 335–51. Hillsdale, N.J.: Lawrence Erlbaum Associates.

———. 1989. Sex differences in human mate preferences: Evolutionary hypotheses tested in 37 cultures. *Behav. Brain Sci.* 12:1–49.

Buss, L. W. 1981. Group living, competition and the evolution of cooperation in a sessile invertebrate. *Science* 213:1012–14.

———. 1982. Somatic cell parasitism and the evolution of somatic tissue compatibility. *Proc. Natl. Acad. Sci. USA* 79:5337–41.

Bygott, J. D., B. C. R. Bertram, and J. P. Hanby. 1979. Male lions in large coalitions gain reproductive advantages. *Nature* 282:839–41.

Cagle, F. R. 1944. Home range, homing behavior, and migration in turtles. *U. Mich. Mus. Zool. Misc. Publ.* 61:1–34.

Calahan, C. M., and C. Gliddon. 1985. Genetic neighbourhood sizes in *Primula vulgaris*. *Heredity* 54:65–70.

Caley, M. J. 1987. Dispersal and inbreeding avoidance in muskrats. *Anim. Behav.* 35:1225–33.

Callan, H. G., and H. Spurway. 1951. A study of meiosis in interracial hybrids of the newt, *Triturus cristatus*. *J. Genet.* 50:235–49.

Campbell, C. A. 1978. Genetic divergence between populations of *Thais lamellosa* (Gmelin). In *Marine Organisms: Genetics, Ecology and Evolution*, ed. B. Battaglia and J. A. Beardmore, 157–70. New York: Plenum Press.

Campbell, D. R. 1991. Comparing pollen dispersal and gene flow in a natural plant population. *Evolution* 45:1965–68.

Campbell, D. R., and J. L. Dooley. 1992. The spatial scale of genetic differentiation in a hummingbird-pollinated plant: Comparison with models of isolation by distance. *Am. Nat.* 139:735–48.

Campbell, D. R., and N. M. Waser. 1987. The evolution of plant mating systems: Multilocus simulations of pollen dispersal. *Am. Nat.* 129:593–609.

Campbell, R. B. 1986. The interdependence of mating structure and inbreeding depression. *Theor. Popul. Biol.* 30:232–44.

Cancino, J. M., and R. N. Hughes. 1988. The zooidal polymorphism and astogeny of *Celleporella hyalina* (Bryozoa: Cheilostomata). *J. Zool.* (Lond.) 215:167–81.

Capanna, E., A. Gropp, H. Winking, G. Noack, and M. V. Civitelli. 1976. Robertsonian metacentrics in the mouse. *Chromosoma* 58:341–53.

Carlton, J. T. 1982. The historical biogeography of *Littorina littorea* on the Atlantic coast of North America, and implications for the interpretation of the structure of New England intertidal communities. *Malacol. Rev.* 15(1–2):146.

Carr, A. 1967. *So Excellent a Fishe: A Natural History of Sea Turtles*. Garden City: Natural History Press.

———. 1975. The Ascension Island green turtle colony. *Copeia* 1975: 547–55.

Carr, A., and M. H. Carr. 1972. Site fixity in the Carribean green turtle. *Ecology* 53:425–29.

Carré, D., and C. Carré. 1990. Complex reproductive cycle in *Eucheilota paradoxica* (Hydrozoa: Leptomedusae): Medusae, polyps and frustules produced from medusa stage. *Mar. Biol.* 104:303–10.

Carson, H. L. 1958. Response to selection under different conditions of recombination in *Drosophila*. *Cold Spring Harb. Symp. Quant. Biol.* 23:291–305.

———. 1959. Genetic conditions which promote or retard the formation of species. *Cold Spring Harb. Symp. Quant. Biol.* 24:87–105.

———. 1968. The population flush and its genetic consequences. In *Population Biology and Evolution*, ed. R. C. Lewontin, 123–37. New York: Syracuse University Press.

———. 1971. Speciation and the founder principle. *Stadler Genet. Symp.* 3:51–70.

———. 1975. The genetics of speciation at the diploid level. *Am. Nat.* 109:83–92.

———. 1982. Speciation as a major reorganization of polygenic balances. In *Mechanisms of Speciation*, ed. C. Barigozzi, 411–33. New York: Alan R. Liss.

———. 1986. Sexual selection and speciation. In *Evolutionary Processes and Theory*, ed. S. Karlin and E. Nevo, 391–409. Orlando, Fla.: Academic Press.

Carson, H. L., W. E. Johnson, P. S. Nair, and F. M. Sene. 1975. Allozymic and chromosomal similarity in two *Drosophila* species. *Proc. Natl. Acad. Sci. USA* 72:4521–25.

Carson, H. L., and A. R. Templeton. 1984. Genetic revolutions in relation to speciation phenomena: The founding of new populations. *Ann. Rev. Ecol. Syst.* 15:97–131.

Carter, P. A., and W. B. Watt. 1988. Adaptation at specific loci. V. Metabolically adjacent enzyme loci may have very distinct experiences of selective pressures. *Genetics* 119:913–24.

Carvalho, G. R. 1989. Microgeographic genetic differentiation and dispersal capacity in the intertidal isopod, *Jaera albifrons* Leach. In *Reproduction, Genetics and Distributions of Marine Organisms*, ed. J. S. Ryland and P. A. Tyler, 265–71. Fredensborg, Denmark: Olsen & Olsen.

Case, S. M., P. G. Haneline, and M. F. Smith. 1975. Protein variation in several species of *Hyla. Syst. Zool.* 24:281–95.

Casebolt, D. B., R. V. Henrickson, and D. W. Hird. 1985. Factors associated with birth rate and live birth rate in multi-male breeding groups of rhesus monkeys. *Am. J. Primatol.* 8:289–97.

Casper, B. B. 1988. Evidence for selective embryo abortion in *Cryptantha flava. Am. Nat.* 132:318–26.

Caswell, H. 1985. The evolutionary demography of clonal reproduction. In *Population Biology and Evolution of Clonal Organisms*, ed. J. B. C. Jackson, L. W. Buss, and R. E. Cook, 187–224. New Haven: Yale University Press.

Cavallin, H. 1973. Incest. *Sexual Behavior*, Feb., 19–22.

Cavener, D. R., and M. T. Clegg. 1981. Multigenic response to ethanol in *Drosophila melanogaster. Evolution* 35:1–10.

Cesaroni, D., G. Allegrucci, M. Caccone, M. Cobilli Sbordoni, E. De Matthaeis, M. Di Rao, and B. Sbordoni. 1981. Genetic variability and divergence between populations and species of *Nesticus* cave spiders. *Genetica* 56:81–92.

Chaffee, C., and D. R. Lindberg. 1986. Larval biology of early Cambrian molluscs: The implications of small body size. *Bull. Mar. Sci.* 39(2): 536–49.

Chakraborty, R. 1980. Gene-diversity analysis in nested subdivided populations. *Genetics* 96:721–26.

————. 1981. The distribution of the number of heterozygous loci in an individual in natural populations. *Genetics* 98:461–66.

Chakraborty, R., and M. Nei. 1977. Bottleneck effects on average heterozygosity and genetic distance with the stepwise mutation model. *Evolution* 31(2):347–56.

Chaouat, G., J.-P. Kolb, S. Chaffaux, M. Riviere, D. Lankar, I. Athanassakis, D. Green, and T. G. Wegmann. 1987. The placenta and the survival of the fetal allograft. In *Immunoregulation and Fetal Survival*, ed. T. J. Gill III and T. G. Wegmann, 239–51. Oxford: Oxford University Press.

Chaouat, G., J.-P. Kolb, N. Kiger, M. Stanislawski, and T. G. Wegmann. 1985. Immunologic consequences of vaccination against abortion in mice. *J. Immunol.* 134:1594–98.

Chapais, B. 1983. Reproductive activity in relation to male dominance and the likelihood of ovulation in rhesus monkeys. *Behav. Ecol. Sociobiol.* 12:215–28.

Chapais, B., and Mignault, C. 1991. Homosexual incest avoidance among females in captive Japanese macaques. *Am. J. Primatol.* 23:171–83.

Chapman, R. W. 1989. Spatial and temporal variation of mitochondrial DNA haplotype frequencies in the striped bass (*Morone saxatilis*) 1982-year class. *Copeia* 1989: 344–48.

————. 1990. Mitochondrial DNA analysis of striped bass populations in Chesapeake Bay. *Copeia* 1990: 355–66.

Charlesworth, B. 1980a. The cost of sex in relation to mating system. *J. Theor. Biol.* 84:655–72.

————. 1980b. *Evolution in Age-structured Populations*. Cambridge: Cambridge University Press.

————. 1988. The evolution of mate choice in a fluctuating environment. *J. Theor. Biol.* 130:191–204.

Charlesworth, B., and D. Charlesworth. 1976. An experiment on recombinational load in *Drosophila melanogaster*. *Genet. Res.* 25:267–74.

————. 1987. The effect of investment in attractive structures on allocation to male and female functions in plants. *Evolution* 41:948–68.

Charlesworth, D. 1982. On the nature of the self-incompatibility locus in homomorphic and heteromorphic systems. *Am. Nat.* 119:732–35.

————. 1985. Distribution of dioecy and self-incompatibility in angiosperms. In *Evolution: Essays in Honor of John Maynard Smith*, ed. P. J. Greenwood, P. H. Harvey, and M. Slatkin, 237–68. London: Cambridge University Press.

Charlesworth, D., and B. Charlesworth. 1979. The evolution and breakdown of S-allele systems. *Heredity* 43:41–55.

————. 1987. Inbreeding depression and its evolutionary consequences. *Annu. Rev. Ecol. Syst.* 18:237–68.

Charlesworth, D., M. T. Morgan, and B. Charlesworth. 1990. Inbreeding depression, genetic load and the evolution of outcrossing rates in a multilocus system with no linkage. *Evolution* 44:1469–89.

Charnov, E. L. 1979. Simultaneous hermaphroditism and sexual selection. *Proc. Natl. Acad. Sci. USA* 76:2480–84.

————. 1982. *The Theory of Sex Allocation*. Princeton, N.J.: Princeton University Press.

————. 1987. On sex allocation and selfing in higher plants. *Evol. Ecol.* 1:30–36.

Chelazzi, G., and F. Francisci. 1979. Movement patterns and homing behavior of *Testudo hermanni* Gmelin (Reptilia, Testudinidae). *Monitore Zool. Ital.* 13:105–27.

Cheney, D. L. 1981. Intergroup encounters among free-ranging vervet monkeys. *Folia Primatol.* 35:124–46.

Cheney, D. L., and R. M. Seyfarth. 1983. Nonrandom dispersal in free-ranging vervet monkeys: Social and genetic consequences. *Am. Nat.* 122:392–412.

Cheney, D. L., R. M. Seyfarth, S. J. Andelman, and P. C. Lee. 1988. Reproductive success in vervet monkeys. In *Reproductive Success*, ed. T. H. Clutton-Brock, 384–402. Chicago: University of Chicago Press.

Cheney, D. L., R. M. Seyfarth, and B. Smuts. 1986. Social relationships and social cognition in non-human primates. *Science* 234:1361–66.

Chepko-Sade, B. D., and Z. T. Halpin, eds. 1987. *Mammalian Dispersal Patterns: The Effects of Social Structure on Population Genetics*. Chicago: University of Chicago Press.

Chepko-Sade, B. D., W. M. Shields, J. Berger, Z. T. Halpin, W. T. Jones, L. L. Rogers, J. P. Rood, and A. T. Smith. 1987. The effects of dispersal and social structure on effective population size. In *Mammalian Dispersal Patterns*, ed. B. D. Chepko-Sade and Z. T. Halpin, 287–321. Chicago: University of Chicago Press.

Chesser, R. K., and R. J. Baker. 1986. On factors affecting the fixation of chromosomal rearrangements and neutral genes: Computer simulations. *Evolution* 40:625–32.

Chia, F.-S. 1976. Sea anemone reproduction: Patterns and adaptive radiations. In *Coelenterate Ecology and Behavior*, ed. G. O. Mackie, 261–70. New York: Plenum Press.

Chia, F.-S., and B. J. Crawford. 1973. Some observations on gametogenesis, larval development and substratum selection of the sea pen *Ptilosarcus guerneyi*. *Mar. Biol.* 23:73–82.

Child, A. R. 1984. Biochemical polymorphism in Charr (*Salvelinus alpinus* L.) from three Cumbrian lakes. *Heredity* 53:249–57.

Chourrout, D. 1988. Induction of gynogenesis, triploidy and tetraploidy in fish. In *ISI Atlas of Science: Animal and Plant Sciences*, 65–70.

Christein, D. R., S. I. Guttman, and D. H. Taylor. 1979. Heterozygote deficiencies in a breeding population of *Bufo americanus* (Bufonidae: Anura): The test of a hypothesis. *Copeia* 1979: 498–502.

Christenson, T. E. 1984. Behaviour of colonial and solitary spiders of the theridid species *Anelosimus eximius*. *Anim. Behav.* 32:725–34.

Christian, J. J. 1961. Phenomenon associated with population density. *Proc. Natl. Acad. Sci. USA* 47:428–49.

Christiansen, F. B., O. Frydenberg, A. O. Glydenholm, and V. Simonsen. 1974. Genetics of *Zoarces* populations. VI. Further evidence, based on age group samples, of a heterozygote deficit in the ESTIII polymorphism. *Hereditas* 77:225–36.

Christy, J. H. 1987. Competitive mating, mate choice and mating associations of brachyuran crabs. *Bull. Mar. Sci.* 41(2):177–91.

Chuang, S.-H. 1977. Larval development in *Discinisca* (inarticulate brachiopod). *Am. Zool.* 17:39–53.

Cieza de Leon, P. de. 1959. *The Incas.* Norman: University of Oklahoma Press.

Clarke, A. G. 1971. The effects of maternal pre-immunization on pregnancy in the mouse. *J. Reprod. Fert.* 24:369–75.

Clarke, B. C. 1979. The evolution of genetic diversity. *Proc. Roy. Soc.* (Lond.), ser. B, 205:453–74.

Clegg, M. T., and R. W. Allard. 1973. Viability versus fecundity selection in the slender wild oat, *Avena barbata* L. *Science* 181:667–68.

Clegg, M. T., J. F. Kidwell, and C. R. Horch. 1980. Dynamics of correlated genetic systems. V. Rates of decay of linkage disequilibria in experimental populations of *Drosophila melanogaster*. *Genetics* 94:217–34.

Cloney, R. A. 1987. Phylum Urochordata, Class Ascidiacea. In *Reproduction and Development of Marine Invertebrates of the Northern Pacific Coast*, ed. M. F. Strathmann, 607–39. Seattle: University of Washington Press.

Clutton-Brock, T. H. 1989. Female transfer and inbreeding avoidance in social mammals. *Nature* 337:70–72.

———, ed. 1988. *Reproductive Success.* Chicago: University of Chicago Press.

Clutton-Brock, T. H., and P. H. Harvey. 1977. Primate ecology and social organization. *J. Zool.* (Lond.). 183:1–39.

Coates, A. G., and J. B. C. Jackson. 1987. Clonal growth, algal symbiosis and reef formation by corals. *Paleobiology* 13:363–78.

Cockburn, A., M. P. Scott and D. J. Scotts. 1985. Inbreeding avoidance and male-biased natal dispersal in *Antechinus* spp. (Marsupialia: Dasyuridae). *Anim. Behav.* 33:908–15.

Cockerham, C. C. 1956. Effects of linkage on the covariances between relatives. *Genetics* 41:138–41.

Cockerham, C. C., and B. S. Weir. 1968. Sib mating with two linked loci. *Genetics* 60:629–40.

———. 1977a. Digenic descent measures for finite populations. *Genet. Res.* (Camb.) 30:121–47.

———. 1977b. Quadratic analyses of reciprocal crosses. *Biometrics* 33:187–203.

Cohen, J. 1977. *Statistical Power Analysis for the Behavioral Sciences.* New York: Academic Press.

Cohen, S. 1990. Outcrossing in field populations of two species of self-fertile ascidians. *J. Exp. Mar. Biol. Ecol.* 140:147–58.

Cohen, T., N. Bloch, Y. Flum, M. Kadar, and E. Goldschmidt. 1963. School attainments in an immigrant village. In *Genetics of Migrant and Isolate Populations*, ed. E. Goldschmidt. New York: Williams and Wilkins.

Cohn, V. H., M. A. Thompson, and G. P. Moore. 1984. Nucleotide sequence comparison of the Adh gene in three drosophilids. *J. Mol. Biol.* 20:31–37.

Cole, S., F. R. Hainsworth, A. C. Kamil, and L. L. Wolf. 1982. Spatial learning as an adaptation in hummingbirds. *Science* 217:655–57.

Coles, J. F., and D. P. Fowler. 1976. Inbreeding in neighboring trees in two white spruce populations. *Silvae Genet.* 25:29–34.

Colvin, J. D. 1986. Proximate causes of male emigration at puberty in rhesus macaques. In *The Cayo Santiago Macaques*, ed. R. G. Rawlins and M. J. Kessler, 131–58. Albany: State University of New York Press.

Connor, J. L., and M. J. Bellucci. 1979. Natural selection resisting inbreeding depression in captive wild housemice (*Mus musculus*). *Evolution* 33:929–40.

Cook, P. L. 1985. Bryozoa from Ghana—a preliminary survey. *Koninklijk Museum voor Midden—Afrika (Tevuren, België), Zoologische Wetenschappen—Annals* 238:1–315.

Cooper, D. C., and R. A. Brink. 1940. Partial self-incompatibility and the collapse of fertile ovules as factors affecting seed formation in alfalfa. *J. Agric. Res.* 60:453–72.

Cooper, E. L. 1961. Growth of wild and hatchery strains of brook trout. *Trans. Am. Fish. Soc.* 90:424–38.

Coppens d'Eeckenbrugge, G., M. Ngendahayo, and B. P. Louant. 1986. Intra- and interspecific incompatibility in *Brachiaria ruziziensis* Germain et Evard (Panicoideae). In *Biotechnology and Ecology of Pollen*, ed. D. L. Mulcahy, G. B. Mulcahy, and E. Ottaviano, 257–64. New York: Springer.

Cords, M. 1987. Forest guenons and patas monkeys: Male-male competition in one-male groups. In *Primate Societies*, ed. B. B. Smuts, D. L. Cheney, R. M. Seyfarth, T. T. Struhsaker, and R. W. Wrangham, 98–111. Chicago: University of Chicago Press.

Cornell, T. J., K. A. Berven, and G. J. Gamboa. 1989. Kin recognition by tadpoles and froglets of the wood frog *Rana sylvatica*. *Oecologia* 78:312–16.

Cothran, E. G., R. Chesser, M. H. Smith, and P. E. Johns. 1983. Influences of genetic variability and maternal factors on fetal growth in white-tailed deer. *Evolution* 37:282–91.

Coyne, J. A. 1983. Genetic basis of differences in genital morphology among three sibling species of *Drosophila*. *Evolution* 37:1101–18.

———. 1984. Genetic basis of male sterility in hybrids between two closely related species of *Drosophila*. *Proc. Natl. Acad. Sci. USA* 81:4444–47.

———. 1985. Genetic studies of three sibling species of *Drosophila* with relationship to theories of speciation. *Genet. Res.* 46:169–92.

Coyne, J. A., and B. Charlesworth. 1989. Genetic analysis of x-linked sterility in hybrids between three sibling species of *Drosophilia*. *Heredity* 62:97–106.

Craddock, E. M., and W. E. Johnson. 1979. Genetic variation in Hawaiian *Drosophila*. V. Chromosomal and allozymic diversity in *Drosophila silvestris* and its homosequential species. *Evolution* 33:137–55.

Craig, J. L., and I. G. Jamieson. 1988. Incestuous mating in a communal bird. *Am. Nat.* 131:58–70.

Crawford, T. J. 1984a. The estimation of neighbourhood parameters for plant populations. *Heredity* 52(2): 273–83.

————. 1984b. What is a population? In *Evolutionary Ecology*, ed. B. Shorrocks, 135–73. Oxford: Blackwell Scientific Publications.

Crespi, B. J. 1988. Adaptation, compromise and constraint: The development, morphometrics, and behavioral basis of a fighter-flier polymorphism in male *Hoplothrips karnyii* (Insecta: Thysanoptera). *Behav. Ecol. Sociobiol.* 23:93–104.

Crisp, D. J. 1978. Genetic consequences of different reproductive strategies in marine invertebrates. In *Marine Organisms: Genetics, Ecology and Evolution*, ed. B. Battaglia and J. A. Beardmore, 257–73. New York: Plenum Press.

Crockett, C. M., and J. F. Eisenberg. 1987. Howlers: Variations in group size and demography. In *Primate Societies*, ed. B. B. Smuts, D. L. Cheney, R. M. Seyfarth, T. T. Struhsaker, and R. W. Wrangham, 54–68. Chicago: University of Chicago Press.

Crosby, J. L. 1970. The evolution of genetic discontinuity: Computer models of the selection of barriers to interbreeding between species. *Heredity* 25:253–97.

Crossman, E. J. 1990. Reproductive homing in meskellunge, *Esox masquinongy*. *Can. J. Fish. Aquat. Sci.* 47:1803–12.

Crothers, J. H. 1980. Further observations on the growth of the common dog-whelk, *Nucella lapillus* (L.), in the laboratory. *J. Molluscan Stud.* 46:181–85.

————. 1981. On variation in *Nucella lapillus* (L.): Shell shape in populations from the Solway Firth. *J. Molluscan Stud.* 47:11–16.

Crow, J. F. 1948. Alternative hypotheses of hybrid vigor. *Genetics* 33:447–87.

————. 1970. Genetic loads and the cost of natural selection. In *Mathematical Topics in Population Genetics*, ed. K.-I. Kojima, 128–77. Berlin: Springer-Verlag.

————. 1988. *The Importance of Recombination*. In *The Evolution of Sex: An Examination of Current Ideas*, ed. R. E. Michod and B. R. Levin, 56–73. Sunderland, Mass.: Sinauer Associates.

Crow, J. F., and M. Kimura. 1970. *An Introduction to Population Genetics Theory*. New York: Harper and Row.

————. 1979. Efficiency of truncation selection. *Proc. Natl. Acad. Sci. USA* 76:396–99.

Crow J. F., and N. E. Morton. 1955. Measurement of gene frequency drift in small populations. *Evolution* 9:202–14.

Crow, J. F., and M. J. Simmons. 1983. The mutation load in *Drosophila*. In *The Genetics and Biology of Drosophila*, vol. 3c, ed. M. Ashburner, H. L. Carson, and J. N. Thompson, Jr., 1–35. New York: Academic Press.

Crowcroft, P. 1966. *Mice All Over*. Chester Springs, Pa.: Dufour Editions, Inc.

Crowe, L. K. 1971. The polygenic control of outbreeding in *Borago officinalis*. *Heredity* 27:111–18.

Crozier, R. H. 1971. Heterozygosity and sex determination in haplo-diploidy. *Am. Nat.* 105:399–412.

Crozier, R. H. 1976. Why male-haploid and sex-linked genetic systems seem to have unusually sex-limited genetic loads. *Evolution* 30:623–24.

———. 1977. Evolutionary genetics of the Hymenoptera. *Annu. Rev. Entomol.* 22:263–88.

———. 1985. 1. 3. 4. Adaptive consequences of male-haploidy. In *Spider Mites: Their Biology, Natural Enemies and Control,* ed. W. Helle and M. W. Sabelis, 201–22. Amsterdam: Elsivier.

Cruden, R. W. 1988. Temporal dioecism: Systematic breadth, associated traits, and temporal patterns. *Bot. Gaz.* 149:1–15.

Crump, M. L. 1986. Homing and site fidelity in a neotropical frog, *Atelopus varius* (Bufonidae). *Copeia* 1986: 438–44.

Crumpacker, D. W. 1967. Genetic loads in maize (*Zea mays* L.) and other cross-fertilized plants and animals. *Evol. Biol.* 1:306–423.

Cruzan, M. B. 1989a. Pollen tube attrition in *Erythronium grandiflorum. Am. J. Bot.* 76:562–70.

———. 1989b. Post-pollination selection in *Erythronium grandiflorum.* Ph.D. dissertation, State University of New York.

———. 1990. Pollen donor interactions during pollen tube growth in *Erythronium grandiflorum. Am. J. Bot.* 77:116–22.

Curry, R. L., and P. R. Grant. 1990. Galapagos mocking-birds: Territorial cooperative breeding in a climatically variable environment. In *Cooperative Breeding in Birds,* ed. P. B. Stacey and W. D. Koening, 289–331. Cambridge: Cambridge University Press.

da Cunha, A. B., and T. Dobzhansky. 1954. A further study of chromosomal polymorphism in *Drosophila willistoni* in relation to environment. *Evolution* 8:119–34.

da Cunha, A. B., H. Burla, and T. Dobzhansky. 1950. Adaptive chromosomal polymorphism in *Drosophila willistoni. Evolution* 4:212–35.

Daly, J. C., and J. L. Patton. 1986. Growth, reproduction, and sexual dimorphism in *Thomomys bottae* pocket gophers. *J. Mammal.* 67:256–65.

Daly, M., and M. Wilson. 1983. *Sex, Evolution, and Behavior,* 2d ed. Boston: Willard Grant Press.

———. 1988. *Homicide.* New York: Aldine.

———. 1990. Is parent-offspring conflict sex-linked? Freudian and Darwinian models. *J. Pers.* 58:163–90.

Daniell, A., and N. D. Murray. 1986. Effects of inbreeding in the budgerigar *Melopsittacus undulatus* (Aves: Psittacidae). *Zoo Biol.* 5:233–38.

Danzmann, R. G., M. M. Ferguson, and F. W. Allendorf. 1987. Heterozygosity and oxygen-consumption rates as predictors of growth and developmental rate in rainbow trout. *Physiol. Zool.* 60:211–20.

———. 1988. Heterozygosity and components of fitness in a strain of rainbow trout. *Biol. J. Linn. Soc.* 39:285–304.

Darchen, R. 1965. Ethologie d'une araignee sociale, *Agelena consociata* Denis. *Biologica Gabonica* 1:117–46.

———. 1968. Ethologie d' *Achaearanea disparata* Denis (Araneae: Theridiidae), araignee sociale du Gabon. *Biologica Gabonica* 1:5–25.

———. 1978. Les essaimages de l'araignee sociale, *Agelena consociata* Denis, dans la foret gabonaise. *C. R. Hebd. Seances Acad. Sci.*, ser. D., *Sci. Nat.* 287:1035–37.

Darlington, C. D. 1960. Cousin marriage and the evolution of the breeding system in man. *Heredity* 14:297–332.

Darlington, C. D., and K. Mather. 1949. *The Elements of Genetics.* London: Allen and Unwin.

Darwin, C. 1859. *On the Origin of Species.* London: John Murray.

————. 1868. *The Variation of Animals and Plants under Domestication.* London: John Murray.

————. 1876. *The Effects of Crossing and Self-fertilization in the Vegetable Kingdom.* London: John Murray.

————. 1877. *The Different Forms of Flowers on Plants of the Same Species.* London: John Murray.

Darwin, G. 1875a. Marriages between first cousins in England and their effects. *J. Stat. Soc.* 38:153–84.

————. 1875b. Note on the marriages of first cousins. *J. Stat. Soc.* 38:344–48.

Dauer, D. M., C. A. Maybury, and R. M. Ewing. 1981. Feeding behavior and general ecology of several spionid polychaetes from the Chesapeake Bay. *J. Exp. Mar. Biol. Ecol.* 54:21–38.

Davies, M. S., and R. W. Snaydon. 1976. Rapid population differentiation. III. Measures of selection pressures. *Heredity* 36:59–66.

Davis, A. R., and A. J. Butler. 1989. Direct observations of larval dispersal in the colonial ascidian *Podoclavella moluccensis* Sluiter: Evidence for closed populations. *J. Exp. Mar. Biol. Ecol.* 127(2): 189–203.

Dawson, R. S. 1972. *Imperial China.* London: Hutchinson.

Day, A. J., and B. L. Bayne. 1988. Allozyme variation in populations of the dogwhelk *Nucella lapillus* (Prosobranchia: Muricacea) from the south west peninsula of England. *Mar. Biol.* 99:93–100.

de Boer, L. E. M. 1982. Karyological problems in breeding owl monkeys, *Aotus trivirgatus. International Zoo Yearbook* 22:119–24.

DeFries, J. C., and G. E. McClearn. 1972. Behavioral genetics and the fine structure of mouse populations: A study in microevolution. In *Evolutionary Biology*, vol. 5, ed. T. Dobzhansky, M. K. Hecht, and W. C. Steere, 279–91. New York: Appleton-Century-Crofts.

Dempster, E. R. 1955. Maintenance of genetic heterogeneity. *Cold Spring Harb. Symp. Quant. Biol.* 20:25–32.

Denniston, C. 1974. An extension of the probability approach to genetic relationships: One locus. *Theor. Popul. Biol.* 6:58–75.

Denny, M. W. 1988. *Biology and the Mechanics of the Wave-Swept Environment.* Princeton, N.J.: Princeton University Press.

Denny, M. W., and M. F. Shibata. 1989. Consequences of surf-zone turbulence for settlement and external fertilization. *Am. Nat.* 134:859–89.

Densmore, L. D., J. W. Wright, and W. M. Brown. 1985. Length variation and heteroplasmy are frequent in mitochondrial DNA from parthenogenetic and bisexual lizards (genus *Cnemidophorus*). *Genetics* 110:689–707.

DeRosa, C. T., and D. H. Taylor. 1980. Homeward orientation mechanisms in three species of turtles (*Trionyx spinifer, Chrysemys picta,* and *Terrapene carolina*). *Behav. Ecol. Sociobiol.* 7:15–23.

Dethier, M. N. 1980. Tidepools as refuges: Predation and the limits of the harpacticoid copepod *Tigriopus californicus* (Baker). *J. Exp. Mar. Biol. Ecol.* 42:99–111.

Devi, R. R., N. A. Rao, and A. H. Bittles. 1981. Consanguinity, fecundity and post-natal mortality in Karnatka, South India. *Ann. Hum. Biol.* 8:469–72.

Devlin, B., K. Roeder, and N. C. Ellstrand. 1988. Fractional paternity assignment: Theoretical development and comparison to other methods. *Theor. Appl. Genet.* 76:369–80.

Dewey, S. E., and J. S. Heywood. 1988. Spatial genetic structure in a population of *Psychotria nervosa.* I. Distribution of genotypes. *Evolution* 42:834–38.

de Wreede, R. E., and T. Klinger. 1988. Reproductive strategies in Algae. In *Plant Reproductive Ecology: Patterns and Strategies,* ed. J. Lovett Doust and L. Lovett Doust, 267–84. New York: Oxford University Press.

Dewsbury, D. A. 1982. Avoidance of incestuous breeding between siblings in two species of *Peromyscus* mice. *Biol. Behav.* 7:157–69.

Dickinson, H. G., and D. Lewis. 1973. Cytochemical and ultrastructural differences between intraspecific compatible and incompatible pollinations in *Raphanus. Proc. Roy. Soc.* (Lond.), ser. B, 183:21–38.

Diehl, W. J., P. M. Gaffney, and R. K. Koehn. 1986. Physiological and genetic aspects of growth in the mussel *Mytilus edulis.* I. Oxygen consumption, growth, and weight loss. *Physiol. Zool.* 59:201–11.

Diehl, W. J., P. M. Gaffney, J. H. McDonald, and R. K. Koehn. 1985. Relationship between weight standardized oxygen consumption and multiple-locus heterozygosity in the marine mussel *Mytilus edulis* L. (Mollusca). In *Proceedings of the 19th European Marine Biology Symposium,* ed. P. Gibbs, 531–36. Cambridge: Cambridge University Press.

Dimont, M. I. 1962. *Jews, God and History.* Bergenfield, N.J.: Signet Books, New American Library.

Distant, W. L. 1898. Zoological rambles in the Transvaal. *Zoologist* 2:249–60.

Dittus, W. P. J. 1988. Group fission among wild toque macaques as a consequence of female resource competition and environmental stress. *Anim. Behav.* 36:1626–45.

Dobson, F. S. 1982. Competition for mates and predominant juvenile male dispersal in mammals. *Anim. Behav.* 30:1183–92.

Dobson, F. S., and W. T. Jones. 1985. Multiple causes of dispersal. *Am. Nat.* 126:855–58.

Dobzhansky, T. 1936. Studies on hybrid sterility. II. Localization of sterility factors in *Drosophila pseudoobscura* hybrids. *Genetics* 21:113–35.

———. 1941. *Genetics and the Origin of Species,* 2d ed. New York: Columbia University Press.

———. 1951. *Genetics and the Origin of Species.* New York: Columbia University Press.

———. 1957. Genetics of natural populations. XXVI. Chromosomal variability in island and continental populations of *Drosophila willistoni* from Central America and the West Indies. *Evolution* 11:280–93.

———. 1970. *Genetics of the Evolutionary Process.* New York: Columbia University Press.

———. 1974. Genetic analysis of hybrid sterility within the species *Drosophila pseudoobscura. Hereditas* 77:81–88.

Doherty, P. J. 1979. A demographic study of a subtidal population of the New Zealand articulate brachiopod *Terebratella inconspicua. Mar. Bio.* 52:331–42.

Dole, J. W. 1971. Dispersal of recently metamorphosed leopard frogs, *Rana pipiens. Copeia* 1971: 221–28.

———. 1972. Homing and orientation of displaced toads, *Bufo americanus*, to their home sites. *Copeia* 1972: 151–58.

Don, J., and R. R. Avtalion. 1988. Production of F_1 and F_2 diploid gynogentic tilapias and analysis of the "Hertwig curve" obtained using ultraviolet irradiated sperm. *Theor. Appl. Genet.* 76:253–59.

Dressler, R. L. 1981. *The Orchids: Natural History and Classification.* Cambridge, Mass.: Harvard University Press.

Drickamer, L. C., and S. H. Vessey. 1973. Group changing in free-ranging male rhesus monkeys. *Primates* 14:359–68.

Dube, R., and M. Hebert. 1988. Sexual abuse of children under 12 years of age: A review of 511 cases. *Child Abuse Negl.* 12:321–30.

DuBois, C. 1944. *The People of Alor.* Minneapolis: University of Minnesota Press.

Duerden, J. E. 1902. Aggregated colonies in madreporarian corals. *Am. Nat.* 36:461–71.

———. 1904. The coral *Siderastrea radians* and its postlarval development. *Publ. Carnegie Inst. Wash.* 20:1–129.

Dumas, C., and R. B. Knox. 1983. Callose and determination of pistil viability and incompatibility. *Theor. Appl. Genet.* 67:1–10.

Dunbar, R. I. M. 1984. *Reproductive Decisions.* Princeton, N.J.: Princeton University Press.

Dunlap, D. G., and C. K. Scatterfield. 1982. Habitat selection in larval anurans: Early experience and substrate pattern selection in *Rana pipiens. Devel. Psychobiol.* 18:37–58.

Dunn, D. F. 1975a. Gynodioecy in an animal. *Nature* 253:528–29.

———. 1975b. Reproduction of the externally brooding sea anemone *Epictis Epiactis prolifera* Verrill, 1869. *Biol. Bull.* 148:199–218.

———. 1977. Dynamics of external brooding in the sea anemone *Epiactis prolifera. Mar. Biol.* 39:41–49.

———. 1982. Cnidaria. In *Synopsis and Classification of Living Organisms*, vol. 1, ed. S. P. Parker, 669–705. New York: McGraw-Hill.

Dunn, D. F., F.-S. Chia, and R. Levine. 1980. Nomenclature of *Aulactinia* (= *Bunodactis*), with description of *Aulactinia incubans* n.sp. (Coelenterata: Actiniaria), an internally brooding sea anemone from Puget Sound. *Can. J. Zool.* 58:2071–80.

Dunn, L. C. 1957. Evidence of evolutionary forces leading to the spread of lethal genes in wild populations of house mice. *Proc. Natl. Acad. Sci. USA* 43:158–63.

East, E. M., and D. F. Jones. 1919. *Inbreeding and Outbreeding: Their Genetic and Sociological Significance.* Philadelphia: J. B. Lippincott.

Easteal, S. 1985. The ecological genetics of introduced populations of the giant toad *Bufo marinus.* II. Effective population size. *Genetics* 110:107–22.

———. 1986. The ecological genetics of introduced populations of the giant toad *Bufo marinus.* IV. Gene flow estimated from admixture in Australian populations. *Heredity* 56:145–56.

Easteal, S., and R. B. Floyd. 1986. The ecological genetics of introduced populations of the giant toad, *Bufo marinus* (Amphibia: Anura): Disperal and neighbourhood size. *Biol. J. Linn. Soc.* 27:17–45.

Eberhard, W. G. 1979. Rate of egg production by tropical spiders in the field. *Biotropica* 11:292–300.

———. 1990. Animal genitalia and female choice. *Am. Scientist* 78:134–41.

Eckelbarger, K. J. 1986. Vitellogenic mechanisms and the allocation of energy to offspring in polychaetes. *Bull. Mar. Sci.* 39(2):426–43.

Eckert, K. L., S. A. Eckert, T. W. Adams, and A. D. Tucker. 1989. Inter-nesting migrations by leatherback sea turtles (*Dermochelys coriacea*) in the West Indies. *Herpetologica* 45:190–94.

Eernisse, D. J. 1988. Reproductive patterns in six species of *Lepidochitona* (Mollusca: Polyplacophora) from the Pacific coast of North America. *Biol. Bull.* 174:287–302.

Ehrlich, P. R., and P. H. Raven. 1969. Differentiation of populations. *Science* 165:1228–32.

Elbadry, E. A., and M. S. F. Tawfik. 1966. Life cycle of the mite *Adaetylidium* sp. (Acarina: Pyemotidae), a predator of thrips eggs in the United Arab Republic. *Ann. Entomol. Soc.* 59:458–61.

Elisbao, T., and N. Freire-Maia. 1984. Inbreeding effect on mordibity. II. Analysis of a third survey including and excluding infant-juvenile mortality among Brazilian whites and negroes. *Am. J. Hum. Genet.* 18:387–90.

Elliott, L. F., P. W. Waser, G. F. McCracken, N. E. Link, and M. K. Gustin. 1989. Genetic variation in two populations of *Dipodomys spectabilis. J. Mammal.* 70:852–55.

Ellis-Quinn, B. A., and C. A. Simon. 1989. Homing behavior of the lizard *Sceloporus jarrovi. J. Herpetol.* 23:146–52.

Ellstrand, N. C. 1984. Multiple paternity within the fruits of the wild radish, *Raphanus sativus. Am. Nat.* 123:819–28.

Ellstrand, N. C., and D. L. Marshall. 1985. Inter-population gene flow by pollen in wild radish, *Raphanus sativus. Am. Nat.* 126:606–16.

Ellstrand, N. C., A. M. Torres, and D. A. Levin. 1978. Density and the rate of apparent outcrossing in *Helianthus annuus* (Asteraceae). *Syst. Bot.* 3:403–7.

El-Shafei, A., P. S. S. Rao, and A. K. Sandhu. 1986. Congential malformation and consanguinity. *Aust. N.Z. J. Obstet. Gynecol.* 26:168–76.

Elston, R. C., and K. Lange. 1976. The genotypic distribution of relatives of homozygotes when consanguinity is present. *Ann. Hum. Genet.* 39:493–96.

Ember, M. 1975. On the origin and extension of the incest taboo. *Behav. Sci. Res.* 10:249–81.

Emigh, T. H., and E. Pollak. 1979. Fixation probabilities and effective population numbers in diploid populations with overlapping generations. *Theor. Popul. Biol.* 15:86–107.

Emlen, S. T. 1969. Homing ability and orientation in the painted turtle *Chrysemys picta marginata*. *Behaviour* 33:58–76.

Emlen, S. T., and L. W. Oring. 1977. Ecology, sexual selection and the evolution of mating systems. *Science* 197(4300): 215–23.

Emlen, S. T., and P. H. Wrege. 1986. Forced copulations and intra-specific parasitism: Two costs of social living in the white-fronted bee-eater. *Ethology* 71:2–29.

Emson, R. H. 1986. Life history patterns in rock pool animals. In *The Ecology of Rocky Coasts*, ed. P. G. Moore and R. Seed, 220–22. New York: Columbia University Press.

Emson, R. H., and R. G. Crump. 1979. Description of a new species of *Asterina* (Asteroidea), with an account of its ecology. *J. Mar. Biol. Assoc. UK* 59:77–94.

Emson, R. H., P. V. Mladenov, and I. C. Wilkie. 1985. Patterns of reproduction in small Jamaican brittle stars: Fission and brooding predominate. In *The Ecology of Coral Reefs*, ed. M. L. Reaka, 87–100. Symposia Series for Undersea Research, NOAA's Undersea Research Program, vol. 3(1).

Emson, R. H., and I. C. Wilkie. 1980. Fission and autotomy in echinoderms. *Oceanogr. Mar. Biol. Annu. Rev.* 18:155–250.

Endler, J. A. 1977. *Geographic Variation, Speciation and Clines*. Princeton, N.J.: Princeton University Press.

———. 1978. A predator's view of animal color patterns. In *Evolutionary Biology*, vol. 11, ed. M. K. Hecht, W. C. Steere, and B. Wallace, 319–64. New York: Plenum Press.

———. 1979. Gene flow and life history patterns. *Genetics* 93:263–84.

———. 1980. Natural selection on color patterns in *Poecilia reticulata*. *Evolution* 34:76–91.

———. 1983. Natural and sexual selection on color patterns in poeciliid fishes. *Environ. Biol. Fish.* 9:173–90.

———. 1986. *Natural Selection in the Wild*. Princeton, N.J.: Princeton University Press.

———. 1988. Sexual selection and predation risk in guppies. *Nature* 332:593–94.

Engstrom, N. A. 1982. Brooding behavior and reproductive biology of a subtidal Puget Sound sea cucumber, *Cucumaria lubrica* (Clark, 1901) (Echinodermata: Holothuroidea). In *Echinoderms: Proceedings of the International Conference, Tampa Bay*, ed. J. M. Lawrence, 447–50. Rotterdam: A. A. Balkema.

Ennos, R. A., and M. T. Clegg. 1982. Effect of population substructuring on estimates of outcrossing rate in plant populations. *Heredity* 48:283–92.

Enomoto, T. 1978. On social preferences in sexual behavior of Japanese monkeys (*Macaca fuscata*). *J. Hum. Evol.* 7:283–93.

Epperson, B. K., and M. T. Clegg. 1986. Spatial autocorrelation analysis of flower color polymorphisms within substructured populations of morning glory (*Ipomoea purpurea*). *Am. Nat.* 128:840–58.

———. 1987. First-pollination primacy and pollen selection in the morning glory *Ipomoea purpurea*. *Heredity* 58:5–14.

Ernst, C. H. 1970. Homing ability in the painted turtle *Chrysemys picta* (Schneider). *Herpetologica* 26:399–403.

Eshel, I., and W. D. Hamilton. 1984. Parent-offspring correlation in fitness under fluctuating selection. *Proc. Roy. Soc.* (Lond.), ser. B, 222:1–14.

Estep, D. Q., K. Nieuwenhuijsen, K. E. M. Bruce, K. J. de Neef, P. A. Walters III, S. C. Baker, and A. K. Slob. 1988. Inhibition of sexual behavior among subordinate stumptail macaques, *Macaca arctoides*. *Anim. Behav.* 36:854–64.

Evans, P. G. H. 1987. Electrophoretic variability of gene products. In *Avian Genetics*, ed. F. Cooke and P. A. Buckley, 105–62. London: Academic Press.

Ewens, W. J. 1969. *Population Genetics*. London: Methuen and Co., Ltd.

Ewens, W. J., P. J. Brockwell, J. M. Gani, and S. I. Resnick. 1987. Minimum viable population size in the presence of catastrophes. In *Viable Populations for Conservation*, ed. M. E. Soulé, 59–68. Cambridge: Cambridge University Press.

Ewert, M. 1969. Seasonal movements of the toads *Bufo americanus* and *B. cognatus* in northwestern Minnesota. Ph.D. thesis, University of Minnesota.

Fadlallah, Y. H. 1983a. Population dynamics and life history of a solitary coral, *Balanophyllia elegans*, from central California. *Oecologia* 58:200–207.

———. 1983b. Sexual reproduction, development and larval biology in scleractinian corals: A review. *Coral Reefs* 2:129–50.

Fadlallah, Y. H., R. H. Karlson, and K. P. Sebens. 1984. A comparative study of sexual reproduction in three species of Panamanian zoanthids (Coelenterata: Anthozoa). *Bull. Mar. Sci.* 35(1):80–89.

Fadlallah, Y. H., and J. S. Pearse. 1982a. Sexual reproduction in solitary corals: Overlapping oogenic and brooding cycles, and benthic planulas in *Balanophyllia elegans*. *Mar. Biol.* 71:223–31.

———. 1982b. Sexual reproduction in solitary corals: Synchronous gametogenesis and broadcast spawning in *Paracyathus stearnsii*. *Mar. Biol.* 71:233–39.

Falconer, D. S. 1981. *Introduction to Quantitative Genetics*. 2d ed. New York: Longman.

———. 1989. *Introduction to Quantitative Genetics*. 3d ed. Burnt Mill, Harlow: Longman.

FAO/UNEP. 1981. Conservation of the genetic resources of fish: Problems and recommendations. *Report of the Expert Consultation on the Genetic Resources of Fish*, Rome, June 9–13, 1980. FAO Fisheries Technical Paper (217):43 p.

Farr, J. A. 1977. Male rarity or novelty, female choice behavior, and sexual selection in the guppy, *Poecilia reticulata* Peters (Pisces: Poeciliidae). *Evolution* 31:162–68.

———. 1980. Social behavior patterns as determinants of reproductive success in the guppy, *Poecilia reticulata* Peters (Pisces: Poeciliidae): An experimental study of the effects of intermale competition, female choice, and sexual selection. *Behaviour* 74:38–91.

———. 1983. The inheritance of quantitative fitness traits in guppies, *Poecilia reticulata* (Pisces: Poeciliidae). *Evolution* 37:1193–1209.

Farr, J. A., and K. Peters. 1984. The inheritance of quantitative fitness traits in guppies, *Poecilia reticulata* (Pisces: Poeciliidae). II. Tests for inbreeding effects. *Heredity* 52:285–96.

Farrant, P. 1985. Reproduction in the temperate Australian soft coral *Capnella gaboensis*. *Proc. V Int. Coral Reef Cong.* 4:319–24.

Farris, M. A., and J. B. Mitton. 1984. Population density, outcrossing rate and heterozygote superiority in ponderosa pine. *Evolution* 38:1151–54.

Fauchald, K. 1983. Life diagram patterns in benthic polychaetes. *Proc. Biol. Soc. Wash.* 96(1): 160–77.

Faulk, W. P., A. Temple, R. E. Lovins, and N. Smith. 1973. Antigens of human trophoblasts: A working hypothesis for their role in normal and abnormal pregnancies. *Proc. Natl. Acad. Sci. USA* 75:1947–51.

Fautin, D. G., and F.-S. Chia. 1986. Revision of sea anemone genus *Epiactis* (Coelenterata: Actiniaria) on the Pacific coast of North America, with descriptions of two new brooding species. *Can. J. Zool.* 64:1665–74.

Feder, J. L., M. H. Smith, R. K. Chesser, M. J. Godt, and K. Asbury. 1984. Biochemical genetics of mosquitofish. II. Demographic differentiation of populations in a thermally altered reservoir. *Copeia* 1984: 108–19.

Feldman, M. 1972. Selection for linkage modification. I. Random mating populations. *Theor. Popul. Biol.* 3:324–46.

Feldman, M. W., and F. B. Christiansen. 1984. Population genetic theory and the cost of inbreeding. *Am. Nat.* 123:642–53.

Feldman, M. W., F. B. Christiansen, and L. D. Brooks. 1980. Evolution of recombination in a constant environment. *Proc. Natl. Acad. Sci. USA* 83:4824–27.

Fell, P. E. 1974. Porifera. In *Reproduction of Marine Invertebrates*, vol. 1, ed. A. C. Giese and J. S. Pearse, 51–132. New York: Academic Press.

Felsenstein, J. 1971. Inbreeding and variance effective numbers in populations with overlapping generations. *Genetics* 68:581–97.

———. 1975. A pain in the torus: Some difficulties with models of isolation by distance. *Am. Nat.* 109(967): 359–68.

———. 1981. Skepticism towards Santa Rosalia, or why are there so few kinds of animals? *Evolution* 35:124–38.

———. 1988. Sex and the evolution of recombination. In *The Evolution of Sex*, ed. R. E. Michod and B. R. Levin. Sunderland, Mass.: Sinauer Associates.

Felsenstein, J., and S. Yokoyama. 1976. The evolutionary advantage of recombination. II. Individual selection for recombination. *Genetics* 83:845–59.

Fenster, C. B. 1991. Gene flow in *Chamaecrista fasciculata* (Leguminosae). I. Gene dispersal. *Evolution* 45:398–409.

———. 1988. Gene flow and population differentiation in *Chamaecrista fasciculata* (Leguminosae). Ph.D. dissertation, University of Chicago, Chicago, Ill.

Fenster, C. B., and V. L. Sork. 1988. Effect of crossing distance and male parent on *in vivo* pollen tube growth in *Chamaecrista fasciculata*. *Am. J. Bot.* 75:1898–1903.

Ferguson, A. 1980. *Biochemical Systematics and Evolution*. New York: John Wiley and Sons.

Ferguson, A., and J. B. Taggart. 1991. Genetic differentiation among the sympatric trout (*Salmo trutta*) populations of Lough Melvin, Ireland. *Biol. J. Linn. Soc.* 43: 221–37.

Ferguson, G. W. 1971. Observations on the behavior and interactions of two sympatric *Sceloporus* in Utah. *Am. Midl. Nat.* 86:190–96.

Ferguson, M. M., R. G. Danzmann, and F. W. Allendorf. 1985. Developmental divergence among hatchery strains of rainbow trout (*Salmo gairdneri*). II. Hybrids. *Can. J. Genet. Cytol.* 27:298–307.

Figueroa, F., M. Golubic, D. Nizetic, and J. Klein. 1985. Evolution of mouse major histocompatibility complex genes borne by *t* chromosomes. *Proc. Natl. Acad. Sci. USA* 82:2819–23.

Finklehor, D. 1978. Psychological, cultural and family factors in incest and family sexual abuse. *J. Marriage Fam. Couns.* 4:41–49.

———. 1979. *Sexually Victimized Children*. New York: The Free Press.

Fischer, E. A. 1981. Sexual allocation in a simultaneously hermaphroditic coral reef fish. *Am. Nat.* 117(1):64–82.

Fisher, R. A. 1930. *The Genetical Theory of Natural Selection*. Oxford: Oxford University Press.

———. 1941. Average excess and average effect of a gene substitution. *Ann. Eugen.* 11:53–63.

———. 1958. *The Genetical Theory of Natural Selection*, 2d ed. New York: Dover.

Fitch, W. M., and W. R. Atchley. 1985. Evolution in inbred strains of mice appears rapid. *Science* 228:1169–75.

Fitch, W. M., and E. Margoliash. 1970. The usefulness of amino acid and nucleotide sequences in evolutionary studies. *Evol. Biol.* 4:67–109.

FitzGerald, G. J., and N. van Havre. 1987. The adaptive significance of cannibalism in sticklebacks (Gasterosteidae: Pisces). *Behav. Ecol. Sociobiol.* 20:125–28.

Flinn, M. 1981. Uterine vs. agnatic kinship variability and associated cousin marriage preferences: An evolutionary biological analysis. In *Natural Selection and Social Behavior*, ed. R. D. Alexander and D. W. Tinkle, 439–75. New York: Chiron Press.

Foerster, R. E. 1936. The return from the sea of sockeye salmon (*Oncorhynchus nerka*) with special reference to percentage survival, sex proportions and progress of migration. *J. Fish. Res. Board Can.* 3:26–42.

Fogden, M. P. L. 1972. The seasonality and population dynamics of equatorial forest birds in Sarawak. *Ibis* 114:307–43.

Foltz, D. W., and J. L. Hoogland. 1983. Genetic evidence of outbreeding in the black-tailed prairie dog (*Cynomys ludovicianus*). *Evolution* 37:273–81.

Foote, C. J., and P. A. Larkin. 1988. The role of male choice in the assortative mating of anadromous and nonanadromous sockeye salmon (*Oncorhynchus nerka*). *Behaviour* 106:43–62.

Foote, C. J., C. C. Wood, and R. E. Withler. 1989. Biochemical genetic comparison of sockeye salmon and kokanee, the anadromous and nonanadromous forms of *Oncorhyncus nerka*. *Can. J. Fish. Aquat. Sci.* 46:149–58.

Ford, C. E., Jr. 1964. Reproduction in the aggregating sea anemone, *Anthopleura elegantissima*. *Pacific Sci.* 18:138–45.

Ford, N. L. 1983. Variation in mate fidelity in monogamous birds. In *Current Ornothology*, vol. 1, ed. R. F. Johnston, 329–56. New York: Plenum Press.

Forster Blouin, S., and M. Blouin. 1988. Inbreeding avoidance behaviors. *Trends Ecol. Evol.* 3(9):230–33.

Fortes, M. 1936. Kinship, incest and exogamy of the Northern Territories of the Gold Coast. In *Custom is King*, ed. L. H. D. Buxton, 239–59. London: Hutchinson.

———. 1949. Time and social structure. In *Social Structure: Essays Presented to A. R. Radcliffe-Brown*, ed. M. Fortes, 36–50. London: Oxford University Press.

Fox, R. 1980. *The Red Lamp of Incest*. New York: E. P. Dutton.

Francis, L. 1988. Cloning and aggression among sea anemones (Coelenterata: Actiniaria) of the rocky shore. *Biol. Bull.* 174:241–53.

Frank, S. A. 1987. Demography and sex ratio in social spiders. *Evolution* 41:1267–81.

Frank, S. A., and M. Slatkin. 1990. Evolution in a variable environment. *Am. Nat.* 136:244–60.

Frankel, O. H., and M. E. Soulé. 1981. *Conservation and Evolution*. Cambridge: Cambridge University Press.

Frankel, R. 1983. *Heterosis: Reappraisal of Theory and Practice*. Berlin: Springer-Verlag.

Franklin, E. C. 1972. Genetic load in loblolly pine. *Am. Nat.* 106:262–65.

Fraser, F. C., and C. J. Biddle. 1976. Estimating the risks for offspring of first cousin matings. *Am. J. Hum. Genet.* 28:522–26.

Fredholm, M., and B. Kristensen. 1987. Further evidence of segregation distortion in the SLA system in Danish Landrace pigs. *Anim. Genet.* 18(Suppl.)1: 25–26.

Frei, O. M., C. W. Stuber, and M. M. Goodman. 1986. Use of allozymes as genetic markers for predicting performance in maize single cross hybrids. *Crop Sci.* 26:37–42.

Freire-Maia and T. Elisbao. 1984. Inbreeding effect on morbidity. III. A review of the world literature. *Am. J. Hum. Genet.* 18:391–400.

Freud, S. 1913. *Totem and Taboo*. New York: Vintage Books.

———. 1939. *Moses and Monotheism*. New York: Alfred A. Knopf.

———. 1955. *Moses and Monotheism*. New York: Vintage Books.

Frichtman, H. K. 1974. The planula of the stylasterine hydrocoral *Allopora petrograpta* Fisher: Its structure, metamorphosis and development of the primary cyclostem. *Proc. II Int. Coral Reef Symp.* 2:245–25

Fried, K., and A. M. Bittles. 1974. Some effects on the offspring of uncle-niece marriage in the Moroccan Jewish community in Jerusalem. *Am. J. Hum. Genet.* 26:65–72.

Fry, W. G. 1971. The biology of larvae of *Ophlitaspongia seriata* from two North Wales populations. In *Proceedings of the Fourth European Marine Biology Symposium*, ed. D. J. Crisp, 155–78. Cambridge: Cambridge University Press.

Fryxell, P. A. 1957. Mode of reproduction of higher plants. *Bot. Rev.* 23:135–233.

Fujino, K., and T. Kang. 1968. Transferrin groups of tunas. *Genetics* 59:79–91.

Fujio, Y. 1982. A correlation of heterozygosity with growth rate in the Pacific oyster, *Crassostrea gigas*. *Tohoku J. Agric. Res.* 33:66–75.

Fukuda, F. 1988. Influence of artificial food supply on population parameters and dispersal in the Hakone T troop of Japanese macaques. *Primates* 29:477–92.

Furnier, G. R., P. Knowles, M. A. Clyde, and B. P. Dancik. 1987. Effects of avian seed dispersal on the genetic structure of whitebark pine populations. *Evolution* 41:607–12.

Futuyma, D. J. 1986. *Evolutionary Biology*, 2d ed. Sunderland, Mass.: Sinauer Associates.

Fyfe, J. L. 1957. Relational incompatibility in diploid and tetraploid lucerne. *Nature* 179:591–92.

Fyfe, J. L., and N. T. J. Bailey. 1951. Plant breeding studies in leguminous forage crops. I. Natural cross-breeding in winter beans. *J. Agric. Sci.* 41:371–78.

Gaines, M. S., and L. R. McClenaghan, Jr. 1980. Dispersal in small mammals. *Annu. Rev. Ecol. Syst.* 11:163–96.

Gajardo, G. M., and J. A. Beardmore. 1989. Ability to switch reproductive mode in *Artemia* is related to maternal heterozygosity. *Mar. Ecol. Prog. Ser.* 55:191–95.

Galen, C., R. C. Plowright, and J. D. Thomson. 1985. Floral biology and regulation of seed set in the lily, *Clintonia borealis*. *Am. J. Bot.* 72:1544–52.

Galiana, A., F. J. Ayala, and A. Moya. 1989. Flush-crash experiments in *Drosophilia*. In *Evolutionary Biology of Transient Unstable Populations*, ed. A. Fontdevila, 58–73. Berlin: Springer-Verlag.

Gall, G. A. E. 1987. Inbreeding. In *Population Genetics and Fishery Management*, ed. N. Ryman and F. Utter, 47–87. Seattle: University of Washington Press.

Ganeshaiah, K. N., R. Uma Shaanker, and G. Shivashankar. 1986. Stigmatic inhibition of pollen grain germination: Its implications for frequency distribution of seed number in pods of *Leucaena leucocephala* (Lam) de Wit. *Oecologia* 70:568–72.

Garcilaso, de la Vega. 1871. *Royal Commentaries of the Yncas 1609–1617*. London: Hakluya Society.

Garton, D. W. 1984. Relationship between multiple locus heterozygosity and physiological energetics of growth in the estuarine gastropod *Thais haemastoma*. *Physiol. Zool.* 57:530–43.

Garton, D. W., R. K. Koehn, and T. M. Scott. 1984. Multiple-lows heterozygosity and the physiological energetics of growth in the coot clam, *Mulinia lateralis*, from a natural population. *Genetics* 108:445–55.

Gatz, A. J. 1981. Non-random mating by size in American toads, *Bufo americanus*. *Anim. Behav.* 29:1004–12.

Gaulin, S. J. C., and A. Schlegel. 1980. Paternal confidence and paternal investment: A cross-cultural test of a sociobiological hypothesis. *Ethol. Sociobiol.* 1:301–9.

Gebhard, P. H., C. V. Christenson, J. H. Gagnon, and W. B. Pomeroy. 1965. *Sex Offenders: An Analysis of Types*. New York: Harper and Row.

Gee, J. M. 1963. Pelagic life of *Spirorbis* larvae. *Nature* 198(4885):1109–10.

Gee, J. M., and G. B. Williams. 1965. Self- and cross-fertilization in *Spirorbis borealis* and *Spirorbis pagenstecheri*. *J. Mar. Biol. Assoc. UK* 45:275–85.

Gerrodette, T. 1981. Dispersal of the solitary coral *Balanophyllia elegans* by demersal planular larvae. *Ecology* 62(3):611–19.

Ghiselin, M. T. 1974. *The Economy of Nature and the Evolution of Sex*. Berkeley: University of California Press.

Gibbons, J. W. 1968. Reproductive potential, activity, and cycles in the painted turtle, *Chrysemys picta*. *Ecology* 49:399–409.

Gibbs, H. L., and P. R. Grant. 1989. Inbreeding in Darwin's medium ground finches (*Geospiza fortis*). *Evolution* 43:1273–84.

Gilbert, D., C. Packer, A. E. Pusey, J. C. Stephens, and S. J. O'Brien. 1991. Analytical DNA fingerprinting in lions: Parentage, genetic diversity and kinship. *J. Hered.* 82:378–86.

Gill, D. E. 1978. Effective population size and interdemic migration rates in a meta-population of the red-spotted newt, *Notophthalmus viridescens* (Rafinesque). *Evolution* 32:839–49.

Gillespie, H. H., and M. Turelli. 1989. Genotype-environment interactions and the maintenance of polygenic variation. *Genetics* 121:129–38.

Gillespie, J. 1973. Polymorphism in random environments. *Theor. Popul. Biol.* 4:193–95.

Gillespie, J. H. 1978. A general model to account for enzyme variation in natural populations. V. The SAS-CFF model. *Theor. Popul. Biol.* 14:1–45.

Gillespie, L. L., and J. B. Armstrong. 1981. Suppression of first cleavage in the Mexican axolotl (*Ambystoma mexicanum*) by heat shock or hydrostatic pressure. *J. Exp. Zool.* 218:441–45.

Gilpin, M. E., and M. E. Soulé. 1986. Minimum viable populations: The processes of species extinctions. In *Conservation Biology: The Science of Scarcity and Diversity*, ed. M. E. Soulé, 13–34. Sunderland, Mass.: Sinauer Associates.

Ginsburger-Vogel, T., and H. Charniaux-Cotton. 1982. Sex determination. In *The Biology of Crustacea*, vol. 2, ed. L. G. Abele, 257–81. New York: Academic Press.

Ginzburg, L. R. 1979. Why are heterozygotes often superior in fitness? *Theor. Popul. Biol.* 15:264–67.

———. 1983. *Theory of Natural Selection and Population Growth*. Menlo Park, Calif.: Benjamin/Cummings.

Gittins, S. P. 1983. Population dynamics of the common toad (*Bufo bufo*) at a lake in mid-Wales. *J. Anim. Ecol.* 52:981–88.

Gjedrem, T. 1985. Improvement of productivity through breeding schemes. *GeoJournal* 10:233–41.

Gjerde, B. 1988. Complete diallele cross between six inbred groups of rainbow trout, *Salmo gairdneri*. *Aquaculture* 75:71–87.

Gjerde, B., K. Gunnes, and T. Gjedrem. 1983. Effect of inbreeding on survival and growth in rainbow trout. *Aquaculture* 34:327–32.

Glander, K. E. 1980. Reproduction and population growth in free-ranging mantled howling monkeys. *Am. J. Phys. Anthropol.* 53:25–36.

Glass, N. L., S. J. Vollmer, C. Staben, J. Grotelueschen, R. L. Metzenberg, and

C. Yanofsky. 1988. DNAs of the two mating-type alleles of *Neurospora crassa* are highly dissimilar. *Science* 241:570–73.

Glesener, R. R., and Tilman, D. 1978. Sexuality and the components of environmental uncertainty: Clues from geographical parthenogenesis in terrestrial animals. *Am. Nat.* 112:659–73.

Glinski, T. H., and C. O'Neil Krekorian. 1985. Individual recognition in free-living adult male desert iguanas, *Dipsosaurus dorsalis*. *J. Herpetol.* 19:541–44.

Gohar, H. A. F. 1940. The development of some Xeniidae (Alcyonaria) (with some ecological aspects). *Publ. Mar. Biol. Stat. Ghardaqa (Red Sea)* 3:27–79.

Goldizen, A. W., and Terborgh, J. 1989. Demography and dispersal patterns of a tamarin population: Possible causes of delayed breeding. *Am. Nat.* 134:208–24.

Goldschmidt, R. 1940. *The Material Basis of Evolution.* New Haven: Yale University Press.

Goodall, J. 1986. *The Chimpanzees of Gombe: Patterns of Behavior.* Cambridge, Mass.: Harvard University Press.

Goodman, D. 1987. The demography of chance extinction. In *Viable Populations for Conservation,* ed. M. E. Soulé, 11–34. Cambridge: Cambridge University Press.

Gordon, L., and P. O'Keefe. 1984. Incest as a form of family violence: Evidence from historical case records. *J. Marriage Fam.* 46:27–34.

Gorman, C. G., Y. J. Kim, and C. E. Taylor. 1977. Genetic variability in irradiated and control populations of *Cnemidophorous tigris* (Sauria: Teiidae) from Mercury, Nevada with a discussion of genetic variability in lizards. *Theor. Appl. Genet.* 49:9–14.

Gorman, W. L., and M. S. Gaines. 1987. Patterns of genetic variation in the cricket frog, *Acris crepitans,* in Kansas. *Copeia* 1987: 352–60.

Gosling, E. M. 1984. The systematic status of *Mytilus galloprovincialis* in western Europe: A review. *Malacologia* 25(2): 551–68.

Gottlieb, L. D. 1981. Electrophoretic evidence and plant populations. *Prog. Phytochem.* 7:1–46.

Goundie, T. R., and S. H. Vessey. 1986. Survival and dispersal of young white-footed mice born in nest boxes. *J. Mammal.* 67:53–60.

Gouzoules, S. 1984. Primate mating systems, kin associations and cooperative behavior: Evidence for kin recognition? *Yearbook Phys. Anthropol.* 27:99–134.

Grant, B. R., and P. R. Grant. 1989. *Evolutionary Dynamics of a Natural Population: The Large Cactus Finch of the Galapagos.* Chicago: University of Chicago Press.

Grant, P. R. 1986. *Ecology and Evolution of Darwin's Finches.* Princeton, N.J.: Princeton University Press.

Grant, P. R., and B. R. Grant. 1992. Demography and the genetically effective size of two populations of Darwin's finches. *Ecology* 73:766–84.

Grant, V. 1975. *Genetics of Flowering Plants.* New York: Columbia University Press.

Grant, W. S., and F. M. Utter. 1988. Genetic heterogeneity on different geographic scales in *Nucella lamellosa* (Prosobranchia: Thaididae). *Malacologia* 28:275–87.

Grant, W. S., et al. 1987. Lack of genetic stock discretion in Pacific cod (*Gadus macrocephalus*). *Can. J. Fish. Aquat. Sci.* 44:490–98.

Grassle, J. F., and J. P. Grassle. 1977. Temporal adaptations in sibling species of *Capitella*. In *Ecology of Marine Benthos*, ed. B. C. Coull, 177–89. Columbia: University of South Carolina Press.

———. 1978. Life histories and genetic variation in marine invertebrates. In *Marine Organisms: Genetics, Ecology and Evolution*, ed. B. Battaglia and J. A. Beardmore, 347–64. New York: Plenum Press.

Grassle, J. P., and J. F. Grassle. 1976. Sibling species in the marine pollution indicator, *Capitella capitata* (Polychaeta). *Science* 192:567–69.

Gray, R. H. 1984. Effective breeding size and the adaptive significance of color polymorphism in the cricket frog (*Acris crepitans*) in Illinois, U.S.A. *Amphibia-Reptilia* 5:101–7.

Green, M. C. 1975. The laboratory mouse, *Mus musculus*. In *Handbook of Genetics*, vol. 4: *Vertebrates of Genetic Interest*, ed. R. C. King, 203–41. New York: Plenum Press.

Green, M. M. 1988. Mobile DNA elements and spontaneous gene mutation. In *Eukaryotic Transposable Elements as Mutagenic Agents*, ed. M. E. Lambert, J. F. McDonald, and I. B. Weinstein, 41–50. Cold Spring Harbor, N.Y.: Cold Spring Harbor Press.

Green, R. H., S. M. Singh, B. Hicks, and J. M. McCuaig. 1983. An arctic intertidal population of *Macoma balthica* (Mollusca: Pelecypoda): Genotypic and phenotypic components of population structure. *Can. J. Fish. Aquat. Sci.* 40:1360–71.

Greenwood, P. J. 1980. Mating systems, philopatry and dispersal in birds and mammals. *Anim. Behav.* 28:1140–62.

———. 1983. Mating systems and the evolutionary consequences of dispersal. In *The Ecology of Animal Movement*, ed. I. R. Swingland and P. J. Greenwood, 116–31. Oxford: Clarendon Press.

———. 1987. Inbreeding, philopatry and optimal outbreeding in birds. In *Avian Genetics*, ed. F. Cooke and P. A. Buckley, 207–22. London: Academic Press.

Greenwood, P. J., P. H. Harvey and C. M. Perrins. 1978. Inbreeding and dispersal in the great tit. *Nature* 271:52–54.

———. The role of dispersal in the great tit (*Parus major*): The causes, consequences and heritability of natal dispersal. *J. Anim. Ecol.* 48:123–42.

Gregory, P. T. 1984. Communal denning in snakes. In *Vertebrate Ecology and Systematics: A Tribute to Henry S. Fitch*, ed. R. A. Siegel, L. E. Hunt, J. L. Knight, L. Malaret, and N. L. Zuschlag, 57–75. Lawrence: Univ. of Kansas Museum of Natural History.

Grieg, J. C. 1979. Principles of genetic conservation in relation to wildlife management in Southern Africa. *S. Afr. J. Wildl. Res.* 9:57–78.

Griffin, A. R., G. F. Moran, and Y. J. Fripp. 1987. Preferential outcrossing in *Eucalyptus regnans* F. Muell. *Aust. J. Bot.* 35:465–75.

490

REFERENCES

Grigg, R. W. 1979. Reproductive ecology of two species of gorgonian corals: Relations to vertical and geographical distribution. In *Reproductive Ecology of Marine Invertebrates*, ed. S. E. Stancyk, 41–59. Columbia: University of South Carolina Press.

Gromov, V. S. 1981. Social organization of the family groups of clawed bird *Meriones unguiculatus* in natural colonies. *Zoologicheskii Zhurnal* 60:1683–94.

Groot, C., T. P. Quinn, and T. J. Hara. 1986. Responses of migrating adult sockeye salmon (*Oncorhynchus nerka*) to population-specific odours. *Can. J. Zool.* 64:926–32.

Grosberg, R. K. 1987. Limited dispersal and proximity-dependent mating success in the colonial ascidian *Botryllus schlosseri*. *Evolution* 41(2): 372–84.

———. 1988. The evolution of allorecognition specificity in clonal invertebrates. *Q. Rev. Biol.* 63(4): 377–412.

———. 1991. Sperm-mediated gene flow and the genetic structure of a population of the colonial ascidian *Botryllus schlosseri*. *Evolution* 45(1): 130–42.

Grosberg, R. K., and J. F. Quinn. 1986. The genetic control and consequences of kin recognition by the larvae of a colonial marine invertebrate. *Nature* 322(6078): 456–59.

———. 1988. The evolution of allorecognition specificity. In *Invertebrate Histo-recognition*, ed. R. K. Grosberg, D. Hedgecock, and K. Nelson, 157–67. New York: Plenum Press.

Grove, K. F. 1983. A cryptic stylar outcrossing mechanism in an autogamous tropical herb. *Bull. Ecol. Soc. Am.* 64:115.

Grubb, J. C. 1973. Olfactory orientation in *Bufo woodhousei fowleri*, *Pseudacris clarki* and *Pseudacris streckri*. *Anim. Behav.* 21:726–32.

Grula, J. W., and O. R. Taylor, Jr. 1980. The effect of X-chromosome inheritance on mate-selection behavior in the sulfur butterflies, *Colias eurytheme* and *C. philodice*. *Evolution* 34:688–95.

Gummere, G. R., P. J. McCormick, and D. Bennett. 1986. The influence of genetic background and the homologous chromosome 17 on *t*-haplotype transmission ratio distortion in mice. *Genetics* 114:235–45.

Gusinde, M. 1937. *The Yaghan: The Life and Thought of the Water Nomads of Cape Horn*. HRAF translation. Modling bei Wien: Anthropos-Bibliothek.

Gustavsson, I. 1969. Cytogenetics, distribution and phenotypic effects of a translocation in Swedish cattle. *Hereditas* 63:168–69.

Guttman, S. I., and K. G. Wilson. 1973. Genetic variation in the genus Bufo. I. An extreme degree of transferrin and albumin polymorphism in a population of the American toad (*Bufo americanus*). *Biochem. Genet.* 8:329–40.

Hadfield, M. G., E. A. Kay, M. U. Gillette, and M. C. Lloyd. 1972. The Vermetidae (Mollusca: Gastropoda) of the Hawaiian Islands. *Mar. Biol.* 12(1): 81–98.

Hafner, J. C., D. J. Hafner, J. L. Patton, and M. F. Smith. 1983. Contact zones and the genetics of differentiation in the pocket gopher *Thomomys bottae* (Rodentia: Geomyidae). *Syst. Zool.* 32:1–20.

Hagen, D. W. 1967. Isolating mechanisms in threespine sticklebacks (*Gasterosteus*). *J. Fish. Res. Board Can.* 24:1637–92.

Haig, D., and M. Westoby. 1988a. Inclusive fitness, seed resources, and maternal care. In *Plant Reproductive Ecology: Patterns and Strategies*, ed. J. Lovett Doust and L. Lovett Doust, 60–79. New York: Oxford University Press.

———. 1988b. On limits to seed production. *Am. Nat.* 131:757–59.

Haigh, G. R. 1983. The effects of inbreeding and social factors on the reproduction of young female *Peromyscus maniculatus bairdii*. *J. Mammal.* 64:48–54.

Haldane, J. B. S. 1924. The mathematical theory of natural and artificial selection. Part II. The influence of partial self-fertilisation, inbreeding, assortative mating, and selective fertilisation on the composition of Mendelian populations, and on natural selection. *Proc. Camb. Phil. Soc. Biol. Sci.* 1:158–63.

———. 1926. A mathematical theory of natural and artificial selection. Part III. *Proc. Camb. Phil. Soc.* 23:363–72.

———. 1932. *The Causes of Evolution*. Ithaca, N.Y.: Cornell University Press.

———. 1957. The cost of natural selection. *J. Genet.* 55:511–24.

Hamilton, W. D. 1964a. The genetical evolution of social behaviour, I. *J. Theor. Biol.* 7:1–16.

———. 1964b. The genetical evolution of social behaviour II. *J. Theor. Biol.* 7:1–16.

———. 1966. The moulding of senescence by natural selection. *J. Theor. Biol.* 12:12–45.

———. 1967. Extraordinary sex ratios. *Science* 156:477–88.

———. 1971. Selection of selfish and altruistic behavior in some extreme models. In *Man and Beast: Comparative Social Behavior*, vol. 2, ed. J. F. Eisenberg and W. S. Dillon, 59–91. Washington, D.C.: Smithsonian Press.

———. 1972. Altruism and related phenomena, mainly in social insects. *Annu. Rev. Ecol. Syst.* 3:192–232.

———. 1975. Innate social aptitudes of man: An approach from evolutionary genetics. In *Biosocial Anthropology*, ed. R. Fox, 133–55. New York: Wiley.

———. 1978. Evolution and diversity under bark. *Roy. Entomol. Soc. Symp.* 9:154–75.

———. 1979. Wingless and fighting males in fig wasps and other insects. In *Sexual Selection and Reproductive Competition in Insects*, ed. M. S. Blum and N. A. Blum, 167–220. New York: Academic Press.

———. 1980. Sex versus non-sex versus parasites. *Oikos* 35:282–90.

———. 1982. Pathogens as causes of genetic diversity in their host populations. In *Population Biology of Infectious Diseases*, R. M. Anderson and R. M. May. New York: Springer.

———. 1987. Kinship, recognition, disease and intelligence: Constraints of social evolution. In *Animal Societies: Theories and Facts*, ed. Y. Ito, J. L. Brown, and J. Kikkawa, 82–102. Tokyo: Japan Scientific Societies Press.

———. 1990. Memes of Haldane and Jayakar in a theory of sex. *J. Genetics* 69:17–32.

Hamilton, W. D., R. Axelrod, and R. Tanese. 1990. Sexual reproduction as an adaptation to resist parasites. *Proc. Natl. Acad. Sci. USA* 87:3566–73.

Hamilton, W. D., and R. M. May. 1977. Dispersal in a stable habitat. *Nature* 269:578–81.

Hamilton, W. D., and M. Zuk. 1982. Heritable true fitness and bright birds: A role for parasites? *Science* 218:384–87.

Hamilton, W. J. III, and J. B. Bulger. N.d. The relationship of dominance rank to competitive ability at age of natal transfer in male baboons. *Behav. Ecol. Sociobiol.* In press.

Hammerberg, C., and J. Klein. 1975. Linkage disequilibrium between H–2 and *t* complexes in chromosome 17 of the mouse. *Nature* 258:296–99.

Hammerstein, P., and G. A. Parker. 1987. Sexual selection: Games between the sexes. In *Dahlem Conference: Sexual Selection, Testing the Alternatives*, ed. J. Bradbury and M. Andersson. New York: Wiley.

Hamrick, J. L. 1982. Plant population genetics and evolution. *Am. J. Bot.* 69:1685–93.

———. 1983. The distribution of genetic variation within and among natural plant populations. In *Genetics and Conservation*, ed. C. M. Schonewald-Cox, S. M. Chambers, B. MacBryde, and L. Thomas. Menlo Park, Calif.: Benjamin/Cummings.

Hamrick, J. L., Y. B. Linhart, and J. B. Mitton. 1979. Relationships between life history characteristics and electrophoretically-detected genetic variation in plants. *Annu. Rev. Ecol. Syst.* 10:173–200.

Hanby, J. P., and J. D. Bygott. 1987. Emigration of subadult lions. *Anim. Behav.* 35:161–69.

Hanken, J., and P. W. Sherman. 1981. Multiple paternity in Belding's ground squirrel litters. *Science* 212:351–53.

Hannaford Ellis, C. J. 1983. Patterns of reproduction in four *Littorina* species. *J. Molluscan Stud.* 49:98–106.

Harder, L. D., J. D. Thomson, M. B. Cruzan, and R. S. Unnasch. 1985. Sexual reproduction and variation in floral morphology in an ephemeral vernal lily, *Erythronium americanum*. *Oecologia* 67:286–91.

Harding, J., R. W. Allard, and D. G. Smeltzer. 1966. Population studies in predominantly self-pollinated species. IX. Frequency dependent selection in *Phaseolus lunatus*. *Proc. Natl. Acad. Sci. USA* 56:99–104.

Harrigan, J. F. 1972. The planula larva of *Pocillopora damicornis:* Lunar periodicity of swarming and substratum selection behavior. Ph.D. thesis, University of Hawaii, Honolulu.

Harris, D. L. 1964. Genotypic covariances between relatives. *Genetics* 50:1319–48.

Harris, H. 1966. Enzyme polymorphisms in man. *Proc. Roy. Soc. (Lond.),* ser. B, 164:298–310.

Harrison, P. L., R. C. Babcock, G. D. Bull, J. K. Oliver, C. C. Wallace, and B. L. Willis. 1984. Mass spawning in tropical reef corals. *Science* 223(4641):1186–89.

Harrison, R. G. 1989. Animal mitochondrial DNA as a genetic marker in population and evolutionary biology. *Trends Ecol. Evol.* 4:6–11.

Harrison, R. G., S. F. Wintermeyer, and T. M. Odell. 1983. Patterns of genetic variation within and among gypsy moth, *Lymantria dispar* (Lepidoptera: Lymantriidae), populations. *Ann. Entomol. Soc. Am.* 766:52–56.

Hartl, D. L. 1971. Some aspects of natural selection in arrhenotokous populations. *Am. Zool.* 11:309–25.

———. 1977. Mechanisms of a case of genetic coadaptation in populations of *Drosophila melanogaster*. *Proc. Natl. Acad. Sci. USA* 74:324–28.

Hartl, D. L., and A. G. Clark. 1989. *Principles of Population Genetics*, 2d ed. Sunderland, Mass.: Sinauer Associates.

Hartman, W. D. 1982. Porifera. In *Synopsis and Classification of Living Organisms*, vol. 1, ed. S. P. Parker, 640–66. New York: McGraw-Hill.

Harvey, P. H., R. D. Martin, and T. H. Clutton-Brock. 1987. Life histories in comparative perspective. In *Primate Societies*, ed. B. B. Smuts, D. L. Cheney, R. M. Seyfarth, T. T. Struhsaker, and R. W. Wrangham, 181–96. Chicago: University of Chicago Press.

Harvey, P. H., and K. Ralls. 1986. Do animals avoid incest? *Nature* 320:575–76.

Haskins, C. P., E. F. Haskins, J. J. A. McLaughlin, and R. E. Hewitt. 1961. Polymorphism and population structure in *Lebistes reticulatus*. In *Vertebrate Speciation*, ed. W. F. Blair, 320–95. Austin: Univ. of Texas Press.

Hasler, A. D., and A. T. Scholz. 1983. *Olfactory Imprinting and Homing in Salmon*. Berlin: Springer-Verlag.

Hassell, M. P., J. K. Waage, and R. M. May. 1983. Variable parasitoid sex ratio and their effect on host-parasitoid dynamics. *J. Anim. Ecol.* 52:889–904.

Hauenschild, C. 1954. Genestische und entwicklungsphysiologische Untersuchungen über Intersexualität und Gewebevertraglichkeit bei *Hydractinia echinata* Flemm. (Hydroz. Bougainvill.). *Roux' Archiv für Entwicklungsmechanik* 147:1–41.

Hawkins, A. J. S., B. L. Bayne, and A. J. Day. 1986. Protein turnover, physiological energetics and heterozygosity in the blue mussel, *Mytilus edulis:* The basis of variable age-specific growth. *Proc. Roy. Soc.* (Lond.), ser. B, 229:161–76.

Hawkins, A. J. S., B. L. Bayne, A. J. Day, J. Rusin, and C. M. Worrall. 1989. Genotype-dependent interrelations between energy metabolism, protein metabolism and fitness. In *Reproduction, Genetics and Distributions of Marine Organisms*, ed. J. S. Ryland and P. A. Tyler, 283–92. Fredensborg, Denmark: Olsen and Olsen.

Hay, D. E., and J. D. McPhail. 1975. Mate selection in threespine sticklebacks (*Gasterosteus*). *Can. J. Zool.* 53:441–50.

Hayaki, H. 1985. Copulation of adolescent male chimpanzees, with special reference to the influence of adult males, in the Mahale National Park, Tanzania. *Folia Primatol.* 44:148–60.

Hedgecock, D. 1978. Population subdivision and genetic divergence in the red-bellied newt, *Taricha rivularis*. *Evolution* 32:271–86.

———. 1982. Genetic consequences of larval retention: Theoretical and methodological aspects. In *Estuarine Comparisons*, ed. V. S. Kennedy, 553–68. New York: Academic Press.

———. 1986. Is gene flow from pelagic larval dispersal important in the adaptation and evolution of marine invertebrates? *Bull. Mar. Sci.* 39(2): 550–64.

Hedgecock, D., M. L. Tracey, and K. Nelson. 1982. Genetics. In *The Bio-*

logy of Crustacea, vol. 2, ed. L. G. Abele, 283–403. New York: Academic Press.

Hedrick, P. W. 1981. The establishment of chromosomal variants. *Evolution* 35:322–32.

———. 1984. Is there an inbreeding optimum? *Zoo Biol.* 3:167–69.

———. 1985. Inbreeding and selection in natural populations. In *Population Genetics in Forestry*, ed. H. R. Gregorius, 71–91. Lecture Notes in Biomathematics 60. New York: Springer-Verlag.

———. 1986. Genetic polymorphism in heterozygenous environments: A decade later. *Annu. Rev. Ecol. Syst.* 17:535–66.

Hedrick, P. W., and C. C. Cockerham. 1986. Partial inbreeding: Equilibrium heterozygosity and the heterozygosity paradox. *Evolution* 40:856–61.

Hedrick, P. W., M. E. Ginevan, and E. P. Ewing. 1976. Genetic polymorphism in heterogenous environments. *Annu. Rev. Ecol. Syst.* 7:1–32.

Hedrick, P. W., S. Jain, and L. Holden. 1978. Multilocus systems in evolution. *Evol. Biol.* 11:101–84.

Hegner, R. E., S. T. Emlen, and N. J. Demong. 1982. Spatial organization of the white-fronted bee-eater. *Nature* 298:264–66.

Helle, W. 1965. Inbreeding depression in an arrhenotokous mite (*Tetranychus utricae* Koch). *Entomol. Exp. Appl.* 8:299–304.

Heller, R., and J. Maynard Smith. 1978. Does Muller's ratchet work with selfing? *Genet. Res.* 34:289–93.

Hellman, E. W., and J. N. Moore. 1983. Effect of genetic relationship to pollinizer on fruit, seed, and seedling parameters in highbush and rabbiteye blueberries. *J. Am. Soc. Hort. Sci.* 108:401–5.

Hemelaar, A. S. M. 1981. Age determination of male *Bufo bufo* (Amphibia, Anura) from the Netherlands, based on year rings in phalanges. *Amphibia-Reptila* 3/4:223–33.

Hendler, G. 1975. Adaptational significance of the patterns of ophiuroid development. *Am. Zool.* 15:691–715.

Hendler, G., and B. S. Littman. 1986. The ploys of sex: Relationships among the mode of reproduction, body size and habitats of coral-reef brittlestars. *Coral Reefs* 5:31–42.

Hendler, G., and R. W. Peck. 1988. Ophiuroids off the deep end: Fauna of the Belizean fore-reef slope. In *Echinoderm Biology*, ed. R. D. Burke, P. V. Mladenov, P. Lambert, and R. L. Parsley, 411–19. Rotterdam: A. A. Balkema.

Hepper, P. G. 1986. Kin recognition: Functions and mechanisms. A review. *Biol. Rev.* 61:63–93.

Hepper, P. G., and B. Waldman. 1992. Embryonic olfactory learning in frogs. *Q. J. Exp. Psychol.* B. 44:179–97.

Hershkovitz, P. 1977. *Living New World Monkeys*, vol. 1. Chicago: University of Chicago Press.

———. 1984. Taxonomy of the squirrel monkey genus *Saimiri* (Cebidae, Platyrrhini): A preliminary report with description of a hitherto unnamed form. *Am. J. Primatol.* 7:155–210.

Herskowitz, I. 1988. A regulatory hierachy for cell specialization in yeast. *Nature* 342:749–57.

————. 1989. Life cycle of the budding yeast *Saccharomyces cerevisiae. Microbiol. Rev.* 52:536–53.

Heslop-Harrison, J. 1975. Incompatibility and the pollen-stigma interaction. *Annu. Rev. Plant Physiol.* 26:403–25.

————. 1983. Self-incompatibility: Phenomenology and physiology. *Proc. Roy. Soc.* (Lond.), ser. B, 218:371–95.

Hess, H., B. Bingham, S. Cohen, R. K. Grosberg, W. Jefferson, and L. Walters. 1988. The scale of genetic differentiation in *Leptosynapta clarki* (Heding), an infaunal brooding holothuroid. *J. Exp. Mar. Biol. Ecol.* 122:187–94.

Hessing, M. B. 1989. Differential pollen tube success in *Geranium caespitosum. Bot. Gaz.* 150:404–10.

Heusser, H. 1968. Die Lebensweise der Erdkröte, *Bufo bufo* L: Wanderungen und Sommerquartiere. *Revue Suisse de Zoologie* 75:928–82.

————. 1969. Die Lebensweise der Erdkröte, *Bufo bufo* (L); Das Orientierungsproblem. *Revue Suisse de Zoologie* 76:443–518.

Heyward, A. J., and R. C. Babcock. 1986. Self- and cross-fertilization in scleractinian corals. *Mar. Biol.* 90:191–95.

Hickey, D. A., and T. McNeilly. 1975. Competition between metal tolerant and normal plant populations: A field experiment on normal soil. *Evolution* 29:458–64.

Highsmith, R. C. 1985. Floating and algal rafting as potential dispersal mechanisms in brooding invertebrates. *Mar. Ecol. Prog. Ser.* 25:169–79.

Hilbish, T. J., and K. M. Zimmerman. 1988. Genetic and nutritional control of the gametogenic cycle in *Mytilus edulis. Mar. Biol.* 98:223–28.

Hill, D. A. 1986. Seasonal differences in the spatial relations of adult male rhesus macaques. In *The Cayo Santiago Macaques,* ed. R. G. Rawlins and M. J. Kessler, 159–71. Albany: State University of New York Press.

————. 1987. Social relationships between adult male and female rhesus macaques: Sexual consortships. *Primates* 28:439–56.

Hill, J. L. 1974. *Peromyscus:* Effects of early pairing on reproduction. *Science* 186:1042–44.

Hill, W. G., and A. Robertson. 1966. The effect of linkage on limits to artificial selection. *Genet. Res.* 8:269–94.

Hintz, R. L., and T. J. Foose. 1982. Inbreeding, mortality and sex ratio in gaur (*Bos gaurus*) under captivity. *J. Hered.* 73:297–98.

Hirth, H. F. 1966. The ability of two species of snakes to return to a hibernaculum after displacement. *S.W. Nat.* 11:49–53.

Hoagland, K. E. 1978. Protandry and the evolution of environmentally-mediated sex change: A study of the Mollusca. *Malacologia* 17(2):365–91.

Hoeck, H. N. 1982. Population dynamics, dispersal and genetic isolation in two species of hyrax (*Heterohyrax brucei* and *Procavia johnstoni*) on habitat islands in the Serengeti. *Z. Tierpsychol.* 59:177–210.

————. 1989. Demography and competition in hyrax. A 17 years study. *Oecologia* 89:353–60.

Hoffmann, R. J. 1986. Variation in contributions of asexual reproduction to the genetic structure of populations of the sea anemone *Metridium senile. Evolution* 40(2):357–65.

———. 1987. Short-term stability of genetic structure in populations of the sea anemone *Metridium senile*. *Mar. Biol.* 93:499–507.

Hogenboom, N. G. 1975. Incompatibility and incongruity: Two different mechanisms for the non-functioning of intimate partner relationships. *Proc. Roy. Soc.* (Lond.), ser. B, 188:361–75.

Holekamp, K. E. 1984a. Dispersal in ground-dwelling sciurids. In *The Biology of Ground-dwelling Squirrels*, ed. J. O. Murie and G. R. Michener, 297–320. Lincoln: University of Nebraska Press.

———. 1984b. Natal dispersal in Belding's ground squirrels (*Spermophilus beldingi*). *Behav. Ecol. Sociobiol.* 16:21–30.

———. 1986. Proximal causes of natal dispersal in Belding's ground squirrels (*Spermophilus beldingi*). *Ecological Monographs* 56:365–91.

Holekamp, K. E., and P. W. Sherman. 1989. Why male ground squirrels disperse. *Am. Sci.* 77:232–39.

Holmes, W. G., and P. W. Sherman. 1983. Kin recognition in animals. *Am. Sci.* 71:46–55.

Holsinger, K. E. 1988a. The evolution of self-fertilization in plants: Lessons from population genetics. *Acta Oecologica: Oecologia Plantarum* 9(1): 95–102.

———. 1988b. Inbreeding depression doesn't matter: The genetic basis of mating-system evolution. *Evolution* 42(6): 1235–44.

———. 1991. Mass-action models of plant mating systems: The evolutionary stability of mixed mating systems. *Am. Nat.* 138:606–22.

Holsinger, K. E., M. W. Feldman, and F. B. Christiansen. 1984. The evolution of self-fertilization in plants: A population genetic model. *Am. Nat.* 124:446–53.

Holtsford, T. P. 1984. Breeding system, fruiting pattern, and optimal outcrossing distance of a Sierra Nevada lily. M.S. thesis, University of California, Irvine.

Honeycutt, R. L., S. V. Edwards, K. Nelson, and E. Nevo. 1987. Mitochondrial DNA variation and the phylogeny of African mole rats (Rodentia: Bathyergidae). *Syst. Zool.* 36: 280–92.

Hoogland, J. L. 1982. Prairie dogs avoid extreme inbreeding. *Science* 215:1639–41.

———. 1992. Levels of inbreeding among prairie dogs. *Am. Nat.* 139:591–602.

Hopkins, K. 1980. Brother-sister marriage in Roman Egypt. *Soc. Comp. Stud. Soc. Hist.* 22:303–54.

Horne, E. A., and R. G. Jaeger. 1988. Territorial pheromones of female red-backed salamanders. *Ethology* 78:143–52.

Horrall, R. M. 1981. Behavioral stock-isolating mechanisms in Great Lakes fishes with special reference to homing and site imprinting. *Can. J. Fish. Aquat. Sci.* 38:1481–96.

Houde, A. E. 1988a. Sexual selection in guppies called into question. *Nature* 333:711.

———. 1988b. The effects of female choice and male-male competition on the mating success of male guppies. *Anim. Behav.* 36:888–96.

Howard, J. G., M. Bush., and D. E. Wildt. 1991. Teratospermia in domestic cats compromises penetration of zona-free hamster ova and cat zonae pellucidae. *J. Androl.* 12:36–45.

Howard, R. D. 1988. Sexual selection on male body size and mating behaviour in American toads, *Bufo americanus*. *Anim. Behav.* 36:1796–1808.

Howard, W. E. 1949. Dispersal, amount of inbreeding, and longevity in a local population of prairie deermice on the George Reserve, southern Michigan. *Contrib. Lab. Vert. Biol. Univ. Mich.* 43:1–52.

Howe, H. F., and L. C. Westley. 1986. Ecology of pollination and seed dispersal. In *Plant Ecology*, ed. M. J. Crawley, 185–215. Oxford: Blackwell Scientific Publications.

Hoy, M. A. 1977. Inbreeding in the arrhenotokous predator *Metaseiulus occidentalis* ([Nesbitt] Acari: Phytoseiidae). *Int. J. Acar.* 3(2):117–21.

Hsu, T. C., and S. H. Hampton. 1970. Chromosomes of Callithricidae with special reference to an XX/'XO' sex chromosome system in Goeldi's marmoset (*Callimico goeldii* Thomas 1904). *Folia Primatol.* 13:183–95.

Hubbell, T. H., and R. M. Norton. 1978. The systematics of the cave-crickets of the North American tribe Hadenoecini. *Misc. Pub. Mus. Zool. Univ. Mich.* 156:1–124.

Hubby, J. L., and R. C. Lewontin. 1966. A molecular approach to the study of genic heterozygosity in natural populations. I. The number of alleles at different loci in *Drosophila pseudoobscura*. *Genetics* 54:577–94.

Huettel, M. D., and G. L. Bush. 1972. The genetics of host selection and its bearing on sympatric speciation in *Procecidochares* (Diptera: Tephritidae). *Entomol. Exp. Appl.* 15:465–80.

Huffman, M. A. 1987. Consort intrusion and female mate choice in Japanese macaques (*Macaca fuscata*). *Ethology* 75:221–34.

Hughes, K. W., and R. K. Vickery, Jr. 1974. Patterns of heterosis and crossing barriers from increasing genetic distance between populations of the *Mimuluus luteus* complex. *J. Genet.* 61:235–45.

Hughes, R. G. 1977. Aspects of the biology and life-history of *Nemertesia antennina* (L.) (Hydrozoa: Plumulariidae). *J. Mar. Biol. Assoc. UK* 57:641–57.

Hughes, R. N. 1978. The biology of *Dendropoma corallinaceum* and *Serpulorbis natalensis*, two South African vermetid gastropods. *Zool. J. Linn. Soc.* 64:111–27.

———. 1986. *A Functional Biology of Marine Gastropods*. Baltimore: Johns Hopkins University Press.

———. 1987. The functional ecology of clonal animals. *Funct. Ecol.* 1:63–69.

———. 1989. *A Functional Biology of Clonal Animals*. London: Chapman and Hall.

Hughes, R. N., and D. J. Roberts. 1981. Comparative demography of *Littorina rudis*, *L. nigrolineata* and *L. neritoides* on three contrasted shores in North Wales. *J. Anim. Ecol.* 50:251–68.

Hughes, T. P., and J. B. C. Jackson. 1985. Population dynamics and life histories of foliaceous corals. *Ecol. Monogr.* 55(2): 141–66.

Hunkapiller, T., H. Huang, L. Hood, and J. H. Campbell. 1982. The impact of modern genetics on evolutionary theory. In *Perspectives on Evolution*, ed. R. Milkman, 164–89. Sunderland, Mass.: Sinauer Associates.

Hutter, P., and M. Ashburner. 1987. Genetic rescue of inviable hybrids between *Drosophila melanogaster* and its sibling species. *Nature* 327:331–33.

Huxley, J. 1942. *Evolution: The Modern Synthesis*. London: Allen and Unwin.

Hyman, L. H. 1940. *The Invertebrates*, vol. 1, *Protozoa through Ctenophora*. New York: McGraw-Hill.

———. 1955. *The Invertebrates*, vol. 4, *Echinodermata*. New York: McGraw-Hill.

———. 1967. *The Invertebrates*, vol. 6, *Mollusca I*. New York: McGraw-Hill.

IPPL. 1982. 71,500 monkey lives in danger. *IPPL Newsletter* 9:2–6.

Itani, J. 1975. Twenty years with Mount Takasaki monkeys. In *Primate Utilization and Conservation*, ed. G. Bermant and D. G. Lindburg, 101–25. New York: John Wiley and Sons.

Itoigawa, N., K. Negayama, and K. Kondo. 1981. Experimental study on sexual behavior between mother and son in Japanese monkeys (*Macaca fuscata*). *Primates* 22:494–502.

Jablonski, D., and R. A. Lutz. 1983. Larval ecology of marine benthic invertebrates: Paleobiological implications. *Biol. Rev.* 58:21–89.

Jackson, J. B. C. 1968. Bivalves: Spatial and size-frequency distributions of two intertidal species. *Science* 161:479–80.

———. 1977. Competition on marine hard substrata: The adaptive significance of solitary and colonial strategies. *Am. Nat.* 111:743–67.

———. 1979. Morphological strategies of sessile animals. In *Biology and Systematics of Colonial Organisms*, ed. G. Larwood and B. R. Rosen, 499–555. London: Academic Press.

———. 1985. Distribution and ecology of clonal and aclonal benthic invertebrates. In *Population Biology and Evolution of Clonal Organisms*, ed. J. B. C. Jackson, L. W. Buss, and R. E. Cook, 297–355. New Haven: Yale University Press.

———. 1986. Modes of dispersal of clonal benthic invertebrates: Consequences for species' distributions and genetic structure of local populations. *Bull. Mar. Sci.* 39(2): 588–606.

Jackson, J. B. C., and A. G. Coates. 1986. Life cycles and evolution of clonal (modular) animals. *Phil. Trans. Roy. Soc.* (Lond.), ser. B, 313:7–22.

Jackson, W. M. 1988. Can individual differences in history of dominance explain development of linear dominance hierarchies? *Ethology* 79:71–77.

Jacquard, A. 1974. *The Genetic Structure of Populations*. New York: Springer.

———. 1975. Inbreeding: One word, several meanings. *Theor. Popul. Biol.* 7:469–95.

Jacson, C. C., and K. J. Joseph. 1973. Life history, bionomics and behaviour of the social spider *Stegodyphus sarasinorum*. *Insectes Soc.* 20:189–203.

Jaeger, R. G., and W. F. Gergits. 1979. Intra- and interspecific communication in salamanders through chemical signals on the substrate. *Anim. Behav.* 27:150–56.

Jaeger, R. G., J. M. Goy, M. Tarver, and C. E. Marquez. 1986. Salamander territoriality: Pheromonal markers as advertisement by males. *Anim. Behav.* 34:860–64.

Jain, S. K. 1976. The evolution of inbreeding in plants. *Annu. Rev. Ecol. Syst.* 7:469–95.

Jain, S. K., and A. D. Bradshaw. 1966. Evolutionary divergence among adjacent plant populations. I. The evidence and its theoretical analysis. *Heredity* 20:407–41.

Jambunathan, N. S. 1905. The habits and life history of the social spider (*Stegodyphus sarasinorum* Karsch). *Smithsonian Misc. Coll.* 47:365–72.

James, D. A. 1965. Effects of antigenic dissimilarity between mother and foetus on placental size in mice. *Nature* 205:613–14.

———. 1967. Some effects of immunological factors on gestation in mice. *J. Reprod. Fert.* 14:265–75.

Jameson, D. L. 1957. Population structure and homing responses in the Pacific tree frog. *Copeia* 1957: 221–28.

Jamieson, I. G. 1989. Levels of analysis or analyses at the same level. *Anim. Behav.* 37:696–97.

Janson, K. 1983. Selection and migration in two distinct phenotypes of *Littorina saxatilis* in Sweden. *Oecologia* 59:58–61.

———. 1985. A morphologic and genetic analysis of *Littorina saxatilis* (Prosobranchia) from Venice, and on the problem of *saxatilis-rudis* nomenclature. *Biol. J. Linn. Soc.* 24:51–59.

———. 1987a. Allozyme and shell variation in two marine snails (*Littorina*, Prosobranchia) with different dispersal abilities. *Biol. J. Linn. Soc.* 30:245–56.

———. 1987b. Genetic drift in small and recently founded populations of the marine snail *Littorina saxatilis*. *Heredity* 58:31–37.

Janson, K., and R. D. Ward. 1984. Microgeographic variation in allozyme and shell characters in *Littorina saxatilis* Olivi (Prosobranchia: Littorinidae). *Biol. J. Linn. Soc.* 22:289–307.

Jarvis, J. U. M. 1981. Eusociality in a mammal: Cooperative breeding in naked mole rat colonies. *Science* 212:571–73.

Jeffreys, A. J. 1987. Highly variable minisatellites and DNA fingerprints. *Biochem. Soc. Trans.* 15:309–17.

Jeffreys, A. J., N. J. Royle, V. Wilson, and Z. Wong. 1988. Spontaneous mutation rates to new length alleles at tandem-repetitive hypervariable loci in human DNA. *Nature* 332:278–80.

Jeffreys, A. J., V. Wilson, and S. L. Thein. 1985. Individual-specific "fingerprints" of human DNA. *Nature* 316:76–79.

Jennison, B. L. 1979. Gametogenesis and reproductive cycles in the sea anemone *Anthopleura elegantissima* (Brandt, 1835). *Can. J. Zool.* 57:403–11.

Jinks, J. L. 1983. Biometrical genetics of heterosis. In *Heterosis, Reappraisal of Theory and Practice*, ed. R. Frankel. New York: Springer.

Johannesson, K. 1988. The paradox of Rockall: Why is a brooding gastropod (*Littorina saxatilis*) more widespread than one having a planktonic larval dispersal stage (*L. littorea*)? *Mar. Biol.* 99:507–13.

Johnson, C. N. 1986. Sex-biased philopatry and dispersal in mammals. *Oecologia* 69:626–27.

Johnson, D. L. E. 1978. Genetic differentiation in two members of the *Drosophilia athabasca* complex. *Evolution* 32:798–811.

Johnson, F. M., G. T. Roberts, R. K. Sharma, F. Chasalow, R. Zweidinger, A. Morgan, R. W. Hendren, and S. E. Lewis. 1981. The detection of mutants in mice by electrophoresis: Results of a model induction experiment with procarbazine. *Genetics* 97:113–24.

Johnson, M. S. 1988. Founder effects and geographic variation in the land snail *Theba pisana*. *Heredity* 61:133–42.

Johnson, M. S., and J. L. Brown. 1980. Genetic variation among trait groups and apparent absence of close inbreeding in grey-crowned babblers. *Behav. Ecol. Sociobiol.* 7:93–98.

Johnson, M. S., B. Clarke, and J. Murray. 1977. Genetic variation and reproductive isolation in *Partula*. *Evolution* 31:116–26.

Johnson, W. E., and R. K. Selander. 1971. Protein variation and systematics in kangaroo rats (Genus *Dipodomys*). *Syst. Zool.* 20:377–405.

Jokiel, P. L. 1984. Long distance dispersal of reef corals by rafting. *Coral Reefs* 3:113–16.

———. 1990. Transport of reef corals into the Great Barrier Reef. *Nature* 347:665–67.

Jolly, A. 1985. *The Evolution of Primate Behavior*. 2d ed. New York: Macmillan.

Jones, C. S., C. M. Lessells, and J. R. Krebs. 1991. Helpers-at-the nest in European bee-eaters (*Merops apiaster*): A genetic analysis. In *DNA Fingerprinting*, ed. T. Burke, G. Dolf, A. Jeffreys, and R. Wolff, 169–92. Basel: Birkhauser.

Jones, W. T. 1984. Natal philopatry in bannertailed kangaroo rats. *Behav. Ecol. Sociobiol.* 15:151–55.

———. 1986. Survivorship in philopatric and dispersing kangaroo rats (*Dipodomys spectabilis*). *Ecology* 67:202–7.

———. 1987. Dispersal patterns in kangaroo rats (*Dipodomys spectabilis*). In *Mammalian Dispersal Patterns: The Effects of Social Structure on Population Genetics*, ed. B. D. Chepko-Sade and Z. T. Halpin, 119–27. Chicago: University of Chicago Press.

Jones, W. T., P. M. Waser, L. F. Elliott, N. E. Link, and B. B. Bush. 1988. Philopatry, dispersal, and habitat saturation in the banner-tailed kangaroo rat, *Dipodomys spectabilis*. *Ecology* 69:1466–73.

Joseph J. Peters Institute. 1979. *Rape Victim Study*. Philadelphia: Joseph J. Peters Institute.

Julian, V., and C. Mohr. 1980. Father-daughter incest: Profile of the offender. *National Study on Child Abuse and Neglect*. Denver: American Humane Association.

Kaestner, A. 1970. *Invertebrate Zoology*, vol. 3, *Crustacea*. New York: Interscience.

Kahler, A. L., C. O. Gardner, and R. W. Allard. 1984. Nonrandom mating in experimental populations of maize. *Crop Sci.* 24:350–54.

Kalezić, M. L., and N. Tucić. 1984. Genic diversity and population genetic structure of *Triturus vulgaris* (Urodela: Salamandridae). *Evolution* 38:389–401.

Kamil, L. J. Inbreeding depression and IQ. *Psychol. Bull.* 87:469–78.

Karlin, S. 1968. Equilibrium behaviour of population genetics models with nonrandom mating. Part I. Preliminaries and special mating systems. *J. Appl. Prob.* 5:231–313.

Karlson, R. H. 1981. Reproductive patterns in *Zoanthus* spp. from Discovery Bay, Jamaica. *Proc. IV Int. Coral Reef Symp.* 2:699–704.

———. 1983. Disturbance and monopolization of a spatial resource by *Zoanthus sociatus* (Coelenterata, Anthozoa). *Bull. Mar. Sci.* 33(1): 118–31.

———. 1986. Disturbance, colonial fragmentation and size-dependent life history variation in two coral reef cnidarians. *Mar. Ecol. Prog. Ser.* 28:245–49.

———. 1988a. Growth and survivorship of clonal fragments in *Zoanthus solanderi* Lesueur. *J. Exp. Mar. Biol. Ecol.* 123:31–39.

———. 1988b. Size-dependent growth in two zoanthid species: A contrast in clonal strategies. *Ecology* 69(4): 1219–32.

Kaufman, I., A. L. Peck, and C. K. Taquiri. 1954. The family constellation and overt incestuous relations between father and daughter. *Am. J. Orthopsychol.* 24:266–79.

Kawanaka, K. 1973. Intertroop relationships among Japanese monkeys. *Primates* 14:113–59.

Keane, B. 1990a. Dispersal and inbreeding avoidance in the white-footed mouse, *Peromyscus leucopus*. *Anim. Behav.* 40:143–52.

———. 1990b. The effect of relatedness on reproductive success and mate choice in the white-footed mouse, *Peromyscus leucopus*. *Anim. Behav.* 39:264–73.

Kelly, M., J. Burke, M. Smith, A. Klar, and D. Beach. 1988. Four mating-type genes control sexual differentiation in the fission yeast. *EMBO J.* 7:1537–47.

Kempe, R. S., and C. H. Kempe. 1978. *Child Abuse.* Cambridge, Mass.: Harvard University Press.

Kendall-Tackett, K. A. 1987. Perpetrators and their acts: Data from 365 adults molested as children. *Child Abuse Negl.* 11:237–45.

Kenrick, J. 1986. A method for estimating self-incompatibility. In *Pollination '86,* ed. E. G. Williams, R. B. Knox, and D. Irvine, 116–20. Melbourne, Australia: University of Melbourne Press.

Keough, M. J. 1984. Kin-recognition and the spatial distribution of larvae of the bryozoan *Bugula neritina* (L.). *Evolution* 38(1):142–47.

Keough, M. J., and H. Chernoff. 1987. Dispersal and population variation in the bryozoan *Bugula neritina*. *Ecology* 68(1): 199–210.

Kerr, W. E. 1976. Population studies in Hymenoptera. 2. Sex-limited genes. *Evolution* 30:94–99.

Kerster, H. W. 1964. Neighborhood size in the rusty lizard, *Sceloporus olivaceus*. *Evolution* 18:445–57.

Kesseli, R. V., and S. K. Jain. 1985. Breeding systems and population structure in *Limnanthes*. *Theor. Appl. Genet.* 71:292–99.

Keverling Buisman, A., and R. van Weeren. 1982. Breeding and management of Przewalski horses in captivity. In *Breeding Przewalski Horses in Captivity for Release into the Wild*, ed. J. Bouman, I. Bouman, and A. Groeneveld, 77–160. Rotterdam: Foundation for the Preservation and Protection of the Przewalski Horse.

Khoury, M. J., B. H. Cohen, E. L. Diamond, G. A. Chase, and V. A. McKusick.

1987. Inbreeding and pre-reproductive mortality in the older order Amish. III. Direct and indirect effects of inbreeding. *Am. J. Epidemiol.* 125:473–83.

Kiester, A. R., C. W. Schwartz, and E. R. Schwartz. 1982. Promotion of gene flow by transient individuals in an otherwise sedentary population of box turtles (*Terrapene carolina triunguis*). *Evolution* 36:617–19.

Kimura, M. 1959. Conflict between self-fertilization and outbreeding in plants. *Annu. Rep. Natl. Inst. Genet. Jpn.* 9:87–88.

———. 1983. *The Neutral Theory of Molecular Evolution.* Cambridge: Cambridge University Press.

Kimura, M., and J. F. Crow. 1978. Effect of overall phenotypic selection on genetic change at individual loci. *Proc. Natl. Acad. Sci. USA* 75:6168–71.

Kimura, M., and T. Ohta. 1971. *Theoretical Aspects of Population Genetics.* Princeton, N.J.: Princeton University Press.

Kincaid, H. L. 1976a. Effects of inbreeding on rainbow trout populations. *Trans. Am. Fish. Soc.* 105:273–80.

———. 1976b. Inbreeding in rainbow trout (*Salmo gairdneri*). *J. Fish. Res. Board Can.* 33:2420–26.

———. 1983. Inbreeding in fish populations used for aquaculture. *Aquaculture* 33:215–27.

King, M. C., and A. C. Wilson. 1975. Evolution at two levels: Molecular similarities and biological differences between humans and chimpanzees. *Science* 188:107–16.

King, P. S. 1987. Macro- and microgeographic structure of a spatially subdivided beetle species in nature. *Evolution* 41:401–16.

Kinomura, K., and K. Yamauchi. 1987. Fighting and mating behaviors of dimorphic males in the ant *Cardiocondyla wroughtoni*. *J. Ethol.* 5:75–81.

Kirkendall, L. R. 1983. The evolution of mating systems in bark and ambrosia beetles (Coleoptera: Scolytidae and Platypodidae). *Zool. J. Linn. Soc.* 77:293–352.

Kirkpatrick, M. 1986. The handicap mechanism of sexual selection does not work. *Am. Nat.* 127:222–40.

———. 1987. Sexual selection by female choice in polygynous animals. *Annu. Rev. Ecol. Syst.* 18:43–70.

Kirpichnikov, V. S. 1981. *Genetic Bases of Fish Selection.* Berlin: Springer-Verlag.

Klein, J., and F. Figueroa. 1981. Polymorphism of the mouse *H–2* loci. *Immunol. Rev.* 60:23–57.

Klekowski, E. J. 1988. *Mutation, Developmental Selection, and Plant Evolution.* New York: Columbia University Press.

Knight, A. J., R. N. Hughes, and R. D. Ward. 1987. A striking example of the founder effect in the mollusc *Littorina saxatilis. Biol. J. Linn. Soc.* 32:417–26.

Knight, S. E., and D. M. Waller. 1987. Genetic consequences of outcrossing the cleistogamous annual, *Impatiens capensis* I: Population genetic structure. *Evolution* 41:969–78.

Knowlton, N., J. C. Lang, and B. D. Keller. 1989. Fates of staghorn coral isolates on hurricane-damaged reefs in Jamaica: The role of predators. *Proc. VI Int. Coral Reef Symp.* 2:83–88.

Knox, R. B. 1984. Pollen-pistil interactions. In *Cellular Interactions: Encyclopedia of Plant Physiology*, n.s., vol. 17, ed. H. F. Linskins and J. Heslop-Harrison, 508–608. Berlin: Springer-Verlag.

Kodric-Brown, A. 1977. Reproductive success and the evolution of breeding territories in pupfish (*Cyprinodon*). *Evolution* 31:750–66.

Koehn, R. K. 1983. Biochemical genetics and adaptation in molluscs. In *The Mollusca*, vol. 2, *Environmental Biochemistry and Physiology*, ed. P. W. Hochachka, 305–30. New York: Academic Press.

————. 1987. The importance of genetics to physiological ecology. In *New Directions in Ecological Physiology*, ed. M. E. Feder, A. F. Bennett, W. W. Burggren, and R. B. Huey, 170–88. Cambridge: Cambridge University Press.

Koehn, R. K., W. J. Diehl, and T. M. Scott. 1988. The differential contribution by individual enzymes of glycolysis and protein catabolism to the relationship between heterozygosity and growth rate in the coot clam, *Mulinia lateralis*. *Genetics* 118:121–30.

Koehn, R. K., and P. M. Gaffney. 1984. Genetic heterozygosity and growth rate in *Mytilus edulis*. *Mar. Biol.* 82:1–7.

Koehn, R. K., R. Milkman, and J. B. Mitton. 1976. Population genetics of marine pelecypods. IV. Selection, migration and genetic differentiation in the blue mussel *Mytilus edulis*. *Evolution* 30:2–32.

Koehn, R. K., R. I. E. Newell, and F. Immermann. 1980. Maintenance of an aminopeptidase allele frequency cline by natural selection. *Proc. Natl. Acad. Sci. USA* 77(9): 5385–89.

Koehn, R. K., and S. E. Shumway. 1982. A genetic/physiological explanation for differential growth rate among individuals of the American oyster, *Crassostrea virginica* (Gmelin). *Mar. Biol. Lett.* 3:35–42.

Koehn, R. K., F. J. Turano, and J. B. Mitton. 1973. Population genetics of marine pelecypods. II. Genetic differences in microhabitats of *Modiolus demissus*. *Evolution* 27:100–105.

Koehn, R. K., and G. C. Williams. 1978. Genetic differentiation without isolation in the American eel, *Anguilla rostrata*. II. Temporal stability of geographic patterns. *Evolution* 32:624–37.

Koehn, R. K., A. J. Zera, and J. G. Hall. 1983. Enzyme polymorphism and natural selection. In *Evolution of Genes and Proteins*, ed. M. Nei and R. K. Koehn, 115–36. Sunderland, Mass.: Sinauer Associates.

Koenig, W. D., and R. L. Mumme. 1987. *Population Ecology of the Cooperatively Breeding Acorn Woodpecker*. Princeton, N.J.: Princeton University Press.

Koenig, W. D., R. L. Mumme, and F. A. Pitelka. 1984. The breeding system of the acorn woodpecker in central coastal California. *Z. Tierpsychol.* 65:289–308.

Koenig, W. D., and F. A. Pitelka. 1979. Relatedness and inbreeding avoidance: Counterploys in the communally nesting acorn woodpecker. *Science* 206:1103–5.

Kojis, B. L., and N. J. Quinn. 1981. Aspects of sexual reproduction and larval development in the shallow water hermatypic coral, *Goniastrea australensis* (Edwards and Haime, 1857). *Bull. Mar. Sci.* 31(3):558–73.

Kölreuter, J. G. 1761. *Vorläufige Nachricht von einingen das Geschlecht der Pflanzen betreffenden Versuchen und Beobachtungen.* Leipzig: Gleditschischen Handlung.

Komatsu, M., Y. T. Kano, H. Yoshizawa, S. Akabane, and C. Oguro. 1979. Reproduction and development of the hermaphroditic sea-star, *Asterina minor* Hayashi. *Biol. Bull.* 157:258–74.

Kondrashov, A. S. 1982. Selection against harmful mutations in large sexual and asexual populations. *Genet. Res.* 40:325–32.

———. 1985. Deleterious mutations as an evolutionary factor. II. Facultative apomixis and selfing. *Genetics* 111:635–53.

Kondrashov, A. S., and J. F. Crow. 1988. King's formula for the mutation load with epistatis. *Genetics* 120:853–56.

Koptur, S. 1984. Outcrossing and pollinator limitation of fruit set: Breeding systems of neotropical *Inga* trees (Fabaceae: Mimosoideae). *Evolution* 38:1130–43.

Kott, P. 1974. The evolution and distribution of Australian tropical Ascidiacea. *Proc. II Int. Coral Reef Symp.* 1:405–23.

Krafft, B. 1966. Etude du comportement sociale de l'araignee *Agelena consociata* Denis. *Biologica Gabonica* 2:235–50.

———. 1969. Quelques remarques sur les phenomenes sociaux ches la Araignees avec une etude particuliere d'*Agelena consociata* Denis. *Bull. Mus. Natl. Hist. Nat.* Suppl. 1:70–75.

———. 1970. Contribution a la biologie et a l'ethologie d'Agelena consociata Denis (Araignee du sociale du Gabon): Premiere Partie. *Biologica Gabonica* 3:199–301.

Kramer, G., and R. Mertens. 1938. Rassenbildung bei west-istrianischen inseleidechsen in Abhangigkeit von isolierungsalter und arealgrosse. *Archiv für naturgeschichte* (N.F.) 7:189–234.

Krear, H. R. 1965. An ecological and ethological study of the pika (*Ochotona princeps saxitilis* Bangs) in the Front Range of Colorado. Ph.D. dissertation, University of Colorado, Boulder.

Krekorian, C. O. 1977. Homing in the desert iguana, *Dipsosaurus dorsalis*. *Herpetologica* 33: 123–27.

Krimbas, C. B., and M. Loukas. 1980. The inversion polymorphism of *Drosophila subobscura*. *Evol. Biol.* 12:163–234.

Kristiansson, A. C., and J. D. McIntyre. 1976. Genetic variation in chinook salmon (*Oncorhynchus tshawytscha*) from the Columbia River and three Oregon coastal rivers. *Trans. Am. Fish. Soc.* 105:620–23.

Kruckeberg, A. 1957. Variation in infertility of hybrids between isolated populations of the serpentine species *Streptanthus glandulosus* Hook. *Evolution* 11:185–211.

Kruuk, H. 1972. *The Spotted Hyena.* Chicago: University of Chicago Press.

Kuhn, T. S. 1970. *The Structure of Scientific Revolutions.* 2d ed. Chicago: University of Chicago Press.

Kuhnlein, U., D. Zadworny, Y. Dawe, R. W. Fairfull, and J. S. Gavora. 1990. Assessment of DNA fingerprinting: Development of a calibration curve using defined strains of chickens. *Genetics* 125:161–65.

Kullman, E., St. Nawabi, and W. Zimmerman. 1971. Neue Ergebnisse zur Bruthbiologie cribellater Spinnen aus Afghanistan und der Serengeti. *Zeitscrift Kölner Zoo* 14:87–108.

Kumar, P., R. A. Pai, and M. S. Swaminathan. 1967. Consanguineous marriages and the genetic load due to lethal genes in Kerala. *Ann. Hum. Genet.* 31:141–47.

Kummer, H. 1978. On the value of social relationships to nonhuman primates: A heuristic scheme. *Soc. Sci. Inf.* 17:687–705.

Kurland, J. A. 1979. Paternity, mother's brother, and human sociality. In *Evolutionary Biology and Human Social Behavior: An Anthropological Perspective*, ed. N. A. Chagnon and W. G. Irons, 145–80. North Scituate, Mass.: Duxbury Press.

Kynard, B. E. 1978. Breeding behavior of a lacustrine population of threespine sticklebacks (*Gasterosteus aculeatus* L.). *Behaviour* 67:178–207.

L'Abée-Lund, J. H., and L. A. Vøllestad. 1985. Homing precision of roach *Rutilus rutilus* in Lake Arungen, Norway. *Environ. Biol. Fish.* 13:235–39.

Lacaze-Duthiers, H. de. 1897. Faune du Golfe du Lion. *Arch. Zool. Exp. Gen.* (ser. 3) 5:1–249.

Lacey, E. A., and P. W. Sherman. 1991. Social organization of naked mole-rat colonies: Evidence for divisions of labor. In *The Biology of the Naked Mole-Rat*, ed. P. W. Sherman, J. U. M. Jarvis, and R. D. Alexander, 275–336. Princeton, N.J.: Princeton University Press.

Lambert, G. 1968. The general ecology and growth of a solitary ascidian, *Corella willmeriana*. *Biol. Bull.* 135(2): 296–307.

Lambert, G., C. C. Lambert, and D. P. Abbott. 1981. *Corella* species in the American Pacific Northwest: Distinction of *C. inflata* Huntsman, 1912 from *C. willmeriana* Herdman, 1898 (Ascidiacea, Phlebobranchia). *Can. J. Zool.* 59:1493–1504.

Lande, R. 1979. Effective deme sizes during long-term evolution estimated from rates of chromosomal rearrangement. *Evolution* 33:234–51.

———. 1981. Models of speciation by sexual selection on polygenic traits. *Proc. Natl. Acad. Sci. USA* 78:3721–25.

Lande, R., and G. F. Barrowclough. 1987. Effective population size, genetic variation, and their use in population management. In *Variable Populations for Conservation*, ed. M. E. Soulé, 87–123. Cambridge: Cambridge University Press.

Lande, R., and D. W. Schemske. 1985. The evolution of self-fertilization and inbreeding depression in plants. I. Genetic models. *Evolution* 39:24–40.

Lank, D. B., P. Mineau, R. F. Rockwell, and F. Cooke. 1989. Intraspecific nest parasitism and extra-pair copulation in lesser snow geese. *Anim. Behav.* 37:4–89.

Larson, A., D. B. Wake, and K. P. Yanev. 1984. Measuring gene flow among populations having high levels of genetic fragmentation. *Genetics* 106: 293–308.

Latter, B. D. H., and A. Robertson. 1962. The effects of inbreeding and artificial selection on reproductive fitness. *Genet. Res.* 3:110–38.

Law, R., and D. H. Lewis. 1983. Biotic environments and the maintenance of sex: Some evidence from mutualistic symbiosis. *Biol. J. Linn. Soc.* 20:249–76.

Lazenby-Cohen, K. A., and A. Cockburn. 1988. Lek promiscuity in a semelparous mammal, *Antechinus stuartii* (Marsupialia: Dasyuridae)? *Behav. Ecol. Sociobiol.* 22:195–202.

Leach, E. R. 1961. *Rethinking Anthropology.* London: Athlone Press.

Leary, R. F., F. W. Allendorf, and K. L. Knudsen. 1989. Genetic differences among rainbow trout spawned on different days within a single season. *Progressive Fish-Culturist* 51:10–19.

Ledig, F. T., R. P. Guries, and B. A Bonefield. 1983. The relation of growth to heterozygosity in pitch pine. *Evolution* 37:1227–38.

Lee, A. K., and A. Cockburn. 1985. *Evolutionary Ecology of Marsupials.* Cambridge: Cambridge University Press.

Lee, T. D., and F. A. Bazzaz. 1982. Regulation of fruit maturation pattern in an annual legume, *Cassia fasciculata. Ecology* 63:1374–88.

Lerner, I. M. 1954. *Genetic Homeostasis.* Edinburgh: Oliver and Boyd.

Lertzman, K. P. 1981. Pollen transfer: Processes and consequences. M.S. thesis, University of British Columbia, Vancouver.

Lessios, H. A. 1988. Mass mortality of *Diadema antillarum* in the Caribbean: What have we learned? *Annu. Rev. Ecol. Syst.* 19:371–93.

Lester, L. J., and R. K. Selander. 1979. Population genetics of haplo-diploid insects. *Genetics* 92:1329–45.

Lévi, C. 1956. Étude des *Halisarca* de Roscoff: Embryologie et systematique des Démosponges. *Arch. Zool. Exp. Gen.* 93:1–184.

Lévi-Strauss, C. 1969. *The Elementary Structure of Kinship.* Boston: Beacon.

Levin, D. A. 1975. Pest pressure and recombination systems in plants. *Am. Nat.* 109:437–51.

———. 1978. Genetic variation in annual *Phlox:* Self compatible versus self incompatible species. *Evolution* 32:245–63.

———. 1981. Dispersal versus gene flow in plants. *Ann. Mo. Bot. Gard.* 68:232–53.

———. 1984. Inbreeding depression and proximity-dependent crossing success in *Phlox drummondii. Evolution* 38:116–27.

———. 1986. Breeding structure and genetic variation. In *Plant Ecology,* ed. M. J. Crawley, 217–51. Oxford: Blackwell Scientific Publications.

———. 1988. The paternity pool of plants. *Am. Nat.* 132:309–17.

Levin, D. A., and H. W. Kerster. 1974. Gene flow in seed plants. *Evol. Biol.* 7:139–220.

Levin, D. A., and L. Watkins. 1984. Assortative mating in *Phlox. Heredity* 53:595–602.

Levin, L. A. 1984a. Life history and dispersal patterns in a dense infaunal polychaete assemblage: Community structure and response to disturbance. *Ecology* 65(4): 1185–1200.

———. 1984b. Multiple patterns of development in *Streblospio benedicti* Webster (Spionidae) from three coasts of North America. *Biol. Bull.* 166:494–508.

Levinson, J. R., and H. O. McDevitt. 1976. Murine *t* factors: An association between alleles at *t* and *H–2. J. Exp. Med.* 144:834–939.

Levinton, J. S., and R. K. Koehn. 1976. Population genetics of mussels. In *Marine Mussels: Their Ecology and Physiology,* ed. B. L. Bayne, 357–84. Cambridge: Cambridge University Press.

Levitan, D. R. 1989. Life history and population consequences of body size regulation in the sea urchin *Diadema antillarum* Philippi. Ph.D. dissertation, University of Delaware, Newark.

———. 1991. Influence of body size and population density on fertilization success and reproductive output in a free-spawning invertebrate. *Biol. Bull.* 181:261–68.

Lewis, D. 1949. Incompatibility in flowering plants. *Biol. Rev.* 24:472–96.

Lewis, D., and L. K. Crowe. 1954. Structure of the incompatibility gene. IV. Types of mutations in *Prunus avium* L. *Heredity* 8:357–63.

Lewis, D., S. C. Verma and M. I. Zuberi. 1988. Gametophytic-sporophytic incompatibility in the Cruciferae *Raphanus sativus. Heredity* 61:355–66.

Lewis, H. 1962. Catastrophic selection as a factor in speciation. *Evolution* 16:257–71.

———. 1966. Speciation in flowering plants. *Science* 152:167–72.

———. 1974a. Settlement and growth factors influencing the contagious distribution of some Atlantic reef corals. *Proc. II Int. Coral Reef Symp.* 2:201–6.

———. 1974b. The settlement behaviour of planulae larvae of the hermatypic coral *Favia fragum* (Esper). *J. Exp. Mar. Biol. Ecol.* 15:165–72.

Lewis, K. R., and B. John. 1959. Breakdown and restoration of chromosome stability following inbreeding in a locust. *Chromosoma* 10:589–618.

Lewontin, R. C. 1962. Interdeme selection controlling a polymorphism in the house mouse. *Am. Nat.* 96:65–78.

———. 1974. *The Genetic Basis of Evolutionary Change.* New York: Columbia University Press.

———. 1985. Population genetics. *Annu. Rev. Genet.* 19:81–102.

Lewontin, R. C., and L. C. Dunn. 1960. The evolutionary dynamics of a polymorphism in the house mouse. *Genetics* 45:705–22.

Lewontin, R. C., and J. L. Hubby. 1966. A molecular approach to the study of genic heterozygosity in natural populations. II. Amount of variation and degree of heterozygosity in natural populations of *Drosophila pseudoobscura. Genetics* 54:595–609.

Libby, W. J., and S. H. Strauss. 1987. Allozyme heterosis in radiata pine is poorly explained by overdominance. *Am. Nat.* 130:879–90.

Licht, L. E. 1976. Sexual selection in toads (*Bufo americanus*). *Can. J. Zool.* 54:1277–84.

Lidicker, W. Z., Jr. 1976. Social behaviour and density regulation in house mice living in large enclosures. *J. Anim. Ecol.* 45:677–97.

Lidicker, W. Z., Jr., and J. L. Patton. 1987. Patterns of dispersal and genetic structure in populations of small rodents. In *Mammalian Dispersal Patterns: The Effects of Social Structure on Population Genetics,* ed. B. D. Chepko-Sade and Z. T. Halpin, 144–61. Chicago: University of Chicago Press.

Lindburg, D. G. 1969. Rhesus monkeys: Mating season mobility of adult males. *Science* 166:1176–78.

Lindelius, R. 1980. Effects of parental consanguinity on mortality and reproductive function. *Hum. Hered.* 30:185–91.

Linhart, Y. B., and J. B. Mitton. 1985. Relationships among reproduction, growth rate and protein heterozygosity in ponderosa pine. *Am. J. Bot.* 72:181–84.

Linhart, Y. B., J. B. Mitton, K. B. Sturgeon, and M. L. Davis. 1981. Genetic variation in space and time in a population of Ponderosa pine. *Heredity* 46:407–26.

Lloyd, D. G. 1979. Some reproductive factors affecting the selection of self-fertilization in plants. *Am. Nat.* 113(1): 67–79.

———. 1980. Sexual strategies in plants. I. An hypothesis of serial adjustment of maternal investment during one reproductive season. *New Phytol.* 86:69–79.

———. 1988. Benefits and costs of biparental and uniparental reproduction in plants. In *The Evolution of Sex: An Examination of Current Ideas,* ed. R. E. Michod and B. R. Levin, 233–52. Sunderland, Mass.: Sinauer Associates.

Lloyd, D. G., and C. J. Webb. 1986. The avoidance of interference between the presentation of pollen and stigmas in angiosperms. I. Dichogamy. *N.Z. J. Bot.* 24:135–62.

Loekle, D. M., D. M. Madison, and J. J. Christian. 1982. Time dependency and kin recognition of cannibalistic behavior among poeciliid fishes. *Behav. Neural Biol.* 35:315–18.

Logan, A., and J. P. A. Noble. 1971. A Recent shallow-water brachiopod community from the Bay of Fundy. *Maritime Sediments* 7(2):85–91.

Loiselle, P. V. 1983. Filial cannibalism and egg recognition by males of the primitively custodial teleost *Cyprinodon macularius californiensis* Girard (Atherinomorpha: Cyprinodontidae). *Ethol. Sociobiol.* 4:1–9.

Lomnicki, A. 1978. Individual differences between animals and the natural regulation of their numbers. *J. Anim. Ecol.* 47:461–75.

Loncke, D. J., and M. E. Obbard. 1977. Tag success, dimensions, clutch size and nesting site fidelity for the snapping turtle, *Chelydra serpentina* (Reptilia, Testudines, Chelydridae) in Algonquin Park, Ontario, Canada. *J. Herpetol.* 11:243–44.

Longwell, A. C. 1976. Review of genetic and related studies on commercial oysters and other pelecypod mollusks. *J. Fish. Res. Board Can.* 33:1100–1007.

Longwell, A. C., and S. S. Stiles. 1973. Gamete cross incompatibility and inbreeding in the commercial American oyster, *Crassostrea virginica* Gmelin. *Cytologia* 38:521–33.

Lord, E. M., and K. J. Eckard. 1984. Incompatibility between the dimorphic flowers of *Collomia grandiflora,* a cleistogamous species. *Science* 223:695–96.

Loveless, M. D., and J. L. Hamrick. 1984. Ecological determinants of genetic structure in plant populations. *Annu. Rev. Ecol. Syst.* 15:65–95.

Lovett Doust, L. 1981. Population dynamics and local specialization in a clonal perennial (*Ranunculus repens*). II. The dynamics of leaves, and a reciprocal transplant replant experiment. *J. Ecol.* 69:757–68.

Loya, Y. 1976a. The Red Sea coral *Stylophora pistillata* is an *r* strategist. *Nature* 259(5543): 478–80.

———. 1976b. Settlement, mortality and recruitment of a Red Sea scleractinian coral population. In *Coelenterate Ecology and Behavior*, ed. G. O. Mackie, 89–100. New York: Plenum.

Lubet, P., G. Prunus, M. Masson, and D. Bucaille. 1984. Recherches expérimentales sur l'hybridation de *Mytilus edulis* L. et *M. galloprovincialis* Lmk (Mollusques Lamellibranches). *Bull. Soc. Zool. Fr.* 109:87–98.

Lubin, Y. D. 1986. Courtship and alternative mating strategies in a social spider. *J. Arachnology* 14:239–57.

Lubin, Y. D., and R. H. Crozier. 1985. Electrophoretic evidence for population differentiation in a social spider *Achaeranea wau* (Theridiidae). *Insectes Sociaux* 32:297–304.

Lubin, Y. D., and M. H. Robinson. 1982. Dispersal by swarming in a social spider. *Science* 216:319–21.

Lucas, J. S., W. J. Nash, and M. Nishida. 1985. Aspects of the evolution of *Acanthaster planci* (L) (Echinodermata, Asteroidea). *Proc. V Int. Coral Reef Cong.* 5:327–32.

Lundqvist, A. 1975. Complex self-incompatibility systems in angiosperms. *Proc. Roy. Soc.* (Lond)., ser. B, 188:235–45.

Luyten, P. H., and N. R. Liley. 1991. Sexual selection and competitive mating success of male guppies (*Poecilia reticulata*) from four Trinidad populations. *Behav. Ecol. Sociobiol.* 28:329–36.

Lynch, C. B. 1977. Inbreeding effects upon animals derived from a wild population of *Mus musculus*. *Evolution* 31:526–37.

Lynch, M. 1988a. Design and analysis of experiments on random drift and inbreeding depression. *Genetics* 120:791–807.

———. 1988b. Estimation of relatedness by DNA fingerprinting. *Mol. Biol. Evol.* 5:584–99.

———. 1988c. The rate of polygenic mutation. *Genet. Res.* 51:137–48.

———. 1990. The similarity index and DNA fingerprinting. *Mol. Biol. Evol.* 7:478–84.

———. 1991. The genetic interpretation of inbreeding depression and outbreeding depression. *Evolution* 45:622–29.

Lyons, E. E., N. M. Waser, M. V. Price, J. Antonovics, and A. F. Motten. 1989. Sources of variation in plant reproductive success and implications for concepts of sexual selection. *Am. Nat.* 134:409–33.

McCabe, S. 1983. FBD marriage: Further support for the Westermarck hypothesis of the incest taboo? *Am. Anthropol.* 85:50–69.

McCall, C., T. Mitchell-Olds, and D. M. Waller. 1988. Might a plant that often selfs have an optimal out-crossing distance? *Bull. Ecol. Soc. Am.* 69:223.

———. 1989. Fitness consequences of outcrossing in *Impatiens capensis:* Tests of the frequency dependent and sib competition models. *Evolution* 43:1075–84.

————. 1991. The distance between mates affects seedling characters in a population of *Impatiens capensis* (Balsaminaceae). *Am. J. Bot.*, 78:964–70.

Macartney, J. M., P. T. Gregory, and K. W. Larsen. 1988. A tabular survey of data on movements and home ranges of snakes. *J. Herpetol.* 22:61–73.

McCracken, G. F. 1987. Genetic structure of bat social groups. In *Recent Advances in the Study of Bats*, ed. M. B. Fenton, P. Racey, and J. M. V. Rayner, 281–98. Cambridge: Cambridge University Press.

McCracken, G. F., and J. W. Bradbury. 1981. Social organization and kinship in the polygynous bat *Phyllostomus hastatus*. *Behav. Ecol. Sociobiol.* 8:11–34.

Macdonald, D., ed. 1984. *The Encyclopedia of Mammals*. New York: Facts on File Publications.

McDonald, J. F. 1989. The potential evolutionary significance of retroviral-like transposable elements in peripheral populations. In *Evolutionary Biology of Transient Unstable Populations*, ed. A. Fontdevila, 190–205. Berlin: Springer-Verlag.

McDonald, J. H., and R. K. Koehn. 1988. The mussels *Mytilus galloprovincialis* and *M. trossulus* on the Pacific coast of North America. *Mar. Biol.* 99:111–18.

McEuen, F. S. 1987. Phylum Echinodermata, Class Holothuroidea. In *Reproduction and Development of Marine Invertebrates of the Northern Pacific Coast*, ed. M. F. Strathmann, 574–96. Seattle: University of Washington Press.

McEuen, F. S., B. L. Wu, and F. S. Chia. 1983. Reproduction and development of *Sabella media*, a sabellid polychaete with extratubular brooding. *Mar. Biol.* 76:301–9.

McFadien-Carter, M. 1979. Scaphopoda. In *Reproduction of Marine Invertebrates*, vol. 5, ed. A. C. Giese and J. S. Pearse, 95–111. New York: Academic Press.

McFarland Symington, M. 1988. Demography, ranging patterns, and activity budgets of black spider monkeys (*Ateles paniscus chamek*) in the Manu National Park, Peru. *Am. J. Primatol.* 15:45–67.

McGavin, M. 1978. Recognition of conspecific odors by the salamander *Plethodon cinereus*. *Copeia* 1978: 356–58.

McGrew, W. C., and McLuckie, B. C. 1986. Philopatry and dispersion in the cotton-top tamarin, *Saguinas (o.) oedipus*: An attempted laboratory simulation. *Int. J. Primatol.* 7:399–420.

McIntyre, J. A., and W. P. Faulk. 1983. Recurrent spontaneous abortion in human pregnancy: Results of immunogenetical, cellular and humoral studies. *Am. J. Reprod. Immunol.* 4:165–70.

McKaye, K. R., and G. W. Barlow. 1976. Competition between color morphs of the midas cichlid, *Cichlasoma citrinellum*, in Lake Jiloa, Nicaragua. In *Investigations of the Ichthyofauna of Nicaraguan Lakes*, ed. T. B. Thorson, 465–75. Lincoln: School of Life Sciences, University of Nebraska-Lincoln.

McKenna, J. J. 1979. The evolution of allomothering behavior among colobine monkeys: Function and opportunism in evolution. *Am. Anthropol.* 81:818–40.

Mackie, G. L. 1984. Bivalves. In *The Mollusca*, vol. 7, ed. A. S. Tompa, N. H. Verdonk, and J. A. M. van den Biggelaar, 351–418. Orlando: Academic Press.

Mackie, G. O. 1974. Locomotion, flotation, and dispersal. In *Coelenterate Biology*, ed. L. Muscatine and H. M. Lenhoff, 313–57. New York: Academic Press.

McKinney, F. K., and J. B. C. Jackson. 1989. *Bryozoan Evolution*. Boston: Unwin Hyman.

McKinney, F., K. M. Cheng, and D. Bruggers. 1984. Sperm competition in apparently monogamous birds. In *Sperm Competition and the Evolution of Animal Mating Systems*, ed. R. L. Smith, 523–45. New York: Academic Press.

McLaren, A. 1975. Antigenic disparity: Does it affect placental size, implantation or population genetics? In *Immunology of Trophoblast*, ed. R. G. Edwards, C. W. S. Howe, and M. H. Johnson, 255–73. Cambridge: Cambridge University Press.

McPhee, H. C., E. Z. Russel, and J. Zeller. 1931. An inbreeding experiment with poland china swine. *J. Hered.* 22:393–403.

McPheron, B. A., C. D. Jorgenson, and S. H. Berlocher. 1988. Low genetic variability in a Utah cherry-infesting population of the apple maggot, *Rhagoletis pomonella*. *Entomol. Exp. Appl.* 46:155–60.

McVey, M. E., R. G. Zahary, D. Perry, and J. MacDougal. 1981. Territoriality and homing behavior in the poison dart frog (*Dendrobates pumilio*). *Copeia* 1981: 1–8.

Madison, D. M. 1969. Homing behaviour of the red-cheeked salamander, *Plethodon jordani*. *Anim. Behav.* 17:25–39.

———. 1975. Intraspecific odor preferences between salamanders of the same sex: Dependence on season and proximity of residence. *Can. J. Zool.* 53:1356–61.

Maisch, H. 1972. *Incest.* New York: Stein and Day.

Makaveev, T., I. Venev, and M. Baulov. 1978. Investigations of activity level and polymorphisms of some blood enzymes in farm animals with different growth energy. II. Correlations between homo- and heterozygosity of some protein and enzyme phenotypes and fattening ability and slaughter indices in various breeds of fattened pigs. *Genet. Sel.* 10:229–36.

Makov, E., and A. H. Bittles. 1986. On the choice of mathematical models for the estimation of lethal gene equivalents in man. *Heredity* 57:377–80.

Malecot, G. 1969. *The Mathematics of Heredity*, rev., ed. and trans. D. M. Yermanos. San Francisco: W. H. Freeman.

Malinowski, B. 1929. *The Sexual Life of Savages in Northwestern Malanesia*. London: Routledge.

Mallet, A. L., and L. E. Haley. 1983. Effects of inbreeding on larval and spat performance in the American oyster. *Aquaculture* 33:229–35.

Malti and K. R. Shivanna. 1985. The role of the pistil in screening compatible pollen. *Theor. and Appl. Genet.* 70:684–86.

Manachenko, G. P. 1981. Allozymic variation in *Araneus ventricosus* (Arachnida, Araneae). *Isozyme Bull.* 14:78.

Mantel, N. 1967. The detection of disease clustering and a generalized regression approach. *Cancer Res.* 27:209–20.

Manzolillo, D. L. 1982. Intertroop transfer by adult male *Papio anubis*. Ph.D. dissertation, University of California, Los Angeles.

Marcallo, F. A., N. Freire-Maia, J. B. C. Azvedo, and I. A. Simoes. 1964. Inbreeding effect on mortality and morbidity in South Brazilian populations. *Ann. Hum. Gen.* 27:203–17.

Marliave, J. B. 1986. Lack of planktonic dispersal of rocky intertidal fish larvae. *Trans. Am. Fish. Soc.* 115:149–54.

Marsh, C. W. 1979. Comparative aspects of social organization in the Tana River red colobus, *Colobus badius rufomitratus*. *Z. Tierpsychol.* 51:337–62.

Marshall, D. L., and N. C. Ellstrand. 1988. Effective mate choice in wild radish: Evidence for selective seed abortion and its mechanisms. *Am. Nat.* 131:739–56.

Marshall, G. A. K. 1898. Notes on the South African social spiders. *Zoologist* 2:417–22.

Martin, F. W. 1963. Distribution and interrelationships of incompatibility barriers in the *Lycopersicon hirsutum* Humb. and Bonpl. complex. *Evolution* 17:519–28.

———. 1970. Pollen germination on foreign stigmas. *Bull. Torrey Bot. Club* 97:1–6.

Martin Chavez, E. 1986. Gametogenesis and origin of planulae in the hermatypic coral *Pocillopora damicornis*. *Hawaii Inst. Mar. Biol. Tech. Rep.* 37:193–205.

Maruyama, T., and P. A. Fuerst. 1985. Population bottlenecks and nonequilibrium models in population genetics. II. Number of alleles in a small population that was formed by a recent bottleneck. *Genetics* 11:675–89.

Maturo, F. 1991. Self-fertilization in gymnolaemate Bryozoa. In *Bryozoaires Actuels et Fossiles: Bryozoa, Living and Fossil*, ed. F. P. Bigey and J.-L. d'Hondt, 572. *Bull. Soc. Sci. Nat. Ouest Fr., Mem. HS. 1.*

May, R. M. 1979. When to be incestuous. *Nature* 279:192–94.

Maynard, E. A. 1934. The aquatic migration of the toad, *Bufo americanus* Le Conte. *Copeia* 1934: 174–77.

Maynard Smith, J. 1956. Fertility, mating behaviour and sexual selection in *Drosophila subobscura*. *J. Genet.* 54:261–79.

———. 1968. "Haldane's dilemma" and the rate of evolution. *Nature* 29:1114–16.

———. 1971. What use is sex? *J. Theor. Biol.* 30:319–35.

———. 1974. Recombination and the rate of evolution. *Genetics* 78:299–304.

———. 1977. The sex habit in plants and animals. In *Measuring Selection in Natural Populations*, ed. F. B. Christiansen and T. M. Fenchel, 315–31. Berlin: Springer-Verlag.

———. 1978. *The Evolution of Sex*. Cambridge: Cambridge University Press.

———. 1982. *Evolution and the Theory of Games*. New York: Cambridge University Press.

———. 1983. Current controversies in evolutionary biology. In *Dimensions of Darwinism*, ed. M. Grene, 273–86. Cambridge: Cambridge University Press.

———. 1984. The ecology of sex. In *Behavioural Ecology: An Evolutionary Ap-

proach, ed. J. R. Krebs and N. B. Davies, 201–21. Oxford: Blackwell Scientific Publications.

———. 1988. The evolution of recombination. In *The Evolution of Sex: An Examination of Current Ideas,* ed. R. E. Michod and B. R. Levin, 106–25. Sunderland, Mass.: Sinauer Associates.

Mayr, E. 1942. *Systematics and the Origin of Species.* New York: Columbia University Press.

———. 1954. Change of genetic environment and evolution. In J. Huxley, A. C. Hardy, and E. B. Ford, *Evolution as a Process.* London: Allen and Unwin.

———. 1963. *Animal Species and Evolution.* Cambridge, Mass.: Harvard University Press.

———. 1976. *Evolution and the Diversity of Life.* Cambridge, Mass.: Harvard University Press.

Mayr, E., and W. B. Provine, eds. 1980. *The Evolutionary Synthesis: Perspectives on the Unification of Biology.* Cambridge, Mass.: Harvard University Press.

Meagher, T. R. 1986. Analysis of paternity within a natural population of *Chamaelirium luteum.* I. Identification of most-likely fathers. *Am. Nat.* 128:199–215.

Mehlman, P. 1986. Male intergroup mobility in a wild population of the Barbary macaque (*Macaca sylvanus*), Ghomaran Rif Mountains, Morocco. *Am. J. Primatol.* 10:67–81.

Melnick, D. J. 1988. The genetic structure of a primate species: Rhesus macaques and other cercopithecine monkeys. *Int. J. Primatol.* 9:195–231.

Melnick, D. J., and M. C. Pearl, 1987. Cercopithecines in multimale groups: Genetic diversity and population structure. In *Primate Societies,* ed. B. B. Smuts, D. L. Cheney, R. M. Seyfarth, T. T. Struhsaker, and R. W. Wrangham, 121–34. Chicago: University of Chicago Press.

Mendelsohn, H. 1980. Observations on a captive colony of *Iguana iguana.* In *Reproductive Biology and Diseases of Captive Reptiles,* ed. J. B. Murphy and J. T. Collins, 119–23. Lawrence: Society for the Study of Amphibians and Reptiles.

Merrell, D. J. 1968. A comparison of the estimated size and the "effective size" of breeding populations of the leopard frog, *Rana pipiens. Evolution* 22:274–83.

Merritt, R. B., W. H. Kroon, D. A. Wienski, and K. A. Vincent. 1984. Genetic structure of natural populations of the red-spotted newt, *Notophthalmus viridescens. Biochem. Genet.* 22:669–86.

Metraux, A. 1965. *The Incas.* London: Studio Vista, Ltd.

Meylan, A. B., B. W. Bowen, and J. C. Avise. 1990. A genetic test of the natal homing versus social facilitation models for green turtle migration. *Science* 248:724–27.

Michod, R. E. 1979. Genetical aspects of kin selection: Effects of inbreeding. *J. Theor. Biol.* 81:223–33.

———. 1980. Evolution of interactions in family structured populations: Mixed mating models. *Genetics* 96:275–96.

————. 1982. The theory of kin selection. *Annu. Rev. Ecol. Syst.* 13:23–55.

Michod, R. E., and W. W. Anderson. 1979. Measures of genetic relationship and the concept of inclusive fitness. *Am. Nat.* 114:637–47.

Michod, R. E., and W. D. Hamilton. 1980. Coefficients of relatedness in sociobiology. *Nature* 288:694–97.

Michod, R. E., and B. R. Levin, eds. 1988. *The Evolution of Sex: An Examination of Current Ideas.* Sunderland, Mass.: Sinauer Associates.

Michod, R. E., and M. J. Sanderson. 1985. Behavioural structure and the evolution of cooperation. In *Evolution: Essays in Honour of John Maynard Smith,* ed. J. J. Greenwood, P. H. Harvey, and M. Slatkin. Cambridge: Cambridge University Press.

Milkman, R. 1978. Selection differentials and selection coefficients. *Genetics* 88:391–403.

————. 1982. Toward a unified selection theory. In *Perspectives on Evolution,* ed. R. Milkman, 105–18. Sunderland, Mass.: Sinauer Associates.

Millar, J. S. 1971. Breeding of the pika in relationship to the environment. Ph.D. dissertation, University of Alberta, Edmonton.

————. 1972. Timing of breeding of pikas in southwestern Alberta. *Can. J. Zool.* 50:665–69.

Millar, J. S., and F. C. Zwickel. 1972. Determination of age, age structure, and mortality of the pika, *Ochotona princeps* (Richardson). *Can. J. Zool.* 50: 229–32.

Millar, R. H. 1971. The biology of ascidians. *Adv. Mar. Biol.* 9:1–100.

Mishler, B. D. 1988. Reproductive ecology in Bryophytes. In *Plant Reproductive Ecology,* ed. J. Lovett Doust and L. Lovett Doust, 285–328. Oxford: Oxford University Press.

Missakian, E. A. 1973. Geneological mating activity in free-ranging groups of rhesus monkeys (*Macaca mulatta*) on Cayo Santiago. *Behaviour* 45:225–41.

Mitchell-Olds, T. 1986. Quantitative genetics of survival and growth in *Impatiens capensis. Evolution* 40:107–16.

Mitton, J. B. 1989. Physiological and demographic variation associated with allozyme variation. In *Isozymes in Plant Biology,* ed. D. Soltis and P. Soltis, 127–45. Portland, Ore.: Dioscorides Press.

Mitton, J. B., and M. Boyce. In preparation. The evolution of female choice in a constant environment.

Mitton, J. B., C. Carey, and T. D. Kocher. 1986. The relation of enzyme heterozygosity to standard and active oxygen consumption and body size of tiger salamanders, *Ambystoma tigrinum. Physiol. Zool.* 59:574–82.

Mitton, J. B., and M. C. Grant. 1984. Associations among protein heterozygosity, growth rate, and developmental homeostasis. *Annu. Rev. Ecol. Syst.* 15:479–99.

Mitton, J. B., and R. M. Jeffers. 1989. The genetic consequences of mass selection for growth rate in Engelmann spruce. *Silvae Genet.* 38:6–12.

Mitton, J. B., and R. K. Koehn. 1975. Genetic organization and adaptive response of allozymes to ecological variables in *Fundulus heteroclitus. Genetics* 79:97–111.

Mitton, J. B., and B. A. Pierce. 1980. The distribution of individual heterozygosity in natural populations. *Genetics* 95:1043–54.

Mitton, J. B., and R. Thornhill. In preparation. The advantage of a highly heterozygous mate and the evolution of female choice.

Mladenov, P. V., and R. H. Emson. 1988. Density, size structure and reproductive characteristics of fissiparous brittle stars in algae and sponges: Evidence for interpopulational variation in levels of sexual and asexual reproduction. *Mar. Ecol. Prog. Ser.* 42:181–94.

Moav, R., and G. W. Wohlfarth. 1968. Genetic improvement of yield in carp. In *FAO Fisheries Reports*, No. 44, vol. 4. ed. T. V. R. Pillay, 12–29. Proceedings of the FAO World Symposium on Warm-water Pond Fish Culture. Rome: Food and Agriculture Organization of the United Nations.

Moll, R. H., J. H. Lonnquist, J. VelezFourtuno, and E. C. Johnson. 1965. The relationship of heterosis and genetic divergence in maize. *Genetics* 52:139–44.

Montalvo, A. M. 1992. Relative success of self and outcross pollen after mixed- and single-donor pollinations in *Aquilegia caerulea*. *Evolution*. In press.

Moore, J. 1984. Female transfer in primates. *Int. J. Primatol.* 5:537–89.

———. 1988. Primate dispersal (letter). *Trends Ecol. Evol.* 3:144–45.

———. 1992. Dispersal, nepotism, and primate social behavior. *Int. J. Primatol.* 13:357–78.

Moore, J., and Ali, R. 1984. Are dispersal and inbreeding avoidance related? *Anim. Behav.* 32:94–112.

———. 1985. Inbreeding and dispersal: Reply to Packer (1985). *Anim. Behav.* 33:1367–69.

Mork, J., and G. Sundnes. 1985. O-group cod (*Gadus morhua*) in captivity: Differential survival of certain genotypes. *Helgolander Meeresuntersuchungen* 39:63–70.

Morse, D. E., N. Hooker, A. N. C. Morse, and R. A. Jensen. 1988. Control of larval metamorphosis and recruitment in sympatric agariciid corals. *J. Exp. Mar. Biol. Ecol.* 116:193–217.

Mortimer, J. A., and A. Carr. 1987. Reproduction and migrations of the Ascension Island green turtle (*Chelonia mydas*). *Copeia* 1987: 103–13.

Morton, N. E. 1959. Empirical risks in consanguineous marriages: Birth weight, gestation time and measurements of infants. *Am. J. Hum. Genet.* 10:344–49.

———. 1960. The mutational load due to detrimental genes in man. *Am. J. Hum. Genet.* 12:348–64.

———. 1978. Effect of inbreeding on IQ and mental retardation. *Proc. Natl. Acad. Sci. USA* 75:3906–8.

Morton, N. E., C. S. Chang, and M. Mi. 1967. *Genetics of Interracial Crosses in Hawaii*. New York: S. Karger.

Morton, N. E., J. F. Crow, and H. J. Muller. 1956. An estimate of the mutational damage in man from data on consanguineous marriages. *Proc. Nat. Acad. Sci. USA* 42:855–63.

Motro, U. 1991. Avoiding inbreeding and sibling competition: The evolution of sexual dimorphism for dispersal. *Am. Nat.* 137:108–15.

Mowbray, J. F. 1987. A controlled trial of immunotherapy with paternal cells in recurrent spontaneous abortion. In *Immunoregulation and Fetal Survival*, ed. T. J. Gill III, T. G. Wegmann, and E. Nisbet-Brown, 300–308. Oxford: Oxford University Press.

Mowbray, J. F., C. Gibbings, H. Liddell, P. W. Reginald, J. L. Underwood, and R. W. Beard. 1985. Controlled trial of treatment of recurrent spontaneous abortion by immunisation with paternal cells. *Lancet* i:941–43.

Mrakovčič, M., and L. E. Haley. 1979. Inbreeding depression in the zebra fish, *Brachydanio rerio* (Hamilton Buchanan). *J. Fish Biol.* 15:323–27.

Mukai, T., L. E. Mettler, and S. Chigusa. 1971. Linkage disequilibrium in a local population of *Drosophila melanogaster. Proc. Natl. Acad. Sci. USA* 68:1065–69.

Mulcahy, D. L. 1984. Self-compatibility, self-incompatibility and Stebbin's rule. In *Pollination '84*, ed. E. G. Williams and R. B. Knox, 48–52. Melbourne, Australia: University of Melbourne.

Mulcahy, D. L., P. Curtis, and A. A. Snow. 1983. Pollen competition in a natural population. In *Handbook of Experimental Pollination Biology*, ed. C. E. Jones and R. J. Little, 330–37. New York: Van Nostrand Reinhold.

Mulcahy, D. L., and G. B. Mulcahy. 1983. Gametophytic self-incompatibility re-examined. *Science* 220:1247–51.

Mulcahy, G. B., and D. L. Mulcahy. 1982. The two phases of growth of *Petunia hybrida* pollen tubes through compatible styles. *J. Palynol.* 18:1–3.

Müller, H. 1893. *The Fertilisation of Flowers.* Trans. W. D'Arcy Thompson. London: Macmillan.

Muller, H. J. 1964. The relation of recombination to mutational advance. *Mutat. Res.* 1:2–9.

Mumme, R. L., W. D. Koenig, R. M. Zink, and J. A. Marten. 1985. Genetic variation and parentage in a Californian population of acorn woodpeckers. *Auk* 102:305–12.

Murdock, G. P. 1949. *Social Structure.* New York: Macmillan.

Murdy, W. H., and M. E. B. Carter. 1988. Regulation of the timing of pollen germination by the pistil in *Talinum mengesii* (Portulaceae). *Am. J. Bot.* 74:1888–92.

Murphy, J. B., J. E. Rehg, P. F. A. Maderson, and W. B. McCrady. 1987. Scutellation and pigmentation defects in a laboratory colony of Western diamondback rattlesnakes (*Crotalus atrox*): Mode of inheritance. *Herpetologica* 43:292–300.

Murphy, P. A. 1981. Celestial compass orientation in juvenile American alligators (*Alligator mississippiensis*). *Copeia* 1981: 638–43.

Murray, B. G. 1967. Dispersal in vertebrates. *Ecology* 48:975–78.

———. 1984. A demographic theory on the evolution of mating systems as exemplified by birds. *Evol. Biol.* 18:71–140.

Murray, M. G. 1987. The closed environment of the fig receptacle and its influence on male conflict in the Old World fig wasp, *Philotrypesis pilosa. Anim. Behav.* 35:488–506.

Murray, M. G., and R. Gerrard. 1984. Conflict in the neighbourhood: Models where close relatives are in direct competition. *J. Theor. Biol.* 111:237–46.

Murray, R. D., and E. O. Smith. 1983. The role of dominance and intrafamilial bonding in the avoidance of close inbreeding. *J. Hum. Evol.* 12:481–86.

Muus, B. J. 1967. The fauna of Danish estuaries and lagoons: Distribution and ecology of dominating species in the shallow reaches of the mesohaline zone. *Meddelelser fra Danmarks Fiskeri-og Havundersøgelser NY Serie* 5(1): 1–316.

Myles, T. G., and W. L. Nutting. 1988. Termite eusocial evolution: A reexamination of Bartz's hypothesis and assumptions. *Q. Rev. Biol.* 63:1–23.

Nadeau, J. H., E. K. Wakeland, D. Gotze, and J. Klein. 1981. The population genetics of the *H–2* polymorphism in European and North African populations of the house mouse (*Mus musculus* L.). *Genet. Res.* 37:17–31.

Nadel, S. F. 1947. *The Nuba.* London: Oxford University Press.

Nakamura, R. R., and M. L. Stanton. 1989. Embryo growth and seed size in *Raphanus sativus:* Maternal and paternal effects *in vivo* and *in vitro. Evolution* 43:1435–43.

Nasrallah, J. B., T.-H Kao, C.-H. Chen, M. L. Goldberg, and M. E. Nasrallah. 1987. Amino acid sequence of glycoproteins encoded by three alleles of the *S*-locus of *Brassica oleracea. Nature* 326:617–19.

Nasrallah, J. B., T.-H Kao, M. L. Goldberg, and M. E. Nasrallah. 1985. A cDNA clone encoding an *S*-locus-specific glycoprotein from *Brassica oleracea. Nature* 318:263–67.

Nasrallah, M. E., and J. B. Nasrallah. 1986. Molecular biology of self-incompatibility in plants. *Trends Genet.* 2:239–44.

Naylor, A. F. 1962. Mating systems which could increase heterozygosity for a pair of alleles *Am. Nat.* 96:51–60.

Nei, M. 1975. *Molecular Population Genetics and Evolution.* Amsterdam: North-Holland.

———. 1978. Estimation of average heterozygosity and genetic distance from a small number of individuals. *Genetics* 89:583–90.

———. 1980. Stochastic theory of population genetics and evolution. In *Symposium on Mathematical Models in Biology,* ed. C. Barigozzi, 17–47. Berlin: Springer-Verlag.

———. 1987. *Molecular Evolutionary Genetics.* New York: Columbia University Press.

Nei, M., T. Maruyama, and R. Chakraborty. 1975. The bottleneck effect and genetic variability in populations. *Evolution* 29(1): 1–10.

Nelder, J. A. 1988. Letter to the editor. *Am. Sci.* 76:431–32.

Nelson, K., and M. Soulé. 1987. Genetical conservation of exploited fishes. In *Population Genetics and Fishery Management,* ed. N. Ryman and F. Utter, 345–68. Seattle: University of Washington Press.

Nettancourt, D. de. 1977. *Incompatibility in Angiosperms.* Berlin: Springer-Verlag.

Nevo, E. 1978. Genetic variation in natural populations: Patterns and theory. *Theor. Popul. Biol.* 13:121–77.

Nevo, E., and A. Beiles. 1991. Genetic diversity and ecological heterogeneity in amphibian evolution. *Copeia* 1991: 565–92.

Nevo, E., A. Beiles, and R. Ben-Shlomo. 1983. The evolutionary significance of genetic diversity: Ecological, demographic and life history correlates. In *Evolutionary Dynamics of Genetic Diversity*, ed. G. S. Mani, 174–206. Heidelberg: Springer-Verlag.

Nevo, E. and S. Y. Yang. 1979. Genetic diversity and climatic determinants of tree frogs in Israel. *Oecologia* 41: 47–63.

Newsweek. 1981. Incest epidemic. November 30:68.

Newton, I., ed. 1989. *Lifetime Reproduction in Birds*. London: Academic Press.

Nichols, J. T. 1939. Range and homing of individual box turtles. *Copeia* 1939: 125–27.

Nishida, M., and J. S. Lucas. 1988. Genetic differences between geographic populations of the Crown-of-thorns starfish throughout the Pacific region. *Mar. Biol.* 98:359–68.

Nishida, T. 1966. A sociological study of solitary male monkeys. *Primates* 7:141–204.

Nishihira, M. 1967. Dispersal of the larvae of a hydroid, *Sertularella miurensis*. *Bull. Mar. Biol. Station Asamushi* 13(1): 49–56.

———. 1968. Dynamics of natural populations of epiphytic Hydrozoa with special reference to *Sertularella miurensis* Stechow. *Bull. Mar. Biol. Station Asamushi* 13(2): 103–24.

Noakes, D. L. G., S. Skúlason, and S. S. Snorrason. 1989. Alternative life history styles in salmonine fishes with emphasis on arctic charr, *Salvelinus alpinus*. In *Alternative Life History Styles of Animals*, ed. M. N. Bruton, 329–46. Dordrecht: Kluwer Academic Publishers.

Noble, J. P. A., and A. Logan. 1981. Size-frequency distributions and taphonomy of brachiopods: A Recent model. *Palaeogeography, Palaeoclimatology, Palaeoecology* 36:87–105.

Nonacs, P. 1986. Ant reproductive strategies and sex allocation theory. *Q. Rev. Biol.* 61:1–21.

Nordeng, H. 1971. Is the local orientation of anadromous fishes determined by pheromones? *Nature* 233: 411–13.

Norikoshi, K., and N. Koyama. 1975. Group shifting and social organization among Japanese monkeys. In *Proceedings from the Symposia of the Fifth Congress of the International Primatological Society*, ed. S. Kondo, M. Kawai, A. Ehara, and S. Kawamura, 43–61. Tokyo: Japan Science Press.

Norris, K. J., and J. K. Blakey. 1989. Evidence for cuckoldry in the great tit *Parus major*. *Ibis* 131:436–42.

Nozawa, K., T. Shotake, Y. Kawamoto, and Y. Tanabe. 1982. Population genetics of Japanese monkeys: II. Blood protein polymorphisms and population structure. *Primates* 23:252–71.

Nunney, L. 1989. The maintenance of sex by group selection. *Evolution* 43:245–57.

Nyholm, K.-G. 1949. On the development and dispersal of *Acthenaria actinia* with special reference to *Halcampa duodecimcirrata*, M. Sars. *Zoologiska Bidrag Uppsala* 27:467–505.

Oates, J. F. 1982. Coat color aberrations in *Presbytis johnii*: A founder effect? *Primates* 23:307–11.

Obbard, M. E., and R. J. Brooks. 1981. A radio-telemetry and mark-recapture study of activity in the common snapping turtle, *Chelydra serpentina*. *Copeia* 1981: 630–37.

Ober, C. L., A. O. Martin, J. L. Simpson, W. W. Hauck, D. B. Amos, D. D. Kostyu, M. Fotino, and F. H. Allen, Jr. 1983. Shared HLA antigens and reproductive performance among Hutterites. *Am. J. Hum. Genet.* 35:994–1004.

O'Brien, S. J., J. S. Martenson, C. Packer, L. Herbst, V. de Vos, P. Joslin, J. Ott-Joslin, D. E. Wildt, and M. Bush. 1987. Biochemical genetic variation in geographic isolates of African and Asiatic lions. *Natl. Geographic Res.* 3: 114–24.

Ó Foighil, D. 1987. Cytological evidence for self-fertilization in *Lasaea subviridis* (Galeommatacea, Bivalvia). *Int. J. Invert. Reprod. Dev.* 12:83–90.

Ó Foighil, D., and C. Thiriot-Quiévreux. 1991. Ploidy and pronuclear interaction in northeastern Pacific *Lasaea* clones (Mollusca: Bivalvia). *Biol. Bull.* 181:222–31.

Ohnishi, O. 1982. Population genetics of cultivated common buckwheat, *Fagopyrum esculentum* Moench. I. Frequency of chlorophyll deficient mutants in Japanese populations. *Jpn. J. Genet.* 57:623–39.

Ohta, A. T. 1980. Coadaptive gene complexes in incipient species of Hawaiian *Drosophila*. *Am. Nat.* 115:121–32.

Ohta, T., and C. C. Cockerham. 1974. Detrimental genes with partial selfing and effects on a neutral locus. *Genet. Res.* 23:191–200.

Oka, H. 1970. Colony specificity in compound ascidians. In *Profiles of Japanese Science and Scientists*, ed. H. Yukawa, 195–206. Tokyo: Kodansha.

Oldham, R. S. 1966. Spring movements in the American toad, *Bufo americanus*. *Can. J. Zool.* 44:63–100.

———. 1967. Orienting mechanisms of the green frog, *Rana clamitans*. *Ecology* 48:477–91.

Oliver, C. G. 1979. Genetic differentiation and hybrid viability within and between some lepidoptera species. *Am. Nat.* 114:681–94.

Oliver, J. H. 1962. A mite parasitic in the cocoons of earthworms. *J. Parasitol.* 48:120–23.

Oliver, J. K., and B. L. Willis. 1987. Coral-spawn slicks in the Great Barrier Reef: Preliminary observations. *Mar. Biol.* 94:521–29.

Olney, P. J. S., P. Ellis, and B. Sommerfelt, eds. 1988. Studbooks and world registers for rare species of wild animals in captivity. *International Zoo Yearbook* 27:482–90.

Olsén, K. H. 1989. Sibling recognition in juvenile Arctic charr, *Salvelinus alpinus* (L.). *J. Fish Biol.* 34:571–81.

Ono, R. D., J. D. Williams, and A. Wagner. 1983. *Vanishing Fishes of North America*. Washington, D.C.: Stone Wall Press.

Orr, H. A. 1987. Genetics of male and female sterility in hybrids of *Drosophila pseudoobscura* and *D. persimilis*. *Genetics* 116:555–63.

———. 1989. Genetics of sterility in hybrids between two species of *Drosophilia*. *Evolution* 43:180–89.

Orr, J., J. P. Thorpe, and M. A. Carter. 1982. Biochemical genetic confirmation of the asexual reproduction of brooded offspring in the sea anemone *Actinia equina*. *Mar. Ecol. Prog. Ser.* 7:227–29.

Ostarello, G. L. 1973. Natural history of the hydrocoral *Allopora californica* Verrill (1866). *Biol. Bull.* 145:548–64.

———. 1976. Larval dispersal in the subtidal hydrocoral *Allopora californica* Verrill (1866). In *Coelenterate Ecology and Behavior*, ed. G. O. Mackie, 331–37. New York: Plenum Press.

Østerbye, U. 1975. Self-incompatibility in *Ranunculus acris* L. *Hereditas* 80:91–112.

Østergård, H., B. Kristensen, and S. Andersen. 1989. Investigations in farm animals of associations between the MHC system and disease resistance and fertility. *Livest. Prod. Sci.* 22:49–67.

Ottaviano, E., M. Sari-Gorla, and D. L. Mulcahy. 1980. Pollen tube growth rates in *Zea mays:* Implications for genetic improvement of crops. *Science* 210:437–38.

Ottaway, J. R. 1979. Population ecology of the intertidal anemone *Actinia tenebrosa*. II. Geographical distribution, synonymy, reproductive cycle and fecundity. *Aust. J. Zool.* 27:273–90.

Overal, W. L., and P. R. Ferreira da Silva. 1982. Population dynamics of the quasisocial spider *Anelosimus eximius* (Araneae: Theridiidae). In *The Biology of the Social Insects: Proceedings of the Ninth Congress of the International Union for the Study of Social Insects*, ed. M. E. Breed, C. D. Michener, and H. E. Evans, 181–82. Boulder: Westview Press.

Packer, C. 1979. Inter-troop transfer and inbreeding avoidance in *Papio anubis*. *Anim. Behav.* 27:1–36.

———. 1985. Dispersal and inbreeding avoidance. *Anim. Behav.* 33:676–78.

———. 1986. The ecology of sociality in felids. In *Ecological Aspects of Social Evolution*, ed. D. I. Rubenstein and R. W. Wrangham, 429–51. Princeton, N.J.: Princeton University Press.

Packer, C., D. Gilbert, A. E. Pusey, and S. J. O'Brien. 1991. A molecular genetic analysis of kinship and cooperation in African lions. *Nature* 351: 562–65.

Packer, C., L. Herbst, A. E. Pusey, J. D. Bygott, J. P. Hanby, S. J. Cairns, and M. Borgerhoff-Mulder. 1988. Reproductive success of lions. In *Reproductive Success*, ed. T. H. Clutton-Brock, 363–83. Chicago: University of Chicago Press.

Packer, C., and A. E. Pusey. 1979. Female aggression and male membership in troops of Japanese macaques and olive baboons. *Folia Primatol.* 31: 212–18.

———. 1982. Cooperation and competition within male lions: Kin selection or game theory? *Nature* 296:740–42.

———. 1983. Male takeovers and female reproductive parameters: A simulation of oestrus synchrony in lions (*Panthera leo*). *Anim. Behav.* 31:334–40.

———. 1984. Infanticide in carnivores. In *Infanticide: Comparative and Evolutionary Perspectives*, ed. G. Hausfater and S. B. Hrdy, 31–42. New York: Aldine.

Packer, C., A. E. Pusey, H. Rowley, D. A. Gilbert, J. Martenson, and S. J.

O'Brien. 1991. Case study of a population bottleneck: Lions of the Ngorongoro Crater. *Conserv. Biol.* 5:219–30.

Pain, J. 1964. Premieres observations sur lesespeces nouvelle d'araignees sociales. *Biologica Gabonica* 1:47–48.

Paine, R. T. 1963. Ecology of the brachiopod *Glottidia pyramidata*. *Ecol. Monogr.* 33(3): 187–213.

Palanichamy, S., and T. J. Pandian. 1983. Incubation, hatching, and yolk utilization in the eggs of the orb weaving spider *Cyrtophora cicatrosa* (Araneae; Araneidae). *Proc. Indian Acad. Sci.* 92:369–74.

Palmer, A. R. 1984. Species cohesiveness and genetic control of shell color and form in *Thais emarginata* (Prosobranchia, Muricacea): Preliminary results. *Malacologia* 25(2):477–91.

Pamilo, P. 1984. Genetic relatedness and evolution of insect sociality. *Behav. Ecol. Sociobiol.* 15:241–48.

Pamilo, P., and R. H. Crozier. 1982. Measuring genetic relatedness in natural populations: Methodology. *Theor. Popul. Biol.* 21:171–93.

Pandey, K. K. 1956. Mutations of self-incompatibility alleles in *Trifolium pratense* and *T. repens*. *Genetics* 41:327–43.

Pandey, K. K. 1969. Elements of the s-gene complex. V. Interspecific cross-compatibility relationships and theory of the evolution of the s complex. *Genetica* 40:447–74.

Parker, G. A. 1974. Assessment strategy and the evolution of animal conflicts. *J. Theor. Biol.* 47:223–43.

Parker, H., and S. Parker. 1986. Father-daughter sexual abuse: An emerging perspective. *Am. J. Orthopsychol.* 56:531–49.

Parker, S. 1976. The pre-cultural basis of the incest taboo: Toward a biosocial theory. *Am. Anthropol.* 78:285–305.

Parker, S. P. 1982. *Synopsis and Classification of Living Organisms*. New York: McGraw-Hill.

Parker, W. S. 1984. Immigration and dispersal of slider turtles *Pseudemys scripta* in Mississippi farm ponds. *Am. Midl. Nat.* 112:280–93.

Parker, W. S., and E. R. Pianka. 1976. Ecological observations of the leopard lizard (*Crotaphytus wislizeni*) in different parts of its range. *Herpetologica* 32: 95–114.

Parsons, T. 1954. The incest taboo in relation to social structure and the socialization of the child. *Br. J. Sociol.* 5:101–7.

Partridge, L. 1983. Non-random mating and offspring fitness. In *Mate Choice*, ed. P. Bateson, 227–56. Cambridge: Cambridge University Press.

———. 1988. The rare-male effect: What is its evolutionary significance? *Phil. Trans. Roy. Soc.* (Lond.), ser. B, 319:525–39.

Pastner, C. M. 1986. The Westermarck hypothesis and first cousin marriage: The cultural modification of negative sexual imprinting. *J. Anthropol. Res.* 24:573–86.

Paterniani, E. 1969. Selection for reproductive isolation between two populations of maize, *Zea mays* L. *Evolution* 23:534–47.

Patterson, J. T., and T. Dobzhansky. 1945. Incipient reproductive isolation between two subspecies of *Drosophila pallidipennis*. *Genetics* 30:429–38.

Patton, J. L. 1985. Population structure and the genetics of speciation in pocket gophers, genus *Thomomys. Acta Zool. Fenn.* 170:109–14.

Patton, J. L., and J. H. Feder. 1981. Microspatial genetic heterogeneity in pocket gophers: Non-random breeding and drift. *Evolution* 35:912–20.

Patton, J. L., and S. W. Sherwood. 1982. Genome evolution in pocker gophers (genus *Thomomys*) I. Heterochromatin variation and speciation potential. *Chromosoma* 85:149–62.

———. 1983. Chromosome evolution and speciation in rodents. *Annu. Rev. Ecol. Syst.* 14:139–58.

Patton, J. L., and M. F. Smith. 1981. Molecular evolution in *Thomomys:* Phyletic systematics, paraphyly, and rates of evolution. *J. Mammal.* 62:493–500.

Patton, J. L., and S. Y. Yang. 1977. Genetic variation in *Thomomys bottae* pocket gophers: Macrogeographic patterns. *Evolution* 31:697–720.

Paul, A., and J. Kuester. 1985. Intergroup transfer and incest avoidance in semifree-ranging Barbary macaques (*Macaca sylvanus*) at Salem (FRG). *Am. J. Primatol.* 8:317–22.

Pawlik, J. R. 1988. Larval settlement and metamorphosis of two gregarious sabellariid polychaetes: *Sabellaria alveolata* compared with *Phragmatopoma californica. J. Mar. Biol. Assoc. UK* 68:101–24.

Payne, R. B., L. L. Payne, and I. Rowley. 1988. Kin and social relationships in splendid fairy-wrens: Recognition by song in a cooperative bird. *Anim. Behav.* 36:1341–51.

Pearse, J. S. 1979. Polyplacophora. In *Reproduction of Marine Invertebrates*, vol. 5, ed. A. C. Giese and J. S. Pearse, 27–85. New York: Academic Press.

Penney, D. F., and E. G. Zimmerman. 1976. Genetic divergence and local population differentiation by random drift in the pocket gopher genus *Geomys. Evolution* 30:473–83.

Pennington, B. J. 1979. Enzyme genetics in taxonomy: Diagnostic enzyme loci in the spider genus *Meta. Bull. Br. Arachnol. Soc.* 4:377–92.

Pennington, J. T. 1985. The ecology of fertilization of echinoid eggs: The consequences of sperm dilution, adult aggregation and synchronous spawning. *Biol. Bull.* 169:417–30.

Pereira, M. E., and M. L. Weiss. 1991. Female mate choice, male migration and the threat of infanticide in ringtailed lemurs. *Behav. Ecol. Sociobiol.* 28: 141–52.

Peterman, R. M. 1990. The importance of reporting statistical power: The forest decline and acidic deposition example. *Ecology* 71:2024–27.

Peters, E. C. 1978. Effects of long-term exposure of the coral *Manicina areolata* (Linné, 1758) to water soluble extracts. M.S. thesis, University of South Florida, St. Petersburg.

Petras, M. L. 1967a. Studies of natural populations of *Mus.* I. Biochemical polymorphisms and their bearing on breeding structure. *Evolution* 21:259–74.

———. 1967b. Studies of natural populations of *Mus.* II. Polymorphisms at the T locus. *Evolution* 21:466–78.

Pettibone, M. H. 1982. Annelida. In *Synopsis and Classification of Living Organisms,* vol. 2, ed. S. P. Parker, 1–61. New York: McGraw-Hill.

Pfennig, D. W. 1990. "Kin regognition" among spadefoot tadpoles: A side effect of habitat selection? *Evolution* 44:785–98.

Philippi, T., and J. Seger. 1989. Hedging one's evolutionary bets, revisited. *Trends Ecol. Evol.* 4:41–44.

Philipsen, M., and B. Kristensen. 1985. Preliminary evidence of segregation distortion in the SLA system. *Anim. Blood Groups Biochem. Genet.* 16:125–33.

Pierce, B. A., and J. B. Mitton. 1982. Allozyme heterozygosity and growth in the tiger salamander, *Ambystoma tigrinum. J. Hered.* 73:250–53.

Piron, R. D. 1978. Spontaneous skeletal deformities in the zebra danio (*Brachydanio rerio*) bred for fish toxicity tests. *J. Fish Biol.* 13:79–83.

Plummer, M. V. 1990. Nesting movements, nesting behavior, and nest sites of green snakes (*Opheodrys aestivus*) revealed by radiotelemetry. *Herpetologica* 46:190–95.

Pollak, E. 1987. On the theory of partially inbreeding finite populations. I. Partial selfing. *Genetics* 117:353–60.

———. 1988. On the theory of partially inbreeding finite populations. II. Partial sib mating. *Genetics* 120:303–11.

Pomiankowski, A. 1987. The costs of choice in sexual selection. *J. Theor. Biol.* 128:195–218.

———. 1988. The evolution of female mate preference for male genetic quality. In *Oxford Surveys in Evolutionary Biology*, vol. 5, ed. P. H. Harvey and L. Partridge, 136–84. Oxford: Oxford University Press.

Porter, C. A., and J. W. Sites, Jr. 1985. Normal disjunction in Robertsonian heterozygotes from a highly polymorphic lizard population. *Cytogenet. Cell Genet.* 39:250–57.

———. 1987. Evolution of *Sceloporus grammicus* complex (Sauria: Iguanidae) in central Mexico. II. Studies on rates of non-disjunction and the occurrence of spontaneous chromosomal mutations. *Genetica* 75:131–44.

Porter, K. R. 1972. *Herpetology.* Philadelphia: W. B. Saunders.

Potswald, H. E. 1968. The biology of fertilization and brood protection in *Spirorbis (Laeospira) morchi. Biol. Bull.* 135:208–22.

Powell, J. R. 1978. The founder-flush speciation theory: An experimental approach. *Evolution* 32:465–74.

Power, D. M. 1979. Evolution in peripheral isolated populations: *Carpodacus* finches on the California islands. *Evolution* 33:834–47.

Powers, D. A. 1987. A multidisciplinary approach to the study of genetic variation within species. In *New Directions in Ecological Physiology*, ed. M. E. Feder, A. F. Bennett, W. W. Burggren, and R. B. Huey, 102–30. Cambridge: Cambridge University Press.

Price, G. R. 1970. Selection and covariance. *Nature* 227:529–31.

———. 1972. Extension of covariance selection mathematics. *Ann. Hum. Genet.* 35:485–90.

Price, M. V., and N. M. Waser. 1979. Pollen dispersal and optimal outcrossing in *Delphinium nelsoni. Nature* 277:294–97.

———. 1982. Population structure, frequency-dependent selection, and the maintenance of sexual reproduction. *Evolution* 36:35–43.

Pugh, S. R., and R. H. Tamarin. 1988. Inbreeding in a population of meadow voles, *Microtus pennsylvanicus*. *Can. J. Zool.* 66:1831–34.

Punzo, F. 1976. The effects of early experience on habitat selection in tadpoles of the Malayan painted frog, *Kaloula pulchra* (Anura: Microhylidae). *J. Bombay Nat. Hist. Soc.* 73:270–77.

Pusey, A. E. 1980. Inbreeding avoidance in chimpanzees. *Anim. Behav.* 28:543–52.

———. 1987. Sex-biased dispersal and inbreeding avoidance in birds and mammals. *Trends Ecol. Evol.* 2:295–99.

———. 1988. Primate dispersal (reply to Moore). *Trends Ecol. Evol.* 3:145–46.

Pusey, A. E., and C. Packer. 1987a. The evolution of sex-biased dispersal in lions. *Behaviour* 101:275–310.

———. 1987b. Philopatry and dispersal. In *Primate Societies*, ed. B. B. Smuts, D. L. Cheney, R. M. Seyfarth, T. T. Struhsaker, and R. W. Wrangham, 250–66. Chicago: University of Chicago Press.

———. 1993. Infanticide in lions: Consequences and counter-strategies. In *Infanticide and Parental Care*, ed. S. Parmigiani, B. Svare, and F. vom Saal. London: Harwood Academic. In press.

Pyefinch, K. A., and F. S. Downing. 1949. Notes on the general biology of *Tubularia larynx* Ellis & Solander. *J. Mar. Biol. Assoc.* 28:21–43.

Quattro, J. M., and R. C. Vrijenhoek. 1989. Fitness differences among remnant populations of the endangered Sonoran topminnow. *Science* 245:976–78.

Queller, D. C. 1983. Kin selection and conflict in seed maturation. *J. Theor. Biol.* 100:153–72.

Queller, D. C., and K. F. Goodnight. 1989. Estimating relatedness using genetic markers. *Evolution* 43:258–75.

Quiatt, D. 1988. Regulation of mating choice in nonhuman primates. In *Human Mating Patterns*, ed. C. G. N. Mascie-Taylor and A. J. Boyce, 133–52. Cambridge: Cambridge University Press.

Quinn, T. P. 1984. Homing and straying in Pacific salmon. In *Mechanisms of Migration in Fishes*, ed. J. D. McCleave, G. P. Arnold, J. J. Dodson, and W. H. Neill, 357–62. New York: Plenum Press.

———. 1985. Homing and the evolution of sockeye salmon (*Oncorhynchus nerka*). In *Migration: Mechanisms and Adaptive Significance*, ed. M. A. Rankin. *Contributions in Marine Science* (Suppl.) 27:353–66.

Quinn, T. P., and C. A. Busack. 1985. Chemosensory recognition of siblings in juvenile coho salmon (*Oncorhynchus kisutch*). *Anim. Behav.* 33:51–56.

Quinn, T. P., and T. J. Hara. 1986. Sibling recognition and olfactory sensitivity in juvenile coho salmon (*Oncorhynchus kisutch*). *Can. J. Zool.* 64:921–25.

Quinn, T. P., and R. F. Tallman. 1987. Seasonal environmental predictability and homing in riverine fishes. *Environ. Biol. Fish.* 18:155–59.

Quinn, T. P., and G. M. Tolson. 1986. Evidence of chemically mediated population recognition in coho salmon (*Oncorhynchus kisutch*). *Can. J. Zool.* 64:84–87.

Quinn, T. W., J. S. Quinn, F. Cooke, and B. N. White. 1987. DNA marker analysis detects multiple maternity and paternity in single broods of the lesser snow goose. *Nature* 326:392–94.

Quinsey, V. L., T. C. Chaplin, and W. F. Carrington. 1979. Sexual preference among incestuous and non-incestuous child molesters. *Behav. Ther.* 10:562–65.

Rabenold, K. N. 1985. Cooperation in breeding by non-reproductive wrens: Kinship, reciprocity, and demography. *Behav. Ecol. Sociobiol.* 17:1–17.

Rabenold, P. P., K. N. Rabenold, W. H. Piper, J. Haydock, and S. W. Zack. 1990. Shared paternity revealed by genetic analysis in cooperatively breeding tropical wrens. *Nature* 348:538–40.

Radha Rama Devi, A. N. Apaji Rao, and A. H. Bittles. 1981. Consanguinity, fecundity and post-natal mortality in Karnatka, India. *Ann. Hum. Biol.* 8:469–72.

Raff, J. W., and R. B. Knox. 1982. Pollen tube growth in *Prunus avium*. In *Pollination '82*, ed. E. G. Williams, R. B. Knox, J. H. Gilbert, and P. Bernhardt, 123–34. Melbourne, Australia: University of Melbourne.

Raff, R. A., J. A. Anstrom, J. E. Chin, K. G. Field, M. T. Ghiselin, D. J. Lane, G. J. Olsen, N. R. Pace, A. L. Parks, and E. C. Raff. 1987. Molecular and developmental correlates of macroevolution. In *Development as an Evolutionary Process*, ed. R. A. Raff and E. C. Raff, 109–38. New York: Alan R. Liss.

Raff, R. A., and T. C. Kaufman. 1983. *Embryos, Genes and Evolution*. New York: Macmillan.

Ralls, K., and J. Ballou. 1982a. Effect of inbreeding on juvenile mortality in some small mammal species. *Lab. Anim.* 16:159–66.

———. 1982b. Effects of inbreeding on infant mortality in captive primates. *Int. J. Primatol.* 3:491–505.

———. 1983. Extinction: Lessons from zoos. In *Genetics and Conservation: A Reference for Managing Wild Animal and Plant Populations*, ed. C. M. Schonewald-Cox, S. M. Chambers, B. MacBryde, and W. L. Thomas, 164–84. Menlo Park, Calif.: Benjamin/Cummings.

Ralls, K., J. D. Ballou, and A. Templeton. 1988. Estimates of lethal equivalents and the cost of inbreeding in mammals. *Conserv. Biol.* 2:185–93.

Ralls, K., K. Brugger, and J. Ballou. 1979. Inbreeding and juvenile mortality in small populations of ungulates. *Science* 206:1101–3.

Ralls, K., K. Brugger, and A. Glick. 1980. Deleterious effects of inbreeding in a herd of captive Dorcas gazelle. *International Zoo Yearbook* 20:137–46.

Ralls, K., P. H. Harvey, and A. M. Lyles. 1986. Inbreeding in natural populations of birds and mammals. In *Conservation Biology: The Science of Scarcity and Diversity*, ed. M. E. Soulé, 35–56. Sunderland, Mass.: Sinauer Associates.

Ramsey, P. R., and J. M. Wakeman. 1987. Population structure of *Sciaenops ocellatus* and *Cynoscion nebulosus* (Pisces: Sciaenidae): Biochemical variation, genetic subdivision and dispersal. *Copeia* 1987: 682–95.

Rao, P. S., and S. G. Inbaraj. 1977. Inbreeding effects on human reproduction in Tamil Nadu of south India. *Ann. Hum. Genet.* 41:87–97.

————. 1979. Trends in human reproductive wastage in relation to long-term practice of inbreeding. *Ann. Hum. Genet.* 42:401–13.

Rasmussen, D. I. 1968. Genetics. In *Biology of Peromyscus (Rodentia)*, ed. J. A. King, 340–72. Special Publication 2, American Society of Mammalogists.

Rasmussen, E. 1973. Systematics and ecology of the Isefjord marine fauna (Denmark), with a survey of the eelgrass (*Zostera*) vegetation and its communities. *Ophelia* 11(1–2): 1–507.

Rawlins, R. G., and M. J. Kessler. 1986. Demography of the free-ranging Cayo Santiago macaques (1976–1983). In *The Cayo Santiago Macaques: History, Behavior and Biology*, ed. R. G. Rawlins and M. J. Kessler, 47–72. Albany: State University of New York Press.

Read, A. F., and P. H. Harvey. 1988. Genetic relatedness and the evolution of animal mating patterns. In *Human Mating Patterns*, ed. C. G. Mascie-Taylor and A. J. Boyce, 115–33. Cambridge: Cambridge University Press.

Record, R. G., and E. Armstrong. 1975. The influence of the birth of a malformed child on the mother's future reproduction. *Br. J. Preventative Med.* 29:267–73.

Reddy, P. G. 1985. Effects of inbreeding on mortality: A study among three South Indian communities. *Hum. Biol.* 57:47–59.

Redford, D. B. 1984. *Akhenaten—The Heretic King*. Princeton, N.J.: Princeton University Press.

Redmond, A. M., L. E. Robbins, and J. Travis. 1989. The effects of pollination distance on seed set in three populations of *Amianthium muscaetoxicum* (Liliaceae). *Oecologia* (Berlin) 79:260–64.

Reed, C. G. 1987a. Phylum Brachiopoda. In *Reproduction and Development of Marine Invertebrates of the Northern Pacific Coast*, ed. M. F. Strathmann, 486–93. Seattle: University of Washington Press.

————. 1987b. Phylum Bryozoa. In *Reproduction and Development of Marine Invertebrates of the Northern Pacific Coast*, ed. M. F. Strathmann, 494–510. Seattle: University of Washington Press.

Rees, W. J. 1957. Evolutionary trends in the classification of capitate hydroids and medusae. *Bull. Br. Mus. (Nat. Hist.) Zool.* 4:455–534.

Reese, J. G. 1980. Demography of European mute swans in Chesapeake Bay. *Auk* 97:449–64.

Reeve, H. K., D. F. Westneat, W. A. Noon, P. W. Sherman, and C. F. Aquadro. 1990. DNA "fingerprinting" reveals high levels of inbreeding in colonies of the eusocial naked mole-rat. *Proc. Natl. Acad. Sci. USA* 87:2496–2500.

Reh, W., and A. Seitz. 1990. The influence of land use on the genetic structure of populations of the common frog *Rana temporaria. Biol. Conserv.* 54:239–50.

Reik, W., A. Collick, M. L. Norris, S. C. Barton, and M. A. Surani. 1987. Genomic imprinting determines the methylation of parental alleles in transgenic mice. *Nature* 328:248–51.

Reimer, J. D., and M. L. Petras. 1967. Breeding structure of the house mouse, *Mus musculus*, in a population cage. *J. Mammal.* 48:88–99.

Reisenbichler, R. R. 1988. Relation between distance transferred from natal stream and recovery rate for hatchery coho salmon. *N.A. J. Fish. Manage.* 8:172–74.

Rettenmeyer, C. W. 1962. The diversity of arthropods found with Neotropical army ants and observations on the behavior of representative species. *Proc. North Central Branch, Am. Assoc. Economic Entomol.* 17:14–15.

Rice, W. R. 1985. Disruptive selection on habitat preference and the evolution of reproductive isolation: An exploratory experiment. *Evolution* 39:645–56.

Rich, S. S., A. E. Bell, and S. P. Wilson. 1979. Genetic drift in small populations of *Tribolium. Evolution* 33:579–84.

Richard, A. F. 1987. Malagasy prosimians: Female dominance. In *Primate Societies,* ed. B. B. Smuts, D. L. Cheney, R. M. Seyfarth, T. T. Struhsaker, and R. W. Wrangham, 25–33. Chicago: University of Chicago Press.

Richards, A. J. 1986. *Plant Breeding Systems.* Boston: George Allen and Unwin.

Richdale, L. E. 1957. *A Population Study of Penguins.* Oxford: Clarendon Press.

Richmond, R. H. 1987. Energetics, competency, and long-distance dispersal of planula larvae of the coral *Pocillopora damicornis. Mar. Biol.* 93:527–33.

————. 1990. Relationships among reproductive mode, biogeographic distribution patterns and evolution in scleractinian corals. *Adv. Invert. Reprod.* 5:317–22.

Richmond, R. H., and C. L. Hunter. 1990. Reproduction and recruitment of corals: Comparisons among the Caribbean, the tropical Pacific and the Red Sea. *Mar. Ecol. Prog. Ser.* 60:185–203.

Ricker, W. E. 1972. Heredity and environmental factors affecting certain salmonid populations. In *The Stock Concept in Pacific Salmon,* ed. R. C. Simon and P. A. Larkin, 19–160. H. R. MacMillan Lectures in Fisheries. Vancouver: University of British Columbia.

Riechert, S. E. 1982. Spider interaction strategies: Communication vs. coercion. In *Spider Communication: Mechanisms and Ecological Significance,* ed. P. N. Witt and J. Rovner. Princeton, N.J.: Princeton University Press.

————. 1985. Why do some spiders cooperate? *Agelena consociata,* a case study. *Fla. Entomol.* 17:105–16.

Riechert, S. E., R. Roeloffs, and A. C. Echternacht. 1986. The ecology of the cooperative spider *Agelena consociata* in equatorial Africa. *J. Arachnol.* 14:175–91.

Rinkevich, B., and Y. Loya. 1979a. The reproduction of the Red Sea coral *Stylophora pistillata.* I. Gonads and planulae. *Mar. Ecol. Prog. Ser.* 1:133–44.

————. 1979b. The reproduction of the Red Sea coral *Stylophora pistillata.* II. Synchronization in breeding and seasonality of planulae shedding. *Mar. Ecol. Prog. Ser.* 1:145–52.

Rinkevich, B., and I. L. Weissman. 1989. Variation in the outcomes following chimera formation in the colonial tunicate *Botryllus schlosseri. Bull. Mar. Sci.* 45(2): 213–27.

Ritland, K. 1983. Estimation of mating systems. In *Isozymes in Plant Genetics and Breeding,* ed. S. D. Tanksley and T. J. Orton. New York: Elsevier.

————. The effective proportion of self-fertilization with consanguineous matings in inbred populations. *Genetics* 106:139–52.

————. 1985. The genetic mating structure of subdivided populations. I. Openmating model. *Theor. Popul. Biol.* 27:51–74.

————. 1990. Inferences about inbreeding depression based upon changes of the inbreeding coefficient. *Evolution* 44:1230–41.

Ritland, K., and F. R. Ganders. 1985. Variation in the mating system of *Bidens menziesii* (Asteraceae) in relation to population substructure. *Heredity* 55:235–44.

————. 1987a. Covariation of selfing rates with parental gene fixation indices within populations of *Mimulus guttatus*. *Evolution* 41:760–71.

————. 1987b. Crossability of *Mimulus guttatus* in relation to components of gene fixation. *Evolution* 41:722–86.

Ritland, K., and S. K. Jain. 1981. A model for the estimation of outcrossing rate and gene frequencies using *n* independent loci. *Heredity* 47:35–52.

Roach, D. A., and R. D. Wulff. 1987. Maternal effects in plants. *Annu. Rev. Ecol. Syst.* 18:209–35.

Robbins, L. W., G. D. Hartman, and M. H. Smith. 1987. Dispersal, reproductive strategies, and the maintenance of genetic variability in mosquitofish, *Gambusia affinis*. *Copeia* 1987:156–64.

Robbins Leighton, D. 1987. Gibbons: Territoriality and monogamy. In *Primate Societies*, ed. B. B. Smuts, D. L. Cheney, R. M. Seyfarth, T. T. Struhsaker, and R. W. Wrangham, 135–45. Chicago: University of Chicago Press.

Roberts, D. F., and B. Bonne. 1981. Reproduction and inbreeding among the Samaritans. *Soc. Biol.* 20:64–70.

Robertson, D. R. 1973. Field observations on the reproductive behaviour of a pomacentrid fish, *Acanthochromis polyacanthus*. *Z. Tierpsychol.* 32:319–24.

Robinson, H. F., and R. E. Comstock. 1955. Analysis of genetic variability in corn with reference to possible effects of selection. *Cold Spring Harb. Symp. Quant. Biol.* 20:127–36.

Robinson, J. G. 1988a. Demography and group structure in wedge-capped capuchin monkeys, *Cebus olivaceus*. *Behaviour* 104:202–32.

————. 1988b. Group size in wedge-capped capuchin monkeys *Cebus olivaceus* and the reproductive success of males and females. *Behav. Ecol. Sociobiol.* 23:187–97.

Rockwell, R. F., and G. F. Barrowclough. 1987. Gene flow and the genetic structure of populations. In *Avian Genetics*, ed. F. Cooke and P. A. Buckley, 223–55. London: Academic Press.

Rodda, G. H. 1984. Homeward paths of displaced juvenile alligators as determined by radiotelemetry. *Behav. Ecol. Sociobiol.* 14:241–46.

Rodger, J. C., and B. L. Drake. 1987. The enigma of the fetal graft. *Am. Sci.* 75:51–57.

Rodhouse, P. G., and P. M. Gaffney. 1984. Effect of heterozygosity on metabolism during starvation in the American oyster *Crassostrea virginica*. *Mar. Biol.* 80:179–88.

Rodhouse, P. G., J. H. McDonald, R. I. E. Newell, and R. K. Koehn. 1986.

Gamete production, somatic growth and multiple locus heterozygosity in *Mytilus edulis* L. *Mar. Biol.* 90:209–14.

Roeloffs, R. M., and S. E. Riechert. 1988. Dispersal and population genetic structure of the cooperative spider, *Agelena consociata* in west African rainforest. *Evolution* 42:173–83.

Ronen, A., I. Ronen, and E. Goldschmidt. 1963. Marriage systems. In *The Genetics of Migrant and Isolate Populations*, ed. E. Goldschmidt. New York: William and Wilkins.

Rood, J. L. 1987. Migration and dispersal among dwarf mongooses. In *Mammalian Dispersal Patterns: The Effects of Social Structure on Population Genetics*, ed. B. D. Chepko-Sade and Z. T. Halpin, 85–103. Chicago: University of Chicago Press.

Roscoe, J. 1911. *The Baganda: An Account of Their Native Customs and Beliefs*. London: Macmillian.

Rosenberger, A. L., and K. B. Strier. 1989. Adaptive radiation of the ateline primates. *J. Hum. Evol.* 18:717–50.

Rossi, L. 1975. Sexual races in *Cereus pedunculatus* (Boad). *Pubbl. Staz. Zool. Napoli* 39(suppl.): 462–70.

Rothenbuhler, N. 1964. Behaviour genetics of nest cleaning in honeybees. I. *Anim. Behav.* 12:578–83.

Rottenberg, D. 1977. *Finding Our Fathers: A Guidebook to Jewish Genealogy*. New York: Random House.

Roughgarden, J. 1979. *Theory of Population Genetics and Evolutionary Ecology: An Introduction*. New York: Macmillan.

———. 1989. The evolution of marine life cycles. In *Mathematical Evolutionary Theory*, ed. M. W. Feldman, 270–300. Princeton, N.J.: Princeton University Press.

Rowell, T. E. 1988. Beyond the one-male group. *Behaviour* 104:189–201.

Rowley, I., and M. G. Brooker. 1986. The response of small insectivorous birds to fire in heathlands. In *Nature Conservation: The Role of Remnants of Native Vegetation*, ed. D. A. Saunders, G. W. Arnold, A. A. Burbidge, and A. J. M. Hopkins, 211–18. Sydney: Surrey Beatty and Sons.

Rowley, I., and E. M. Russell. 1989. Lifetime reproductive success in the splendid fairy-wren. In *Lifetime Reproduction in Birds*, ed. I. Newton, 233–52. London: Academic Press.

———. 1990a. Splendid fairy-wrens: Demonstrating the importance of longevity. In *Cooperative Breeding in Birds*, ed. P. B. Stacey and W. D. Koenig, 3–30. Cambridge: Cambridge University Press.

———. 1990b. Philandering—A mixed mating strategy in the splendid fairy wren. *Behav. Ecol. and Sociobiol.* 27:431–37.

———. 1991. Demography of passerines in the temperate southern hemisphere. In *Bird Population Studies: Relevance to Conservation and Management*, ed. C. M. Perrins, J. D. Lebreton, and G. M. Hirons, 22–44. Oxford: Oxford University Press.

Rowley, I., E. M. Russell, and M. G. Brooker. 1986. Inbreeding: Benefits may outweight costs. *Anim. Behav.* 34:939–41.

Roychoudhury, A. K., and K. S. Sankhala. 1979. Inbreeding in white tigers. *Proceedings of the Indian Academy of Sciences* 88B Part 1, No. 5:311–23.

Ruas, P. M., and N. Freire-Maia. 1984. Inbreeding effects on morbidity. I. Three analyses (one with cousin and sib controls) of two surveys among Brazilian whites and negroes. *Am. J. Hum. Genet.* 18:381–400.

Rudran, R. 1979. The demography and social mobility of a red howler (*Alouatta seniculus*) population in Venezuela. In *Vertebrate Ecology in the Northern Neotropics*, ed. J. F. Eisenberg, 107–26. Washington, D.C.: Smithsonian Institution Press.

Ruiz de Elvira, M.-C., and J. G. Herndon. 1986. Disruption of sexual behaviour by high ranking rhesus monkeys (*Macaca mulatta*). *Behaviour* 96:227–40.

Rumiz, D. I. 1990. *Alouatta caraya*: Population density and demography in Northern Argentina. *Am. J. Primatol.* 21:279–94.

Russell, D. E. 1986. *The Secret Trauma*. New York: Basic Books.

Russell, E. M., and I. Rowley. 1992. Philopatry or dispersal: Competition for territory vacancies in the splendid fairy-wren. *Anim. Behav.* In press.

Rutherford, J. C. 1973. Reproduction, growth and mortality of the holothurian *Cucumaria pseudocurata*. *Mar. Biol.* 22:167–76.

Ryan, M. J. 1985. *The Tungara Frog, A Study in Sexual Selection and Communication*. Chicago: University of Chicago Press.

————. 1990. Sexual selection, sensory systems, and sensory exploitation. *Oxford Surveys in Evolutionary Biology* 7:157–95.

Ryland, J. S. 1970. *Bryozoans*. London: Hutchinson University Library.

————. 1974. Behaviour, settlement and metamorphosis of bryozoan larvae: A review. *Thalassia Jugoslavica* 10(1/2):239–62.

————. 1982. Bryozoa. In *Synopsis and Classification of Living Organisms*, vol. 2, ed. S. P. Parker, 743–69. New York: McGraw-Hill.

Ryland, J. S., and J. D. D. Bishop. 1990. Prevalence of cross-fertilization in the hermaphroditic compound ascidian *Diplosoma listerianum*. *Mar. Ecol. Prog. Ser.* 61:125–32.

Ryman, N. 1970. A genetic analysis of recapture frequencies of released young of salmon (*Salmo salar* L.). *Hereditas* 65:159–60.

————. 1983. Patterns of distribution of biochemical genetic variation in salmonids: Differences between species. *Aquaculture* 33:1–21.

Ryman, N., F. W. Allendorf, and G. Stahl. 1979. Reproductive isolation with little genetic divergence in sympatric populations of brown trout (*Salmo trutta*). *Genetics* 92:247–62.

Sabbadin, A. 1971. Self- and cross-fertilization in the compound ascidian *Botryllus schlosseri*. *Dev. Biol.* 24:379–91.

————. 1978. Genetics of the colonial ascidian, *Botryllus schlosseri*. In *Marine Organisms: Genetics, Ecology, and Evolution*, ed. B. Battaglia and J. A. Beardmore, 195–209. New York: Plenum Press.

————. 1982. Formal genetics of ascidians. *Am. Zool.* 22:765–73.

Sade, D. S. 1968. Inhibition of son-mother mating among free-ranging rhesus monkeys. *Sci. Psychoanal.* 12:18–38.

St. Louis, V. L., and J. C. Barlow. 1988. Genetic differentiation among ancestral and introduced populations of the Eurasian tree sparrow (*Passer montanus*). *Evolution* 42:266–76.

Sammarco, P. W., and J. C. Andrews. 1988. Localized dispersal and recruitment in Great Barrier Reef corals: The Helix experiment. *Science* 239:1422–24.

Samollow, P. B., and M. E. Soulé. 1983. A case of stress-related heterozygote superiority in nature. *Evolution* 37:646–49.

Sankaranarayanan, K. 1988. Mobile genetic elements, spontaneous mutations and the assessment of genetic radiation hazards in man. In *Eukaryotic Transposable Elements as Mutagenic Agents*, ed. M. E. Lambert, J. F. McDonald, and I. B. Weinstein, 319–36. Cold Spring Harbor, N.Y.: Cold Spring Harbor Press.

Sanz, C. 1945. Pollen-tube growth in intergeneric pollinations on *Datura stramonium*. *Proc. Natl. Acad. Sci. USA* 31:361–67.

Sarvetnick, N., H. S. Fox, E. Mann, P. E. Mains, R. W. Elliott, and L. M. Silver. 1986. Nonhomologous pairing in mice heterozygous for a *t* haplotype can produce recombinant chromosomes with duplications and deletions. *Genetics* 113:723–34.

Sastry, A. N. 1979. Pelecypoda (excluding Ostreidae). In *Reproduction of Marine Invertebrates*, vol. 5, ed. A. C. Giese and J. S. Pearse, 113–292. New York: Academic Press.

Saunders, L. C., and E. M. Lord. 1989. Directed movement of latex particles in the gynoecia of three species of flowering plants. *Science* 243:1606–8.

Sausman, K. A. 1984. Survival of captive-born *Ovis canadensis* in North American zoos. *Zoo Biol.* 3:111–21.

Schaal, B. A., and D. A. Levin. 1976. The demographic genetics of *Liatris cylindracea* Michx. *Am. Nat.* 110:191–206.

Schaller, G. B. 1972. *The Serengeti Lion*. Chicago: University of Chicago Press.

Scharff, J. W. 1981. Free enterprise and the ghetto family. *Pyschol. Today* (March): 41–48.

Scheerer, P. D., G. H. Thorgaard, F. W. Allendorf, and K. L. Knudsen. 1986. Androgenetic rainbow trout produced from inbred and outbred sperm sources show similar survival. *Aquaculture* 57:289–98.

Scheltema, R. S. 1971. Larval dispersal as a means of genetic exchange between geographically separated populations of shallow-water benthic marine gastropods. *Biol. Bull.* 140(2): 284–322.

———. 1986a. Long-distance dispersal by planktonic larvae of shoal-water benthic invertebrates among central Pacific islands. *Bull. Mar. Sci.* 39(2):241–56.

———. 1986b. On dispersal and planktonic larvae of benthic invertebrates: An eclectic overview and summary of problems. *Bull. Mar. Sci.* 39(2): 290–322.

Scheltema, R. S., I. P. Williams, M. A. Shaw, and C. Loudon. 1981. Gregarious settlement by the larvae of *Hydroides dianthus* (Polychaeta: Serpulidae). *Mar. Ecol. Prog. Ser.* 5:69–74.

Schemske, D. W. 1978. Evolution of reproductive characteristics in *Impatiens* (Balsaminaceae): The significance of cleistogamy and chasmogamy. *Ecology* 59:596–613.

———. 1984. Population structure and local selection in *Impatiens pallida* (Balsamaceae), a selfing annual. *Evolution* 38:817–32.

Schemske, D. W., and R. Lande. 1985. The evolution of self-fertilization and inbreeding depression in plants. II. Empirical observations. *Evolution* 39(1): 41–52.

Schemske, D. W., and L. P. Pautler. 1984. The effects of pollen composition on fitness components in a neotropical herb. *Oecologia* 62:31–36.

Schmidt, H. 1967. A note on the sea-anemone *Bunodactis verrucosa* Pennant. *Pubbl. Staz. Zool. Napoli* 35:252–53.

Schmitt, J., and J. Antonovics. 1986. Experimental studies on the evolutionary significance of sexual reproduction. IV. Effect of neighbor relatedness and aphid infestation on seedling performance. *Evolution* 40:830–36.

Schmitt, J., and D. Ehrhardt. 1987. A test of the sib competition hypothesis for outcrossing advantage in *Impatiens capensis*. *Evolution* 41:579–90.

———. 1990. Enhancement of inbreeding depression by dominance and suppression in *Impatiens capensis*. *Evolution* 44:269–78.

Schmitt, J., and S. E. Gamble. 1990. The effect of distance from the parental site on offspring performance in *Impatiens capensis:* A test of the local adaptation hypothesis. *Evolution* 44:2022–30.

Schnee, B. K., and D. M. Waller. 1986. Reproductive behavior of *Amphicarpaea bracteata* (Leguminosae), an amphicarpic annual. *Am. J. Bot.* 73:376–86.

Schoen, D. J. 1982. The breeding system of *Gilia achilleifolia:* Variation in floral characteristics and outcrossing rate. *Evolution* 36:352–60.

Schoen, D. J., and M. T. Clegg. 1984. Estimation of mating system parameters when outcrossing events are correlated. *Proc. Natl. Acad. Sci. USA* 81:5258–62.

Schou, O., and M. Philipp. 1983. An unusual heteromorphic incompatibility system. II. Pollen tube growth and seed sets following compatible and incompatible crossings within *Anchusa officinalis* L. (Boraginaceae). In *Pollen: Biology and Implications for Plant Breeding*, ed. D. L. Mulcahy and E. Ottaviano, 219–27. New York: Elsevier.

Schroeder, P. C., and C. O. Hermans. 1975. Annelida: Polychaeta. In *Reproduction of Marine Invertebrates*, vol. 3, ed. A. C. Giese and J. S. Pearse, 1–213. New York: Academic Press.

Schubert, C. A., L. M. Ratcliffe, and P. T. Boag. 1989. A test of inbreeding avoidance in the zebra finch. *Ethology* 82:265–74.

Schull, W. J. 1958. Empirical risks in consanguineous marriages: Sex ratio, malformation and viability. *Am. J. Hum. Genet.* 10:294–343.

———, ed. 1963. *Genetic Selection in Man*. Ann Arbor: University of Michigan Press.

Schull, W. J., H. Nagano, M. Yamamoto, and J. Komatus. 1970. The effects of parental consanguinity and inbreeding in Hirado, Japan. I. Stillbirths and pre-reproductive mortality. *Am. J. Hum. Genet.* 22:239–62.

Schull, W. J., and J. V. Neel. 1965. *The Effects of Inbreeding in Japanese Children*. New York: Harper and Row.

Schulz, B., F. Banuett, M. Dahl, R. Schlesinger, W. Schäfer, T. Martin, I. Herskowitz, and R. Kahmann. 1990. The *b* alleles of *U. maydis*, whose combinations program pathogenic development, code for polypeptides containing a homeodomain-related motif. *Cell* 80:295–306.

Schwaegerle, K. E., and B. A. Schaal. 1979. Genetic variability and founder effect in the pitcher plant *Sarracenia purpurea* L. *Evolution* 33:1210–18.

Schwartz, O. A., and K. B. Armitage. 1980. Genetic variation in social mammals: The marmot model. *Science* 207:665–67.

Scofield, V. L., J. M. Schlumpberger, and I. L. Weissman. 1982. Colony specificity in the colonial tunicate *Botryllus* and the origins of vertebrate immunity. *Am. Zool.* 22:783–94.

Scott, P. E. 1989. Ecological consequences of variation in pollinator availability: Ocotillo, carpenter bees, and hummingbirds in two deserts. Ph.D. dissertation, Louisiana State University, Baton Rouge.

Scribner, K. T., J. E. Evans, S. J. Morreale, M. H. Smith, and J. W. Gibbons. 1986. Genetic divergence among populations of the yellow-bellied slider turtle (*Pseudemys scripta*) separated by aquatic and terrestrial habitats. *Copeia* 1986: 691–700.

Seavey, S. R., and K. S. Bawa. 1986. Late-acting self-incompatibility in angiosperms. *Bot. Rev.* 52:195–219.

Sebens, K. P. 1982. Asexual reproduction in *Anthopleura elegantissima* (Anthozoa: Actiniaria): Seasonality and spatial extent of clones. *Ecology* 63(2):434–44.

————. 1983a. The larval and juvenile ecology of the temperate octocoral *Alcyonium siderium* Verrill. I. Substratum selection by benthic larvae. *J. Exp. Mar. Biol. Ecol.* 71:73–89.

————. 1983b. The larval and juvenile ecology of the temperate octocoral *Alcyonium siderium* Verrill. II. Fecundity, survival, and juvenile growth. *J. Exp. Mar. Biol. Ecol.* 72:263–85.

————. 1983c. Population dynamics and habitat suitability of the intertidal sea anemones *Anthopleura elegantissima* and *A. xanthogrammica*. *Ecol. Monogr.* 53(4): 405–33.

————. 1983d. Settlement and metamorphosis of a temperate soft-coral larva (*Alcyonium siderium* Verrill): Induction by crustose algae. *Biol. Bull.* 165:286–304.

————. 1986. Spatial relationships among encrusting marine organisms in the New England subtidal zone. *Ecol. Monogr.* 56(1): 73–96.

Seed, R. 1976. Ecology. In *Marine Mussels: Their Ecology and Physiology*, ed. B. L. Bayne, 13–65. Cambridge: Cambridge University Press.

Seemanova, E. 1971. A study of children of incestuous matings. *Hum. Hered.* 21:108–28.

Seger, J., and W. D. Hamilton. 1988. Parasites and sex. In *The Evolution of Sex: An Examination of Current Ideas*, ed. R. E. Michod and B. R. Levin. Sunderland, Mass.: Sinauer Associates.

Seibt, U., and W. Wickler. 1988. Bionomics and social structure of "Family Spiders" of the genus *Stegodyphus*, with special reference to the African species *S. dumicola* and *S. mimosarum* (Arachnida, Eeresida). *Verh. naturwiss. Ver. Hamburg (NF)* 30:256–301.

Selander, R. K. 1970. Biochemical polymorphism in populations of the house mouse and old-field mouse. *Symp. Zool. Soc. Lond.* 26:73–91.

———. 1976. Genetic variation in natural populations. In *Molecular Evolution*, ed. F. J. Ayala. Sunderland, Mass.: Sinauer Associates.

Seligman, B. Z. 1932. The incest barrier: Its role in social organization. *Br. J. Psychol.* 22:250–76.

Seligman, C. G., and B. Z. Seligman. 1932. *Tribes of the Nilotic Sudan*. London: George Routledge and Sons.

Sella, G. 1985. Reciprocal egg trading and brood care in a hermaphroditic polychaete worm. *Anim. Behav.* 33:938–44.

———. 1990. Sex allocation in the simultaneously hermaphroditic polychaete worm *Ophryotrocha diadema*. *Ecology* 71:27–32.

Sellmer, G. 1967. Functional morphology and ecological life history of the gem clam, *Gemma gemma* (Eulamellibranchia; Veneridae). *Malacologia* 5:137–223.

Sene, F. M., and H. L. Carson. 1977. Genetic variation in Hawaiian *Drosophila*. IV. Allozymic similarity between *D. silverstris* and *D. heteroneura* from the island of Hawaii. *Genetics* 86:187–98.

Serradilla, J. M., and F. J. Ayala. 1983. Alloprocoptic selection: A mode of natural selection promoting polymorphism. *Proc. Natl. Acad. Sci. USA* 80:2022–25.

Shaffer, H. B. 1984a. Evolution in a paedomorphic lineage. I. An electrophoretic analysis of the Mexican ambysomatid salamanders. *Evolution* 38:1194–1206.

———. 1984b. Evolution in a paedomorphic lineage. II. Allometry and form in the Mexican ambystomatid salamanders. *Evolution* 38:1207–18.

Shaffer, M. 1987. Minimum viable populations: Coping with uncertainity. In *Viable Populations for Conservation*, ed. M. E. Soulé, 69–86. Cambridge: Cambridge University Press.

Shaklee, J. B., and C. S. Tamaru. 1977. Biochemical and morphological evidence of sibling species of bonefish, *Albula vulpes*. *Am. Zool.* 17:973.

Shami, S. A., and Zahida. 1982. Study of consanguineous marriages in the population of Lahore, Punjab, Pakistan. *Biologia* 28:1–15.

Shapiro, D. Y. 1983. On the possibility of kin groups in coral reef fishes. In *The Ecology of Deep and Shallow Coral Reefs*, ed. M. L. Reaka, 39–45. Washington, D.C.: United States Dept. of Commerce.

Shapiro, D. Y., D. A. Hensley, and R. S. Appeldoorn. 1988. Pelagic spawning and egg transport in coral-reef fishes: A skeptical overview. *Environ. Biol. Fish.* 22:3–14.

Shapovalov, L., and A. C. Taft. 1954. The life histories of the steelhead rainbow trout (*Salmo gairdneri gairdneri*) and silver salmon (*Oncorhynchus kisutch*). California Dept. of Fish and Game, Fish Bulletin (98): 375 pp.

Sharp, P. L. 1973. Behaviour of the pika (*Ochotona princeps*) in the Kananaskis region of Alberta. M. S. thesis, University of Alberta, Edmonton.

Sharp, P. M. 1984. The effect of inbreeding on competitive male mating ability in *Drosophila melanogaster*. *Genetics* 106:601–12.

Shaw, R. G. 1986. Response to density in a wild population of the perennial herb *Salvia lyrata*: Variation among families. *Evolution* 40:492–505.

Shepher, J. 1971. *Self-imposed Incest Avoidance and Exogamy in Second Generation Kibbutz Adults*. New York: University Microfilms.

———. 1983. *Incest: A Biosocial View*. New York: Academic Press.

Sheridan, M., and R. H. Tamarin. 1986. Kinships in a natural meadow vole population. *Behav. Ecol. Sociobiol.* 19:207–11.

Sherman, P. W. 1977. Nepotism and the evolution of alarm calls. *Science* 197:1246–53.

———. 1980. The limits of ground squirrel nepotism. In *Sociobiology: Beyond Nature/Nurture?* ed. G. W. Barlow and J. Silverberg, 505–44. Boulder: Westview Press.

———. 1989. The clitoris debate and levels of analysis. *Anim. Behav.* 37:697–98.

Sherman, P. W., J. U. M. Jarvis, and R. D. Alexander, eds. 1991. *The Biology of the Naked Mole-rat*. Princeton, N.J.: Princeton University Press.

Sherman, P. W., and M. L. Morton. 1984. Demography of Belding's ground squirrels. *Ecology* 65:1617–28.

———. 1988. Extra-pair fertilizations in mountain white-crowned sparrows. *Behav. Ecol. Sociobiol.* 22:413–20.

Sherman, P. W., T. D. Seeley, and H. K. Reeve. 1988. Parasites, pathogens and polyandry in social Hymenoptera. *Am. Nat.* 131:602–10.

Sherwood, S. W., and J. L. Patton. 1982. Genome evolution in pocket gophers (genus *Thomomys*) II. Variation in cellular DNA content. *Chromosoma* 85:163–79.

Shick, J. M. 1991. *A Functional Biology of Sea Anemones*. London: Chapman and Hall.

Shick, J. M., R. J. Hoffmann, and A. N. Lamb. 1979. Asexual reproduction, population structure, and genotype-environment interactions in sea anemones. *Am. Zool.* 19:699–713.

Shick, J. M., and A. N. Lamb. 1977. Asexual reproduction and genetic population structure in the colonizing sea anemone *Haliplanella luciae*. *Biol. Bull.* 153:604–17.

Shields, W. M. 1979. Philopatry, inbreeding and the adaptive advantages of sex. Ph.D. dissertation, Ohio State University, Columbus.

———. 1982. *Philopatry, Inbreeding, and the Evolution of Sex*. Albany: State University of New York Press.

———. 1983. Optimal inbreeding and the evolution of philopatry. In *The Ecology of Animal Movement*, ed. I. R. Swingland and P. J. Greenwood, 132–59. Oxford: Clarendon Press.

———. 1987. Dispersal and mating systems: Investigating their causal connections. In *Mammalian Dispersal Patterns: The Effects of Social Structure on Popu-*

lation Genetics, ed. B. D. Chepko-Sade and Z. T. Halpin, 3–24. Chicago: University of Chicago Press.

————. 1988. Sex and adaptation. In *The Evolution of Sex: An Examination of Current Ideas*, ed. R. E. Michod and B. R. Levin, 253–69. Sunderland, Mass.: Sinauer Associates.

Shin, H.-S., J. Stavnezer, D. Bennett, and K. Artzt. 1982. Genetic structure and origin of *t* haplotypes of mice, analyzed with H-2 cDNA probes. *Cell* 29:969–76.

Shoemaker, A. H. 1982. The effect of inbreeding and management on propagation of pedigree leopards. *International Zoo Yearbook* 22:198–206.

Shoemaker, J. S., and D. M. Waller. N.d. Genetic consequences of outcrossing in the cleistogamous annual, *Impatiens capensis* IV. Microgeographic population structure. Manuscript.

Short, R. V., A. C. Chandley, R. C. Jones, and W. R. Allen. 1974. Meiosis in interspecific equine hybrids. II. The Przewalski horse/domestic horse hybrid. *Cytogenet. Cell Genet.* 13:465–78.

Shull, G. H. 1948. What is "heterosis"? *Genetics* 33:439–46.

Sigg, H., A. Stolba, J. J. Abegglen, and V. Dasser. 1982. Life history of hamadryas baboons: Physical development, infant mortality, reproductive parameters and family relationships. *Primates* 23:473–87.

Silver, L. M. 1982. Genomic analysis of the H-2 complex region associated with mouse *t* haplotypes. *Cell* 29:961–68.

————. 1985. Mouse *t* haplotypes. *Annu. Rev. Genet.* 19:179–208.

Silver, L. M., M. Hammer, H. Fox, J. Garrels, M. Bucan, B. Herrmann, A.-M. Frischauf, H. Lehrach, H. Winking, F. Figueroa, and J. Klein. 1987. Molecular evidence for the rapid propagation of mouse *t* haplotypes from a single, recent, ancestral chromosome. *Mol. Biol. Evol.* 4:473–82.

Simanek, D. E. 1978. Genetic variability and population structure of *Poecilia latipinna*. *Nature* 276:612–14.

Simmons, M. J., and J. F. Crow. 1977. Mutations affecting fitness in *Drosophila* populations. *Annu. Rev. Genet.* 11:49–78.

Singh, S. M. 1982. Enzyme heterozygosity associated with growth at different developmental stages in oysters. *Can. J. Genet. Cytol.* 24:451–58.

Singh, S. M., and E. Zouros. 1978. Genetic variation associated with growth rate in the American oyster (*Crassostrea virginica*). *Evolution* 32:342–53.

Singleton, G. R., and D. A. Hay. 1983. The effect of social organization on reproductive success and gene flow in colonies of wild house mice, *Mus musculus*. *Behav. Ecol. Sociobiol.* 12:49–56.

Sinsch, U. 1987. Orientation behaviour of toads (*Bufo bufo*) displaced from the breeding site. *J. Comp. Physiol. A* 161:715–27.

————. 1990. Migration and orientation in anuran amphibians. *Ethol. Ecol. Evol.* 2:65–79.

Sjögren, P. 1991. Genetic variation in relation to demography of peripheral pool frog populations (*Rana lessonae*). *Evol. Ecol.* 5:248–71.

Skibinski, D. O. F. 1983. Natural selection in hybrid mussel populations. In

Protein Polymorphism: Adaptive and Taxonomic Significance, ed. G. S. Oxford and D. Rollinson, 283–98. London: Academic Press.

Slater, P. J. B., and F. A. Clements. 1981. Incestuous mating in zebra finches. *Z. Tierpsychol.* 57:201–8.

Slatis, H. M. 1960. An analysis of inbreeding in the European bison. *Genetics* 45:275–87.

———. 1975. *Research in Zoos and Aquariums.* Washington, D.C.: National Academy of Sciences.

Slatis, H. M., and R. E. Hoene. 1961. The effect of consanguinity on the distribution of continuously variable characteristics. *Am. J. Hum. Genet.* 13:28–31.

Slatis, H. M., R. H. Reis, and R. E. Hoene. 1958. Consanguineous marriages in the Chicago region. *Am. J. Hum. Genet.* 6:444–64.

Slatkin, M. 1975. Gene flow and selection in a cline. *Genetics* 75:733–56.

———. 1981. Estimating levels of gene flow in natural populations. *Genetics* 99:323–35.

———. 1985. Gene flow in natural populations. *Annu. Rev. Ecol. Syst.* 16:393–430.

———. 1987. Gene flow and the geographic structure of natural populations. *Science* 236:787–92.

Slovenko, R. 1979. Legal briefs: Incest. *SEICUS Report* 7:4–5.

Small, M. F. 1984. Aging and reproductive success in female *Macaca mulatta.* In *Female Primates: Studies by Women Primatologists,* ed. M. F. Small, 249–59. New York: Alan R. Liss.

Smith, A. T. 1974a. The distribution and dispersal of pikas: Consequences of insular population structure. *Ecology* 55:1112–19.

———. 1974b. The distribution and dispersal of pikas: Influences of behavior and climate. *Ecology* 55:1368–76.

———. 1978. Comparative demography of pikas (*Ochotona*): Effect of spatial and temporal age-specific mortality. *Ecology* 59:133–39.

———. 1980. Temporal changes in insular populations of the pika (*Ochotona princeps*). *Ecology* 60:8–13.

———. 1987. Population structure of pikas: Dispersal versus philopatry. In *Mammalian Dispersal Patterns: The Effects of Social Structure on Population Genetics,* ed. B. D. Chepko-Sade and Z. T. Halpin, 128–42. Chicago: University of Chicago Press.

Smith, A. T., and B. L. Ivins. 1983a. Colonization in a pika population: Dispersal versus philopatry. *Behav. Ecol. Sociobiol.* 13:37–47.

———. 1983b. Reproductive tactics of pikas: Why have two litters? *Can. J. Zool.* 61:1551–59.

———. 1984. Spatial relationships and social organization in adult pikas: A facultatively monogamous mammal. *Z. Tierpsychol.* 66:289–308.

———. 1987. Temporal separation between philopatric juvenile pikas and their parents limits behavioural conflict. *Anim. Behav.* 35:1210–14.

Smith, A. T., H. J. Smith, X. G. Wang, X. Yin, and J. Liang. 1986. Social behavior of the steppe-dwelling black-lipped pika. *Natl. Geographic Res.* 2:57–74.

Smith, A. T., and X. G. Wang. 1991. Social relationships of adult black-lipped pikas (*Ochotona curzoniae*). *J. Mammal.* 72:231–47.

Smith, D. G. 1982. Inbreeding in three captive groups of rhesus monkeys. *Am. J. Phys. Anthropol.* 58:447–51.

———. 1986a. Inbreeding in the maternal and paternal lines of four captive groups of rhesus monkeys (*Macaca mulatta*). In *Current Perspectives in Primate Biology*, ed. D. M. Taub and F. A. King, 214–25. New York: Van Nostrand Reinhold Company.

———. 1986b. Incidence and consequences of inbreeding in three captive groups of rhesus macaques (*Macaca mulatta*). In *Primates: The Road to Self-Sustaining Populations*, ed. K. Benirschke, 857–74. New York: Springer.

Smith, D. G., F. W. Lorey, J. Suzuki, and M. Abe. 1987. Effect of outbreeding on weight and growth rate of captive infant rhesus macaques. *Zoo Biol.* 6:201–12.

Smith, D. G., and M. F. Small. 1982. Selection and the transferrin polymorphism in rhesus monkeys (*Macaca mulatta*). *Folia Primatol.* 37:127–36.

Smith, D. G., and S. Smith. 1988. Parental rank and the reproductive success of natal rhesus males. *Anim. Behav.* 36:554–62.

Smith, D. R. 1986. Population genetics of the cooperative spider *Anelosimus eximius* (Araneae, Theridiidae). *J. Arachnol.* 14:201–17.

Smith, E. B. 1968. Pollen competition and relatedness in *Haplopappus* section Isopappus. *Bot. Gaz.* 129:371–73.

Smith, J. N. M. 1988. Determinants of lifetime reproductive success in the song sparrow. In *Reproductive Success*, ed. T. H. Clutton-Brock, 154–72. Chicago: University of Chicago Press.

Smith, M. F., J. L. Patton, J. C. Hafner, and D. J. Hafner. 1983. *Thomomys bottae* pocket gophers of the central Rio Grande Valley, New Mexico: Local differentiation, gene flow and historical biogeography. *Occasional Papers of the Museum of Southwestern Biology* 2:1–6.

Smith, M. H., and R. K. Chesser. 1981. Rationale for conserving genetic variation of fish gene pools. *Ecol. Bull.* 34:13–26.

Smith, M. H., K. T. Scribner, J. D. Hernandez, and M. C. Wooten. 1989. Demographic, spatial, and temporal genetic variation in *Gambusia*. In *Ecology and Evolution of Livebearing Fishes (Poeciliidae)*, ed. G. K. Meffe and F. F. Snelson, Jr., 235–57. Englewood Cliffs, N.J.: Prentice Hall.

Smith, M. W., M. H. Smith, and R. K. Chesser. 1983. Biochemical genetics of mosquitofish. I. Environmental correlates, and temporal and spatial heterogeneity of allele frequencies within a river drainage. *Copeia* 1983: 182–93.

Smith, P. J. 1987. Homozygous excess in sand flounder, *Rhombosolea plebeia*, produced by assortative mating. *Mar. Biol.* 95:489–92.

Smith, P. J., and R. I. C. C. Francis. 1984. Glucosephosphate isomerase genotype frequences, homozygous excess and size relationships in the sand flounder *Rhombosolea plebeia*. *Mar. Biol.* 79:93–98.

Smith, P. J., and Y. Fujio. 1982. Genetic variation in marine teleosts: High variability in habitat specialists and low variability in habitat generalists. *Mar. Biol.* 69:7–20.

Smith, R. I. 1950. Embryonic development in the viviparous nereid polychaete, *Neanthes lighti* Hartman. *J. Morphol.* 87:417–65.

Smith, R. S., and F. G. Whoriskey, Jr. 1988. Multiple clutches: Female three-spine sticklebacks lose the ability to recognize their own eggs. *Anim. Behav.* 36:1838–39.

Smith, S. G., and D. R. Wallace. 1971. Allelic sex determination in a lower hymenopteran *Neodiprion nigroscutum. Midd. Can. J. Genet. Cytol.* 13:617–21.

Smouse, P. E. 1986. The fitness consequences of multiple locus heterozygosity under the multiplicative overdominance and inbreeding depression models. *Evolution* 40:946–57.

Smuts, B. B. 1985. *Sex and Friendship in Baboons.* New York: Aldine.

———. 1987. Sexual competition and mate choice. In *Primate Societies,* ed. B. B. Smuts, D. L. Cheney, R. M. Seyfarth, T. T. Struhsaker, and R. W. Wrangham, 385–99. Chicago: University of Chicago Press.

Smyth, C. A., and J. L. Hamrick. 1987. Realized gene flow via pollen in artificial populations of musk thistle, *Carduus nutans* L. *Evolution* 41:613–19.

Snedecor, G. W., and W. G. Cochran. 1967. *Statistical Methods.* Ames: Iowa State University Press.

Snow, D. W., and A. Lill. 1974. Longevity records for some neo-tropical land-birds. *Condor* 76:262–67.

Sobrevila, C. 1988. Effects of distance between pollen donor and pollen recipient on fitness components in *Espeletia schultzii. Am. J. Bot.* 75:701–24.

Sokal, R. R., and F. J. Rohlf. 1981. *Biometry,* 2d ed. San Francisco: W. H. Freeman and Co.

Sokal, R. R., and N. L. Oden. 1978. Spatial autocorrelation in biology. II. Some biological implications and four applications of evolutionary and ecological interest. *Biol. J. Linn. Soc.* 10:229–49.

Sommer, V., and L. S. Rajpurohit. 1989. Male reproductive success in harem troops of Hanuman langurs (*Presbytis entellus*). *Int. J. Primatol.* 10:293–317.

Sorenson, F., and R. S. Miles. 1982. Inbreeding depression in height, height growth, and survival of Douglas-fir, ponderosa pine and noble fir to 10 years of age. *For. Sci.* 28:283–92.

Sorenson, F. C., and T. C. White. 1988. Effect of natural inbreeding on variance structure in tests of wind-pollination Douglas-fir progenies. *For. Sci.* 34:102–18.

Soulé, M. 1973. The epistasis cycle: A theory of marginal populations. *Annu. Rev. Ecol. Syst.* 4:165–87.

———. 1976. Allozyme variation: Its determinants in space and time. In *Molecular Evolution,* ed. F. J. Ayala, 60–70. Sunderland, Mass.: Sinauer Associates.

———. 1980. Thresholds for survival: Maintaining fitness and evolutionary potential. In *Conservation Biology: An Evolutionary-Ecological Perspective,* ed. M. E. Soulé and B. A. Wilcox, 151–69. Sunderland, Mass.: Sinauer Associates.

———. 1986. *Conservation Biology: The Science of Scarcity and Diversity.* Sunderland, Mass.: Sinauer Associates.

Soulé, M. E., and B. A. Wilcox, eds. 1980. *Conservation Biology: An Evolutionary-Ecological Perspective.* Sunderland, Mass.: Sinauer Associates.

Spencer, W. P. 1944. The genetic basis of differences between two species of *Drosophila*. *Am. Nat.* 78:183–88.

Spieth, H. T. 1949. Sexual behavior and isolation in *Drosophila*. I. The interspecific mating behavior of species of the *willistoni* group. *Evolution* 3:67–81.

Spight, T. M. 1974. Sizes of populations of a marine snail. *Ecology* 55(4): 712–29.

Spiro, M. E. 1958. *Children of the Kibbutz*. Cambridge, Mass.: Harvard University Press.

Spoecker, P. D. 1967. Movements and seasonal activity cycles of the lizard *Uta stansburiana stejnegeri*. *Am. Midl. Nat.* 77:484–94.

Sprague, G. F. 1983. Heterosis in maize: Theory and practice. In *Heterosis: Reappraisal of Theory and Practice*, ed. R. Frankel. Berlin: Springer-Verlag.

Stabell, O. B. 1984. Homing and olfaction in salmonoids: A critical review with special reference to the Atlantic salmon. *Biol. Rev.* 59:333–88.

Stack, C. 1975. *All Our Kin: Strategies for Survival in a Black Community*. New York: Harper and Row.

Ståhl, G. 1987. Genetic population structure of Altantic salmon. In *Population Genetics and Fishery Management*, ed. N. Ryman and F. Utter, 121–40. Seattle: University of Washington Press.

Stam, P. 1983. The evolution of reproductive isolation in closely adjacant plant populations through differential flowering time. *Heredity* 50:105–18.

Stanford, C. 1991. Social dynamics of intergroup encounters in the capped langur (*Presbytis pileata*). *Am. J. Primatol.* 25:35–47.

Stanton, M. L., A. A. Snow, S. N. Handel, and J. Bereczky. 1989. The impact of a flower-color polymorphism on mating patterns in experimental populations of wild radish (*Raphanus raphanistrum* L.). *Evolution* 43:335–46.

Statistical Abstracts of the U. S. 1979. Washington, D.C.: U. S. Government Printing Office.

Stead, A. D., I. N. Roberts, and H. G. Dickinson. 1979. Pollen-pistil interaction in *Brassica oleracea*. Events prior to pollen germination. *Planta* 146:211–16.

Stearns, S. C. 1987. The selection-arena hypothesis. In *The Evolution of Sex and its Consequences*, ed. S. C. Stearns, 337–49. Basel: Birkhauser Verlag.

Stebbins, G. L., Jr. 1950. *Variation and Evolution in Plants*. New York: Columbia University Press.

———. 1957. Self-fertilization and population variability in the higher plants. *Am. Nat.* 91: 337–54.

———. 1970. Adaptive radiation in Angiosperms. I. Pollination mechanisms. *Annu. Rev. Ecol. Syst.* 1:307–26.

Stenseth, N. C. 1983. Causes and consequences of dispersal in small mammals. In *The Ecology of Animal Movement*, ed. I. R. Swingland and P. J. Greenwood, 63–101. Oxford: Clarendon Press.

Stephenson, A. G. 1981. Flower and fruit abortion: Proximate causes and ultimate functions. *Annu. Rev. Ecol. Syst.* 12:253–79.

Stephenson, A. G., and J. A. Windsor. 1986. *Lotus corniculatus* regulates offspring quality through selective fruit abortion. *Evolution* 40:453–58.

Stephenson, T. A. 1931. Development and the formation of colonies in *Pocillo-*

pora and *Porites*. Part 1. *British Museum (Natural History) Great Barrier Reef Expedition 1928–29 Scientific Reports* 3(3): 113–34.

Stevenson, A. C., H. A. Johnston, M. I. P. Stewart, and D. R. Golding. 1966. *Congenital Malformations: A Report of a Study of Series of Consecutive Births in 24 Centres.* Geneva: World Health Organization.

Stewart, K. J., and A. H. Harcourt. 1987. Gorillas: Variation in female relationships. In *Primate Societies*, ed. B. B. Smuts, D. L. Cheney, R. M. Seyfarth, T. T. Struhsaker, and R. W. Wrangham, 155–64. Chicago: University of Chicago Press.

Stewart, R. J., A. Francoeur, and R. Loiselle. 1986. Fighting males in the ant genus *Cardiocondyla*. In *Abstracts of the 10th International Congress of the Union for Study of Social Insects*, ed. J. Eder and H. Rembold, 174. Munich: Verlag.

Stickel, L. F. 1989. Home range behavior among box turtles (*Terrapene c. carolina*) of a bottomland forest in Maryland. *Journal of Herpetology* 23:40–44.

Stoddart, J. A. 1983. Asexual production of planulae in the coral *Pocillopora damicornis*. *Mar. Biol.* 76:279–84.

———. 1984. Genetical structure within populations of the coral *Pocillopora damicornis*. *Mar. Biol.* 81:19–30.

———. 1986. Biochemical genetics of *Pocillopora damicornis* in Kaneohe Bay, Oahu, Hawaii. *Hawaii Inst. Mar. Biol. Tech. Rep.* 37:133–50.

———. 1988. Coral populations fringing islands: Larval connections. *Aust. J. Mar. Freshw. Res.* 39:109–15.

Stoddart, J. A., R. C. Babcock, and A. J. Heyward. 1988. Self-fertilization and maternal enzymes in the planulae of the coral *Goniastrea favulus*. *Mar. Biol.* 99:489–94.

Stoddart, J. A., and R. Black. 1985. Cycles of gametogenesis and planulation in the coral *Pocillopora damicornis*. *Mar. Ecol. Prog. Ser.* 23:153–64.

Stoner, G., and F. Breden. 1988. Phenotypic differentiation in female preference related to geographic variation in male predation risk in the Trinidad guppy (*Poecilia reticulata*). *Behav. Ecol. Sociobiol.* 22:285–91.

Stout, A. B. 1938. The genetics of incompatibilities in homomorphic flowering plants. *Bot. Rev.* 9:275–369.

Stouthamer, R., R. F. Luck, and W. D. Hamilton. 1990. Antibiotics cause parthenogenetic *Trichogramma* (Hymenoptera/Trichogrammatidae) to revert to sex. *Proc. Natl. Acad. Sci. USA* 87:2424–27.

Strathmann, M. F. 1987. *Reproduction and Development of Marine Invertebrates of the Northern Pacific Coast.* Seattle: University of Washington Press.

Strathmann, M. F., and D. J. Eernisse. 1987. Phylum Mollusca, Class Polyplacophora. In *Reproduction and Development of Marine Invertebrates of the Northern Pacific Coast*, ed. M. F. Strathmann, 205–19. Seattle: University of Washington Press.

Strathmann, R. R. 1974. The spread of sibling larvae of sedentary marine invertebrates. *Am. Nat.* 108:29–44.

———. 1978. The evolution and loss of feeding larval stages of marine invertebrates. *Evolution* 32(4): 894–906.

————. 1981. On barriers to hybridization between *Strongylocentrotus droebachiensis* (O. F. Müller) and *S. pallidus* (G. O. Sars). *J. Exp. Mar. Biol. Ecol.* 55:39–47.

————. 1985. Feeding and non-feeding larval development and life-history evolution in marine invertebrates. *Annu. Rev. Ecol. Syst.* 16:339–61.

————. 1990. Why life histories evolve differently in the sea. *Am. Zool.* 30:197–207.

Strathmann, R. R., and M. F. Strathmann. 1982. The relationship between adult size and brooding in marine invertebrates. *Am. Nat.* 119(1): 91–101.

Strathmann, R. R., M. F. Strathmann, and R. H. Emson. 1984. Does limited brood capacity link adult size, brooding and simultaneous hermaphroditism? A test with the starfish *Asterina phylactica*. *Am. Nat.* 123(6): 796–818.

Stratton, D. A., M. B. Cruzan, and J. D. Thomson. 1985. Pollen tube growth rate and outcrossing distance in *Erythronium grandiflorum*. *Am. J. Bot.* 72:866.

Streisinger, G., C. Walker, N. Dower, D. Knauber, and F. Singer. 1981. Production of clones of homozygous diploid zebra fish (*Brachydanio rerio*). *Nature* 291:293–96.

Strier, K. B. 1990. New World primates, new frontiers: Insights from the woolly spider monkey, or muriqui (*Brachyteles arachnnoides*). *Int. J. Primatol.* 11:7–19.

Strijbosch, H., P. T. J. C. van Rooy, and L. A. C. J. Voesenek. 1983. Homing behaviour of *Lacerta agilis* and *Lacerta vivipara* (Sauria, Lacertidae). *Amphibia-Reptilia* 4:43–47.

Ström, R. 1977. Brooding patterns of bryozoans. In *Biology of Bryozoans*, ed. R. M. Woollacott and R. L. Zimmer, 23–55. New York: Academic Press.

Strum, S. C. 1982. Agonistic dominance in male baboons: An alternative view. *Int. J. Primatol.* 3:175–202.

————. 1987. *Almost Human*. New York: Random House.

Suchanek, T. H. 1978. The ecology of *Mytilus edulis* L. in exposed rocky intertidal communities. *J. Exp. Mar. Biol. Ecol.* 31:105–20.

————. 1986. Mussels and their role in structuring rocky shore communities. In *The Ecology of Rocky Coasts*, ed. P. G. Moore and R. Seed, 70–96. New York: Columbia University Press.

Sugiyama, Y. 1976. Life history of male Japanese monkeys. *Adv. Stud. Behav.* 7:255–84.

Sugiyama, Y., and H. Ohsawa. 1982. Population dynamics of Japanese monkeys with special reference to the effect of artificial feeding. *Folia Primatol.* 39:238–63.

Sullivan, B. K. 1992. Sexual selection and calling behavior in the American toad (*Bufo americanus*). *Copeia* 1992:1–7.

Surani, M. A. H. 1987. Evidences and consequences of differences between maternal and paternal genomes during embryogenesis in the mouse. In *Experimental Approaches to Mammalian Embryonic Development*, ed. J. Rossant and R. A. Pedersen, 401–35. Cambridge: Cambridge University Press.

Svensson, L. 1988. Inbreeding, crossing and variation in stamen number in *Scleranthus annuus* (Caryophyllaceae), a selfing annual. *Evol. Trends Plants* 2:31–37.

———. 1990. Distance-dependent regulation of stamen number in crosses of *Scleranthus annuus* (Caryophyllaceae) from a discontinuous population. *Am. J. Bot.* 77:889–96.

Swingland, I. R., and P. J. Greenwood, eds. 1983. *The Ecology of Animal Movement*. Oxford: Clarendon Press.

Symons, D. 1979. *The Evolution of Human Sexuality*. Oxford: Oxford University Press.

Szmant, A. M. 1986. Reproductive ecology of Caribbean reef corals. *Coral Reefs* 5:43–54.

Tabachnick, W. J. 1977. Geographic variation of five biochemical polymorphisms in *Notophthalmus viridescens*. *J. Hered.* 86:117–22.

Tabachnick, W. J., and D. K. Underhill. 1972. Serum transferrin and serum albumin polymorphism in two populations of the red-spotted newt, *Notophthalmus viridescens viridescens*. *Copeia* 1972: 525–28.

Taborsky, M. 1985. Breeder-helper conflict in a cichlid fish with broodcare helpers: An experimental analysis. *Behaviour* 95:45–75.

———. 1987. Cooperative behaviour in fish: Coalitions, kin groups and reciprocity. In *Animal Societies: Theories and Facts*, ed. Y. Ito, J. L. Brown, and J. Kikkawa, 229–37. Tokyo: Japan Scientific Societies Press.

Taborsky, M., and D. Limberger. 1981. Helpers in fish. *Behav. Ecol. Sociobiol.* 8:143–45.

Takahata, Y. 1982a. Social relations between adult males and females of Japanese monkeys in the Arashiyama B Troop. *Primates* 23:1–23.

———. 1982b. The socio-sexual behavior of Japanese monkeys. *Z. Tierpsychol.* 59:89–108.

Takayama, S., A. Isogai, C. Tsukamoto, Y. Ueda, K. Hinata, K. Okazaki, and A. Suzuki, 1987. Sequences of *S*-glycoproteins, products of the *Brassica campestris* self-incompatibility locus. *Nature* 326:102–5.

Tamarin, R. H., M. Sheridan, and C. K. Levy. 1983. Determining matrilineal kinship in natural populations of rodents using radionuclides. *Can. J. Zool.* 61:271–74.

Tapper, S. C. 1973. The spatial organisation of pikas (*Ochotona*), and its effect on population recruitment. Ph.D. dissertation, University of Alberta, Edmonton.

Tardent, P. 1963. Regeneration in the Hydrozoa. *Biol. Rev.* 38:293–333.

Tartakovsky, B. 1987. Immune disruption of gestation. In *Immunoregulation and Fetal Survival*, ed. T. J. Gill III and T. G. Wegmann, 233–38. Oxford: Oxford University Press.

Taylor, C., and W. P. Faulk. 1981. Prevention of recurrent abortion with leucocyte transfusions. *Lancet* ii:68–70.

Taylor, C. E., and G. C. Gorman. 1975. Population genetics of a "colonising" lizard: Natural selection for alloyzme morphs in *Anolis grahami*. *Heredity* 35:241–47.

Taylor, V. A. 1981. The adaptive and evolutionary significance of wing polymorphism and parthenogenesis in *Ptinella* Motschulsky (Coleoptera: Ptiliidae). *Ecol. Entomol.* 6:89–98.

Tegelstrom, H., T. Ebenhard, and H. Ryttman. 1983. Rate of karyotype evolution and speciation in birds. *Hereditas* 98:235–39.

Templeton, A. R. 1979. The unit of selection in *Drosophila mercatorum*. II. Genetic revolution and the origin of coadapted genomes in parthenogenetic strains. *Genetics* 92:1265–82.

———. 1980. The theory of speciation via the founder principle. *Genetics* 94:1011–38.

———. 1981. Mechanisms of speciation: A population genetic approach. *Annu. Rev. Ecol. Syst.* 12:23–48.

———. 1986. Coadaptation and outbreeding depression. In *Conservation Biology: The Science of Scarcity and Diversity*, ed. M. E. Soulé, 105–16. Sunderland, Mass.: Sinauer Associates.

———. 1987. Inferences on natural population structure from genetic studies on captive mammalian populations. In *Mammalian Dispersal Patterns: The Effects of Social Structure on Population Genetics*, ed. B. D. Chepko-Sade and Z. T. Halpin, 257–72. Chicago: University of Chicago Press.

Templeton, A. R., H. Hemmer, G. Mace, U. S. Seal, W. M. Shields, and D. S. Woodruff. 1986. Local adaptation, coadaptation, and population boundaries. *Zoo Biol.* 5:115–25.

Templeton, A. R., and B. Read. 1983. The elimination of inbreeding depression in a captive herd of Speke's gazelle. In *Genetics and Conservation: A Reference for Managing Wild Animal and Plant Populations*, ed. C. M. Schonewald-Cox, S. M. Chambers, B. MacBryde, and W. L. Thomas, 241–61. Menlo Park, Calif.: Benjamin/Cummings.

———. 1984. Factors eliminating inbreeding depression in a captive herd of Speke's gazelle (*Gazella spekei*). *Zoo Biol.* 3:177–99.

Tettenborn, U., and A. Gropp. 1970. Meiotic non-disjunction in mice and mouse-hybrids. *Cytogenetics* 9:272–83.

Thayer, C. W. 1977. Recruitment, growth, and mortality of a living articulate brachiopod, with implications for the interpretation of survivorship curves. *Paleobiology* 3(1): 98–109.

———. 1981. Ecology of living brachiopods. In *Lophophorates: Notes for a Short Course*, ed. T. W. Broadhead. University of Tennessee Department of Geological Sciences Studies in Geology 5:110–26.

Thompson, P., and J. W. Sites, Jr. 1986. Comparison of population structure in chromosomally polytypic and monotypic species of *Sceloporus* (Sauria: Iguanidae) in relation to chromosomally-mediated speciation. *Evolution* 40:303–14.

Thorgaard, G. H. 1983. Chromosome set manipulation and sex control in fish. In *Fish Physiology*, vol. 9b, ed. W. S. Hoar, D. J. Randall, and E. M. Donaldson, 405–34. New York: Academic Press.

Thorgaard, G. H., P. D. Scheerer, W. K. Hershberger, and J. M. Myers. 1990. Androgenetic rainbow trout produced using sperm from tetraploid males show improved survival. *Aquaculture* 85:215–21.

Thornhill, N. W. 1987. Rules of mating and marriage pertaining to relatives: An evolutionary biological analysis. Ph.D. dissertation, University of New Mexico, Albuquerque.

————. 1990a. An evolutionary analysis of incest rules. *Ethol. Sociobiol.* 11:113–29.

————. 1990b. The comparative method of evolutionary biology in the study of societies of history. *J. Comp. Sociol.* 27:7–27.

————. 1991. An evolutionary analysis of rules regulating human inbreeding and marriage. *Behav. Brain Sci.* 14:247–93.

————. N.d. Severity of disease and human inbreeding. Manuscript.

Thornhill, N. W., and R. Thornhill. 1984. Review of Joseph Shepher's—Incest: A biosocial perspective. *Ethol. Sociobiol.* 5:211–14.

————. 1987. Rules of mating and marriage pertaining to relatives. In *Psychology and Sociobiology: Ideas, Issues and Applications,* ed. C. Crawford, M. Smith, and C. Krebs, 373–400. Hillsdale, N.J.: Lawrence Erlbaum Associates.

Thornhill, R. 1986. Early history of sexual selection theory. *Evolution* 40:446–47.

Thornhill, R., and J. Alcock. 1983. *The Evolution of Insect Mating Systems.* Cambridge, Mass.: Harvard University Press.

Thornhill, R., and N. W. Thornhill. 1983. Human rape: An evolutionary analysis. *Ethol. Sociobiol.* 4:137–73.

————. 1992. The evolutionary psychology of men's coercive sexuality. *Behav. Brain Sci.* 15:363–421.

Thorson, G. 1946. Reproduction and larval development of Danish marine bottom invertebrates, with special reference to the planktonic larvae in the Sound (Øresund). *Meddelelser fra Kommissionen for Danmarks Fiskeriog Havundersøgelser Serie: Plankton, Kobenhavn* 4(1): 1–523.

————. 1950. Reproduction and larval ecology of marine bottom invertebrates. *Biol. Rev.* 25(1): 1–45.

Tilford, B. 1982. Seasonal rank changes for adolescent and subadult natal males in a free-ranging group of rhesus monkeys. *Int. J. Primatol.* 3:483–90.

Tilson, R. L. 1981. Family formation strategies of Kloss's gibbons. *Folia Primatol.* 35:259–87.

Tinkle, D. W. 1965. Population structure and effective size of a lizard population. *Evolution* 19:569–73.

————. 1967. Home range, density, dynamics, and structure of a Texas population of the lizard *Uta stansburiana.* In *Lizard Ecology: A Symposium,* ed. W. M. Milstead, 5–29. Columbia: University of Missouri Press.

Todd, C. D., J. N. Havenhand, and J. P. Thorpe. 1988. Genetic differentiation, pelagic larval transport and gene flow between local populations of the intertidal marine mollusc *Adalaria proxima* (Alder & Hancock). *Funct. Ecol.* 2:441–51.

Tompkins, R. 1978. Genic control of axolotl metamorphosis. *Am. Zool.* 18:313–19.

Tormes, Y. 1968. *Child Victims of Incest.* Denver: The American Humane Association.

Tranter, P. R. G., D. N. Nicholson, and D. Kinchington. 1982. A description of spawning and post-gastrula development of the cool temperate coral, *Caryophillia smithi*. *J. Mar. Biol. Assoc. UK* 62:845–54.

Trivers, R. L. 1971. The evolution of reciprocal altruism. *Q. Rev. Biol.* 46:35–57.

———. 1985. *Social Evolution*. Menlo Park, Calif.: Benjamin/Cummings.

Turelli, M., and L. Ginzburg. 1983. Should individual fitness increase with heterozygosity? *Genetics* 104:191–209.

Turkington, R., and J. L. Harper. 1979. The growth, distribution, and neighbor relationships of *Trifolium repens* in a permanent pasture. *J. Ecol.* 67:245–54.

Turner, B. J. 1983. Does matrotrophy promote chromosomal evolution in viviparous fishes? *Am. Nat.* 122:152–54.

Turner, J. R. G., M. S. Johnson, and W. F. Eanes. 1979. Contrasted modes of evolution in the same genome: Allozymes and adaptive change in *Heliconius*. *Proc. Natl. Acad. Sci. USA* 76:1924–28.

Tuttle, M. D., and D. Stevenson. 1982. Growth and survival of bats. In *Ecology of Bats*, ed. T. H. Kunz, 105–50. New York: Plenum.

Twitty, V. C. 1961. Experiments on homing behavior and speciation in *Taricha*. In *Vertebrate Speciation*, ed. W. F. Blair, 415–59. Austin: University of Texas Press.

Twitty, V. C., D. Grant, and O. Anderson. 1964. Long distance homing in the newt *Taricha rivularis*. *Proc. Natl. Acad. Sci. USA* 51:51–58.

Tyler-Walters, H., and D. J. Crisp. 1989. The modes of reproduction in *Lasaea rubra* (Montagu) and *L. australis* (Lamarck): (Erycinidae; Bivalvia). In *Reproduction, Genetics and Distributions of Marine Organisms*, ed. J. S. Ryland and P. A. Tyler, 299–308. Fredensborg, Denmark: Olsen & Olsen.

Tyson, J. J. 1984. Evolution of eusociality in diploid species. *Theor. Popul. Biol.* 26:283–95.

Uchida, T., and F. Iwata. 1954. On the development of a brood-caring actinian. *J. Fac. Sci. Hokkaido Univ.*, Ser. 6, *Zoology*, 12(1/2): 220–24.

Uma Shaanker, R., K. N. Ganeshaiah, and K. S. Bawa. 1988. Parent-offspring conflict, sibling rivalry, and brood size patterns in plants. *Annu. Rev. Ecol. Syst.* 19:177–205.

Upton, N. P. D. 1987. Asynchronous male and female life cycles in the sexually dimorphic, harem-forming isopod *Paragnathia formica* (Crustacea: Isopoda). *J. Zool.* 212:677–90.

Uyenoyama, M. K. 1984. Inbreeding and the evolution of altruism under kin selection: Effects on relatedness and group structure. *Evolution* 48(4):778–95.

———. 1985. On the evolution of parthenogenesis: A genetic representation of the "cost of meiosis." *Evolution* 38:87–102.

———. 1986. Inbreeding and the cost of meiosis: The evolution of selfing in populations practicing biparental inbreeding. *Evolution* 40(2): 388–404.

———. 1988a. On the evolution of genetic incompatibility systems: Incompatibility as a mechanism for the regulation of outcrossing distance. In *The Evolution of Sex: An Examination of Current Ideas*, ed. R. E. Michod and B. R. Levin, 212–32. Sunderland, Mass.: Sinauer Associates.

———. 1988b. On the evolution of genetic incompatibility systems. II. Initial increase of strong gametophytic self-incompatibility under partial selfing and half-sib mating. *Am. Nat.* 131:700–722.

———. 1988c. On the evolution of genetic incompatibility systems. III. Introduction of weak gametophytic self-incompatibility under partial inbreeding. *Theor. Popul. Biol.* 34:47–91.

———. 1988d. On the evolution of genetic incompatibility systems. IV. Modification of response to an existing antigen polymorphism under partial selfing. *Theor. Popul. Biol.* 34:347–77.

———. 1989a. Coevolution of the major histocompatibility complex and the *t*-complex in the mouse. I. Generation and maintenance of high-complementarity associations. *Genetics* 121:139–51.

———. 1989b. Coevolution of the major histocompatibility complex and the *t*-complex in the mouse. II. Modification of response to sharing of histocompatibility antigens. *Genetics* 121:153–61.

———. 1989c. On the evolution of genetic incompatibility systems. V. Origin of sporophytic self-incompatibility in response to overdominance in viability. *Theor. Popul. Biol.* 36:339–65.

Uyenoyama, M. K., and B. O. Bengtsson. 1982. Towards a genetic theory for the evolution of the sex ratio. III. Parental and sibling control of brood investment ratio under partial sib-mating. *Theor. Popul. Biol.* 22:43–68.

Uyenoyama, M. K., and D. M. Waller. 1991a. Coevolution of self-fertilization and inbreeding depression. I. Mutation-selection balance at one and two loci. *Theor. Popul. Biol.* 40:14–46.

———. 1991b. Coevolution of self-fertilization and inbreeding depression. II. Symmetric overdominance in viability. *Theor. Popul. Biol.* 40:47–77.

———. 1991c. Coevolution of self-fertilization and inbreeding depression. III. Homozygous lethal mutations at multiple loci. *Theor. Popul. Biol.* 40:173–210.

van den Berghe, P. 1980. Incest and exogamy: A sociobiological reconsideration. *Ethol. Sociobiol.* 1:151–62.

———. 1982. Human inbreeding avoidance: Culture in nature. *Behav. Brain Sci.* 6:91–102.

———. 1987. Incest taboos and avoidance: Some African evidence. In *Psychology and Sociobiology: Ideas, Issues and Applications,* ed. C. Crawford, C. Krebs, and M. Smith, 110–30. Hillsdale, N.J.: Lawrence Erlbaum Associates.

van den Berghe, P., and G. Mesher. 1980. Royal incest and inclusive fitness. *Am. Ethnol.* 7:300–13.

Van Devender, R. W. 1982. Comparative demography of the lizard *Basiliscus basilicus. Herpetologica* 38:189–209.

Van Dijk, H. 1987. A method for the estimation of gene flow parameters from a population structure caused by restricted gene flow and genetic drift. *Theor. Appl. Genet.* 73:724–36.

van Duyl, F. C., R. P. M. Bak, and J. Sybesma. 1981. The ecology of the tropical compound ascidian *Trididemnum solidum.* I. Reproductive strategy and larval behaviour. *Mar. Ecol. Prog. Ser.* 6:35–42.

Van Havre, N., and G. J. FitzGerald. 1988. Shoaling and kin recognition in the threespine stickleback (*Gasterosteus aculeatus* L.) *Biol. Behav.* 13:190–201.

van Noordwijk, A. J. 1987. Quantitative ecological genetics of great tits. In *Avian Genetics*, ed. F. Cooke and P. A. Buckley, 363–80. London: Academic Press.

van Noordwijk, A. J., and W. Scharloo. 1981. Inbreeding in an island population of the great tit. *Evolution* 35:674–88.

van Noordwijk, A. J., P. H. van Tienderen, G. de Jong, and J. H. van Balen. 1985. Genealogical evidence for random mating in a natural population of the great tit (*Parus major* L.). *Naturwiss.* 72:104–6.

van Tienderen, P. H., and A. J. van Noordwijk. 1988. Dispersal, kinship and inbreeding in an island population of the great tit. *J. Evol. Biol.* 1:117–37.

Varvio, S.-L., R. K. Koehn, and R. Väinölä. 1988. Evolutionary genetics of the *Mytilus edulis* complex in the North Atlantic region. *Mar. Biol.* 98:51–60.

Vasil, I. K. 1974. The histology and physiology of pollen germination and pollen tube growth on the stigma and in the style. In *Fertilization in Higher Plants*, ed. H. F. Linskens, 105–18. Amsterdam, Netherlands: North-Holland.

Velikovsky, I. 1960. *Oepidus and Akhenaten*. London: Sedgwick.

Vessey, S. H., and D. B. Meikle. 1987. Factors affecting social behavior and reproductive success of male rhesus monkeys. *Int. J. Primatol.* 8:281–92.

Vetukhiv, M. 1954. Integration of the genotype in local populations of the three species of *Drosophila*. *Evolution* 8:241–51.

———. 1956. Fecundity of hybrids between geographic populations of *Drosophila pseudoobscura*. *Evolution* 10:139–46.

———. 1957. Longevity of hybrids between geographic populations of *Drosophila pseudoobscura*. *Evolution* 11:348–60.

Volkonsky, M. 1940. *Podapolipus diander*, n. sp. acarien heterostygmate parasite de criquet migrateur (*Locusta migratoria* L.). *Archiv Inst. Pasteur Algérie* 18(3): 321–34.

Vollrath, F. 1982. Colony foundation in a social spider. *Z. Tierpsychol.* 60:313–24.

———. 1986. Eusociality and extraordinary sex-ratios in the spider *Anelosimus eximius* (Araneae: Theridiidae). *Behav. Ecol. Sociobiol.* 18:283–87.

Vrijenhoek, R. C. 1985a. Animal population genetics and disturbance: The effects of local extinctions and recolonizations on heterozygosity and fitness. In *The Ecology of Natural Disturbance and Patch Dynamics*, ed. S. T. A. Pickett and P. S. White, 265–85. Orlando: Academic Press.

———. 1985b. Homozygosity and interstrain variation in the self-fertilizing hermaphroditic fish, *Rivulus marmoratus*. *J. Hered.* 76:82–84.

———. 1989. Genotypic diversity and coexistence among sexual and clonal lineages of *Pociliopsis*. In *Speciation and Its Consequences*, ed. D. Otte and J. A. Endler, 386–400. Sunderland, Mass.: Sinauer Associates.

Vrijenhoek, R. C., and S. Lerman. 1982. Heterozygosity and developmental stability under sexual and asexual breeding systems. *Evolution* 36:768–76.

Wade, M. J. 1978a. A critical review of models of group selection. *Q. Rev. Biol.* 53:101–14.

———. 1978b. Kin selection: A classical approach and a general solution. *Proc. Natl. Acad. Sci. USA* 75:6154–58.

———. 1979. The evolution of social interactions by family selection. *Am. Nat.* 113:399–417.

———. 1980. Kin selection: Its components. *Science* 210:665–67.

———. 1985. Soft selection, hard selection, kin selection, and group selection. *Am. Nat.* 125:61–73.

Wade, M. J., and S. J. Arnold. 1980. The intensity of sexual selection in relation to male sexual behaviour, female choice, and sperm precedence. *Anim. Behav.* 28:446–61.

Wade, M. J., and F. Breden. 1981. The effect of inbreeding on the evolution of altruistic behavior by kin selection. *Evolution* 35:844–58.

———. 1987. Kin selection in complex groups. In *Mammalian Dispersal Patterns: The Effects of Social Structure on Population Genetics,* ed. B. D. Chepko-Sade and Z. T. Halpin, 273–83. Chicago: University of Chicago Press.

Wade, T. D. 1979. Inbreeding, kin selection, and primate social evolution. *Primates* 20:355–70.

Wahl, M. 1985a. *Metridium senile:* Dispersion and small scale colonization by the combined strategy of locomotion and asexual reproduction (laceration). *Mar. Ecol. Prog. Ser.* 26:271–77.

———. 1985b. The recolonization potential of *Metridium senile* in an area previously depopulated by oxygen deficiency. *Oecologia* 67:255–59.

Waldman, B. 1981. Sibling recognition in toad tadpoles: The role of experience. *Z. Tierpsychol.* 56:341–58.

———. 1985. Olfactory basis of kin recognition in toad tadpoles. *J. Comp. Physiol. A* 156:565–77.

———. 1987. Mechanisms of kin recognition. *J. Theor. Biol.* 128:159–85.

———. 1989. Do anuran larvae retain kin recognition abilities following metamorphosis? *Anim. Behav.* 37:1055–58.

———. 1991. Kin recognition in amphibians. In *Kin Recognition,* ed. P. G. Hepper. Cambridge: Cambridge University Press.

Waldman, B., J. E. Rice, and R. L. Honeycutt. 1992. Kin recognition and incest avoidance in toads. *Am. Zool.* 32:18–30.

Wallace, B. 1968. *Topics in Population Genetics.* New York: W. W. Norton.

Wallace, B., and M. Vetukhiv. 1955. Adaptive organization of the gene pools of *Drosophila* populations. *Cold Spring Harb. Symp. Quant. Biol.* 20:303–10.

Waller, D. M. 1979. The relative costs of self- and cross-fertilized seeds in *Impatiens capensis* (Balsaminaceae). *Am. J. Bot.* 66:313–20.

———. 1984. Differences in fitness between seedlings derived from cleistogamous and chasmogamous flowers in *Impatiens capensis. Evolution* 38(2):427–40.

———. 1986. Is there disruptive selection for self-fertilization? *Am. Nat.* 128:421–26.

Waller, D. M., and S. E. Knight. 1989. Genetic consequences of outcrossing in the cleistogamous annual, *Impatiens capensis*. II. Outcrossing rates and genotypic correlations. *Evolution* 43(4): 860–69.

Walls, S. C. 1991. Ontogenetic shifts in the recognition of siblings and neighbours by juvenile salamanders. *Anim. Behav.* 42:423–34.

Walls, S. C., and R. E. Roudebush. 1991. Reduced aggression toward siblings as evidence of kin recognition in cannibalistic salamanders. *Am. Nat.* 138:1027–38.

Walters, J. R. 1987. Kin recognition in non-human primates. In *Kin Recognition in Animals*, ed. D. J. C. Fletcher and C. D. Michener, 359–93. New York: John Wiley and Sons.

Wang, X. G., and A. T. Smith. 1988. On the natural winter mortality of the plateau pika (*Ochotona curzoniae*). *Acta Theriologica Sinica* 8:152–56.

———. 1989. Studies on the mating system in plateau pikas (*Ochotona curzoniae*). *Acta Theriologica Sinica* 9:210–15.

Waples, R. S. 1990. Conservation genetics of Pacific salmon. II. Effective population size and the rate of loss of genetic variability. *J. Hered.* 81:267–76.

Wapstra, M., and R. W. M. van Soest. 1987. Sexual reproduction, larval morphology and behaviour in demosponges from the southwest of the Netherlands. In *Taxonomy of Porifera*, ed. J. Vacelet and N. Boury-Esnault, 281–307. Berlin: Springer-Verlag.

Ward, P. D. 1987. *The Natural History of Nautilus*. Boston: Allen & Unwin.

Ward, R. D., N. Billington, and P. D. N. Hebert. 1989. Comparison of allozyme and mitochondrial DNA variation in populations of walleye, *Stizostedion vitreum*. *Can. J. Fish. Aquat. Sci.* 46:2074–84.

Ward, R. D., M. Sarfarazi, C. Azimigarakani, and J. A. Beardmore. 1985. Population genetics of polymorphisms in Cardiff newborn: Relationships between blood group and allozyme heterozygosity and birth weight. *Human Heredity* 35:171–77.

Ward, R. D., T. Warwick, and A. J. Knight. 1986. Genetic analysis of ten polymorphic enzyme loci in *Littorina saxatilis* (Prosobranchia: Mollusca). *Heredity* 57:233–41.

Warneke, M. 1988. *Callimico Goeldii 1988 International Studbook*. Brookfield, Ill.: Chicago Zoological Society.

Warner, R. R., and R. K. Harlan. 1982. Sperm competition and sperm storage as determinants of sexual dimorphism in the dwarf surfperch, *Micrometrus minimus*. *Evolution* 36:44–55.

Waser, N. M. 1983a. Competition for pollination and floral character differences among sympatric plant species: A review of evidence. In *Handbook of Experimental Pollination Biology*, ed. C. E. Jones and R. J. Little, 277–93. New York: Van Nostrand Reinhold.

———. 1983b. The adaptive nature of floral traits: Ideas and evidence. In *Pollination Biology*, ed. L. A. Real, 241–85. New York: Academic Press.

———. 1986. Flower constancy: Definition, cause and measurement. *Am. Nat.* 127:593–603.

———. 1987. Spatial genetic heterogeneity in a population of the montane perennial plant *Delphinium nelsonii. Heredity* 58:249–56.

———. 1988. Comparative pollen and dye transfer by natural pollinators of *Delphinium nelsonii. Funct. Ecol.* 2:41–48.

Waser, N. M., and M. L. Fugate. 1986. Pollen precedence and stigma closure: A mechanism of competition for pollination between *Delphinium nelsonii* and *Ipomopsis aggregata. Oecologia* 70:573–77.

Waser, N. M., and M. V. Price. 1983. Optimal and actual outcrossing in plants and the nature of plant-pollinator interaction. In *Handbook of Experimental Pollination Biology*, ed. C. E. Jones and R. J. Little, 341–59. New York: Van Nostrand Reinhold.

———. 1985. Reciprocal transplant experiments with *Delphinium nelsonii* (Ranunculaceae): Evidence for local adaptation. *Am. J. Bot.* 72:1726–32.

———. 1989. Optimal outcrossing in *Ipomopsis aggregata*: Seed set and offspring fitness. *Evolution* 43:1097–1109.

———. 1991. Outcrossing distance effects in *Delphinium nelsonii*: Pollen loads, pollen tubes and seed set. *Ecology* 72:171–79.

———. 1993. Crossing distance effects on prezygotic performance in plants: An argument for female choice. *Oikos*. In press.

Waser, N. M., M. V. Price, A. M. Montalvo, and R. N. Gray. 1987. Female mate choice in a perennial herbaceous wildflower, *Delphinium nelsonii. Evol. Trends Plants* 1:29–33.

Waser, P. M. 1985. Does competition drive dispersal? *Ecology* 66:1170–75.

———. 1987. A model predicting dispersal distance distributions. In *Mammalian Dispersal Patterns: The Effects of Social Structure on Population Genetics*, ed. B. D. Chepko-Sade and Z. T. Halpin, 251–56. Chicago: University of Chicago Press.

———. 1988. Resources, philopatry, and social interactions among mammals. In *The Ecology of Social Behavior*, ed. C. N. Slobodchikoff, 109–30. New York: Academic Press.

Waser, P. M., S. N. Austad, and B. Keane. 1986. When should animals tolerate inbreeding? *Am. Nat.* 128:529–37.

Waser, P. M., and W. T. Jones. 1983. Natal philopatry among solitary mammals. *Q. Rev. Biol.* 58:355–90.

Wasserthal, L. T., and W. Wasserthal. 1973. Ökologische Bedeutung der Schleimsekretion bei den Planula-Larven der Hydroidengattung *Eudendrium. Mar. Biol.* 22:341–45.

Watanabe, T. K. 1979. A gene that rescues the lethal hybrids between *Drosophila melanogaster* and *D. simulans. Jpn. J. Genet.* 54:325–31.

Watanabe, Y. 1957. Development of *Tethya serica* Lebwohl, a tetraxonian sponge. I. Observations on external changes. *Nat. Sci. Rep. Ochanomizu Univ.* 8(2): 97–104.

———. 1960. Outline of morphological observation on the development of *Tethya serica* Lebwohl. *Bull. Mar. Biol. Stat. Asamushi* 10(2):145–48.

Watt, W. B. 1979. Adaptation at specific loci. I. Natural selection in phosphoglucose isomerase of *Colias* butterflies: Biochemical and population aspects. *Genetics* 87:177–94.

———. 1983. Adaptation at specific loci. II. Demographic and biochemical elements in the maintenance of the *Colias* PGI polymorphism. *Genetics* 103:691–724.

Watt, W. B., P. A. Carter, and S. M. Blower. 1985. Adaptation at specific loci. IV. Differential mating success among glycolytic allozyme genoetypes of colias butterflies. *Genetics* 109:157–75.

Watt, W. B., P. A. Carter, and K. Donohue. 1986. Females' choice of "good genotypes" as mates is promoted by an insect mating system. *Science* 233:1187–90.

Watt, W. B., R. C. Cassin, and M. S. Swan. 1983. Adaptation a specific loci. III. Field behavior and survivorship differences among *Colias* PGI genotypes are predictable from in vitro biochemistry. *Genetics* 103:725–39.

Watts, D. P. 1989. Mountain gorilla life histories, reproductive competition and sociosexual behavior and some implications for captive husbandry. *Zoo Biol.* 9:1–16.

———. 1991. Mountain gorilla reproduction and sexual behavior. *Am. J. Primatol.* 24(3–4): 211–25.

Weatherhead, P. J., and K. A. Boak. 1986. Site infidelity in song sparrows. *Anim. Behav.* 34:1299–1310.

Webb, G. J. W., R. Buckworth, and S. C. Manolis. 1983. *Crocodylus johnstoni* in the McKinlay River Area, N.T. IV. A demonstration of homing. *Aust. Wildl. Res.* 10:403–6.

Webber, H. H. 1977. Gastropoda: Prosobranchia. In *Reproduction of Marine Invertebrates*, vol. 4, ed. A. C. Giese and J. S. Pearse, 1–97. New York: Academic Press.

Weinberg, J. R., and V. R. Starczak. 1988. Morphological differences and low dispersal between local populations of the tropical beach isopod, *Excirolana braziliensis*. *Bull. Mar. Sci.* 42(2): 296–309.

Weinberg, S., and F. Weinberg. 1979. The life cycle of a gorgonian: *Eunicella singularis* (Esper, 1794). *Bijdragen tot de Dierkunde* 48(2): 127–40.

Weinberg, S. K. 1955. *Incest Behavior*. New York: Citadel Press.

Weintraub, J. D. 1970. Homing in the lizard *Sceloporus orcutti*. *Anim. Behav.* 18:132–37.

Weir, B. S. 1990. *Genetic Data Analysis*. Sunderland, Mass: Sinauer Associates.

Weir, B. S., and C. C. Cockerham. 1973. Mixed self and random mating at two loci. *Genet. Res. (Camb.)* 21:247–62.

———. 1984. Estimating F-statistics for the analysis of populations structure. *Evolution* 38:1358–70.

Weisbrot, D. R. 1963. Studies on differences in the genetic architecture of related species of *Drosophilia*. *Genetics* 48:1121–39.

Weller, S. G., and R. Ornduff. 1989. Incompatibility in *Amsinckia grandiflora* (Boraginaceae): Distribution of callose plugs and pollen tubes following inter- and intramorph crosses. *Am. J. Bot.* 76:277–82.

Wells, H. 1979. Self-fertilization: Advantageous or deleterious? *Evolution* 33:252–55.

Wells, M. J., and J. Wells. 1977. Cephalopoda: Octopoda. In *Reproduction of Marine Invertebrates*, vol. 4, ed. A. C. Giese and J. S. Pearse, 291–336. New York: Academic Press.

Werner, D. I., E. M. Baker, E. del C. Gonzalez, and I. R. Sosa. 1987. Kinship recognition and grouping in hatchling green iguanas. *Behav. Ecol. Sociobiol.* 21:83–89.

Werner, R. G. 1979. Homing mechanism of spawning white suckers in Wolf Lake, New York. *N.Y. Fish Game J.* 26:48–58.

Werren, J. H. 1980. Sex ratio adaptations to local mate competition in a parasitic wasp. *Science* 208:1157–59.

———. 1983. Sex ratio evolution under local mate competition in a parasitic wasp. *Evolution* 37:116–24.

———. 1987. Labile sex ratios in wasps and bees. *BioScience* 34:498–506.

Westermarck, E. A. 1891. *The History of Human Marriage*. London: Macmillan.

Westheide, W. 1984. The concept of reproduction in polychaetes with small body size: Adaptations in interstitial species. In *Polychaete reproduction*, ed. A. Fischer and H.-D. Pfannenstiel, 265–87. Stuttgart: Gustav Fischer Verlag.

Westneat, D. F. 1987. Extra-pair fertilization in a predominantly monogamous bird: Genetic evidence. *Anim. Behav.* 35:877–86.

———. 1990. Genetic parentage in the indigo bunting: A study using DNA fingerprinting. *Behav. Ecol. Sociobiol.* 27:67–76.

Westneat, D. F., P. W. Sherman, and M. L. Morton. 1990. The ecology and evolution of extra-pair copulations in birds. In *Current Ornithology*, vol. 7, ed. D. C. Powers, 331–70. New York: Plenum Press.

Westoby, M., and B. Rice. 1982. Evolution of seed plants and inclusive fitness of plant tissues. *Evolution* 36:713–24.

Wetton, J. H., R. E. Carter, D. T. Parkin, and D. Walters. 1987. Demographic study of a wild house sparrow population by DNA fingerprinting. *Nature* 327:147–49.

Wheeler, W. M. 1928. *The Social Insects: Their Origin and Evolution*. London: Kegan Paul, Trench, Trubner.

White, M. J. D. 1968. Models of speciation. *Science* 159:1065–70.

———. 1973. *Animal Cytology and Evolution*. Cambridge: Cambridge University Press.

———. 1978. *Modes of Speciation*. San Francisco: W. H. Freeman and Co.

Whiting, P. W. 1943. Multiple alleles in complementary sex determination of *Habrobracon*. *Genetics* 24:110–11.

Whitkus, R. 1988. Experimental hybridizations among chromosome races of *Carex pachystachya* and the related species *C. macloviana* and *C. preslii* (Cyperaceae). *Syst. Bot.* 13:146–43.

Wickler, W. 1973. Über Koloniegrundung und soziale Bindung von *Stegodyphus mimosarum* Pavesi und anderen sozialen Spinnen. *Tierpsychol.* 32:522–31.

Wickler, W., and U. Seibt. 1983. Monogamy: An ambiguous concept. In *Mate Choice,* ed. P. P. G. Bateson, 33–50. Cambridge: Cambridge University Press.

———. 1986. Aerial dispersal by ballooning in adult *Stegodyphus mimosarum,* a social eresid spider. *Naturwiss.* 73:628–29.

Wiens, D. 1984. Ovule survivorship, brood size, life history, breeding systems, and reproductive success in plants. *Oecologia* 64:47–53.

Wiens, J. A. 1972. Anuran habitat selection: Early experience and substrate selection in *Rana cascadae* tadpoles. *Anim. Behav.* 20:218–20.

Wildt, D., M. Bush, G. L. Goodrowe, C. Packer, A. E. Pusey, J. L. Brown, P. Joslin, and S. J. O'Brien. 1987. Reproductive and genetic consequences of founding isolated lion populations. *Nature* 329:328–31.

Wilkinson, G. S. 1984. Reciprocal food sharing in the vampire bat. *Nature* 308:181–84.

———. 1985. The social organization of the common vampire bat. II. Mating system, genetic structure, and relatedness. *Behav. Ecol. Sociobiol.* 17:123–34.

Wilkinson, G. S., and A. E. M. Baker. 1988. Communal nesting among genetically similar house mice. *Ethology* 77:103–14.

Williams, E. G., V. Kaul, J. L. Rouse, and B. F. Palser. 1986. Overgrowth of pollen tubes in embryo sacs of *Rhododendron* following interspecific pollinations. *Aust. J. Bot.* 34:413–23.

Williams, G. B. 1965. Observations on the behaviour of the planulae larvae of *Clava squamata. J. Mar. Biol. Assoc. UK* 45:257–73.

———. 1976. Aggregation during settlement as a factor in the establishment of coelenterate colonies. *Ophelia* 15(1): 57–64.

Williams, G. C. 1966. *Adaptation and Natural Selection.* Princeton, N.J.: Princeton University Press.

———. 1975. *Sex and Evolution.* Princeton, N.J.: Princeton University Press.

Williams, G. C., R. K. Koehn, and J. B. Mitton. 1973. Genetic differentiation without isolation in the American eel, *Anguilla rostrata. Evolution* 27:192–204.

Williams, G. C., and J. B. Mitton. 1973. Why reproduce sexually? *J. Theor. Biol.* 39:545–54.

Williams, J. G. K., A. R. Kubelik, K. J. Livak, J. A. Rafalski, and S. V. Tingey. 1990. DNA polymorphisms amplified by arbitrary primers are useful as genetic markers. *Nucleic Acids Res.* 18:6531–35.

Williamson, D. I. 1982. Larval morphology and diversity. In *The Biology of Crustacea,* vol. 2, ed. L. G. Abele, 43–110. New York: Academic Press.

Willis, B. L., and J. K. Oliver. 1989. Inter-reef dispersal of coral larvae following the annual mass spawning on the Great Barrier Reef. *Proc. VI Int. Coral Reef Symp.* 2:853–59.

Wills, C. 1978. Rank-order selection is capable of maintaining all genetic polymorphisms. *Genetics* 89:403–14.

———. 1981. *Genetic Variability.* Oxford: Clarendon Press.

Willson, M. F., and N. Burley. 1983. *Mate Choice in Plants: Tactics, Mechanisms, and Consequences.* Princeton, N.J.: Princeton University Press.

Willson, M. F., W. G. Hoppes, D. A. Goldman, P. A. Thomas, P. L. Katusic-

Malmbord, and J. L. Bothwell. 1987. Sibling competition in plants: An experimental study. *Am. Nat.* 129:304–11.

Wilmot, R. L., and C. V. Burger. 1985. Genetic differences among populations of Alaskan sockeye salmon. *Trans. Am. Fish. Soc.* 114:236–43.

Wilson, A. C. 1976. Gene regulation in evolution. In *Molecular Evolution*, ed. F. J. Ayala, 225–34. Sunderland, Mass.: Sinauer Associates.

Wilson, A. C., G. L. Bush, S. M. Case, and M. C. King. 1975. Social structuring of mammalian populations and the rate of chromosomal evolution. *Proc. Natl. Acad. Sci. USA* 72:5061–65.

Wilson, A. C., L. R. Maxson, and V. M. Sarich. 1974. Two types of molecular evolution: Evidence from studies of interspecific hybridization. *Proc. Natl. Acad. Sci. USA* 71:2843–47.

Wilson, D. P. 1968. The settlement behaviour of the larvae of *Sabellaria alveolata* (L.). *J. Mar. Biol. Assoc. UK* 48:387–435.

Wilson, D. S. 1975. A theory of group selection. *Proc. Natl. Acad. Sci. USA* 72:143–46.

———. 1980. *The Natural Selection of Populations and Communities.* Menlo Park, Calif.: Benjamin/Cummings.

Wilson, E. B. 1883. XXIV. The development of *Renilla*. *Phil. Trans. Roy. Soc.* (Lond.) 174:723–815.

Wilson, E. O. 1971. *The Insect Societies.* Cambridge, Mass.: Harvard University Press.

———. 1975. *Sociobiology.* Cambridge, Mass.: Harvard University Press.

Wilson, H. V. 1888. On the development of *Manicina areolata*. *J. Morphol.* 2(2): 191–252.

Wilson, W. H. 1983. Life-history evidence for sibling species in *Axiothella rubrocincta* (Polychaeta: Maldanidae). *Mar. Biol.* 76:297–300.

———. 1991. Sexual reproductive modes in polychaetes: Classification and diversity. *Bull. Mar. Sci.* 48(2): 500–516.

Winemiller, K. O., and D. H. Taylor. 1982. Inbreeding depression in the convict cichlid, *Cichlasoma nigrofasciatum* (Baird and Girard). *J. Fish Biol.* 21:399–402.

Winn, B. E., and B. M. Vestal. 1986. Kin recognition and choice of mates by wild female house mice (*Mus musculus*). *J. Comp. Psychol.* 100:72–75.

Wisely, B. 1958. The development and settling of a serpulid worm, *Hydroides norvegica* Gunnerus (Polychaeta). *Aust. J. Mar. Freshw. Res.* 9:351–61.

———. 1960. Observations on the settling behaviour of larvae of the tubeworm *Spirorbis borealis* Dandin (Polychaeta). *Aust. J. Mar. Freshw. Res.* 12:55–72.

Withler, F. C. 1982. *Transplanting Pacific Salmon.* Canadian Technical Report of Fisheries and Aquatic Sciences 1079, 27 pp.

Wolff, J. O., K. I. Lundy, and R. Baccus. 1988. Dispersal, inbreeding avoidance and reproductive success in white-footed mice. *Anim. Behav.* 36:456–65.

Wool, D. 1987. Differentiation of island populations: A laboratory model. *Am. Nat.* 129:188–202.

Woolfenden, G. E., and J. W. Fitzpatrick. 1978. The inheritance of territory in group breeding birds. *BioScience* 28:104–8.

Woolfenden, G. E., and J. W. Fitzpatrick. 1984. *The Florida Scrub Jay: Demography of a Cooperative-Breeding Bird.* Princeton, N.J.: Princeton University Press.

Wrangham, R. W. 1980. An ecological model of female-bonded primate groups. *Behaviour* 75:262–99.

Wright, J. W., C. Spolsky, and W. M. Brown. 1983. The origin of the partheno-genetic *Cnemidophnous laredoensis* inferred from mitochondrial DNA analysis. *Herpetologica* 39:410–16.

Wright, S. 1921. Systems of mating. II. The effects of inbreeding on the genetic composition of a population. *Genetics* 6:124–43.

———. 1922. Coefficients of inbreeding and relationship. *Am. Nat.* 56:330–38.

———. 1931. Evolution in Mendelian populations. *Genetics* 16:97–159.

———. 1932. The roles of mutation, inbreeding, crossbreeding and selection in evolution. *Proc. VI Int. Cong. Genet.* 1:356–66.

———. 1940. The statistical consequences of Mendelian heredity in relation to speciation. In *The New Systematics*, ed. J. Huxley, 161–83. London: Oxford University Press.

———. 1941. On the probability of fixation of reciprocal translocations. *Am. Nat.* 75:513–22.

———. 1943. Isolation by distance. *Genetics* 28:114–38.

———. 1945. Tempo and mode in evolution: A critical review. *Ecology* 26:414–19.

———. 1946. Isolation by distance under diverse systems of mating. *Genetics* 31:39–59.

———. 1949. Population structure in evolution. *Proc. Am. Phil. Soc.* 93:471–78.

———. 1951. The genetical structure of populations. *Ann. Eugen.* 15:323–54.

———. 1956. Modes of selection. *Am. Nat.* 90:5–24.

———. 1965. The interpretation of population structure by F-statistics with special regard to systems of mating. *Evolution* 19:395–420.

———. 1968. *Evolution and the Genetics of Populations,* vol. 1. Chicago: University of Chicago Press.

———. 1969. *Evolution and the Genetics of Populations,* vol. 2: *The Theory of Gene Frequencies.* Chicago: University of Chicago Press.

———. 1977. *Evolution and the Genetics of Populations,* vol. 3: *Experimental Results and Evolutionary Deductions.* Chicago: University of Chicago Press.

———. 1978. *Evolution and the genetics of populations,* vol. 4: *Variability within and among Natural Populations.* Chicago: University of Chicago Press.

Wulff, J. L. 1986. Variation in clone structure of fragmenting coral reef sponges. *Biol. J. Linn. Soc.* 27:311–30.

Yamaguchi, M. Y. Toshiyuki, H. Nagano, and N. Nakamoto. 1970. Effects of inbreeding on mortality in Fukuaka population. *Am. J. Hum. Genet.* 22:145–59.

Yamazato, K., M. Sato, and H. Yamashiro. 1981. Reproductive biology of an alcyonacean coral, *Lobophytum crassum* Marenzeller. *Proc. IV Int. Coral Reef Symp.* 2:671–78.

Yampolsky, E., and H. Yampolsky. 1922. Distribution of sex forms in phanero-gamic flora. *Bibl. Genet.* 3:1–62.

Yokoyama, S., and M. Nei. 1979. Population dynamics of sex-determining alleles in honey bees and self-incompatibility alleles in plants. *Genetics* 91:609–26.

Young, C. M. 1986. Direct observations of field swimming behavior in larvae of the colonial ascidian *Ecteinascidia turbinata*. *Bull. Mar. Sci.* 39(2): 279–89.

Young, C. M., R. F. Gowan, J. Dalby, Jr., C. A. Pennachetti, and D. Gagliardi. 1988. Distributional consequences of adhesive eggs and anural development in the ascidian *Molgula pacifica* (Huntsman, 1912). *Biol. Bull.* 174:39–46.

Young, S. S. Y. 1967. A proposition on the population dynamics of the sterile *t* alleles in the house mouse. *Evolution* 21:190–98.

Yuhki, N., and S. J. O'Brien. 1990. DNA variation of the mammalian major histocompatibility complex reflects genomic diversity and population history. *Proc. Natl. Acad. Sci. USA* 87:836–40.

Yund, P. O. 1990. An *in situ* measurement of sperm dispersal in a colonial marine hydroid. *J. Exp. Zool.* 253:102–6.

Yund, P. O., and H. M. Parker. 1989. Population structure of the colonial hydroid *Hydractinia* sp. nov. C in the Gulf of Maine. *J. Exp. Mar. Biol. Ecol.* 125:63–82.

Zahavi, A. 1990. Arabian Babblers: The quest for social status in a cooperative breeder. In *Comparative Breeding in Birds,* ed. P. B. Stacey and W. D. Koenig, 103–30. Cambridge: Cambridge University Press.

Zimmer, R. L. 1987. Phylum Phoronida. In *Reproduction and Development of Marine Invertebrates of the Northern Pacific Coast,* ed. M. F. Strathmann, 476–85. Seattle: University of Washington Press.

Zimmer, R. L., and R. M. Woollacott. 1977. Metamorphosis, ancestrulae and coloniality in bryozoan life cycles. In *Biology of Bryozoans,* ed. R. M. Woollacott and R. L. Zimmer, 91–142. New York: Academic Press.

Zimmerman, E. G., C. W. Kilpatrick, and B. J. Hart. 1978. The genetics of speciation in the rodent genus *Peromyscus*. *Evolution* 32:565–79.

Zimmerman, M., and G. H. Pyke. 1988. Pollination ecology of Christmas bells (*Blanfordia nobilis*): Effects of pollen quantity and source on seed set. *Aust. J. Ecol.* 13:93–99.

Zouros, E., and D. W. Foltz. 1987. The use of allelic isozyme variation for the study of heterosis. In *Isozymes: Current Topics in Biological and Medical Research,* vol. 13, ed. M. C. Rattazzi, J. G. Scandalios, and G. S. Whitt, 1–59. New York: Alan R. Liss.

Zouros, E., M. Romero-Dorey, and A. L. Mallet. 1988. Heterozygosity and growth in marine bivalves: Further data and possible explanations. *Evolution* 42(6): 1332–41.

Zouros, E., S. M. Singh, D. W. Foltz, and A. L. Mallet. 1983. Post-settlement viability in the American oyster (*Crassostrea virginica*): An overdominant phenotype. *Genet. Res.* 41:259–70.

Zouros, E., S. M. Singh, and H. E. Miles. 1980. Growth rate in oysters: An overdominant phenotype and possible explanations. *Evolution* 34:856–67.

INDEX

559